COASTS UNDER STRESS

Coasts Under Stress
Restructuring and
Social-Ecological Health

ROSEMARY E. OMMER
AND THE COASTS UNDER STRESS
RESEARCH PROJECT TEAM

McGill-Queen's University Press
Montreal & Kingston • London • Ithaca

© McGill-Queen's University Press 2007

ISBN 978-0-7735-3203-8 (cloth)
ISBN 978-0-7735-3225-0 (paper)

Legal deposit fourth quarter 2007
Bibliothèque nationale du Québec

Printed in Canada on acid-free paper that is 100% ancient forest
free (100% post-consumer recycled), processed chlorine free.

This book has been published with the help of a grant from the
Canadian Federation for the Humanities and Social Sciences,
through the Aid to Scholarly Publications Programme, using funds
provided by the Social Sciences and Humanities Research Council
of Canada. Funding has also been received from the Social Sciences
and Humanities Research Council of Canada and the Natural
Sciences and Engineering Research Council of Canada through the
Coasts Under Stress Research Project.

McGill-Queen's University Press acknowledges the support of the
Canada Council for the Arts for our publishing program. We also
acknowledge the financial support of the Government of Canada
through the Book Publishing Industry Development Program
(BPIDP) for our publishing activities.

Library and Archives Canada Cataloguing in Publication

Ommer, Rosemary E.
  Coasts under stress : restructuring and social-ecological health /
Rosemary E. Ommer and the Coasts Under Stress Research Project
team

Includes bibliographical references and index.
ISBN 978-0-7735-3203-8 (bnd)
ISBN 978-0-7735-3225-0 (pbk)

  1. Social ecology – Atlantic Coast (Canada). 2. Social ecology –
British Columbia – Pacific Coast. 3. Human ecology – Atlantic Coast
(Canada). 4. Human ecology – British Columbia – Pacific Coast.
5. Atlantic Coast (Canada). 6. Pacific Coast (B.C.). 7. Coasts –
Canada. I. Coasts Under Stress (Project) II. Title.

HC113.045 2007   304.20971'09146   C2007-901510-7

Typeset by Jay Tee Graphics Ltd. in 10/12 Baskerville

Contents

Tables and Figures

TABLES

Foreword

The results of the work of Coasts Under Stress (CUS) are to be found in numerous journal articles, two films, one booklet, one book, and four edited collections showing how the various parts of life in coastal communities fit together and how interactive restructuring has generated the risks, threats, and opportunities coastal communities (human and biophysical) confront. Three of the team books are theme-based. One deals with social-ecological knowledge systems and the vital importance and challenges of moving knowledge across disciplinary boundaries, within and between knowledge systems, and from people to researchers to policymakers to students and back to communities, in order to grapple with interactive restructuring and its effects (Lutz and Neis forthcoming). One deals with the relationship between interactive restructuring and power, whether as energy (oil and gas, hydro), as "power over nature constructs," or as power and agency in nature and human communities (Sinclair and Ommer 2006). One deals with the history of health, diet, and nutrition – with a particular focus on the issue of decreasing food security in places where once-stable food webs have suffered radical shock, as have the cultures of human communities that have always been interdependent with now-endangered food sources (Parrish, Turner, and Solberg 2007). There are two publications for special audiences: one for policymakers (Ommer 2006) and one for coastal communities (Ommer et al. 2006). This principal volume was team-written and is, in itself, an experiment in interdisciplinary scholarship. The team planned the volume, contributed their findings to all the sections, commented on the manuscript as it evolved, and approved its final shape.

In all our work, by *environmental restructuring* we mean changes in the environment, usually at large scales, which are thought to be caused, at least in part, by such things as climate change. We take *social restructuring* to mean changes in society at a range of scales that result in, for example, changes in community cohesion, social support, health care delivery, or the availability of educational resources. Such changes include industrial restructuring, which deals with changes in patterns of ownership and control and in work environments, and political restructuring, which deals with shifts in policy regimes. We take *health* to be the capacity to cope with stressors and recognize that people are a part of (not outside) nature. *Social-ecological* health is the capacity of the human-natural world nexus to deal resiliently with change and the stress that it brings (Dolan et al. 2005).

We wish to take this opportunity to thank the Social Sciences and Humanities Research Council of Canada (SSHRC), the Natural Sciences and Engineering Research Council of Canada (NSERC), Memorial University of Newfoundland, and the University of Victoria for major funding of this work, and for ongoing support throughout the lifetime of the project. We owe a debt of gratitude to Yves Mougeot and Katharine Benzekri of SSHRC, along with the various SSHRC officers who assisted us, particularly Jacques Critchley, who got us started, Pierre-François LeFol, who was with us in our "middle period," and Michèle Dupuis, who has seen us through to the end. We also wish to express our gratitude to André Isabelle and Anne Alper of NSERC, whose assistance has likewise been invaluable throughout all the years of our work.

This project could not have been carried out without a dedicated staff, and we here thank Janet Oliver, Carrie Holcapek, Cathy King, Kari Marks, Graeme Bock, Gary Tunnell, Angela Drake, and Moira Wainwright for their hard work, constancy, and continued support through thick and thin. We wish also to thank the other universities whose faculty contributed to our work: the University of British Columbia (and, in particular, the Fisheries Centre and the Department of Geography), Dalhousie University, Saint Mary's University, and the University of New Brunswick. Our heartfelt thanks go to our partners and our advisory boards, named in the appendices to this volume, and to the Centre for Studies in Religion and Society and the Centre for Earth and Ocean Research, both at the University of Victoria, for providing the West Coast part of the team with a home. On the East Coast, Memorial University provided a small building for the use of staff, faculty, and students, while on the West Coast the University of Victoria gave the project director an academic home. We are grateful to both these institutions for their generosity and for their faith in us.

For this, the main publication from the project, the project director owes a particular debt of gratitude to the editorial committee (Dave

Schneider, Steve Bornstein, John Lutz, Barb Neis, Peter Sinclair, and Carrie Holcapek), which helped finalize the text of this volume and she especially wishes to thank Barb Neis and Peter Sinclair, who read the whole volume through several times and made comments and suggestions. The book could not have been completed without them. We also wish to thank two anonymous readers for their insightful comments, Ron Curtis for his careful and thoughtful copyediting, Bob and Gillian McIvor and Bob Cecill for proofreading assistance, and Joan McGilvray and McGill-Queen's University Press for providing a positive publishing experience.

Rosemary E. Ommer, University of Victoria

Abbreviations

ACOA	Atlantic Canada Opportunities Agency
AND	Anglo-Newfoundland Development Corporation
AOI	area of interest
BTF	Back to the Future
CBA	cost-benefit analysis
CCPFA	Canadian Council of Professional Fish Harvesters
CHC	Community Health Council
CHRHC	Community Human Resources and Health Centres
CMA	Coastal Management Area
CNOPB	Canada–Newfoundland Offshore Petroleum Board
CNSOPB	Canada–Nova Scotia Offshore Petroleum Board
COSEWIC	Committee on the Status of Endangered Wildlife in Canada
CUS	Coasts Under Stress Research Project
CVIHSDA	Central Vancouver Island Health Service Delivery Areas
CW	carapace width
CZM	Coastal Zone Management
DFO	Department of Fisheries and Oceans Canada
EA	Enterprise Allocation
ECBC	Enterprise Cape Breton Corporation
ECNASAP	East Coast of North America Strategic Assessment Project
EEM	Environmental Effect Monitoring
EEZ	exclusive economic zone
EI	Employment Insurance
EIA	Environmental Impact Assessment
EIS	Environmental Impact Statement
EPP	Environmental Protection Plan
FAO	Food and Agriculture Organization of the United Nations
FEK	fishers' ecological knowledge

FPU	Fishermen's Protective Union
FRCC	Fisheries Resource Conservation Council
GBSC	Gilbert Bay Steering Committee
GNP	Great Northern Peninsula
GPA	Global Programme of Action for the Protection of the Marine Environment from Land-based Activities
GSC	Geological Survey of Canada
HBC	Hudson's Bay Company
HCCSJ	Health Care Corporation of St John's
HSDA	Health Service Delivery Area
IA	Income Assistance
IBM	Individual-Based Model
ICES	International Council for the Exploration of the Sea
ICNAF	International Commission for the Northwest Atlantic Fisheries
IM	integrated management
IMF	International Monetary Fund
IMMA	International Marine Minerals Society
IQ	individual quota
ITQ	individual transferable quota
IUCN	World Conservation Union
LEK	local ecological knowledge
LMN	Labrador Métis Nation
LOMA	Large Ocean Management Area
MHA	Member of the House of Assembly (provincial legislature)
MOH	Ministry of Health
MPA	Marine Protected Area
MSC	Marine Stewardship Council
NAFO	Northwest Atlantic Fisheries Organization
NARL	North Atlantic Refining Ltd
NAWMP	North American Waterfowl Management Plan
NGO	non-governmental organization
NHA	Northern Health Authority
NHNHA	Newfoundland Hospital and Nursing Home Association
NITA	Northern and Isolation Travel Assistance Program
NPA	National Programme of Action for Protection of the Marine Environment from Land-Based Activities
NPV	net present value
NTA	Newfoundland Teachers' Association
NTL	Newfoundland Transhipment Terminal
NVIHSDA	North Vancouver Island Health Service Delivery Area
NWHSDA	North West Health Service Delivery Area
ORB	optimal restorable biomass

PCCRRHS	Provincial Coordinating Committee for Remote and Rural Health Services
POP	Physicians Outreach Program
PRFD	Prince Rupert Forest District
PSYUS	provincial sustainable yield units
PYLL	potential years of lost life
QCB	Queen Charlotte Basin
RDA	Rural Development Association
RHA	regional health authority
RHB	Regional Health Board
RHDB	Regional Health District Board
RV	research vessel
SAM	social accounting matrix
SARA	Species at Risk Act
SCI	Sustainable Communities Initiative
SEP	Salmonid Enhancement Program
SSP	Strategic Social Plan
TAC	total allowable catch
TAGS	The Atlantic Groundfish Strategy
TB	Tofino Basin
TEK	traditional ecological knowledge
TFL	Tree Farm Licence
TOC	total organic carbon
UI	Unemployment Insurance
UNCLOS	United Nations Convention on the Law of the Sea
VEC	valued ecosystem components
VIHA	Vancouver Island Health Authority
WCOEEA	West Coast Offshore Exploration Environmental Assessment
WCVI	west coast of Vancouver Island
WD	Western diversification
WTO	World Trade Organization
WWF	World Wildlife Fund

PART ONE

How We Got Here: Historical Restructuring and Its Social-Ecological Legacy

1

Introduction – What Stress?
What Coasts?

THE PROBLEM

Rural communities in Canada, and also in many other parts of the developed world, are in crisis. Nowhere is this more true than in the coastal resource-based communities on the East and West Coasts of Canada, where there are now fewer people living than there were one hundred years ago and where the resources and environment that once supported communities are now all but gone. The communities are so diminished that there are now more ghost towns than inhabited ones on the East and West coasts. Changes in natural environments have interacted with political, industrial, and social change to adversely affect the health of the people who live there, their communities, and the natural environments in which they are embedded. Such communities are bellwethers for national and global changes: the flight from rural communities to huge urban agglomerations is not just a Canadian phenomenon, nor is the environmental destruction that is partly driving that migration. Increasingly, scholars, scientists, and social scientists are telling us that we are headed down a pathway to environmental collapse and social chaos unless important changes are made soon. We do not have enough resources left, even if we stop degrading ecosystems now, because already almost two-thirds of our resources are used up (*Guardian Weekly* 2005; Millennium Ecosystem Assessment 2005; Diamond 2005; Right 2004). We need to understand what has happened, what is going on and why ... and then focus on recovery. In short, Canada and the wider world need to make urgent choices. Our accustomed patterns drove us into this mess, and we need new ideas, new ways of thinking, to get us out.

The authors of this book think that the fundamental problem is an inadequate understanding of the highly complex links between social and environmental restructuring and how they interact with the health of people and places. This volume is an attempt to offer new ideas and new ways of thinking based on research projects that have examined these problems as they have developed in all their complexity. We call the approach we have developed "social-ecological health," a new comprehensive analytical framework for understanding social and ecological restructuring and its impact on health in Canada and internationally. This volume and the insights it contains, as well as the other volumes linked to it, result from a bold experiment in bringing an unprecedented diversity of expertise to bear on social and environmental restructuring. Over five years of testing and experimentation, we have crafted a scholarly technique that breaks through disciplinary borders to allow scholars to work effectively together across traditional divides. Not only do our research findings contain new knowledge, a new perspective on coastal issues, and new ways of doing research, but we have generated a new way to make and move knowledge within scholarly teams and, through and from them, to policymakers and communities.

We start with the understanding that today's coastal communities are the product of centuries of *interactive* restructuring between people and natural environments. Interactive restructuring occurs through the interplay of social factors (economics, policy, institutional history) with the biophysical environment. The historical pathways of interaction can be short (as in the case of some nonrenewable resources) or long (as in the case of many renewable resources), but in all cases the human and community health consequences of restructuring can lead to biophysical degradation when risks, costs, and benefits are misaligned and scale asymmetry occurs. Scale asymmetry happens in various ways, depending on which scale is involved.

For our analysis, three kinds of mismatches are crucial: s*patial scale* asymmetries occur when activities appropriate to one geographical level are applied without due consideration at another or when the process is wrong and decisions made at one level pertain to another and people at that other level (usually, but not always, lower down) are not consulted. Fisheries management, for example, occurs at the national scale, which can result in decisions that are taken without adequate local (or international) consultation and that turn out to be inappropriate regionally and locally or ineffective globally. Again, government policy thinking may be directed at "the individual," when "the community" is a more appropriate level of management. We will see examples throughout this book. *Temporal scale* asymmetries occur when (for example) the need to introduce major change slowly is ignored and problems arise from overly rapid change. By the same token, change can come too slowly, as occurred in the case of the East Coast groundfish moratoria, where policy-making did not respond quickly enough to science. *Organizational scale* asymmetries occur when, for example, activities appropriate at the level of the firm are applied to government or community organizations. Working at the appropriate level is of fundamental importance, particularly if the scale of different firms is also considered: they may range from large firms with a global reach to local firms. Firms differ in scale in at least two ways: in the size and in the scope of their activity. Some small firms can have a national or global reach and vice versa, although the typical combination is more symmetrical, some of them being larger than municipalities in numerical terms, and provinces in economic terms. Thus, both practice and policy have to take type and scale of organization into account and consider how they may benefit or be inappropriate for producers whose livelihoods depend on resources. We also consider that the different types of asymmetry tend to go together, decisions made by the wrong organization tending to be for the wrong geographical scale and also for the wrong temporal scale.[1]

Among the sources of misalignment we have uncovered are misdirected flows of benefits that, if they fail to reach resource-dependent regions and generate diversification in them, can lead to overdependence on one resource (a "staple") and a consequent inability to transform the economy when markets falter or the resource is depleted – what is called a "staple trap." The emerging solution to the problem of depleted resources and stressed communities, which we see as a challenge for the recovery of both (albeit in a new form) is multiscale governance, which is distinguished by cross-scale flows of knowledge and cross-scale readjustments of regulatory power.

The team that came together to do the research on which this volume reports has been thinking about the problem of restructuring and social-ecological health for some time. That there was trouble in Canada's coastal communities and marine ecosystems was obvious to us, and we could not ignore the warning signals. In 1992, for example, the East Coast groundfish fishery collapsed, and the subsequent moratoria on fishing threw thirty thousand people out of work. At that time, some of us from a range of disciplines in the natural and social sciences explored how the disaster played out for communities of fish and fishers in a small part of Newfoundland, to try to understand what went wrong and what might be done in the future (Ommer 2002). Subsequently, the salmon fishery on the West Coast of Canada started to experience problems, and we saw therein worrying resemblances to the East Coast situation. We therefore put together a second interdisciplinary research team to look at issues of

Figure 1.1   The Coasts Under Stress study area

ethics and equity in fisheries management on both the East and West Coasts (Coward et al. 2000).

Those two research projects produced some interesting results, not least by bringing home to us the fact that the problems on Canada's coasts extended beyond the fisheries to most resource sectors, renew-

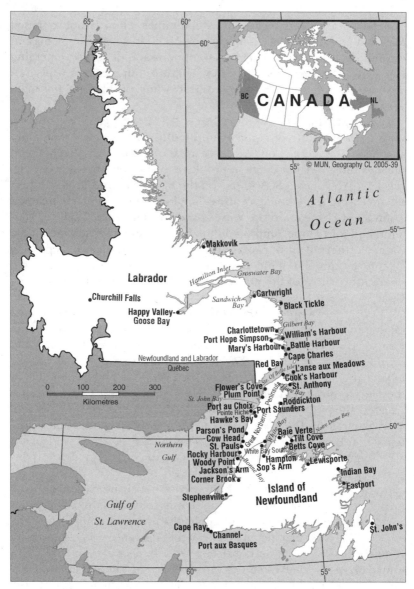

Figure 1.1 (continued)

able and nonrenewable. It was made clear to us that coastal communities had inherited a legacy of increasingly depleted and/or degraded resources (closed mines, ghost towns). Moreover, the extraction industries that remained were concentrated for the most part in the hands of large companies who were themselves challenged competitively.

The result has been distressed communities whose resource-based raisons-d'être have been disappearing as both local environments and community economies change. New options are, of course, emerging (see part 3), but many of them (aquaculture, "underutilized species," offshore oil and gas) could aggravate existing problems if not managed effectively.

That is why the Coasts Under Stress (CUS) team came together to work out a new interdisciplinary and multiscale approach that could analyse what was going on. We chose areas on both coasts that between them would cover the range of problems we needed to assess (see figure 1.1). We selected St Anthony on the East Coast and Prince Rupert on the West as regional centres. Single-resource centres included Tofino and Ucluelet on the West Coast, Port au Choix, Williams Harbour, and Port Hope Simpson on the East. Forestry was studied on the Great Northern Peninsula of Newfoundland as a whole, with a particular concentration on White Bay South, while on the West Coast we considered the whole of the north coast of British Columbia from Haida Gwaii to Ocean Falls. We chose Port Hardy and Port McNeill on the West Coast as examples of mixed economies; we looked at now-defunct mining towns – Anyox on the West Coast, Tilt and Bett's Coves on the East – and we looked at fisheries from Cartwright to Port aux Basques on the East Coast and in the Hecate Strait areas on the West. We looked at oil and gas at Hibernia on the East Coast and its potential in the Hecate Strait and the Tofino Basin on the West Coast. We also selected Haida Gwaii (for archaeological work) and Hartley Bay and Alert Bay (all on the West Coast) to allow us to pay attention to the impact on First Nations communities on that coast, where land claims are of significant import today. There was no equivalent First Nations presence in Newfoundland in the areas where we were working.

In these places, we could look at how our social and natural environments had been changing in an interconnected and complex manner and at an accelerating rate and apply our understanding of how the scale of decision-making matters. Despite extensive investments in resource management and environmental impact assessments, we in Canada have overharvested our groundfish, our salmon, and our forests; our known mineral and energy resources are under pressure, and we continue to degrade our marine, terrestrial, and atmospheric habitats. In coastal areas, as in agricultural areas, our small, rural communities are increasingly in trouble, partly because environmental degradation poses significant problems for people, while our attempts at social responses often exacerbate degradation. These complex interactive processes affect environmental, population, and community health and are in turn affected by them. We suspected that the problem was systemic in cause – a result of policy thrusts and economic imperatives over the long term – rather than simply a result of specific local geographies and resource bases. That is why we chose to work on two apparently very different coasts, unalike in geography and resource base, in timing of settlement (the "old" East Coast and the "young" West Coast, always excepting First Nations), and in culture. We hypothesized that if both coasts were suffering similar impacts, then the problem could reasonably be said to be the result of patterns of human decision-making (hence systemic), which means that it may be fixable. That said, the fundamental problem Canada faces here, both intellectually and practically, is that we have an inadequate understanding of the highly complex links between social and environmental restructuring and how together they interact with the health of people and places. In other words, our work had to pay attention to the *interactive* processes that have been restructuring Canada's biophysical environment, social structure, and the health of its coastal peoples.

RESTRUCTURING

Following a substantial body of literature, we conceptualize what is occurring on our coasts as a process of *restructuring*. In the past, restructuring has been generally recognized to have created rapid and profound social, economic, and political change both in and beyond industrial countries, and the growing literatures on globalization, the dismantling of the welfare state, industrial change, and neoliberalism all speak of the advent of a "new industrial divide." The literature on *social restructuring* has concentrated on regulatory and industrial restructuring through an examination of globalization, technological change, and new industrial divisions and practices, and points out that social restructuring has both global and local dimensions. Globally, "freeing" the operation of economic activity has predominated (Beneria and Lind 1995, 2). The effects can be seen, for example, in shifts in the location of production, as well as in changes in the nature of work and employment (Mackenzie and Norcliffe 1997). Nationally and locally, restructuring has also been associated with the erosion of social safety nets and changes in health and education systems. There has been disagreement, however, on how such restructuring affected people, their communities, and the environment.

The *optimistic view* has been that we have been shifting to a post-industrial society characterized by greater flexibility, more skilled workers, better work coordination, and new approaches to management (Osborne and Gaebler 1992; Piore and Sabel 1984). This shift was to be demonstrated over the longer term by (for example) less state debt, greater economic growth, reduced gaps between rich and poor, increased leisure opportunities, more sophisticated niche market structures, and better food production and distribution as a result of biotechnological innovation (Hart 1996). From a *social determinants of health* perspective, such outcomes from restructuring might have unfortunate short-term negative consequences for some people (lost jobs, reduced incomes, and state support programs), but in the long run it was expected to help solve the problem of poverty and reduce social inequities. With regard to the environment, poor outcomes (it was argued) would improve as a shift away from mass production reduced threats to environmental health as well as environmental threats to human health, while actually creating new kinds of communities, including "virtual communities."

The *pessimistic view* has been that, to the contrary, neoliberal restructuring has deepened the divide between rich and poor people, regions, and countries (Sforza-Roderick, Nova, and Weisbrot 1998), strengthening transnational corporations relative to workers and governments (Bonnano et al. 1994; Friedland 1991; Winson 1992), and destroying communities (Laxer 1995) and, with them, the roots of social integration and social action, while also threatening the health of workers and their families (in Canada as elsewhere: Yalnizyan 1998). Some critics have added that the poverty gap is producing a health gap – Canadians at the bottom of the economic ladder are more likely to die from many different diseases than are the more well-off (Raphael 1999). Such things as work reorganization, capital flight, deregulation, and cuts to public services and social programs, in this view, have affected the health of workers as a consequence of changes in regulations that had been designed to prevent and/or detect work-related health risks and workers' access to public services providing health care, training, and compensation for loss of work, accident, injury, or occupational disease. Restructuring, seen negatively, has become associated with such problems as higher unemployment rates, lower wages and benefits, and poorer working conditions (Amott 1993). "Job flexibility" has become a euphemism for low-income employment without benefits, unions, or much job security (Armstrong 1995; Leach and Winson 1995). The health risks associated with restructuring are claimed to be profound, since strategies such as downsizing and capital flight increase employment uncertainty and unemployment and create psychosocial stress (Mayhew and Quinlan 1999; Picot and Wannell 1993; Witherill and Kolak 1996). The benefits are largely reaped by employers, generally at the expense of workers, primary producers, and the environment, and the health of all of these, while the transfer of power from democratic institutions to private institutions erodes effective democracy. Ecological degradation increases as regulatory constraints weaken (Albo and Roberts 1998; Martin and Schumann 1998). In this book, we investigate the effects of restructuring. In the process, we test the validity of both the positive and the negative view of it, from the perspective of coastal communities.

Such has been the social science literature. "Restructuring" has been used in this literature to refer to periods of relatively rapid, substantive change, what in the biophysical sciences is referred to as "regime shift." Research on such environmental restructuring has usually focused either narrowly on regime shifts within ecosystems (Steele 1998) or more broadly on ecosystem stressors and their potentially devastating consequences for environmental and human health. "Healthy ecosystems" are usually defined as ecosystems that are stable and sustainable, resilient to stressors, and capable of maintaining organization and autonomy over time. Natural disturbances, it is argued, often provide the basis for the revitalization of local ecosystems, while anthropogenic stressors are seen as often resulting in degradation that reduces the capacity of ecosystems to recover, thus permanently affecting their health (Rapport, Costanza, and McMichael 1998). Degradation can worsen under conditions of scientific uncertainty, and/or when politico-economic pressure for continued resource exploitation results in a cycle of resource overexploitation characterized by the so-called "ratchet effect" (Clapp 1998; Ludwig, Hilborn, and Walters 1993).

All these literatures have reflected what we have come to see as an incomplete understanding of the dynamics of environments or the links between ecology, economy, and health. We see restructuring as much more interactive and dynamic and as much less linear or discrete than any of these views understand it, and we think, furthermore, that it has deeper historical roots than any of these literatures recognize. The importance of the temporal dimension is borne out by the work of the eminent Canadian economic historian Harold Innis (e.g., 1954, 1956), who saw post-contact economic development in Canada as grounded in natural-resource ("staple") exploitation. He (and later historians) went so far as to explain Canadian history in terms of the development of regional economies that restructured the economies of "new" and "empty" territories so that they became producers of fish, fur, timber, wheat, and so forth ("staples") for export.

In later work, Mel Watkins (1963, and with Grant 1993) suggested that some regions could continue to shift economic focus (restructure) successfully either by switching resources when the original one failed or by diversifying around the staple base and achieving industrial development. It is argued that such successful restructuring has occurred in various places in Canada in the last two centuries

(Ontario, British Columbia), while other regions have failed to shift staples or engender sufficient diversification around the resource base, becoming stuck in a "staple trap" that has left them economically "backward" (Newfoundland, for example). Some, on the other hand, have been cursed with staples that were considered to be inadequate as instruments of regional growth (Paquet on cod, for example, 1968). More recent staples work, recognizing the narrowness of such an explanation, broadened the approach to deal with issues of agency for local people, the influence of power on the location of economic and social development (Ommer 1991), the importance of the balance between human populations and their environment (Thornton 1979), or more modern developments such as branch plants and transnational corporations (Hayter 2000). In addition, the staples approach, in all of its variants, had one key lacuna: it never dealt with that first, colonial, restructuring, which took the pre-contact world of First Nations and restructured it in the image and likeness of European merchant, and later industrial, capitalism, with consequences that are still being played out across the continent. Nor did it come to grips with environmental restructuring or with issues related to population and community health.

COMPLEX ADAPTIVE SYSTEMS AND SOCIAL-ECOLOGICAL HEALTH

As we began our work, a new literature was emerging that sought to develop new interdisciplinary insights and drew on "vernacular knowledge" in order to gain a better understanding of the dynamics and consequences of social and environmental restructuring (Berkes 1999). This literature saw social and ecological systems as complex, adaptive, and interdependent, as well as highly vulnerable in today's increasingly global postindustrial world. It also argued that these complex adaptive systems, while vulnerable, also contained the germs of how to remain sustainable and resilient, in that their modus operandi protected people and ecosystems even under stressful conditions. There is now a significant and growing literature on ecology, resilience, and complex adaptive systems, as well as an important literature combining analyses of human and ecological systems.

The most notable is the work of Fikret Berkes and his colleagues, whose path-breaking research we can only partially examine here.[2] The seminal paper for all this thinking dates back to 1973, when biologist C.S. ("Buzz") Holling opened a discussion of the importance of what he called "constancy" in ecological systems. He suggested that the *persistence of internal relationships* was what really mattered when a system was "profoundly affected by changes external to it, and continually confronted by the unexpected" (1973, 1). If considered in this way, ecological systems, he argued, benefited from "instability, in the sense of large fluctuations" because they appeared to produce "resilience and a capacity to persist" (15). In introducing the "qualitative" analysis of relationship, rather than the traditional quantitative analysis of presence/absence of component parts, he required that attention be turned from equilibrium as the main feature of healthy ecosystems to *resilience*.

This is the founding premise of what we now refer to as the "new ecology." Since then, Holling (1986) and others (Gunderson and Holling 2002; Berkes, Colding, and Folke 2003; Turner et al. 2003) have all been seeking a new understanding of human and biological cycles and the relationship between them. They speak of "social-ecological systems" (from which we derive our term "social-ecological health") having their own cycles of organization, collapse, and renewal, but point out that such systems may also exhibit other features, such as sudden transitions from one state to another, incremental and slowly emergent change, or change that occurs because of exogenous and rapid disturbance. For social systems that are organized around a particular biophysical system, unanticipated shifts in the biological base are likely to result in periods of social instability and crisis. The reverse is also true – changes in social systems (as in the use of more sophisticated harvesting technology or increased technological intensity), particularly if managed poorly, can induce biophysical change. Because of this interdependence, understanding system-wide feedbacks – how human actions influence ecological systems and how ecological systems in turn alter human behaviour – is crucial (Tengo and Hammer 2003, 133). Scale-dependent interactions also need to be understood. These interactions can be interpreted in the light of several concepts of system complexity, especially that of complex adaptive systems with defined components, such as a community of people, or a population of organisms (Levin 1999), which demonstrate resilience.

Most importantly, however, while work in the new ecology is not "wedded to an equilibrial perspective," as Scoones observes, a great deal of the social science thinking that has built on Holling's original insight has failed to catch up with this adjustment and "has tended to ignore questions of dynamics and variability across time and space, often excluding from the analysis the key themes of uncertainty, dynamics and history" (1999, 480). Such a position is no longer tenable, nor is (Scoones argues) the "nature-culture divide," although it must not be replaced by either determinism or relativism. He calls for work that is concerned with "spatial and temporal dynamics developed in detail and situated analyses of 'people in places,' using in particular historical analysis as a way of explaining environmental change across time and space" (487–90).

We concur. This is exactly the kind of work we have done, except that we go further and also look at social-ecological health and change. Scoones recommends building on an understanding of "environment as both the product of and the setting for human interactions, which link dynamic structural analysis of environmental processes with an appreciation of human agency in environmental transformation" (1999, 490). With this, too, we agree, and indeed our work goes further because we also consider the reciprocities (interdependencies) that exist between human agency and environmental processes. Scoones, however, insists that "the appreciation of complexity and uncertainty in social-ecological systems" may result in "the recognition ... that prediction, management, and control are unlikely, if not impossible" (490). Here, we disagree. We consider this too draconian a conclusion to reach before much more work is done, although we agree with the importance of recognizing uncertainty and complexity.

Fortunately, however, Scoones also claims that in cases where rigidity has given way to flexibility, there might be "major practical consequences for planning, intervention design, and management" (479) as a result. We think this is a wiser interim conclusion than a claim that, given complexity, management may be "impossible."[3] Encouragingly, in the last five years a network of scholars concerned with management and resilience in social-ecological systems has developed. Key players here include, in particular, Fikret Berkes and his colleagues and the authors (most of them part of the Resilience Alliance) whose work is presented in the theoretical book edited by Lance Gunderson and Buzz Holling entitled *Panarchy: Understanding Transformations in Human and Natural Systems* (2002). Berkes, Gunderson, Holling, Folke, and others have explored social-ecological systems in case studies drawn from around the world, and Berkes himself also has a rich collection of work on the Canadian Arctic. Two major books, the first edited by Berkes and Folke (1998) and the second by Berkes, Colding, and Folke (2003) speak to the complex issues of how to link social and ecological systems (1998) and how to "navigate" them (2003). They present cross-scale and cross-disciplinary studies that inform us about the conditions under which resilience is built, and they discuss ways of managing social-ecological systems.

It is this new approach that is detailed and expanded upon in the Gunderson and Holling volume in which Berkes and others in the Resilience Alliance are contributors and which is a major theoretical and structural analysis of complexity, drawing on both natural and social science, although much of the social science is drawn from management and business studies. The book seeks to come to theoretical grips with the thorny issues of complex adaptive systems, change, resilience, cross-scale issues, and the urgent necessity of cross-disciplinary and cross-scale research. Most significantly, it recognizes the importance of responding to uncertainty at a multiplicity of scales, and also across them. The authors say that they "prefer to invent a[nother] term that captures the adaptive and evolutionary nature of adaptive cycles that are nested one within the other and across time scales," because they wish to get away from "rigid top-down" connotations. That is why they call their theoretical system *panarchy*. Panarchy (they describe it more clearly than they actually define it) embraces "biological, ecological and human systems" that "contain a nested set of [the] four-phase adaptive cycles, in which opportunities for periodic reshuffling within levels maintain adaptive opportunity, and the simple interactions across levels maintain integrity." The four phases are characterized by exploitation, conservation, release, and reorganization (Gunderson and Holling 2002, 73–4, and chap. 2).

There is much to ponder here, but it is in its understanding of cross-scale interactions that the book's most useful theoretical insight is to be found. Using the lessons learned in earlier case studies (such as those by Berkes and Holling, among others), *Panarchy* points to the problems of rigidity in management systems at all scales and shows how, in some cases, such rigidities can contribute to meta-system collapse. Rigidity, in this use of the word, is a continued commitment to existing practices in the face of difficulties, rather than seeking a creative new response to changed circumstances. That, in and of itself, is not new. But what is new here is the idea that different levels of a meta-system have different functions that need to be coordinated. Being a theoretical discussion, illustrated only partially by rather limited case studies, the book does not (and cannot) tell us how such coordination should occur. However, it suggests that the top levels of a meta-system should, under uncertainty, provide the slower-moving stabilizing role of assessing and deliberating, and only then acting upon innovative responses to crises in the system. Such highly creative innovative responses, which can generate transformation throughout the meta-system, are most likely to occur at the base of the system, where the scale is small enough for dynamic response to be possible, because experimentation is at a manageable scale.

This is an interesting hypothesis and one that is pertinent to the kinds of social-ecological change that restructuring has created, although it fails to come to grips with important issues of power (see

Sinclair and Ommer 2006). What it does do is speculate on the dynamics of multiscale response differentials, the problem that is at the heart of their conceptual framework for dealing with uncertainty. They hypothesize that creative innovative responses, if successfully filtered across scales and positively assessed at the highest levels, will result in positively creative and (relatively) stable cross-system adjustments. Should, however, rigidities occur in the system, particularly at the higher levels, the results can be catastrophic – and they cite the collapse of northern cod (*Gadus morhua*) as an ecological example, explaining (in broad terms) how institutional rigidities in the human management system failed to cope with uncertainties and change, resulting in both social and ecological disaster (Hutchings, Walters, and Haedrich 1997; Hutchings and Myers 1995). They go further. They suggest that inflexibilities in the system will have cascading effects that can trap lower levels in unproductive stasis, stasis that can, in the long term, lead to systemic failure. The authors speak of the patchiness of resources (well recognized in biology), the importance of cross-scale interactions, and the different kinds of social and ecological transformations that can occur, as well as the different rates of change at different scales. This latter point is crucial, for in "a healthy society, each level is allowed to operate at its own pace, protected from above by slower, larger levels, but invigorated from below by faster, smaller ones" (Gunderson and Holling 2002, 76). In human systems, they assert, failure to allow each level to operate at its own pace will express itself as a "poverty trap" arising from rigidities (such as inequitable power and economic structures, and poor health) in other parts of the system.

Gunderson and Holling's book on panarchy, which encapsulates so much new theoretical thinking, refers to case study examples throughout in a sustained theoretical argument. The examples are drawn in large part, however, either from biology or from the world of business, where concerns over flexibility and rigidity have been uppermost in the entrepreneurial mind for some time now (Cohen and Winn 2007) and have succeeded in altering some business and management practices. The authors speak of the problem of poverty when local systems fail (or are not given the chance) to adapt. They insist that when the various levels do not aid one another, systemic change cannot be responded to appropriately and that the end results are rigidity traps and poverty traps in some levels of the system. These, they conclude, are the hallmarks of maladaptive systems and can eventually bring down the whole structure (see Gunderson and Holling 2002, especially 95–8).

THE COASTS UNDER STRESS APPROACH

This new literature is exceedingly important, and we build on it in our own work, seeking to bring together many strands in the various literatures in the social sciences, humanities, and natural sciences. But we go further, because we add the population health literatures, which are necessary if we are to link the functioning of social-ecological systems to their health and well-being. Our work seeks, therefore, to combine all this new work, along with population health thinking, in such a way as to produce a more holistic view of interactive restructuring and its relationship to the health of people, communities, and the environment. For its part, the population health literature has not paid much attention to the relationship between human/population and community health and environmental degradation, having as its main focus the "social determinants" of health (education, income, employment, gender, work environments, and so on). It has paid even less attention to the relationship between biophysical environments and health, but that relationship is critical to our understanding of the dynamics of coastal communities and health in that context.

While we have drawn on the population health literature to some considerable extent, then, we build into the picture the missing link of environmental health, thereby creating social-ecological health as a "process variable" – that is, a variable that is identified through the way in which restructuring processes (shapes) the interaction between society and environment. There are, we think, connected cascades (causal sequences) of events (*pathways*) between society, the environment, and population and community health that arguably become more visible during periods of restructuring. Restructuring can also enhance the potential for unanticipated health risks, because it can create misalignment between institutional structures and social and environmental realities. This is why it is so important to focus, as we have done, on restructuring. Canada's coasts have been subject to periods of social restructuring (which is experienced in coastal communities) punctuated by periods of rapid regime shift (biological restructuring) within marine ecosystems, with the interaction between these two kinds of restructuring being co-dependent. In some cases, triggers for ecological shifts are relatively remote from the effects, a classic example being the way in which the development of distant water fleets in Europe and elsewhere eventually resulted in the "killer spike" (Hutchings and Myers 1995) for northern cod and other groundfish species in the late 1960s, which was itself the harbinger of the collapse of groundfish in the northwest Atlantic. Triggers can also be close, as when overfishing of the northern Gulf cod stocks by a local fleet of draggers (Palmer and Sinclair 1997) led to their collapse.

More broadly, we should ask if "real" development (which includes maturation and diversification, as opposed to simple growth) was and is being generated around the various staple bases that have come into existence on the East and West Coasts of Canada. To answer that question we must think beyond short-term GDP (gross domestic product) increments to include whatever cultural, social, health, and environmental costs have been involved. Good governance, it seems to us, means knowing as much as possible about the dynamics and potential impacts of social-ecological *interactive* restructuring, using history and comparative research to help anticipate change and uncertainty, in order to minimize social-ecological costs and maintain social-ecological health.

We define health in three categories. First, in the World Health Organization (WHO) definition, human health is "a state of complete physical, mental and social well-being and not merely the absence of disease or infirmity" (WHO 1948).[4] Second, health of communities we take to be the condition of a socio-economic-environmental system where the economic, social, and political components are organized and maintained in such a way as to promote both human and natural environmental well-being so that the community experiences relatively high levels of social support, a culturally acceptable standard of living, less rather than more inequality, and similar benefits that augment individual well-being and provide for low levels of social dysfunction. Third, health of the environment we take to be the result of relatively low levels of human-induced morbidity or mortality of humans or other species, along with relatively low levels of contaminants that induce mortality or morbidity in human or other species.

However, the health of communities and the environment is not merely the absence of morbidity/mortality and social dysfunction. It extends to those interactions of communities with their environment in ways that sustain quality of life and promote resilience in response to stressors. This is "social-ecological health," which includes the cultural, social, and environmental dimensions of health and reflects the link to restructuring by integrating these components into one framework that recognizes that processes operating within social, environmental, and cultural contexts have interdependent, reciprocal, and nonlinear relationships and feedback effects, and also complex causality. More particularly, as our definition implies, resilience is a key idea, and so we seek to explore the idea of resilience, elaborating it to see how it can be achieved across complex adaptive systems that operate across several scales. This multiscale perspective needs to become part of resil-

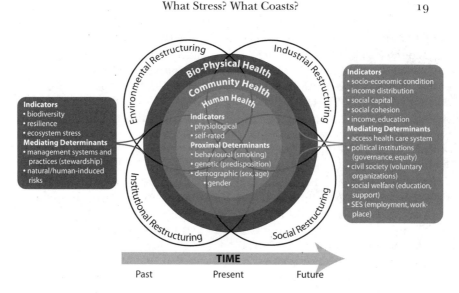

Figure 1.2   The Coasts Under Stress Model (Dolan et al. 2005)

ience thinking and will in the process generate new understanding of what is involved. This should ensure that resilience cannot be understood as any kind of "return to equilibrium" – an understanding that was inherent in Holling's original work (see above – 1973) and that has been part of most biological resilience models but that is increasingly untenable when applied to human (and even biological) systems.

There is no assumption in our work that resilient systems will tend to maintain or restore equilibrium without changing essential relationships among the parts. Instead, we seek to identify, through historical, comparative, interdisciplinary research, the systemic changes that resulted from colonial and postcolonial development in resource-based economies, changes that have flowed through pathways connecting socio-economic, environmental, and political restructuring to social-ecological health. Our conceptual framework (figure 1.2) illustrates these pathways along which various forms of restructuring interact with each other to influence population and community health over time (Dolan et al. 2005). How and why these relationships unfold as they do is significantly affected by the power of various groups of people and organizations to structure courses of action and by the power of other groups to resist. Differences of power among participants in the relationships we examine constitute an important part of our work. Social power (the capacity to control social relationships) can involve a wide range of strategies and the mobilization of relevant resources. This is not to argue that people necessarily achieve the results they intend (as our work clearly shows); nor do we claim that all actions are necessarily consciously goal-directed.[5] Instead, we recognize the wide range of choices that people make and find acceptable, as well as the constraints under which those choices are made (material constraints, knowledge constraints, etc.).

Within the framework of the model, human health (individual/family) is a condition of physical and psychosocial well-being determined by individual predisposition, circumstances, and experiences within particular localities. Aggregate human outcomes, of course, are socially produced, and socio-economic and cultural environments play a significant role in influencing access to health care, environmental exposures (home, work, community), and health behaviour (lifestyle choices, use of health services, options for health promotion) (Adler and Newman 2002). It is generally agreed, for example, that mortality (death) and morbidity (disease) as a measure of health status follow a gradient across socio-economic classes (Evans, Barer, and Marmor 1994). Human health, then, cannot be considered without reference to socially produced attributes such as socio-economic status, gender, education, and culture, and – given that human health is embedded within social, economic, and cultural environments – the restructuring of such attributes has direct implications for health status. Moreover, human, community, and natural environmental health are interdependent, since behavioural and genetic endowments are products of individual biological, environmental, cultural, and social history, which filter community and environmental influences.

We see community health (as we said earlier) as the condition of a socio-economic-environmental system where the economic, social, and political components are organized and maintained in such a way as to promote well-being of both the human and the natural environment. Our framework suggests that healthy communities – those with high levels of trust, social engagement, equitable economic prosperity, for example – are more likely to nurture healthy people. At the same time, the health of a community itself is dependent on investments by individuals, families, and collective organizations (governments) and on the health of the natural environment in which it is embedded. Community-level attributes, then (attributes such as past and current differences in income and education, social capital, social cohesion, and patterns of use/degradation within the natural and built environments), can mediate relationships between human and natural environmental health.

By extension, then, ecological and social resilience and health are interdependent (Folke et al. 1998) and exist together in a particularly powerful relationship in resource-dependent coastal communities

(Adger 2000), since both human and community health are structured within natural environmental conditions (Hancock 1993, 2000).
A healthy natural environment (environmental viability) provides the
wherewithal for human well-being and individual health, and also for
the livelihoods of people in resource-based communities (economic
adequacy). Of course, since there are now very few, if any, relatively
undisturbed natural environments left to us, the present environment
is formed by previous as well as current human activities and management practices. These in turn are mediated by knowledge, culture,
technology, politics, and ownership patterns. Ecologically sensitive and
sustainable management of the natural environment, then, promotes
healthy biophysical systems and maintains healthy ecosystems and thus
the resources upon which humans depend in the long term.

These nested parts of the ecosystem (human, community, biophysical environment) are themselves situated within dynamic and complex
broader environments, of course. Our view of social-ecological health
takes into account the constant change and interactions of the macro
environment and observes that restructuring (whether environmental,
political, economic, or social) provides the context within which
health determinants operate. For example, regime shifts in the biophysical environment can affect human health by influencing the fit
between knowledge systems and the natural and social world (Berkes,
Colding, and Folke 2003), social and environmental policy, social
cohesion, and social capital and support, as well as income, employment, work environments, and nutritional and leisure alternatives.
Broader-scale changes in the form of cumulative human-induced
effects can also affect environmental health by reducing biodiversity
and productivity and by changing disease prevalence and ecosystem
capacity to degrade wastes. Indeed, we can speak of a social-ecological
footprint[6] – the impact that activity in one part of the total system has
on its other constituent parts and the extent and duration of that
impact. That is, our social-ecological definition of health recognizes
that restructuring/health interactions are historically situated, can
change over time and space, and will have uncertain future outcomes.
It also recognizes the need to draw upon various methods, actors,
knowledge systems, and scales of investigation in order to understand
and monitor the relationship between restructuring and the health of
people, communities, and natural environments (Murray et al. 2002).

Our work was designed to allow us to test out these ideas at the level
of two coasts and two regional rural, resource-based, and relatively
remote economies. In this study we show that the key diagnostic features of social-ecological restructuring and health are that they are
(1) interactive processes with (2) bio-physical and human health con-

sequences that have followed on appropriate multi-scale alignments (when successful) and misalignments (when not), and therefore will require (3) multi-scale governance models if social-ecological systems are to operate effectively, promoting creativity and resilience at all levels of the national metasystem.

METHODOLOGY

In this book, we report on this new approach and describe our analysis of the stress that is manifest in the human and biological communities of Canada's East and West Coasts. We do this because we recognize that nowhere has the whole complex adaptive process of coastal social-ecological health and restructuring been thought through in detail before or tested beyond stand-alone case studies. We are attempting to tackle the seemingly intractable decline of coastal communities by using a new methodology put together by natural scientists, social scientists, humanists, and health researchers. It has seven major dimensions. First is *interdisciplinarity*. Integration of the natural and social sciences, as well as a health determinants focus, provides an appreciation, not just of the work of other disciplines, but also of their complexity. This is particularly important when it comes to science-informed resource management, where disciplinary boundaries can mask interactions that may be damaging to both human and biological communities. To be without the input of social scientists – sociologists, anthropologists, and historians, as well as economists and health researchers – is to repeat the old mistake of thinking you can manage (for example) fish through regulations that ignore the realities of stakeholders' lives. That in turn is to ensure that solutions to resource management issues will continue to falter.[7] It is important to understand the range, complexity, and motivations for human behaviour and how these might be successfully addressed if we are to resolve difficult issues. That understanding requires the building of social and natural science partnerships that look jointly at the complexities of resource management. It also means working with social communities and their representatives to create better outcomes.

Coasts Under Stress has taken interdisciplinarity seriously. Members of the team have been working together as biologists, anthropologists, sociologists, biogeochemists, oceanographers, geographers, ethnobotanists, ecologists, education and community health specialists, historians, ethicists, and political scientists.[8] We did this because we were convinced that the problems we face as a consequence of restructuring require the kind of integrated and nuanced analysis that can speak to people *in* their historical and contemporary environments. We also knew that interactions between the various kinds of restructuring, including policy, industrial, and other social changes have crucial and poorly understood consequences. We needed to understand our coasts – the marine environment, the terrestrial one, the substrate (for oil and gas), the littoral, and the watersheds, and also the way these are all linked together ecologically, geographically, and temporally. To do this we had to cut across traditional boundaries and work with new conceptual frameworks and research methods in order to capture the complex (mal)adaptations that occur as human and environmental communities interrelate. This is the only way to understand what has happened to these social-ecological systems, and why.

The second major dimension of our methodology is the *bicoastal* nature of our inquiry. Because the problems we face as a consequence of restructuring require an integrated analysis that can speak to people in their environments, research of this kind needs to be place-specific: as Mackenzie and Norcliffe observe (1997, 6), "Restructuring is a geographically-situated process" where the local situation shapes "larger socio-economic processes into place-specific outcomes." Coasts are interesting places to research. They are edges, boundaries, and/or interfaces, depending on whether one is thinking about them from the inside or outside or across a range of different places, not only geographically but also culturally, politically, and ecologically. Such peripheral (both geographically and politically) places give us different kinds of information from that gained at the continental or political centre. Moreover, they have the great virtue of being national bellwethers for change because of the delicate balance of their component parts, human and ecological.

Culturally, the coastal communities we looked at (see map, figure 1.1) have been an interface between First Nations, colonial, and immigrant cultures. Economically they have been an interface between the formal (colonial and then provincial) economy, the informal subsistence economies, and the political structures of colonial merchant capital and/or foreign and domestic industrial capital, while their people have had to move back and forth between wages (of various kinds) and subsistence. Today they are an interface between First Nations and the dominant culture, between industrial and postindustrial economies, between hinterlands and metropoles, rural and urban. Politically, they are, or have been, on the edge of the US/Canada divide, or the Newfoundland/Quebec divide, and the First Nations/settler divide. Today they are distressed, peripheral, staple-based communities vulnerable to regional and national responses to globalization. Ecologically, they are coastal zones, interfaces between land and sea, marine and land-based ecosystems. Ecologically and culturally, as

Turner, Davidson-Hunt and O'Flaherty (2003) argue, "they exhibit high levels of species richness or biodiversity ... [and] are similarly rich and diverse in cultural terms." As edges, they exist on the edge of the continental landmass, the edge of the continental shelves ... and the edge (margin, periphery) of the continental macro-economy. As boundaries, they are made up of divisions that delineate First Nations' territories, urban and rural environments, resource-based and service-based activities, and places that are "protected" and places that are open to exploitation. Finally, the marine parts of these coasts – the Hecate Strait and the Strait of Belle Isle – are themselves interfaces between different marine ecosystems, providing us with yet another element of complexity.

The third dimension of our methodology is that our work is *community-based*. It is also regionally comparative and informed by larger-scale changes, as is required in complex adaptive systems research. It is place-intensive and time-extensive research that has uncovered the way in which circumstances in the wider economy, which have been building for a long time, have moved Canada towards economic restructuring. Place-extensive (globalization) and time-intensive (competitive response to market demand) economic restructuring have created capital-intensive and labour-extensive strategies for global competitiveness, which have had major impacts on coastal communities, but the effects of such changes at the local, small scale seem to have either passed unnoticed or been dismissed as inevitable.

Coastal communities developed and thrived in economies that were essentially place-intensive and time-extensive. In the formal economies of such places, the task was to exploit the local staple. This was fish ubiquitously on the East Coast to start with and later, in some locations, forests. On the West Coast it was a string of different staples over time, including fish and forests. These staples were exported to metropolitan markets in colonial times and to national and international markets more recently. The formal staple economy was everywhere on the coasts supplemented by an informal economy that effectively subsidized the formal economy by supplementing the very low pay of workers in the staple sector(s) with a means of subsistence in colonial times and with a back-up to the purchase of store products using money obtained through employment (or, more recently, Employment Insurance) in the wage economy. Both formal and informal economies relied on their workers' deep local knowledge of the environment. Both as workers and as residents, people often migrated seasonally to exploit a range of ecological niches (historically) or employment opportunities (historically and today) that, taken together, reflected a yearly round of activities whose purpose was to maintain community and household flexibility during generations of change and uncertainty at the local scale.

Today, however, as flexibility becomes an industrial requirement in the search for global competitiveness, the voices of these small places are barely heard in the clamour for urban renewal and globalization. Their old modus operandi resulted from their need to be flexible and environmentally aware: as such, it was demonstrably sustainable (Ommer and Turner 2004). The kinds of flexibility that allow firms to respond quickly to global production and market demands is dangerous for these social-ecological communities, however, since those kinds of flexibility are built on political, technological, economic, and social changes that enhance flexibility at the level of the postindustrial firm and (sometimes) of governments, but at potentially grave cost to local environments, individuals, and communities. This is because the corporate and institutional flexibilities that have been developed in the last thirty years or so often left coastal communities environmentally damaged and with a specialized infrastructure devoted to staple extraction that is no longer viable. While the staple-extracting corporations can now flexibly shift their operations from country to country, communities are left with the old rigid infrastructures and impoverished environments that make adjustments to restructuring almost impossible. These changes have also supported technologies that permit the global abuse of nature (seen in dying oceans), while companies have responded by searching for techno-fixes (such as biotechnology of various kinds) when nature balks. We will demonstrate these points in detail throughout this book.

The fourth methodological dimension is the use of linked *case studies*. This is the only way to deal systematically and collectively with communities in detail, on two coasts and over time, while still paying attention to the richness and diversity of the habitat, ecosystem, economy, society, culture, and history found in these areas. Of course there are rich literatures associated with the investigation of such themes and places, but when the coasts are studied in this inclusive, complex, and interdisciplinary way, the analysis changes. The questions change and the answers become integrated into a wider picture than single-discipline, single-resource, or single-region analysis can provide.

Case studies allow us to focus on the complexities of one place – the First Nations' (Gitga'at, Coast Tsimshian) settlement of Hartley Bay, for example. They let us "drill down" into one issue – oil and gas impacts, for example, or marine governance. We can explore one form of knowledge – local ecological knowledge (LEK), for example – or one facet of the restructuring crisis, such as its impact on youth in particular places. We can undertake one study of social development (the history of a health care system over an extended period) or one exploration of the pathways between industrial exploitation of a resource and the impact of that exploitation in the coastal zone (log boom debris in British Columbia, for example). Since we undertake case studies within a wider study, however, these kinds of complexities are always considered within the broader concerns of the overarching research question for which these mini-studies provide detailed place-specific and/or temporally extended examinations of particular parts of the restructuring process.

Berkes and Folke (1998) spoke of the need for case studies as the way to get at the complexities of place-specific and temporally dynamic social-ecological interactions and demonstrated that need with a rich array of case studies from around the globe. This study, however, is deliberately not international and not a single-case study. The case studies here are related to one another and examine different aspects of the two regions we studied. Thus, by comparing and contrasting the studies as the evidence demanded, it was possible to get close to the rich and complex totality of lived existence in coastal areas. The case studies covered all the major sectors affected by restructuring, including education, health, environment, and several economic subsectors, while holding "the state" constant, so to speak. The international dimensions of restructuring, which are richly documented in Berkes and Folke (1998), were sacrificed in order to achieve a subnational focus, specifically because this focus permits an examination of the linkages between the different facets that make up life on the East and West Coasts of Canada. This illuminates, not just the impact of restructuring on one place, but also the process of restructuring. If local particularities (kind of resource, weather, culture ...) differ but outcomes are the same or very similar, we can then legitimately start to interrogate the underlying political, ecological, ideological, scientific, and other processes that may be fundamental to, or instrumental in, shaping similarities in the face of difference.

The fifth dimension of our methodology is our particular investigation of *health*, which we have discussed in some detail above. Because this book focuses explicitly on the interactions between social and environmental restructuring and natural environmental, community, and human health, it needs an integrated way of thinking about health as a social-ecological process. Our definition does indeed draw on the contributions from the ecosystem, population, and community health fields of scholarship, capturing the human, social, and environmental dimensions of health, and it reflects the link to restructuring by integrating them into one framework that recognizes that processes operating within social, environmental, and cultural contexts have interdependent, reciprocal, and non-linear relationships, feedback effects, and complex causality. These are the pathways to social-ecological health, and a major part of our task has been to trace these interactive connections between society, ecology, and health.

Our sixth methodological dimension is our focus on *temporal, organizational, and spatial scale, and cross-scale interactions*. The research covered in this book ranges from archaeological investigations that go back as far as ten thousand years to studies at the turn of the twenty-first century and from the local through the regional to the global. This is because full appreciation of how things are demands awareness of both local and larger spatial- and temporal-scale influences on the coasts. It also requires the recognition that if we wish to understand how things are, we need to understand how they got to be the way they now are. Without the organizational, spatial, and temporal dimensions of our research, we could not hope to identify why policies and strategies were developed, how they were implemented, what went wrong, what went right, what was missed or misunderstood, and where the potential for hope now lies for coastal social-ecological communities in the future.

Our development of a social-ecological perspective underlines our conviction that societal processes – political, economic, social, and cultural – over time and at a variety of scales are fundamental, even though under-emphasized in existing integrated models of ecosystem health. Different forms of restructuring are the result of larger-scale processes, past and current, that contribute to reshaping structure-agency, nature-society, and biological and physical relationships. As relationships between individuals and society, between society and nature, and within natural ecosystems change, these different levels of relationships also interact to produce changes to the health of people, communities, and natural environments, as well as interacting to reshape restructuring processes themselves.

Seventh, and last, is our focus on *knowledge* production, knowledge systems, and knowledge transmission. We have discovered in our course of research not only new knowledge and new ways of conceptualizing it but also the range of types of knowledge needed to engage these tangled issues. We have worked with several different knowledge systems: what is called "lay," "vernacular," or "indigenous" knowledge, which comprises both traditional and local knowledge, and what is usually referred to in the literature as "expert" knowledge, by which is meant the kinds of academic knowledge created in the natural and earth sciences, the social sciences, the health and education sciences, and the humanities. We have worked to bring together different types of knowledge, often seen as incommensurable, to provide different perspectives on the same problem.

We have also paid attention to moving knowledge. Most of the problems we chronicle have been known by some people for some time. If there has been no resolution, it is not because the problems are entirely unknown. It is because knowledge about the problems has not penetrated to industry, government, and individuals who have to learn, absorb, and act. We have looked at different mechanisms for getting this knowledge to move and at some of the mechanisms by which it is stalled. Ultimately, our capacity to "know" the world around us is always imperfect and mediated by experience, language, culture, training, memory, mentorship, and the like. Working critically, rigorously, and in a transparent fashion with information from different knowledge systems, while using a comparative, historical approach can help improve our capacity to understand the world and assist in the identification of those processes that constrain such understanding. It becomes clear that all knowledge is, to some degree, a social-ecological product. This lets us imagine alternative pasts and construct alternative futures that, in and of themselves, contribute to the understanding we seek. Taken together, the different kinds of knowledge we have produced, shared, and been offered have illuminated the multifaceted nature of social-ecological restructuring and its impact on the health of people, communities, and their environments.[9]

To summarize, then, our model, with its accompanying methodology, provides an enhanced understanding and explanation of the systematic degradation of the environments and opportunities of our coastal communities, its relationship to human health, and what some possible alternatives might be. Our understanding of our communities' situation is based on the concept of "social-ecological health" (Dolan et al. 2005): we see them as being in a worrisome (and generally declining) state of social-ecological health, as the result of an ongoing process of restructuring. We see that process as involving a complex interaction between people, their communities, their organizations, and their physical environments – an interaction that involves a large number of important factors, some of which (the individual, population, and community health components, in particular) have been neglected.

If we consider the negative aspects of the restructuring process as blockages, rigidities, or wrong turns in decision-making processes, we will gain a richer understanding of what has happened. We can usefully conceptualize these blockages and rigidities in terms of misalignment or asymmetry in the scale of situations, interests, knowledge, and/or power between those making key decisions concerning the exploitation of key resources and those exposed to the consequences of these decisions. By "misalignment," we mean mismatches or divergences in allocation such that benefits go in one direction and risks in another (over both space and time, on occasion), such as when the mercury released from coal burning provides benefits in central North America but risks (such as mercury deposition) in eastern North America.

Specifically, we hypothesize that the greater this mismatch in a decision-making process, the more negative the outcomes are likely to be for the social-ecological health of local communities. By "scale asymmetry," we mean the distribution of risks and benefits to different points or levels in space or time, such as occurs with the concentration of benefits at one location (the centre of a region, nation, or empire) while risks are experienced over multiple locations, which become locked in a staple trap. By "complexity mismatch" (asymmetry) we mean that decision makers have been paying attention to only one dimension of a multiply determined interaction, which results in unintended consequences. This approach involves reducing restructuring to a set of decisions; the approach may be simplistic but it has the advantage of clarity. We note, however, that even if policy-makers accept that mismatches, asymmetries, and inequities have occurred, reversing the downward spiral in which restructuring has generated social-ecological health degradation will still be difficult. One promising way to deal with this problem is to enhance the match/alignment between decisions and their subjects by introducing multilevel modes of decision making, by which we mean decision making that assigns influence to people at multiple levels and that pays attention to the input of factors from multiple levels of time and space and to the impact of decisions at the same multiple levels.

As this volume was being completed, Warren (2005) published his interpretation of "socioecological" systems and reached a similar conclusion about the importance of scale and of misalignment between levels, pointing out that local-level actions are often conducted without adequate understanding of how they are constrained by regional and even global levels of what he labels as population, ecosystem, landscape, and society. Warren brings out the implications for local-level actions concerned with resource management. On the other hand, macrolevel management policy that ignores local variation and needs is also inadequate. Thus the importance of matching institutional forms to specific contexts and scales is emphasized (458).[10]

CONCLUSION

Throughout this book, then, we look at coastal communities' history, their present dilemmas, and their future options in order to identify interactive restructuring, health consequences, scale alignments and

misalignments, and multiscale governance requirements. As we identify the pathways to (or from) social-ecological health, we find negative
activities created through poor restructuring processes that generate
blockages, rigidities, or wrong turns in the pathways to social-ecological
health. We examine these further in terms of the match or mismatch
in scale of situations, interests, knowledge, and/or power between those
making key decisions concerning the exploitation of key resources and
those (people and/or resources) exposed to the consequences of such
decisions, our hypothesis being that the greater the mismatch in a
decision-making process, proves, the more negative the outcomes are
likely to be for social-ecological health. By scale, in this work, we mean
a multiplicity of things – temporal, institutional, and geographic scale
and a mismatch in scale complexity can mean that decision makers pay
attention to only one dimension of a complex interactive effect, resulting in a cascade of unintended consequences. The good news is that,
although many of the pathways we identify here trace a negative downwards spiral, uncovering the match/alignment misfit between decisions and their subjects allows us to suggest how introducing multilevel
modes of decision making could help work towards solutions to such
misalignments in a promising manner.

The story we tell is about vulnerability and resilience – where
resilience came from, where it lies, how endangered it is, and what may
be done to enhance it – in the social-ecological systems of the East and
West Coasts of Canada. Part 1 (chapters 2–4) presents the evidence for
the increasing social-ecological degradation we are postulating, part 2
(chapters 5–11) asks why such accelerating degradation has occurred,
and part 3 (chapters 12–15) considers what needs to be done about it.
Chapter 16 concludes the book.

Part 1 interrogates, using case studies, the history of changes in the
social-ecological structure of the two coasts over time: how their natural resource bases were developed, what happened to them, and the
impact that had on community well-being. It does this by resource sector, inevitably, but always bearing in mind that a key feature of the history of our coasts has been restructuring within and across sectors. The
history of our coasts, like that of other places, is not linear but one in
which there will be environmental, social, cultural, and time variations
of which we will need to be aware. It is also important to remember
that while some communities were focused on one resource, others
had a broader resource base, and yet others were by definition temporary, being based on nonrenewable resources. We also examine the
legacy of the past use of resources. We start with trends over time in
marine ecosystems and the coastal communities that depended, for
the most part, on the sea (chapter 2). In the following chapter (3) we

look at issues in science and the management of fisheries that have to deal with the legacy of scarcity. We then turn to terrestrial ecosystems in the form of coastal forests for an identical examination of their social-ecological past (chapter 4) and in the next chapter (5) look at some of the cross-boundary problems that we inherit from single-staple foci, with their inability to come to grips with interactive ecosystem effects. Chapter 6 focuses on communities created to exploit the substrate, in terms of their exploitation and their health, be that anthropogenic or biophysical.

Part 2 then asks what is the human impact of restructuring and the damage that has been to done social-ecological health. Here we are mindful of the interactive effects of restructuring on many facets of society as a result of the cross-scale and cross-boundary dependencies of coastal communities on the ecologies that they exploit and on which they depend. We also discuss the present-day inheritance of social-ecological restructuring on human society, always mindful of the interconnectedness of resources and coastal communities. We consider the issues of physical and mental human health that coastal-zone social-ecological communities have to confront (chapters 7–11), including overall health care trends (chapter 7); case studies of the lived socio-economic experience and well-being of people, households, and communities (chapters 8 and 9); education and the future prospects for youth (chapter 10); and issues of nutrition and diet (chapter 11), on both coasts. We find that a process of degradation has been masked by discipline-based science, by silo-based policy-making, and by the dynamics of restructuring within and across sectors (intensification, expansion, shifting across sectors – to some degree caused by the unanticipated resilience of coastal people and by unanticipated ecological responses). The process of degradation has also tended to channel that resilience in the direction of extraction, because that is where the jobs are, rather than in the direction of stewardship and recovery. A scientific, industrial, and policy focus on shorter-term temporal scales and large spatial scales within science and a focus on individuals (too many fishermen) rather than on organizational structures and institutions have contributed to this process of accelerating degradation.

We also show that the social-ecological degradation of our coasts is reaching critical proportions overall, although some areas are more degraded than others and in some cases creative, primarily local initiatives have provided important indications of alternative directions and of creative ways to resist degradation and promote recovery of environments, people, and their communities. Too often, these initiatives are having to fight larger policy structures and directions, rather than being nurtured and sustained by this larger environment. The situation on our coasts has become critical, and without focused efforts on social-ecological recovery within coastal areas and coastal communities, it will produce dramatic changes to Canada's coasts that are likely to be associated with ongoing and escalating degradation in the longer term – movement to a new social-ecological threshold – further undermining the resilience and potential for recovery in these areas.

Part 3 then asks what needs to be done. It moves us forward to look at the potential for new opportunities and options, including their problems. Chapter 12 focuses on watershed, terrestrial, and littoral issues, with an emphasis on renewable resources, while chapter 13 considers the problems and potentials of oil and gas development on the West Coast. Our last two data chapters (14 and 15) consider the options for new forms of governance that may be available in the marine, terrestrial, and littoral environments. We conclude (chapter 16) with a discussion of what we have learned and what we think needs to be addressed to ensure the resilience of Canada's coastal communities in the future, mindful that the view from the coasts has relevance for the rest of the country, and indeed the world. Globally, we have not achieved sustainability, and recovery is only marginally even under discussion. What such discussion might entail is dealt with in our conclusion to the book.

2

A Social-Ecological History of Canada's Fisheries

INTRODUCTION: RESEARCHING AND WRITING SOCIAL-ECOLOGICAL FISHERIES HISTORY

In chapter 1 we stated that the key diagnostic features of social-ecological restructuring and health are (1) that they are interactive processes with (2) biophysical and human health consequences that have followed on appropriate multiscale alignments (when successful) and misalignments (when not) and that therefore will require (3) multiscale governance models if social-ecological systems are to operate effectively, promoting creativity and resilience at all levels of the national metasystem. In this chapter[1] and the next we turn to the fisheries and fishing communities of both coasts, to demonstrate that the accelerating degradation of their social-ecological systems (restructuring) has resulted from a long-term interactive process between changes in natural environments and political, industrial, and social change, a process that has over time affected the health of people and communities and the marine ecosystems on which they depend. In this chapter, we construct a social-ecological history of the fisheries on both coasts, since only that history can tell us how degradation came to be, if it has sped up over time, and what the social impacts and responses have been. The following chapter brings the story up to the present, looking at the legacy with which fisheries management now has to deal and how this is being done.

This chapter is an experiment in writing social-ecological history, not a standard history of the fisheries. It focuses on the ecological reconstruction that has occurred in the marine and/or seaboard social-ecological system, along with whatever social restructuring has also happened, in order to establish the historical basis for the present-day situation, which leaves communities of fish and fishers on both

coasts so much at risk of collapse. In order to do this, we have reconstructed – at a range of levels and using three different methodologies
– the key transformations that have taken place in marine ecosystems
on Canada's East and West Coasts. Each approach provides partial
understanding and demonstrates what can be uncovered using that
particular methodology. We record and discuss important changes in
marine food webs and fishing practices and policies and identify some
of the key interactions between these elements that have transformed
marine ecosystems and fishing communities over many decades. Our
three different methodological approaches, taken together, allow us to
paint a picture of what past ecosystems were probably like on both
coasts and then to quantify the changes that have occurred, in the process linking (to some degree and in different ways on each coast)
human and ecosystem history, insofar as they have influenced each
other and social-ecological health. Throughout, as backdrop to our
findings, we have drawn on the existing rich conventional historical
studies of both coasts.

Our first approach has been to assess ecosystem health at the
macroscale and over the long term; because we cannot act effectively
until we know what it is that we have lost, we cannot look forward
without having looked back. We have based our retrospective analysis
on the latest versions of a whole-ecosystem modelling technique
("Ecopath with Ecosim," Christensen and Walters 2004) that employs
data from a wide range of detailed studies and provides a holistic perspective on changes in the ecosystem and its fisheries. The models
employ balanced trophic linkages tuned to past fishery assessments
and survey data to predict changes in biomasses, consumption rates,
diet compositions, and fisheries yields for each species or species
group (e.g., Cox et al. 2002; Kitchell et al. 2002) under past, present,
and alternative fishery and climate regimes (Pitcher and Forrest
2004). This type of ecosystem simulation modelling, which can
express the importance of fishing relative to interspecies predation,
has influenced the recent focus on trophic level as a functional entity,
rather than a sorting tool. In other words, descriptions of the trophic
levels[2] within an ecosystem can be useful to compare the distribution
of feeding at all levels of the food web, and the mean trophic level of
an ecosystem can be used to describe changes within systems over
time (Pauly et al. 1998).

We developed whole-ecosystem mass-balance and simulation models
for marine ecosystems tuned to data on each coast during similar historical periods by using an innovative "Back to the Future" (Pitcher
2005b; Pitcher et al. 2005) technique modelling past marine abundances.[3] The spatial and temporal precision of our ecosystem models is

constrained, however, both for marine ecosystem changes and for fishing activities, in that the models provide slices or progressions through time – our best estimates of what our marine ecosystems must have looked like and how they changed in richness or diversity over the years we selected. What these ecological models cannot do, however, is to tell us very much about how social changes in fisheries have interacted with marine biological changes to affect the health of these ecosystems and of fisheries communities.[4]

On the West Coast, therefore, where First Nations have been fishing for sustenance and tribal exchange for millennia, we executed a *second* approach, which we used to deepen the social part of the social-ecological history of the coast. This work helps to tell us how and why we lose or preserve social-ecological systems over the long term. Identifying some of the human dynamics behind the models meant going to a much finer temporal scale, to get at First Nations' relationship to the coastal marine ecosystem over both pre- and post-contact time.[5] Given the long and complex history of First Nations' interactions with West Coast marine ecosystems, we thought it necessary to carry out a detailed case study (including archaeological and legal history) of interactions between First Nations and coastal ecosystems and of the manner in which their world was reconstructed by European contact and subsequent colonization of the coast. Although time and budget constraints did not allow us to do a parallel analysis of settler history (including local ecological knowledge [LEK], which would have allowed a much finer-grained analysis of people/resource interactions) on the West Coast, the work we carried out reminds us that there is a human, and frequently a political, side to ecosystem change. More work needs to be done to answer such questions as how, for example, the degradation of the salmon resource played out in the lives, economies, and cultures of First Nations and settlers. A full social-ecological history needs all of these pieces. Part of our experiment here has been to identify these gaps and demonstrate the value of that kind of knowledge.

Our third experimental approach (which we carried out only on the East Coast, again because of time and budget constraints) extended our ability to deal with the how and why of social-ecological change by developing new information sources containing the local knowledge of fish harvesters and applying these sources, along with scientific and archival sources, in order to develop a finer-grained (spatial and temporal) picture of marine ecosystem changes and of apparent interactions over time between fish, fisheries, and fisheries policies. Our capacity to study interactive effects of this kind depends on the data and related knowledge we have at our disposal. However, because we are interested in devising a knowledge system that can capture such interactive effects more effectively and because all knowledge is in some sense a social-ecological product, we have also explored the history of science and LEK in relation to multiple fisheries (Alcock, Ings, and Schneider forthcoming). This history has provided us with greater understanding of the social-ecological dynamic associated with knowledge production in different historical contexts and environments. The knowledge systems vary over time and space (Murray and Neis 2004), and they have interacted with each other in different ways in different areas and over our history to influence what fishers, scientists, managers, and policy-makers see in the world around them, how they interpret what they see, and, related to this, human relationships with the environment and with each other.

For our East Coast work, we combined information from fish harvesters with that from archival sources (including census data and International Commission for the Northwest Atlantic Fisheries [ICNAF] and Northwest Atlantic Fishery Organization [NAFO] landings data) and also with research vessel (RV) survey data, in order to reconstruct the history of fish and fisheries between 1950 and the present for the area from Sandwich Bay, Labrador, south through the Labrador Straits to Port aux Basques (southeast Labrador and the northern Gulf) and into White Bay in area 3K (off northeast Newfoundland). Our detailed studies of changes in landings, effort, and stock abundance are helping us to reconstruct changes in coastal marine social-ecological systems, with a particular focus on fish populations and fish harvesting over the past several decades. They have also provided important insights into interactions between changes in fish, fisheries, and fisheries policy over time that might eventually be incorporated into simulation models.

We also carried out some microscale reconstructions of harvester-resource interactions, one of which (that of the St John Bay lobster fishery) used the historical reconstruction of that fishery obtained from LEK and other sources to develop an innovative individual-based model for harvester-lobster interactions that was used to create "what-if" scenarios in order to assess the probable impact on lobster catches of different management initiatives in that fishery (Whalen forthcoming). Thus, the two methodologies complement one another, the former being macrolevel, the latter microlevel.

In this East Coast work, our larger goal has been to test the potential of this kind of reconstruction of environmental history, which relies on multiple and quite different data sources. We have found that it allows us to identify more clearly the dynamics of the social-ecological systems and how and why they have changed; it requires us to discuss these historical reconstructions with fish harvesters and other members of their communities and with policymakers; and it allows us to explore with them the lessons from history and their relevance for current and future options in these areas. Thus, we have built a basis for multilevel discussions with harvesters, other community members, and government on potential strategies for the recovery both of these ecosystems and of the communities that depend on them, taking into account the risks, costs, and benefits associated with such strategies (see also chapters 3 and 14).

TOWARDS A SOCIAL-ECOLOGICAL HISTORY OF WEST COAST FISHERIES

Past Ecosystems

Coastal communities appear to have always lived with the issue of marine ecosystem restructuring.[6] To get a macro perspective on that restructuring, we built mass-balance models whose dates were dictated by important social-ecological historical events on both coasts (Vasconcellos and Pitcher 2002). These models attempted to express biological ecosystem change in all its complexity – predator-prey relations, noncommercial as well as commercial species, indeed the whole food web – and fisheries mortality. We used traditional and local knowledge in drawing up the models and ensured that they simulated biomass changes in response to fishing in a realistic fashion. Interviews seeking information on the changes in abundance of over one hundred kinds of organisms in the Hecate Strait ecosystem were held in Prince Rupert and Haida Gwaii (Ainsworth 2004a; Pitcher, Power, and Wood 2002), and the results for each group of organisms archived in a database (accessible on the web).[7] Interviews were followed up by a community workshop (Power, Haggan, and Pitcher 2004) in which, among other things, elements of the model were improved with local knowledge.

On the West Coast we built these models for important historical times on that coast: the 1750s (pre-contact), the 1900s (pre-trawling), the 1950s, and, of course, the present-day. The 1950s model was tuned to best fit all available stock assessments and survey data and driven forward by annual climate series to 2000. On the East Coast (in area 2J3K3L3NO, which is off the northeast and east coasts of Newfoundland and off Labrador: see figure 1.1) our models provide snapshots of what these ecosystems looked like at several crucial points in their history: in the 1450s (pre-contact), 1900s, 1950s (before distant-water fleets), late 1960s (after peak landings), 1980s (before the cod collapse) and

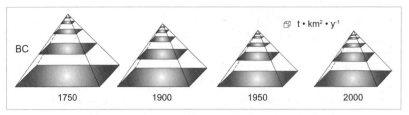

Figure 2.1   Trophic flow pyramids for northern British Columbia from
whole-ecosystem modelling. The height of the pyramid indicates maximum
trophic chain length, each level corresponds to a trophic level in the system,
and surface area of level indicates quantity of biomass flow between trophic
levels (in t•km²•y⁻¹) (Ainsworth et al. 2002).

1990s (after the cod collapse). Models were fitted to all recent stock-
assessment and survey data (e.g., Lilly 2002), although it did not prove
possible to simulate precisely the recent large collapse (Heymans
2003b).

On both coasts, our results from these models indicated a long history of depletion. It is not clear that the first marine restructuring
occurred on contact with Europeans: some restructuring may have
occurred before colonization. It is likely, for example, that the first
human inhabitants of the coast exploited and wiped out the Steller's
sea cow (*Hydrodamalis gigas*; see Pitcher 2004c). That said, figure 2.1
shows the estimated changes in biomass in northern British Columbia
between pre-contact time (1750) and the recent past (2000), while figure 2.2 summarizes the percentage changes in biomass by species. The
picture for salmon is complex. Figure 2.3 shows the reported catches
of Pacific salmon (*Oncorhynchus spp.*) in the Hecate Strait region from
1877 to 1977 (Hewes 1973), with a note that First Nations' catches by
1750 are estimated in our models at 6,400 tonnes (this is one social
aspect of ecological modelling). Here there is a clear cyclical pattern,
but nevertheless the overall trend is also down.

It is not clear what the role of water temperature shifts may have
been (Pitcher and Forrest 2004), but since the largest reductions are
in top predators that are caught for human consumption (as table
food) and since there is continuous decline in overall ecosystem complexity as well, the picture implies significant loss of ecological value
(caused to some significant degree by overfishing: Coleman et al.
2004; Hilborn et al. 2003; Jackson et al. 2001; Pauly et al. 2001) and in
human terms a concomitant loss in social and economic stability. Rose
(2004) suggests, for example, that in the northwest Atlantic, declines
in the 1800s were caused by the Little Ice Age, while those in the 1960s
were the result of overfishing and those in the 1980s were caused by a

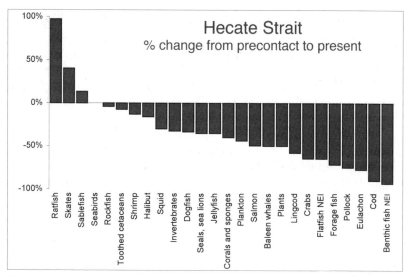

Figure 2.2   Percentage change in biomasses of selected functional groups from pre-contact to the present for northern British Columbia, as calculated using whole-ecosystem simulation models (Ainsworth et al. 2002)

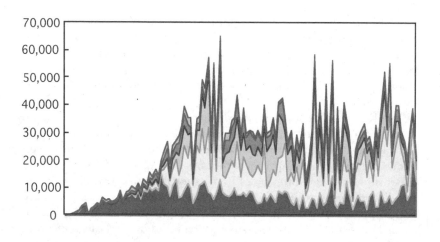

■ Sockeye   □ Pink   ▨ Chum   ■ Coho   ▨ Chinook

Figure 2.3   Reported catches of Pacific salmon in the Hecate Strait region from 1877 to 1977, with a note that First Nations' catches by 1750 are estimated in our models at 6,400 tonnes (Hewes 1973)

combination of climate change and overfishing, which also continues to be instrumental in the present failure of the stocks to recover. We note, furthermore, that the declines are in abundance, diversity, and trophic level over time, indicating substantial decline in ecosystem "health." More positively, we can think of these models in terms of what past abundances it might be possible to restore, an issue that will be revisited in chapter 12.

For more fine detail and an explanation of the interaction of the biophysical with social restructuring, we would be required to do more detailed historical work, the first part of which (our second methodological approach) has been to make a beginning to an explicitly social-ecological history of First Nations on the West Coast.

Towards a Social-Ecological History of First Nations' Fisheries

Early History. As far as we can tell, coastal peoples have always weathered whatever ecosystem changes confronted them. There is a growing literature on ecosystem change and indigenous resilience, most importantly (for our area) that of Lyman and Cannon (2004) and Cannon (1991), which use archaeological data from Namu. Our own archaeological work on Haida Gwaii identified the way of life of indigenous people on those islands and their interactions with the marine ecosystem in the distant past. We investigated two sites, a small part of the rich cultural heritage of the area, and we note in passing that precautions to protect these and similar sites will need to be taken if energy development occurs here in future.

The Kilgii Gwaay site, which we excavated in 2001–2, is over 9,450 (radiocarbon) years old, dating back to about 10,600 years ago. Because it sits in a waterlogged position in the intertidal zone, Kilgii Gwaay has very unusual preservation; not only of stone tools, as is typical of sites of its age, but also of bone, shell, and wood. (The next oldest site on the northwest coast with similar characteristics is a mere 5,500 years old). We found that its ancient inhabitants had significant capacity to exploit maritime resources: halibut (*Hippoglossus stenolepsis*), rockfish (*Sebastes* spp.), ling cod (*Ophiodon elongatus*), seals (harbour seal [*Phoca vitulina*], northern elephant seal [*Mirounga angustirostris*], northern fur seal [*Callorhinus ursinus*], sea lions (Steller's [*Eumetopias jubatus*], California [*Zalophus californianus*]), and sea otters (*Enhydra lutris*), as well as many seabirds, including albatross (*Phoebastria* sp.). Their use of black bear (*Ursus americanus*) was also significant. These resources are nearly identical to those used in historic times and today, and they mirror the present Haida way of life, including manufacture

Figure 2.4   Dr Quentin Mackie with brown bear jaw, Haida Gwaii
(Fedje)

of cordage, wooden wedges, and basketry. The site also contains the
second-oldest evidence on the coast for bone tool construction. The
technological profile and resource use show that people at this very
early date must have been skilled in using watercraft and that they were
able to exploit all the same offshore niches as did their Haida descendants. The unusual preservation at this site makes it a benchmark for
interpretation of many others of the oldest sites, which typically contain only stone tools.

   The second site (K1) has significant faunal remains located in a lime-
stone cave, including bear bones radiocarbon dated from 9,700 to
14,500 years old (back to 17,000 years ago). These natural assemblages of bones include both black bears and brown bears (*Ursus
arctos*), which had not been previously recorded on Haida Gwaii (see

figure 2.4). Because we know that at this time Haida Gwaii was near its glacial maximum, the presence of large terrestrial omnivores whose bone chemistry shows that they did not consume large amounts of marine resources implies either a very early post-glacial route into the area or the presence of a large viable glacial refugium. Taken together, these sites illustrate a highly dynamic environment of changing fauna, flora, and sea levels, as well as the human ability to adapt to such change. These findings, along with the important work at Namu, are vital for our knowledge of the peopling of the Americas and for human ecology in general.

Aboriginal life evolved slowly until contact between Aboriginal communities and Europeans on the Pacific Coast in the late eighteenth century began processes that would lead to massive restructuring of those Aboriginal communities. Seaborne explorers arrived in the 1770s and were followed by traders who returned seasonally for sea otter pelts. The trade was vigorous, frequently violent, and short-lived: once the sea otters disappeared the British and American trading ships did not return. However, the effects of these early years of contact lingered. The uneven distribution of European trade goods, particularly firearms, affected the balance of power between Aboriginal communities, allowing some groups to expand into territories formerly occupied by their neighbours.[8] More severely, the exotic diseases that accompanied the traders and to which Aboriginal peoples had no immunity caused horrendous suffering. Smallpox appears to have arrived on the north coast in the 1770s, spreading throughout the peoples of the northwest coast before the end of the century, and likely resulting in a mortality rate approaching 30 percent (Boyd 1999). Recurring outbreaks followed, as did measles, influenza, and venereal diseases, which caused a massive depopulation (Galois 1996). Current estimates suggest a population decline of as much as 90 percent in the first century after contact, a figure consistent with other regions in the Western Hemisphere (Cole Harris 1997/98).

Land-based trading companies arrived on the Pacific in the early 1800s and established the first permanent non–First Nations presence in the territory. They competed strongly for supplies of furs until 1821, when the Hudson's Bay Company (HBC) took over its principal competitor, the North West Company, securing a trading monopoly (Mackie 1997). The HBC was neither interested nor able to extend its authority much beyond its forts and its trading routes (Cole Harris 1997/98). Nonetheless, many First Nations were now connected through the HBC to international markets and subject to its whims. Since First Nations' fisheries were integral parts of the land-based trade in the first half of the nineteenth century, rather than deploy its labour force to fish in unfamiliar waters, the HBC relied on fish that it purchased from First Nations to provide an important source of food for its workforces in many posts along the coast and in the interior. At some posts, including Fort Langley near the mouth of the Fraser River, the HBC also purchased salmon from First Nations fishers, which it salted and barrelled to export to emerging markets around the Pacific in the 1830s and 1840s (Mackie 1997). Depending on the species of fish and body of water, First Nations fishers deployed a great variety of hooks, spears, traps, weirs, and nets for their catch (Stewart 1977). Fishing was not equally good everywhere, and in many instances families, clans, or villages owned the best locations. Far from being unregulated, as many government officials would later assume, these fisheries were surrounded by rules that governed rights of access and of use.

In the mid-nineteenth century, British interest in the territory turned away from the commercial trading operations of the HBC and towards creating a settler society. Growing settlement transformed the economy from First Nations–European trade towards agriculture, gold mining, forestry, and the fisheries. Instead of working as partners in trade with First Nations, a settler society sought to control resources and to occupy land. In 1846, Britain and the United States extended the forty-ninth parallel as the boundary between their interests in the North American West. These geopolitical decisions, made by imperial powers with the most tentative and contingent understandings of the territories they were dividing, were initially unnoticed by the people living in those territories (Cole Harris 1997/98).

The consequences of such scale asymmetries, however, would soon be felt, and the established trading networks were among the first to be disrupted. The HBC moved its headquarters north from the mouth of the Columbia River to Vancouver Island, where, under a Crown charter in 1849, it began to encourage settlement. To secure the first settlement frontier in British Columbia – the Saanich Peninsula – the senior HBC official and future colonial governor, James Douglas, signed treaties with the First Nations in the early 1850s. He also arranged treaties to secure the land surrounding coal deposits further north on Vancouver Island, but after 1854 the treaty process in British Columbia ended. Thereafter, the question of Aboriginal title was ignored, even suppressed, as the settler society grew (Cail 1974; Tennant 1990; Cole Harris 2002).

The Douglas Treaties are brief two-paragraph agreements in which the signing First Nations purportedly ceded their land to "the white people for ever" in exchange for a one-time monetary payment, protection for their "village sites and enclosed fields," the "liberty to hunt over the unoccupied," and the ability "to carry on [their] fisheries as formerly" (British Columbia 1875). To this point, the fisheries provision in the Douglas Treaties has provided little protection for First Nations' fisheries in British Columbia, unlike similar provisions in the contemporary treaties signed in Washington State – but its meaning continues to be debated and is the subject of continuing litigation.

Early settlers in British Columbia paid little attention to fish until the introduction of canning technology in the early 1870s. Once canning was introduced, British and American capital flooded into the province, followed by European and Japanese fishers and Chinese cannery workers, to exploit the sockeye salmon (*Oncorhynchus nerka*). Canned sockeye salmon could reach markets in eastern North America and Europe, and the industry grew rapidly, first near the mouth of the Fraser River, then on the Skeena and Nass Rivers, and at many other small rivers along the coast. First Nations people participated in this industrial commercial fishery, men in the boats and women in the canneries (Knight 1996). However, the importance of their labour diminished rapidly on the south coast, so that by the early twentieth century they were an insignificant part of the large commercial fleet.

On the north coast they remained crucial participants until the industry consolidated its operations in the south after World War II, but even where there were employment opportunities in the fisheries, Canada did not recognize or protect First Nations ownership. As part of its general refusal to recognize Aboriginal title or negotiate treaties in British Columbia (with the exception of the Douglas Treaties on Vancouver Island and Treaty 8 in the northeast corner of the province), Canada opened First Nations' fisheries to cannery owners.

In place of treaties, the provincial and Dominion governments established a system of Indian reserve allotments, setting aside many small parcels of land, much of it worthless except for its proximity to a fresh- or salt-water fishery. Around Hecate Strait, for example (figure 2.5), the Dominion/provincial reserve commissioners allotted Indian reserves during bursts of activity in 1882 (Masset and Skidegate) and 1891 (Kitkatla). In 1916, the Royal Commission on Indian Affairs for the Province of British Columbia allotted reserves to those groups that had been previously overlooked (Kitasoo and Hartley Bay). The reserves, held by the Dominion in trust for the Indian band to whom it had been allotted, amounted to a small fraction of the total land area in British Columbia. The Royal Commission created the legal boundaries that marked First Nations from non–First Nations land in provincial land registries and courts. Beyond the reserves, First Nations claims were ignored; the vast majority of the province was opened for non–First Nations settlement and development (Cole Harris 2002).

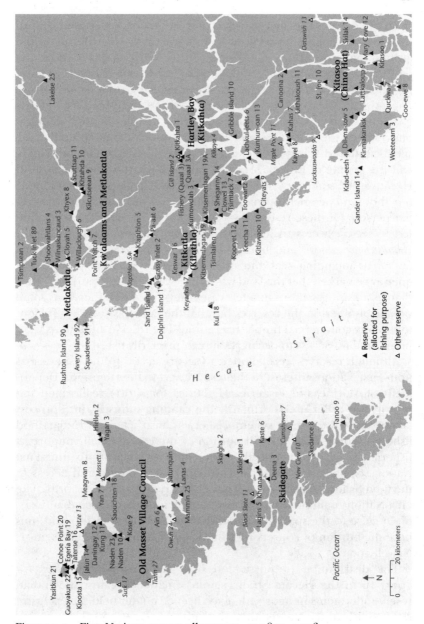

Figure 2.5   First Nations reserve allotments, ca 1891–1916

To the extent that the provincial and Dominion governments justified the small land area set aside as Indian reserve, they did so on the
ground that most Aboriginal peoples in British Columbia, particularly
on the coast, were fishing peoples and did not need a large land base.
To understand the reserve geography in British Columbia, therefore,
one must understand the connections to the fisheries (D.C. Harris
2004). Instead of consolidating indigenous peoples in large contiguous reserves, the Dominion and provincial governments allotted many
small, scattered reserves oriented towards the ocean. For the Haida
and the people of Kitkatla, Hartley Bay, and Kitasoo who live around
Hecate Strait, some 12,660 acres were reserved in 86 parcels for almost
one thousand people: approximately 147 acres per reserve, or 13 acres
per person. Of these reserves, seventy-one (or 83 percent) were allotted to secure access to fisheries, including salmon, halibut, dogfish
(*Squalus acanthias*), and herring (*Clupea pallasii pallasii*), and to sea
mammals including seal and otter. The reserve commissioners frequently remarked that the land was of little value for any purpose other
than as a fishing station. In a few cases they noted that there was some
valuable timber on the reserves but that the land base was insufficient
to support commercial timber harvesting. The reserve allotments were
predicated on access to ocean resources, primarily fish.

Although reserves were allotted to secure access to the fisheries, as
non–First Nations interest in the resource grew, the Dominion Department of Marine and Fisheries (Fisheries for short) reallocated the
resource to other users, primarily the canning industry. In a process
analogous to the reserve system, Fisheries constructed an "Indian food
fishery," setting aside a small portion of the commercially important
fisheries as food fish, but allocating the vast majority to non–First
Nations commercial interests. The department considered access to
the food fishery a privilege, and one that certainly had no priority. The
First Nations' salmon fisheries, most of which were conducted in rivers,
often close to the spawning grounds, had last access to fish and thus
bore the burden of conserving fish for other users (D.C. Harris 2001;
Newell 1993).

The distribution of purse and drag seine salmon licences for
1915/16 in the Hecate Strait region is mapped against the Indian
reserve allotments in figure 2.6. Each licence holder held the exclusive
right to fish in the waters identified in the licence, and the mapping
reveals just how much of the region had been allocated to commercial
fishing interests. Cannery owners held most of the licences; Aboriginal
fishers or communities held none. As the map reveals, many of the licences included waters immediately adjacent to reserves that had been set
aside to secure access to the fishery. Cannery boats were catching the

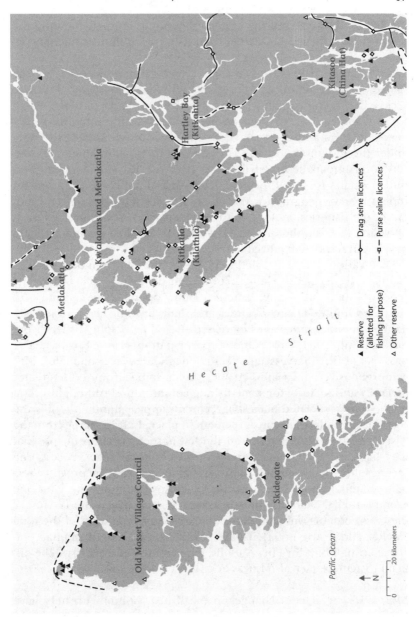

Figure 2.6 Drag and purse seine licence areas in relation to reserve allotments, 1915–16

fish before they entered the rivers, in many cases greatly limiting the opportunities for an Aboriginal food fishery, let alone a commercial fishery.

First Nations fishers worked some of the cannery licences, but the Canadian state, ignoring Aboriginal rights or title claims, assumed access. In this context, First Nation peoples' employment in the industry grew increasingly tenuous, subject as it was to the interests of cannery operators and constrained by their status as wards of the state under the Indian Act. By the 1920s, when the Dominion and provincial governments accepted the Report of the Royal Commission on Indian Affairs for the Province of British Columbia and imposed an Indian reserve geography, Aboriginal peoples were increasingly shut out of the fisheries, looking out from their reserves as other interests harvested the fish, the one resource that, if they had maintained control, might have enabled viable reserve-based economies.

First Nations have struggled to maintain a presence in the commercial fisheries in British Columbia through the twentieth century. The reorganization of many fisheries from open-licence regimes to restricted limited-licence regimes and then, in some cases, to individual transferable quotas, has increased the value of some fisheries but has not enhanced First Nations' participation (Newell 1993). The prospect of fish "harvest agreements," negotiated in conjunction with comprehensive land claims treaties, holds some promise for greater certainty and security for some fishing communities. Although several treaties and associated harvest agreements appear imminent, only the Nisga'a Nation has such an agreement in place. Extrapolating from the Nisga'a Harvest Agreement and the few agreements-in-principle that have been negotiated under the British Columbia Treaty Process, one recent study suggests that if future agreements across the province recognize similar allocations, then in the sockeye salmon fishery (as one example) First Nations' domestic and commercial fisheries under these agreements will amount to approximately one-third of the total (McRae and Pearse 2004). This allocation, which is itself an unacceptable starting point for First Nations, let alone the end goal,[9] will be difficult to achieve (see also Lucas 2004).

Modern History. Recent changes in the British Columbia fishery have not improved the situation. Current restructuring has to be understood as occurring in a time of already severely depleted resources and on the heels of a spiral of downsizing in process since before World War II, albeit accelerated in the last decade. A provincially commissioned study (Gislason, Lam, and Mohan 1996) cites fifteen communities as having been most severely impacted by recent events, and Alert

Bay, on Cormorant Island, northern Vancouver Island, is listed as experiencing the third-greatest employment loss (British Columbia 2000).

The village of Alert Bay was founded with the establishment of a salmon saltery in 1870.[10] The 'Namgis people became the labour force, convinced to move from their homes on the Nimpkish (Gwa'ni) River to a village founded by the settlers next to the plant (Speck 1987). Salmon, as well as herring, was salted and sent to southern Vancouver Island. In 1881 the Alert Bay salmon cannery was built; it was taken over by BC Packers Association in the early 1900s, only to close in 1931, with periodic reopenings after that time. In 1996, 11 percent of total community employment was lost; by 1998 that figure had risen to 20 percent (Vodden 2002). Fishing employment had fallen to less than fifty individuals, 21 percent of conservative 1995 fishing employment estimates (Madill 2004). In addition, Alert Bay has joined the dozens of coastal communities whose processing plants no longer operate. By 1992, management problems, rising interest rates, a poor exchange rate, collapses in the market for chum (*Oncorhynchus keta*) and then sockeye salmon, competitive pressure from major processor BC Packers, and plant inspection requirements were among the reasons given.

Alert Bay (Cormorant Island) has both reserve and municipal communities. About two-thirds of the island's population is of First Nations descent (belonging to 'Namgis or other Kwakwaka'wakw Nations), 50 percent living "on reserve." Misalignment of benefits through insensitive fisheries policies, including the Pacific Salmon Revitalization Strategy, has contributed to significant discrepancies in economic well-being between the two communities. In 1995, for example, average income in the village of Alert Bay was more than double that of the average income of residents living in 'Namgis Reserve R1A. In 1995 also, there were forty-seven registered, resident-owned commercial salmon fishing vessels in Alert Bay. An additional thirteen seine boats were licensed by out-of-town companies but operated by Alert Bay fishermen, for a total of sixty vessels employing approximately 222 people. Over half were operated by members of the 'Namgis First Nation and nearly two-thirds by First Nations residents. An estimated eleven of fourteen licences lost in 1996 (eleven seine, three gillnet) were removed from vessels skippered by First Nations residents. Assuming that these boats employed First Nations crew members, at least fifty of the estimated sixty-three jobs lost (79 percent) belonged to First Nations people: a truly disproportionate impact (Vodden 1999). Crew members, unlike licence holders, were not provided with compensation under the Mifflin Plan.

Like the 'Namgis and other Kwakwaka'wakw peoples of Alert Bay,
Nuu-chah-nulth First Nation communities on the west coast of Vancouver Island (WCVI) have also been disproportionately impacted by the
salmon fleet reduction and other policies. Several relatively remote
fishing-dependent Nuu-chah-nulth communities were also listed in the
Gislason, Lam, and Mohan (1996) report as being among the fifteen
communities in British Columbia most impacted by fleet rationalization. In the early 1990s fishing represented 73 percent of Nuu-chah-
nulth employment. Recent statistics indicate current reliance on the
sales and service sector (Vodden and Kuecks 2003).

The WCVI region is home to approximately forty thousand residents
living in twenty-two small communities. Many are members of the fifteen
First Nations in the region (Day 2004). Once the third largest port on
British Columbia's west coast in terms of tonnage of fish landed, the
municipality of Ucluelet's fishing fleet was drastically reduced as a consequence of the Mifflin Plan. The village subsequently faced reduced
employment at the three local fish plants just as it struggled to cope with
a significant loss of forestry jobs following the 1993 antilogging protests
in Clayoquot Sound. Researchers speaking with people involved with
these sectors found marked concern over the viability of Ucluelet as a
traditionally blue-collar, high-income community. Three fish-processing
plants that ran twenty-four-hour shifts during peak seasons in the early
1990s sit largely idle much of the year. The majority of the region's fish
harvest is now processed in Vancouver instead of in local plants and
more than 50 percent of "a processing workforce that was once one of
the most skilled in the country" has moved out of the area or into other
types of employment (Vodden and Kuecks 2003).

The geoduck clam fishery is among the highest-value fisheries on
the WCVI and is the highest-value invertebrate fishery in British Columbia. There are fifty-five licences coast-wide, twelve of which fished in
the region in 2004, harvesting 864,000 pounds (approximately 22 percent of the British Columbia-wide harvest). Several species of shrimp,
including prawn, are harvested by trap and trawl fisheries. The deep-
water prawn trap fishery became one of WCVI's most valuable fisheries
in the 1990s, accounting for an estimated 6 percent of commercial
trap-caught shrimp landings in British Columbia (1990s average landings were valued at approximately $1.3 million). The shrimp trawl fish-
ery landed an estimated 1,700 tonnes from the region in 2002,
representing 85 percent of the British Columbia-wide harvest (Vodden
2004d; WCVI Aquatic Management Society 2000).

Despite this wealth of resources, few commercial fishing vessels or
licences continue to be locally owned. The small-boat fleet does not

have access to the lucrative (and expensive to enter) fisheries segments, such as halibut, black cod (*Anoplopoma fimbria*), herring, geoduck (*Panopea abrupta*), and prawn, and the local processing workforce has been reduced by at least 50 percent. Day (2004) points out that from 1994 to 2002 licences on WCVI declined by 55 percent, while licensed ownership declined by only 30 percent in urban areas. In total, residents in the region hold only 3 percent of licences in British Columbia's major fisheries, 4 percent of non-quota licences and 2 percent of quota licenses (Ecotrust Canada and Ecotrust 2004). In contrast, of 333 nontransferable commercial licences for clam harvesting in the region, a relatively low-income and labour-intensive fishery, 237 are held by members of area First Nations (Vodden 2004). WCVI residents hold only 180 licences of all other types, while one British Columbia company (Canadian Fishing Company) holds 244 fishing licences. At values as high as $3 million for a geoduck licence, cost is a major barrier (Day 2004). Only two of twelve geoduck licences fishing in the area are locally owned and operated. Ecotrust Canada and Ecotrust (2004) notes that the price of troll and gillnet licences for salmon more than doubled from 1994 to 2002, while the price of halibut quota rose four-fold from 1991 to 2004, pointing out that while capitalization in vessels and equipment may have been reduced through fleet rationalization, overall capital investment rose considerably and is now concentrated in licence and quota values.

The communities of WCVI continue to gain at least some component of their local economy, as well as an important part of their identity, from the fishing industry. Increasingly, the fishery is made up of farmed as well as wild stocks. This is true for both finfish and shellfish. Tofino, which began as a fishing town, just thirty years ago was home to several fish buyers and a locally owned plant handling shrimp, clams, crabs, groundfish, and dogfish. Today the original plant has closed but the area is now one of the main farmed salmon producers in Canada, contributing $44 million annually to the local economy. Two salmon aquaculture processing plants operate in Tofino (Vodden 2004).

Overall, then, the pattern on the West Coast has been one of long-term social-ecological restructuring, both for First Nations and later for local settler families and communities. As all this west coast research shows, First Nations, and now also settler local fisheries, have been significantly reduced or effectively destroyed, and the ecosystem has suffered extensive damage. As a result of serious misalignment of benefits from fisheries policy, no one except the large footloose technologically sophisticated industrial fleets, which are also responsible for significant ecosystem damage, seems to have benefited.

Past Ecosystems

On the East Coast we did not carry out new archaeological research and did only a little work on the history of First Nations (Heymans 2003c). Our ecosystem models, however, were constructed for a portion of the research area and for similar time periods. These models showed that East Coast marine ecosystems have fared much worse than those on the West Coast. East Coast fisheries have seen catastrophic collapses of groundfish stocks such as the northern and Gulf cod stocks (Heymans 2003a; Walters and Maguire 1996). Understanding the factors responsible for variations in abundance of species is essential for sustainable management of the species and of the ecosystem in which they live (MacKenzie et al. 2002). It is even more important for ecological restoration. Our original four mass-balance models for this coast were carried out using the 2J3K3L3NO ecosystem, the area that begins in the north off the south coast of Labrador and extends down and across the Grand Banks. We constructed models representing these ecosystems during the periods around 1450, 1900, 1985, and 1995, based on the same principles as for the West Coast (Heymans 2003a; Rogers 2003). We subsequently added two additional models for the early 1950s and the late 1960s to capture the structure of these ecosystems on the eve of the arrival of the massive distant-water fleets of factory-freezer trawlers and at the time of the so-called killer spike of peak landings in the area of 2J3K3L3NO (Alcock, Ings, and Scheider 2003).

The East Coast models (figures 2.7 and 2.8) show that for most species, as expected, biomass was highest in the 1450 model and decreased over time to the most recent period (1995).[11] Components of total system throughput[12] and its parts (viz., the sum of all the consumption, respirations, production, exports, and flow to detritus) all tend to decrease from their highest values in 1450, through lower values in 1900 and 1985, to the lowest values in 1995. Similarly, the reduction in the trophic level of the catch, net primary production, net systems production, total primary production/total biomass and total biomass excluding detritus are reduced from their highest values in 1450 to the lowest in 1995. In contrast, the total biomass/total throughput increases from 1450 to 1995. We conclude that the Newfoundland ecosystem has been severely stressed over the past century, with impacts appearing to increase over the four time periods.

Since the primary focus of our study is on developments since World War II and in order to have information that fits with our East Coast

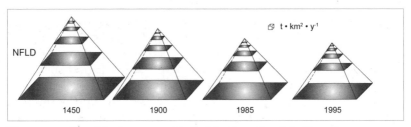

Figure 2.7 Trophic flow pyramids for Newfoundland shelf marine ecosystem. The height of the pyramid indicates maximum trophic chain length, each level corresponds to a trophic level in the system, and surface area of level indicates quantity of biomass flow between trophic levels (in $t \cdot km^2 \cdot y^{-1}$) (Heymans and Pitcher 2002b; Heymans 2003a)

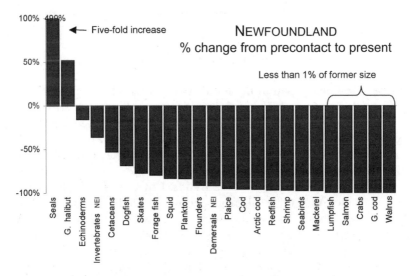

Figure 2.8 Percentage change in biomasses of selected ecosystem groups from pre-contact to present for Newfoundland, as based on whole-ecosystem simulation models (Heymans and Pitcher 2002; Heymans 2003a)

LEK interviews, we decided to construct two additional Ecopath models for 2J3K3L3NO for the period 1954–56, when rapidly expanding distant-water fleets began to fish intensively in this area, and for the period between 1968 and 1970, the years of the killer spike, when northern cod landings, mostly by foreign fleets, reached their peak (Alcock, Ings, and Schneider 2003; Walters and Maguire 1996).

Our two ecosystem models for this period suggest that on the eve of the 1970s, the food webs in the region were still relatively healthy.

Critical decreases in biomass actually occurred throughout the 1970s, after the period when landings peaked. Reduced maturity in the system at the time of the spike in catches in the late 1960s suggests the onset, but not the completion of, an important ecological shift. It is possible that, had Canada moved to a two-hundred-mile exclusive economic zone (EEZ) in 1970, earlier than all other countries, greater resilience might have been maintained. In fact, after the EEZ, Canadian groundfish landings increased substantially until, in 1988, groundfish catches in Newfoundland and Labrador were at 400,000 tonnes. After the collapse in 1993, they were less than 30,000 tonnes, a 90 percent drop. Overall, both East and West Coast models show significant decline in abundance, diversity, and trophic level over these later years, indicating substantial decline in ecosystem "health," reducing the potential gains in wealth and ecological values (such as diversity) that might be expected from any management aimed at restoration.

Towards a Social-Ecological History of Southern Labrador

The ecosystem simulation models discussed earlier in this chapter for both coasts can give us a fairly clear indication of the extent and nature of marine ecosystem degradation over time. However, the spatial and temporal scales in these models are large and can evidently mask the extent and timing of local changes. In addition, they can tell us relatively little about the interactions between people and oceans that produce these long-term, large-scale trends. Consistent with the output from our models, Pauly et al. (1998, 2001) found that for the northern temperate areas (including the Northeast Atlantic), and specifically for both coasts of Canada, mean trophic levels have decreased over the past twenty years. Trophic levels have declined at a rate of 0.1 per decade, while at the same time landings have not increased substantially. Environmental history that seeks to document interactions between fish and fisheries can help us understand more clearly the social-ecological dynamics associated with fishing down at different spatial and temporal scales. This kind of work requires formulating archival and scientific data at appropriate spatial and temporal scales, working with local fishers, collecting their local ecological knowledge of their areas and fisheries, and aggregating that information up to scales that match, as closely as possible, those available in historical and scientific information (Pauly et al. 2001). Such reconstructions may also contribute in very significant ways to recovery strategies, because fine-scale knowledge of habitat and the behaviour of fish and fishers is essential when attempting to protect remnant populations that are vulnerable to rapid depletion and to inform multi-pronged recovery strategies that focus not only on limiting fishing but also on habitat enhancement, the development of stewardship institutions, and maximizing alternative, less exploitative sources of income and employment. Such reconstructions can also provide insights into interactions between policy initiatives and fishing practice (as in the case of the St John Bay lobster fishery; see chapter 8).

The history of European resource extraction begins with the Basque whale fishery in the Strait of Belle Isle, during the second half of the sixteenth century (Tuck and Grenier 1989); followed by an emphasis on the netting of harp seals (*Phoca groenlandica*) and trapping of fur-bearing animals during the French Regime (1702–63); and then on the salmon and cod fisheries after the Treaty of Paris in 1763, which concluded the French Regime (Kennedy 1995). The Treaty of Paris (1783) terminating the American War of Independence confirmed the "liberty" of American fishers to fish the waters off southern Labrador, permitting a fishery that lasted for around a hundred years and brought thousands of men to Labrador each summer to fish (Gosling 1910). During the early British era, the official policy favoured a migratory, or seasonal, "ship fishery" along the southern Labrador coast, stipulating that British vessels provision annually in England and that they return there with their catches of salted cod each fall. Gradually, however, the justification for permanent residence in Labrador out-weighed the older policy prohibiting it, and European (chiefly British) permanent settlement began to be tolerated by government. Finally, three kinds of fisheries from the island of Newfoundland occurred along the Labrador coast during most of the nineteenth and twentieth centuries. These included "stationer" fisher families, who fished from their "room" (fishing station) (Hussey 1981). Two other groups of fishers from Newfoundland were more mobile: the "floater" fishers, chiefly from Conception and Bonavista Bays, who travelled to Labrador in schooners and, later, motorized longliners and fished wherever cod were most plentiful, and the "bankers," chiefly from Newfoundland's south coast, who arrived in Labrador in the fall and fished on schooners well off the coast.

In the cold ocean coastal environment of Newfoundland and Labrador after European settlement and before about 1950, human life was sustained by people's utilization of local resources in the context of a mercantile commercial economy combined with a subsistence way of life based on integrating various seasonal economic activities, long an adaptive device for peripheral societies of the North Atlantic rim. In Labrador, as in western Canada, these interactions were shaped by internal colonial relations between areas (the island of Newfoundland and Labrador) and between First Nations and settler populations. The initial migratory fishery slowly gave way to one that depended on a settled labour force, which made economic sense provided merchants could avoid paying fishers year-round, when the fishing season lasted only four to six months.

Hence the settlers were compelled by necessity to become subsistence producers or occupational pluralists engaged in fishing, gardening, hunting, gathering, and household maintenance, while the merchants developed a truck system to allow them to derive profit from a trade in fish that they owned and that was in effect subsidized by the settler population. Essential consumer goods and fishing supplies that could not be produced locally were delivered by merchants on account to settlers, who paid for them in fish at the end of the season. The household was the local unit of production in both formal and informal sectors. The formal sector functioned successfully only because subsistence production lowered the cost of labour that was contained in the price of fish. The informal subsistence sector could operate successfully only when fish could be exchanged for necessary imported items. Although this total system allowed continuity of life in the outports, it provided little material comfort, security, or cultural resources such as formal schooling. The outports were poor. The amount and value of production in the various communities reveal the complexity of the traditional Newfoundland household economy, its regional differentiations, and the adjustments made to it over time.[13]

The founding European "settlers" in southeast Labrador were all men who came to work for various British fishing companies along the coast, at places like Battle Harbour, Francis Harbour, Venison Islands, Grady, and others.[14] Although many important details remain unknown, it is now increasingly accepted that as in central (Zimmerly 1975) and northern Labrador (Kleivan 1966), many first-generation settler men who chose to remain in southeastern Labrador married the only available women, who were either Inuit or part-Inuit. Children from these unions were physically and culturally mixed, and it is their descendants who now comprise the Labrador Métis Nation, south of the area claimed by the Labrador Inuit Association (Vodden and Kennedy forthcoming).

As the nineteenth century unfolded, settler families along the southeast coast adapted to the marginal nature of individual components of the resource base (as they did all round the coasts of Newfoundland, and indeed all round the North Atlantic rim) through transhumance – the seasonal migration of men, and sometimes their animals and families, from one ecological niche to another. Families wintered in the more sheltered forested bays and coves of Labrador's inner coast, near sources of food, firewood, and water, and then moved each summer to the outer coast's islands and headlands to harvest a succession of resources, including harp seals, migratory waterfowl (principally eider duck [*Somateria mollissima*]), Atlantic salmon [*Salmo salar*], cod, herring, and others. As on the island of Newfoundland, Labrador residents provisioned themselves each spring from merchants who brought them supplies in exchange for their salted cod and salmon, seal skins, and oils, furs, and other local products. In bays such as Sandwich Bay, where salmon were plentiful and could be predictably netted each summer, English trading companies managed the short, lucrative salmon season. Thus, between around 1830 and 1873, the Hunt Company maintained posts on the Eagle and Paradise Rivers in Sandwich Bay; these and other assets were sold to the Hudson's Bay Company (HBC) in 1873, which managed the fishery and retail trade until recent decades. Historically, both the Hunt and Hudson's Bay companies exercised an iron grip on the salmon fishery.

Until the 1920s, most salmon was salted or "pickled" in barrels, although a small salmon canning industry was also introduced by the Hudson's Bay Company in central Labrador in the 1860s (Plaice 1990). Salmon were first purchased fresh for freezing in Sandwich Bay in the 1920s, and this method of processing was prevalent until the federal buy-back of commercial salmon licences a decade or so ago. Labrador's herring fishery had two "booms" throughout history. The first occurred between 1860 and 1880; the second during World War II, when herring oil was used for making explosives (Kennedy 1995, 130–2).

Historically, the summer homes of settlers were located only a short distance from cod fishing grounds, enabling fish to be hand-lined from small boats powered by oar or sail. However, by the twentieth century, declining catches led fishers to fish further from home, employing more gear to catch fewer cod. Indeed, from the 1870s onwards, Newfoundland fisheries, in the face of depletion of local island fisheries, were moving on to Labrador to alleviate their declining catch rates (Hutchings and Myers 1995). Stationary cod traps were invented in 1866 and replaced moveable cod seines in southeastern Labrador around 1900. These traps caught small fish, further weakening the resource.

Meanwhile, offshore (in 2J3K3L3NO) between 1954 and 1956, rapidly expanding distant water fleets had begun to fish intensively in this area, and between 1968 and 1970 (the years of the so-called killer spike) northern cod landings reached their peak. During the early 1950s, most of the landings from this region came from 3L3NO and only 5 percent from southern Labrador (2J). By the late 1950s, the majority of the landings were coming from 2J. As with the historical

expansion of the coastal cod fisheries from the island of Newfoundland to Labrador in the late nineteenth and early twentieth centuries,
in response to declining catch rates, effort changed, though not all of
it shifted north. Spatial and ecological intensification also took place,
coinciding with spatial and ecological expansion in the offshore and
coastal fisheries (Neis and Kean 2003). Thus, by the 1970s (after the
introduction of the midwater trawl by Germany in 1969) effort was
spread across the Grand Banks, off the northeast coast and along the
Labrador coast, while coast and target species had expanded from haddock (*Melanogrammus aeglefinus*), cod, and American plaice (*Hippoglossoides platessoides*) to include redfish (*Sebastes mentella, S. faciatus, and S.
marinus*) and capelin (*Mallotus villosus*). With the extension of the two-
hundred-mile EEZ in 1977 on both coasts of the country, new, increasingly complex science and management initiatives were put in place.[15]
These initiatives failed, however, to ensure recovery of devastated
stocks. Overall, then, the historical picture is one of increasing technology, dwindling stocks, and intensification and expansion of effort contributing to ecosystem change.

Towards a Social-Ecological History of the Northern Gulf

Further south, the history of European exploitation of the marine
resources in the northern Gulf and Strait of Belle Isle, Newfoundland's
west coast and Great Northern Peninsula, starts with the whale fishery
of the Basques (Mannion 1977; Thompson 1961).[16] These whale and
cod fisheries took place in the sixteenth and seventeenth centuries,
and then a subsequent French fishery took over, along with the French
treaty rights on the shore (the "French Shore": Thompson 1961;
Mannion 1977) that persisted until the early twentieth century. There
is strong evidence that even at this date human activities had affected
marine ecosystems in this area. Basques whalers targeted two species of
baleen whales (the right whale [*Eubalaena glacialis*] and the bowhead
whale [*Balaena mysticetus*]) on the Labrador side of the Straits and
appear to have substantially reduced the abundance of these whales by
the end of the sixteenth century. With the destruction of these populations, the Basques shifted effort to concentrate more intensively on
seals, walrus (*Odobenus rosmarus*), and cod.

In 1857 the French gained exclusive fishing rights on the west coast
from Cape St John to the Quirpon Islands, on the north coast from the
Quirpon Islands to Cape Norman and in five specified harbours on the
west coast. With the exception of these harbours, Newfoundlanders
could fish with the French from Cape Norman to Cape Ray, and the
French could fish with the Newfoundlanders in Labrador until the

French Shore question was settled in 1904, when France gave up her rights in Newfoundland (Firestone 1967, 20–1). There were flourishing fisheries – in Port au Choix, for example, in 1857, 120 men and boys caught nine thousand quintals of cod, primarily with bultows (line trawls) set in seventy fathoms of water (Sinclair 1985). At this time, cod seines were also used and were blamed for catching small cod in Ingornachoix Bay, just south of Port au Choix. Other species targeted at this time included seals, herring, and salmon. After 1871, the cod trap spread through the area, and in 1873, lobster canneries came to the northern portion of Newfoundland's West Coast, transforming lobster (*Lebbeus groenlandicus*), which had been largely ignored, into a major commercial resource. Sinclair indicates that at this time fishers abandoned the cod fishery to concentrate on lobsters, with many opening their own family canneries. In 1926 the lobster fishery was closed for three years, and in the 1930s fishers began shipping live lobsters to the United States. In 1941 the last cannery on the St Barbe coast closed. The number of lobster traps fished on the West Coast seems to have been fairly stable until 1965 (65,000 traps reported in 1911, 68,000 in 1945, and 69,000 in 1957), when it increased rapidly to 172,000 (Sinclair 1985, 42).

Lobster was most important to fishers south of Flower's Cove. For those north of Flower's Cove, cod and seals were the primary commercial species. South of Pointe Riche, fishers were more likely to combine fishing with forestry employment. North of Pointe Riche, lobster supplemented their income from cod, salmon, and herring. On the Strait shore, relatively large, stable cod trap crews with hereditary berths fished line trawls in the fall. Other harvesters relied on handlines and line trawl throughout the season (Sinclair 1985).

This merchant-household fishery changed rapidly from the mid-twentieth century, when industrial capitalism and the welfare state became the dominant formal economic forces and as the effects of the distant-water fleets and regional modernization initiatives began the process of dramatically transforming marine ecosystems. The years following Confederation brought crucial change. Investment in the cod fishery expanded in the 1960s, supported by the introduction of fishers' unemployment insurance in 1958 and by a reduction in logging employment for fishers. Despite greater investments in gear, the value of landings per fisher stayed low and was highly unstable. Some were spending more to earn less (Sinclair 1985, 49). The provincial economy modernized rapidly, and a series of "safety-net" devices for the informal sector, through welfare, baby bonus cheques, and unemployment insurance in effect took the place of the old merchant credit subsystem. Cash became the mechanism to maintain the informal sector,

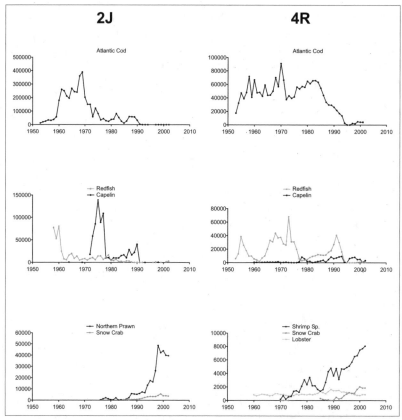

How We Got Here

Figure 2.9   2J and 4R Landings, 1953–2002 (data taken from NAFO 21A
database http://www.nafo.ca/science/frames/research.html)

which was culturally valued, if less critical than before to the functioning of the formal sector. As fish stocks declined, the cash costs of fishing escalated. Fishing was supported by government subsidies, and in
the absence of sustainable management of the stocks, differentiation
on the basis of class and region enhanced divisions and conflict within
fisheries.

From 1953 it is possible to draw on International Commission for
the Northwest Atlantic Fisheries (ICNAF) and North Atlantic Fisheries
Organization (NAFO) landings data to reconstruct landings in southern
Labrador (area 2J) and the northern Gulf (area 4R). For both areas,
these data show the effects on landings of the arrival of the distant-water fleets and the development of Newfoundland's own trawler
fleets. This reconstruction of landings from 4R and 2J (figure 2.9) indicates a pattern similar to that described by Deimling and Liss (1994,

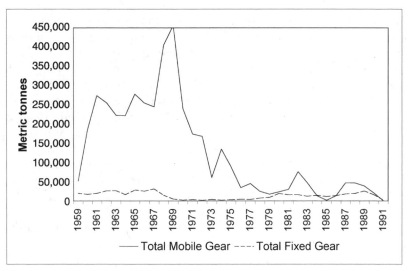

Figure 2.10 Mobile and Fixed Gear Landings, 2J, 1959–91 (data taken from NAFO 21B database http://www.nafo.ca/science/frames/research.html)

60) for landings in the eastern North Pacific. That pattern consists of landings of preferred species increasing over time until they peak and then decline and of diversification in the species targeted by fishers as they shift effort to offset the effects of reductions in landings from their preferred species. A related pattern reveals spatial shifts in landings from inshore and nearshore areas to offshore, deeper-water areas. Finally, these data point to a downward trend in overall landings of all species, which may indicate an overall decline in the productive capacity of marine ecosystems as it relates to commercial species.

As indicated in figure 2.10 the major increase in 2J cod landings came from the rapid expansion of the mobile offshore trawler fleet in this area during the 1960s. Relative to the trawler landings, landings increased slightly and temporarily in the fixed-gear sector in the 1980s and early 1990s, but many of those fishing for cod in this area came from the west and other coasts of Newfoundland and from Quebec.

In the inshore, technology has to improve to keep up with the offshore vessels as much as possible. In 1963, the first of the longliners (decked vessels between thirty-five and sixty-five feet long) arrived in Port au Choix (longliners came to Bonavista and the southwest coast ten years earlier). Longliners, nylon gillnets, and mechanical haulers contributed to a rapid increase in daily landings per vessel. The number of longliners on the northwest coast increased from seventeen in 1966 to forty-eight in 1968 (Sinclair 1985, 60–1). They not only fished locally; some also steamed up to three hundred miles up the Labrador coast in search of cod. They were also used in the seal hunt. During the early 1970s, most longliner skippers converted from gillnets to otter trawl. Sinclair attributes the trigger for the rapid shift to the presence of foreign and Canadian draggers and foreign factory ships in the Gulf, often fishing within only a few miles of the coast. There was little effective regulation of the activities of the draggers, and they interfered substantially with the ability of the smaller boats to fish. In addition, stocks and catch rates were declining. Low prices for fish aggravated the problems of the fleet (64).

Local fishers were encouraged to shift effort over to other species, so some shifted to scallop dredging and others to shrimp dragging. Northern prawn (*Pandalus borealis*) stocks located about twenty miles off Port au Choix were targeted initially, but research and fishing activities soon located more populations. Helped by government subsidies, the shrimp fleet expanded and landings increased from 321,000 pounds from 11 vessels in 1970 to a peak of 5,829,000 pounds from 42 vessels in 1978, declining to 3,984,000 pounds from 36 boats in 1982 (Sinclair 1985, 68). In 1974–75, some shrimp vessels began to shift from shrimp to cod dragging, which could generate more revenue for less effort (Sinclair 1985, 70). Cod dragging required larger and more powerful boats, and the first sixty-five-foot dragger was introduced into the inshore Gulf fleet in 1974. In 1977, 15 local boats decided to go south for the "winter" cod fishery off Port aux Basques. In 1978, 5 headed south and landed on average 700,000 pounds. By 1982, there were 50 draggers prosecuting this fishery; also by 1982, there were 35 shrimp licences and 57 groundfish trawl licences on the northwest coast. There were 221 lobster licences between Trout River and Raleigh and 90 salmon licences. Lobster licences were heavily concentrated below Pointe Riche (Sinclair 1985, 77). The dragger fishery was an elite fishery composed of relatively few skippers and concentrated in few communities. With the draggers came division and conflict that has persisted until today.

As trawlers depleted the cod stocks in the northern Gulf, small boats and longliners from this coast began to travel more frequently to the Labrador coast in search of cod. This pattern culminated in the intensive cod fishery in the Black Tickle area in the late 1980s and early 1990s. The boom ended with the closure of the northern cod fishery in 1992 and in 1994 in the northern Gulf of St Lawrence, but small-boat enterprises were already hurting badly by the mid-1980s. Since that time, some have sold their licences. Small-boat harvesters with lobster licences have struggled to survive from this fishery with sporadic and very limited access to cod and other species, particularly in the area of St John Bay and further north. Larger vessels have often been left with only one or two species, particularly shrimp and, in some areas, snow crab (*Chionoecetes opilio*). Some company-owned vessels now dominate the pelagic fisheries of the northern Gulf. The costs of fishing have escalated relative to incomes, owing, in part, to the downloading of fisheries management costs onto harvesters (see chapter 12).

Since the northern cod moratorium in 1992 and the northern Gulf cod moratorium in 1994, the fishing industry in Newfoundland and Labrador has been transformed from a predominantly cod- and groundfish-based fishery to a shellfish fishery. In 2002, the production value of the fishery reached an all-time high at over $1 billion. The landed value of fish was $515 million, but $421 million of this (82 percent) came from shellfish. Snow crab is now the most commonly processed species and the highest revenue-producing species ($235 million in 2002, compared to $22 million in 1988), but the production of other types of shellfish is also increasing. The landed value of shrimp was $18 million in 1998 but $143 million in 2002 (Newfoundland and Labrador Fisheries and Aquaculture 2002, 2–3).

While the ICNAF and NAFO landings data and existing histories for these areas can give some indication of shifts between sectors in landings and long-term trends in landed value, they cannot tell us much about the deeper spatial and temporal dynamics of these processes. Were there finer-scale spatial and temporal shifts in effort underlying these landings data? Changes in the size of fish (Amorosi, McGovern, and Perdikaris 1994, for Iceland)? Increases in fishing efficiency? Changes in by-catch, discarding, and unreported catches? (Neis and Kean 2003). They also do not tell us about possible interactions between the shifts described; i.e., did the shift to shrimp trawling affect recruitment and eventually landings in other fisheries like Atlantic halibut?

In general, landings data suffer from a number of shortcomings. They tend to be sparse or uneven over time; they are collected in spatial units that often do not fit well with the underlying ecology; they do not include information on by-catch or discarding, are vulnerable to misreporting, and do not take into account changes in effort, including spatial, temporal, and some ecological shifts. Landings are also mediated by changing markets, changes in management, and changes in environmental conditions.

Fish Harvesters' Local Ecological Knowledge and Science

We therefore added the LEK of harvesters to our other data sources to complete our social-ecological history for the period 1950–80 (post-

1980 history is summarized briefly in chapter 8), during which the dynamics of fisheries and fisheries policy became very important. LEK findings provide information on harvesters' recollections of trends in catch rates over time. They also include vital information on how harvesters have responded to environmental change, changing management initiatives, changing market opportunities, and changes in the availability of new technologies by often improving their fishing efficiency (with changes in vessel and engine size, the amount and type of gear used, and fish-finding equipment), changing what they fish for, what they fish with, and where and when they fish, and by changing their crew composition and how they decide how to fish.

From the LEK research that we have done for this project, we see a pattern of spatial, temporal, and ecological intensification in fisheries, as well as a pattern of spatial, temporal, and ecological expansion. Thus, harvesters fish more gear (spatial intensification) in new areas (spatial expansion) more intensively and for longer periods (more hauls per day) on the same and new grounds and target, particularly in the short term, younger and older fish, fish from different stocks or aggregations, and different species (ecological intensification and expansion). Intensification and expansion are often associated with increased conflict, pointing to the presence of a social dimension within fishing down (the food web), although, as indicated by Palmer and Sinclair (1997), cultural dynamics in these small communities can mute overt conflict.

Some harvesters tend to intensify their efforts on local grounds, while others expand to other areas. These processes can have substantial impacts on marine ecosystems, on fish harvesters, and on fishing communities and regions. Overall, the cost of harvesting tends to increase relative to the volume and value of landings, with the result that, in fish harvester households, wealth will tend to flow off the land and onto the sea (out of homes and into vessels). In addition, the relationship between harvesting and home lives will tend to change as harvesters spend more time at sea, away from their families. Other responses to intensification and expansion can include changes in crew composition (such as the increase in women fishing), in skipper-crew relations (changes in the share system), and in the relationship between harvesters and fishing communities (the changes in the companies harvesters sell to and the conditions of sale). These changes will influence occupational health risks and the distribution of wealth within and between generations and communities. Expansion and intensification can also have consequences for fisheries science and management as harvesters shift effort to new species for which the knowledge base for science and management is weaker and where the risks to the species and to the harvesters may differ. Other potential impacts include increased uncertainty and risk taking, and reduced incomes and employment security for some.

CONCLUSION

What we now know, then, is that in Canadian fisheries on both coasts, interactive restructuring has been going on for a very long time indeed. We also know that its legacy up to the 1980s was one of accelerating social fragmentation and conflict, coupled to a pattern of fishing down ecosystems. These changes were exacerbated by management strategies in which both intensification of effort by increasingly larger and technologically more sophisticated fleets and the expansion of fishing into a range of species and (for local fishers) increasingly wider areas offshore were accepted and even encouraged. Initially, while pre-contact exploitation may have been sufficiently severe to do local damage, the overall impact of First Nations' exploitation of the resource was relatively benign, and the legacy (handed down in myth and story) has been one of respect for the resource (see also later chapters).

Post-contact, the picture changes. On both coasts, the resource ceased to be used solely for subsistence purposes, and the pattern became one of accelerating degradation of targeted commercial species. The social impacts of the industrialization of the fisheries in all our East Coast research areas were increasing social fragmentation within northern Gulf communities and increasing pressures by outsiders in some local areas such as the Labrador Straits and the area off Port aux Basques. Fisher-owned longliners, capable of going far offshore and utilizing mobile gear, became a core part of the fishery in many areas, especially the northwest coast. With a labour force of less than 10 percent of the force in the small boat sector, these vessels accounted for about one-third of the cod catch by the 1970s. Foreign and Newfoundland south coast corporate-owned trawlers fished in the area into the 1980s, accompanied in the last decade by owner-operated trawlers. As the overall trawler fleet expanded it took a growing share of the catch, focusing on cod to supply the frozen-fish plants on shore. By 1970, landings by weight from offshore vessels far surpassed those from inshore, and cod processed in freezer plants greatly exceeded the share that was salted. Many sought employment in the plants, where the majority of workers were often women, now wage earners rather than exclusively unpaid household workers. The "glory days" of the 1970s and 1980s entailed frenzied activity and work in communities like Port au Choix and deepening gaps between wealthy and poor fishing families.

Our overall findings suggest that while the Hecate Strait has not witnessed the kind of collapse that occurred in the northwest Atlantic, there needs to be considerable concern about overall West Coast marine ecosystem health and therefore very careful regulation of fisheries. Current regulatory practices have produced acrimony between competing user groups – recreational, First Nations, and several commercial fishers using different gear types (Pinkerton 2000; Walters 1995; and many others). The model of the consequences of fisheries exploitation known as the "fishing down marine food webs" hypothesis (Pauly et al. 1998) describes the effects of a (generic) fishing community shifting effort from a depleted resource at a high level in the food web ("table fish") to another more plentiful one lower down the food web ("trash fish"). These "new" resources may be located in different regions and most likely are located further from shore than the population that was depleted; or they may be a previously underutilized species.

The hypothesis describes a transition from long-lived, higher trophic level, piscivorous bottom fish toward short-lived, low trophic level invertebrates and planktivorous pelagic fish ("the ecological ratchet of depletion," named "Odum's ratchet" [Pitcher 2001]). It is characterized at first by increased landings, then by a transition to stagnating or declining catches. It can be seen through changes identified in spatial, temporal, or ecological dimensions, as well as through changes in intensification and expansion of the fishery (Neis and Kean 2003). Overall, the picture that emerges agrees with our mass balance models: a steady (albeit slight on the West Coast) fishing down the food webs, with higher levels now diminished, overall biomass depleted, and diversity reduced. In other words, the West Coast marine ecosystem's health is endangered, while the East Coast marine ecosystem's health is best described as critical.

Within the last several centuries, but particularly in the post–World War II period, fishery communities and marine ecosystems on the two coasts have changed significantly. For some time now, people living on both coasts, as well as policy-makers and management officials, have been grappling with serious social and environmental challenges that have sometimes been exacerbated by industrial and policy change. The historical reconstructions presented in this chapter suggest that the rate and depth of social and environmental degradation accelerated even faster after World War II. We also see the hallmarks of scale asymmetry in the increasing concentration of power at the national level. This concentration of power, along with inadequate feedback from local areas and misalignment of risks and benefits, resulted in the large technologically sophisticated fleets being permitted to do more and more overfishing, at the expense, ultimately, of the livelihood of local communities. That is, the dependence of those places on a single resource has been more an outcome of commercial, provincial, and national policy than of a highly restricted resource base, particularly on the West Coast. On both coasts, moreover, the benefits of the resource have not been adequately developed at the local level.

In this chapter we have explored applied history up to the 1980s. In the next chapter, we consider the legacy left by this social-ecological history, as well as its insights into where fishing policy and practice went wrong. Together, these two chapters will provide us with a base from which we can understand how policy and practice must change, in order to consider, as we will in chapter fourteen, what future possibilities may exist and what requirements need to be in place for the recovery of fish, fisheries, and fishery-dependent communities in the future.

3

Not Managing for Scarcity: Social-Ecological Issues in Contemporary Fisheries Management and Capture Practices

INTRODUCTION

As chapter 2 made clear, social and environmental restructuring are neither new nor confined to any single resource sector. What is new is the recognition that social-ecological restructuring within and across sectors and on two very different coasts has generated long-term degradation at an accelerating rate. While the social-ecological damage is serious on both coasts, it appears to be most extensive in marine resources and related industries and communities in Newfoundland and Labrador. On the East Coast, degradation has also been more severe towards the north (off Labrador) than in the northern Gulf. This pattern continues to this day – Labrador's cod stocks continue to be in worse shape than those elsewhere, and snow crab stocks are in worse shape than snow crab stocks further south – in the context of relatively elaborate scientific and management infrastructures for fisheries, particularly after the extension of two-hundred-mile EEZs (exclusive economic zones) in the 1970s. The collapse of East Coast groundfish stocks, the historic path towards which we documented in chapter 2, was the result of overfishing in the context of vulnerability due to changing water temperatures, which forced stock collapse and a series of moratoria put in place to allow the stocks to recover. Since then, scale and benefit misalignments have still been occurring. For example, deep-water fleets continue to draw profits from stocks caught elsewhere, further depleting the planet's decreasing ocean resources. In eastern Canada the local emphasis has turned to species other than cod, some of which are now themselves being overfished. It would appear that the lessons that could have been drawn from the collapse of the groundfish stocks have not been learned, and coastal communities continue to rely on an increasingly threatened marine resource base. On the West Coast some salmon stocks are also in danger, quite possibly from overfishing and changing water conditions. While many changes have been made on the East Coast (with 100 percent observer coverage in some fisheries and video camera surveillance systems on many vessels, along with fleet reduction measures, closures, and other conservation measures that have begun to affect some stocks positively), the emphasis on only the ecological side of the problem has created serious problems for small resource-based communities and the small-boat fishery. The lack of socially and ecologically tied analysis and remediation has left many West Coast communities in crisis, while the same lack on the East Coast has created both stock collapse and social distress. This is just not good enough.

Why, in the light of the ongoing East Coast tragedy and the possibility of a human tragedy in the making on the West Coast, have Canadian science and management initiatives failed not only to achieve East Coast stock recovery but indeed even to sustain our fishery resources along with our fishing communities on both coasts? Only when we know what is obstructing productive change can we search for ways to improve matters. We have therefore examined four issues that revolve around scarcity and that we hypothesize are key to identifying the root problems in Canadian fisheries management and practice, particularly with respect to the small-boat fisheries that are a crucial (although not the sole) lifeline of coastal communities. These are (1) misalignments between the changes in fisheries management and the way in which the resultant policies interact with fish harvesting strategies to affect the nature and quality of the information available to fisheries science and hence the accuracy of stock assessments; (2) scale mismatches between the rapid changes in fishing efficiency and practices and their implications for the slower-moving fisheries science and management; (3) the interactive relationship between social and economic power, institutional structures and paradigms, and the practice of science and management; and finally, (4) growing recognition of the mismatch between single-species management and the way in which the marine biophysical environment actually functions, as well as misalignments between data that rely exclusively on our capacity to estimate abundance and establish "appropriate" quotas and the lack of knowledge that still exists on marine ecosystem function. All these issues are, of course, directly related to interactive restructuring, because they point to the management practices that continue to shape fish capture and to alter (i.e., restructure) the social-ecological situation in which coastal communities find themselves.

Current management and capture issues all revolve around the dilemma of how to deal with scarce resources in the face of technological overcapacity and industrial dynamism, inadequate science and associated high levels of uncertainty. As we saw in chapter 2, the misfit between contemporary fishery resources, industrial needs, and science and management capacity has, to a considerable extent, resulted from a history of largely unregulated international capitalist and socialist state expansion. This result has to be linked to resource mismanagement that extends back many decades on the West Coast and arguably centuries on the East (Cadigan 2003). Associated with this mismanagement had been escalating intensification and expansion of fishing effort (not only spatially but also temporally and across different populations, fish sizes, and species) that appears to have been largely invisible to scientists and managers (Neis and Kean 2003; Pauly 2001). We now therefore examine modern fisheries science, practice, and management at both the national and provincial scales, as well as from the perspective of particular fisheries on the East Coast, to see in what ways management and science have helped and/or hindered the resource degradation we have documented.[1] In chapter 9 we return to fisheries, as well as other resource sectors, to examine how they have interacted with industrial, policy, and social restructuring to increase the vulnerability of coastal communities.

STOCK ASSESSMENT ACCURACY

Data Fouling

One vital challenge in getting a handle on ecosystem health is that of calculating the true size and composition of fishing mortality. How well fisheries management can be carried out depends on national, even international, and local participation. In a science-based management system where fisheries are managed primarily on the basis of quotas, accurate knowledge of fishing mortality is a critical but elusive requirement. Given the emphasis on multispecies or ecosystem-based management approaches that is central to the Oceans Act, it is increasingly necessary to quantify discard composition and mortality with a view to understanding the impact of fishing at the stock, population, trophic, and ecosystem level (Borges et al. 2001). Overall, Canadian fisheries management faces problems at the national and international levels, but at the regional level (which is the base on which all ecosystem management must rest) the challenge is to get accurate and precise catch data that reflect the full scope of fishing mortality.[2] As fish stocks decline and effort is directed at more and more species, the challenges of science-based management based on quotas escalate dramatically. In this context, the problem of data fouling caused by such practices as high-grading, discarding, and under-reporting becomes a major challenge for fisheries scientists and managers. If data are distorted through "misreporting" of landings – the failure to obtain an adequate account for activities at sea from harvesters – then the reliability of catch data for the purpose of stock assessments is undermined. That said, how much data fouling actually exists and what the motivations behind it are is hard to establish empirically, because such research is necessarily hindered by the difficulty of obtaining data on behaviour that is either unreported by harvesters (because there is no requirement to report, or because it is illegal) or because the activity is reported but distorted (since it benefits the individual to do so).

Fishing mortalities have always been underestimated in landings data, which record only fish that make it into the formal marketplace. Species and sizes of fish with no commercial value generally do not appear in historical landings data. Also excluded until recently are fish harvested for subsistence purposes and sold locally. Fishery closures and the introduction of management by total allowable catches (TAC) and, more recently, by enterprise allocations and individual quotas have added new sources of data fouling, such as misreporting and concealment of catches, dumping, discarding, and high-grading (Anderson 1972; Copes 2003; C.E. Harris 1998; Angel et al. 1994; Palmer and Sinclair 1997).

The extent of discarding is not well known, but a NAFO document (Kulka 1986) reported that the rate of discarding in the Newfoundland fleet has increased steadily since 1981 and the quantity of cod discarded in 1985 was estimated at 7,897 tonnes. The Keats (1986) report on the inshore fisheries said that the total amount of discarded cod was estimated to have increased from 7.2 percent by number (1.5 percent by weight) in 1981 to 24.4 percent by number (10.7 percent by weight) in 1986. The rate of dumping plus discarding was estimated to have been 8.4 percent for cod in 1985 and may have reached significant levels by 1986.[3] Hutchings (1996) and Hutchings and Ferguson (2000) estimated that the inshore fisheries discarded 5 percent by weight in the early 1980s, and this figure increased to 28 percent in 1989. There are also non-quantified indications of net lining, dumping, and discarding in the fisheries of the northern Gulf associated with the introduction of individual quotas in the 1980s (Palmer and Sinclair 1997). Years before the northern cod collapse and the collapse of the northern Gulf cod, then, extensive discarding and dumping of cod occurred in all sectors of the Canadian fleet (Hache 1989; L. Harris 1990; J.M. Marshall 1990), and it has now been shown that the increased discarding of young fish contributed to the rapidity with which the cod stocks collapsed (Hutchings 1996; Myers et al. 1997).[4]

In 1997, in the auditor general's report, information on Canada's fish stocks was still considered to be inaccurate, due to under-reporting, misreporting, and/or additional fishing mortality caused by unsustainable fishing practices (AGC 1997). In 2003, the Fisheries Resource Conservation Council (FRCC) said that anecdotal reports of cheating were increasing and widespread. Gezelius (2003) compared a Norwegian with a Newfoundland fishing community and described subsistence "poaching" for food as morally accepted in the Newfound-land community, whereas fishing for sale (commercial poaching) was not.

Copes argues that management regulations, especially in individual quota (IQ) regimes, provide incentives for high-grading, discarding, and the reporting of only those fish that are of a high value (1986; 2000), while Angel et al. (1994) report that most regulations associated with quota management per se have not been enforceable,[5] and, indeed, that illegal fishing activities have increased globally as a result of enterprise allocation (EA) and individual transferable quota (ITQ) programs.[6]

Given the context of scarcity in many key commercial species and the problem of endangered and threatened species and continued extensive reliance on management regimes that require accurate information on fishing mortality, we used an East Coast case study to see if we could identify causes and patterns of data fouling and illegal fishing. We gathered information from several sources: the print media, official records of fisheries violations, court records of convictions, and a set of personal interviews with seventeen active and retired fishers who owned inshore vessels (vessels less than 35 feet in length), twenty-six with vessels in the 35–65 foot category, and three in the over 65 foot sector, to see what we could learn about data fouling in the Newfoundland inshore fishery and to see if there was a relationship between fouling and different management regimes.[7] Accepting that, even with good cooperation, fishers' estimates of the amount of fish they discard is likely to be variable and have some bias to underestimation (Saila 1983)[8] and also recognizing that the small size of our sample of interviews would permit us only to begin to suggest cause and effect, we created a questionnaire designed to identify fishing management and catch practices of all kinds that would cause "fouling" in the input data used in stock assessments.

In our analysis, we distinguished between, on the one hand, deliberate distortion of real catch statistics either through under- or over-reporting of catches or through the falsification of documents (illegal) and, on the other hand, unintentional misreporting. Discarding and high-grading, for example, are deliberate illegal activities, while by-catch dumping, quota busting, and price dumping (*sensu* Copes 1986) are routine in certain fisheries and not considered illegal by the fishers. These practices are nonetheless all misreporting and result in data fouling of catch statistics. We therefore asked also about past and present discarding, dumping, by-catch, and misreporting practices, in order to estimate the extent to which fishers were *aware* of the aberrations in their code of practice that could cause data fouling and result in erroneous data, which would then produce incorrect results from the stock assessment models, to the detriment of fishers themselves. Since various activities are overlapping and continuous in practice, we speak here of data fouling to describe the end result of all unreported fishing.

Our questions were applied to all fisheries, not just the now defunct northeast coast cod fishery. Discarding of crab now occurs when the crab is too small to be legally sold.[9] According to the fishers we interviewed, in recent years fish by-catch in the crab fisheries has been almost nil, but "discarding is a big thing" and approximately 38 percent of all (fifty) interviewed fishers said that misreporting of crab was going on in their fisheries. Most fishers (84 percent) interviewed said that they liked the system of IQs and that they liked the way the crab fishery was managed today. Crab is, however, not managed under the FRCC.[10] There are now dockside monitors and graders and, in this system, fishers claim that cheating is difficult. They also approved of the offshore observers.

By-catch from shrimp trawlers consists mostly of small capelin, redfish, and turbot (*Reinhardtius hippoglossoides*). In some fisheries, harvesters are allowed to take a certain percentage of cod even where this fishery is under moratoria. Interviewed harvesters estimated the misreporting of cod to be extremely high, since at the moment there is no "official" quota for cod. Where restricted fisheries are permitted for scientific purposes (the Sentinel Fishery and an Index Fishery) some fishers indicated that under-reporting by weight was rampant. One way under-reporting happens is by recording cod as redfish. Cod were also misreported as turbot and as mackerel (*Scomber scombrus*), herring, or capelin, since these fish were carted by truckloads and not monitored by boatloads. According to one skipper, extensive poaching still goes on because some fishers who live at the poverty line can add to their income this way, fishing and selling cod. This is known locally but goes largely unreported: one respondent expressed his dislike for by-catch management, and two others (retired skippers) said that they were aware that cod sales were still going on, while a third skipper estimated that 30 percent of the cod landed was never reported and that the misreporting of cod was extremely high since there was legally no quota. The results of these interviews with a relatively small sample of fish harvesters provide clear indication that estimates of fishing mortality for cod based on official statistics are underestimating actual mortality.

Discarding is a normal practice in most fisheries and, world-wide, scientists estimate that fishers discarded about 25 percent of what they caught during the 1980s and the early 1990s, about sixty billion pounds each year (Alverson et al. 1994; Alverson 1998). There is now strong evidence that quota-based management does not work, not only because it encourages illegal practices but, more fundamentally, because its very foundations are scientifically suspect. Quota-based management in any fishery assumes *known* fishing and natural fish mortality, and we now know (since the collapse of the East Coast groundfish stocks) that we do not know enough about either natural or fishing mortality. The risks of this management approach are particularly great in contexts of resource scarcity. However, despite all this evidence, as well as the creation of an independent body, the FRCC, which reviews marine scientific advice, the original organizational structure of fisheries management is not only still in place on the East Coast but has been implemented in more fisheries and at finer scales in the years since the collapse of East Coast groundfish stocks (see Schrank, Arnason, and Hanneson [2003] for a complete detailed history), and all major fish stocks and some shellfish stocks are now under enterprise allocation (EA), individual quota (IQ), or individual transferable quota (ITQ) regimes.

In some cases, fisheries that are not under such regimes are also struggling. Thus, while some progress has been made in the management of lobster fisheries based on trap limits and requirements for collaborative integrated management plans and for voluntary "v-notching" (clipping the tail of an egg-bearing female lobster) and some other activities, exploitation rates in lobster fisheries are still generally too high, particularly in areas like St John Bay on Newfoundland's west coast, where the 1980s management regime led to a heavy concentration of licences and hence effort in this relatively small area and where the 1990s management regime has kept these licences and their owners trapped in the area (Whalen 2005).

Related to problems with misreporting, the fish harvester survey also found clear evidence of fishers who continue to have antagonistic feelings towards the DFO (Department of Fisheries and Oceans Canada) and the continuation of 1980s feelings that the DFO is out of touch with realities in fishing communities. In the words of one, "They don't have a clue." As Perry and Ommer (2003) describe it, "Failing to comprehend and deal with ... issues like these creates a rich breeding ground for community rebellion against regulations, promoting various kinds of illicit practices that thwart the legitimate practices of fisheries management." These are issues of mismatched scales of management and ecosystem function, and our results suggest strongly that the situation has been neither clearly perceived nor understood. In consequence, the situation still persists and hence misreporting and illegal activities continue to exist in small Newfoundland communities.

We also asked fishers about their familiarity with national and international programs for sustainable fisheries. More than half (51 percent) of all fishers interviewed were not aware of the Canadian Code of Conduct for Responsible Fishing Operations (Canada Fisheries and Oceans 1998), which is based on the FAO Code of Conduct established in 1995 by the UN (FAO 1995). A copy of the Canadian code is provided by the DFO with each licence. Of those who said they were aware of the code, a great number answered that they had not read it and were not familiar with it. The active fishers were also asked if they were aware of the Marine Stewardship Council (MSC), an independent, global, non-profit organization that was set up to find a solution to the problem of overfishing. First established in 1997 by Unilever, the world's largest buyer of seafood, and in 1999 by the World Wildlife Fund (WWF), the international conservation organization, it became fully independent from both organizations and today is funded by a wide range of organizations, including charitable foundations and corporate organizations around the world. The MSC spent two years developing an environmental standard for sustainable and well-managed fisheries that they put together following worldwide consultation with scientists, fisheries experts, environmental organizations, and other people with a strong interest in preserving fish stocks for the future.[11] They reward environmentally responsible fisheries management and practices with a distinctive "blue product" label: there are now hundreds of products with the label, seven fisheries are now certified and fourteen in progress. Three offices exist (London, England; Seattle, the United States; and Miranda, Australia), but there is no Canadian office, and only two Canadian fish stocks are undergoing the certification process. We thought it important to know the extent to which fishers in Newfoundland were aware of this international standard.

Of thirty-four fishers asked, only 23 percent had heard about the MSC, although all agreed with the idea of eco-labeling, since it would enhance the value of their product. One commented, "Anything that has to do with conservation, I'm in." This general consensus that conservation was a priority was contradicted, to some degree, by the reports of misreporting and discarding.

Calculating Stock Size – Data Sources

The East Coast groundfish moratoria could have marked a turning point in Canadian fisheries management, given the general agreement between federal scientists, policy-makers, industry, and local people that there were serious problems of overharvesting and resource degradation and that new and more effective approaches to science and management were needed. However, an examination of post-moratoria challenges to fisheries science and management suggests that the fisheries policy objective of conservation remains elusive for Canadian fisheries policy-makers.

Fish populations in the world's oceans have been declining at an alarming rate over the past twenty-five years (Sinclair and Murawski 1997). One index of the status ("health") of a fishery ecosystem is an assessment of how every species in the ecosystem's fish community is doing in respect to International Union for the Conservation of Nature and Natural Resources (IUCN) criteria for species at risk. Because marine fishes are widespread and therefore usually exist in large numbers, the only criterion that can apply is decline in abundance. The fish community in an ecosystem is composed of many different species, but it is those with commercial importance that are often of the most interest and therefore the best known. Using scientific survey data, it is possible to assess the status of almost the full fish community. Such data for the northern Gulf of St Lawrence suggest that something like thirty species may fall into the category "threatened" or worse, and the decline of important Newfoundland species such as halibut and cod has also affected the whole human economy (chapters 3, 8, and 12; Hutchings 2000). Clearly, conservation of commercial fisheries has become a problem of conserving fisheries ecosystems, and new approaches beyond conventional single-species management are called for. But noncommercial, and therefore often poorly known, fishes and their interactions within the community are important in an ecosystem context, and changes in their populations should also be viewed with concern. Some of these species – the wolffishes (*Anarhichas* spp.) for example – have become at-risk species because of human activities directed at commercial species (O'Dea and Haedrich 2002).

Thus, examinations of the state of fisheries science in the study areas of the northern Gulf and southern Labrador point to critical gaps in existing knowledge that have been compounded by changes in fisheries and in the location of fish populations for such critical species as Atlantic cod, as well as by cuts to stock assessment science, as government moves to offload fishing costs (including the costs of science and wharves) and responsibilities onto harvesters and their families. For instance, lobster became a much more important fishery in many parts of Newfoundland in the 1990s and is subject to extreme pressure in some areas, but the scientific resources available to monitor the fishery and to assess stock abundance have not expanded to match the need. Scientific resources that do exist appear to have been disproportionately concentrated in some regions relative to others (Murray, Neis, and Bavington 2005). There are similar problems with understaffing in relation to crab and shrimp science where contemporary fishing pressure is intense, but science is in its infancy and scientific resources are very limited.

On the northeast coast and in Labrador most of the cod that survived the overfishing of the 1980s live in the bays rather than in the offshore areas that were the central focus of stock assessment science in the past. Limited fisheries have reduced access to commercial fisheries data, while efforts to develop new indices of abundance that suffer from fewer problems than commercial data (such as sentinel fisheries) are relatively new, provide sparse indices, and are somewhat controversial among harvesters and scientists. Some of the surviving cod may be part of local, bay stocks, as in the case of stocks identified in Trinity and Placentia Bays and in Gilbert Bay, Labrador, but knowledge about the location, life histories, and health of such stocks is limited (Wroblewski, Neis, and Gosse 2005).

Authors of a recent DFO marine environmental review of the estuary and the Gulf of St Lawrence were unable to arrive at unambiguous conclusions regarding human impacts on that system. Just a few examples of the critical gaps they identified in scientific knowledge related to fisheries in this area include knowledge of juvenile habitats that is limited to only a few species or populations; gaps in knowledge about relationships between catchability and abundance for some species; knowledge related to reasons underlying an apparent change in cod migration patterns in the Gulf that call into question existing management areas; biological differences in distribution and reproduction related to the two species of redfish that compose Gulf stocks, *Sebastes faciatus* and *Sebastes mentella;* and the precise impacts on benthic habit, predator-prey relations, and other processes of bottom trawling. In addition, they found that more needs to be known about stock structure for most species, migrations, spawning areas, and predator-prey relations (White and Johns 1997).

The above discussion refers primarily to commercial fisheries, but even less is known about many newly commercialized and non-commercialized fisheries. We have also not been able to take into account gaps in current knowledge about microscale interactions between fisheries and fish populations and the factors responsible for these interactions. Similarly, little is known about the impact of climate change and changes in the timing and volume of freshwater inputs into these areas, as well as microscale currents and tidal effects. It is well known that depleted populations tend to have few year classes and to be particularly susceptible to the effects of environmental fluctuations and of pulse fisheries, but relatively little is known about the way environmental changes have interacted with remnant populations (Hamilton and Haedrich 1999).

Where a history of poor management and overfishing have contributed to the ramping up of effort and efficiency in depleting fisheries and where harvesters are allowed, indeed encouraged, to shift effort over to other, less abundant species about which we have little knowledge, there is a very strong risk of accelerating degradation of fish stocks and overall marine ecosystems even in contexts where management and other factors contribute to the removal of some harvesters and gear (Neis and Kean 2003). These risks will be greater where gears are nonselective and data fouling accelerates due to by-catch, highgrading, and discarding.

Research vessel surveys remain the primary basis for stock assessments. They can provide information on long-term trends in actual abundance of different species. They also generate information on year-class, length at age, and diet. The East Coast of North America Strategic Assessment Project (ECNASAP) database was constructed from research vessel survey data collected in the United States and Canada on living marine resources and their habitats (Brown et al. 1996). For Canada, the ECNASAP data are taken from DFO trawl surveys. All Canadian mobile gear surveys follow a stratified random-sampling scheme based on depth (Frechet et al. 2003). By controlling a range of variables (effort, location, depth, timing) these data provide a statistically valid and reasonably "objective" means of measuring trends in abundance and distribution of certain species over time.

What is needed now for marine fishes is to get an overview of the general situation, and so we have attempted an assessment of the entire Atlantic marine fish fauna. For all Atlantic Canada, we analyzed 266 fish species, and of those, more than 50 percent (140 species) were found in so few years that their status was data deficient. About 18 percent (49 species) had decline rates greater than 50 percent and were therefore endangered, and only slightly more (20 percent, fifty-four species) were "not at risk." Data for the northern Gulf of St Lawrence revealed similar results (figure 3.1). Both "endangered" and "not at risk" categories contained commercial species, but they were of different sorts. In the endangered category, species were often large,

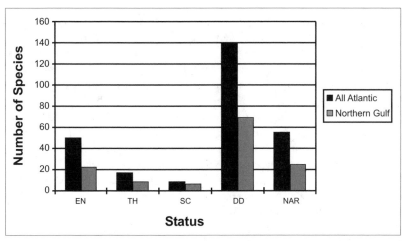

Figure 3.1 The number of species in each of the COSEWIC status categories for all of the western North Atlantic region (N = 266) and for the northern Gulf of St Lawrence alone (N = 129). Based on ECNASAP data. EN = endangered, TH = threatened, SC = special concern (vulnerable), DD = data deficient, NAR = not at risk

long-lived, demersal top predators. In the "not at risk" category, however, the commercial species tended to be small, short-lived, pelagic, and of lower trophic level: e.g., alewife (*Alosa pseudoharengus*), Atlantic herring (*Clupea harengus*), Atlantic mackerel, butterfish (*Peprilus triacanthus*), and capelin. This interesting dichotomy invites further study in detail.

In the northern Gulf, the decline rates of the ten species that fit the "endangered" criterion were significant, well above the 50 percent decline that defines the category, and ranged from about 60 percent to 98 percent. Three species already considered by COSEWIC for Atlantic Canada were endangered in the northern Gulf: Atlantic cod (–75 percent, listed as "special concern" in 1998), Atlantic wolffish, *Anarhichas lupus* (–85 percent, "special concern" in 2000, O'Dea and Haedrich 2002), and spotted wolffish, *Anarhichas minor* (–96 percent, threatened in 2000). Six species were commercially exploited, including haddock (–98 percent), goosefish, *Lophius americanus* (–72 percent), and deepwater redfish, *Sebastes mentella* (–73 percent). Of noncommercial species, sculpins (*Cottidae*) were well-represented in the endangered group: longhorn sculpin, *Myoxocephalus octodecemspinosus* (–61 percent), moustache sculpin, *Triglops murrayi* (–60 percent), sea raven, *Hemitripterus americanus* (–88 percent), and snowflake hookear sculpin,

Artediellus uncinatus (–96 percent). Because of their size and habit, larger sculpins are probably taken regularly in trawl by-catch.

The data deficient category in this analysis results when the species in question was not caught in the first few years or in the last year of the time series. This category is useful in that it focuses attention on how little may actually be known about many marine fish species. For example, it is critical to know the generation time of a species in order to determine the appropriate time period over which abundance change should be calculated. More studies, such as the recent one by Magnússon (2001) on age of the morid *Antimora rostrata*, a deepwater fish that we find has declined about 84 percent in Canadian waters, are much needed. It is also dangerous to extrapolate from a deep-water set of data to shallower waters, where ecosystem dynamics and the social-ecological history may be quite different. More broadly-based research in coastal areas to match that from the deep sea is therefore also needed. That said, the overall picture from our analysis is not an encouraging one, and new approaches beyond conventional single-species management are called for. One such approach would consider restoration of damaged ecosystems over the long run, and we now turn to this.

However, the ECNASAP data do have some shortcomings as a tool for historical reconstruction and for understanding interactions between fishing, fish abundance, and ecosystem change. First, they are somewhat lacking in historical depth. For the Gulf of St Lawrence, for example, data begin in 1970, though there are a total of only five data points for that year. The quantity and comparability of the data generally increase over time. The DFO, for example, draws on data sets that begin in 1984 (Frechet et al. 2003). Likewise, data (eight data points) for Atlantic cod are available for 2J beginning in 1970, two years after the "killer spike" in landings in 1968. Trends in abundance or distribution, in other words, can be compared only against a baseline that represents an already heavily fished ecosystem.

Second, because of obvious cost constraints, trawl surveys are not taken throughout the year. The ECNASAP database contains sporadic data records for certain times of the year, reflecting the original sampling strategy. In the northern Gulf, for example, the longest time series of data are derived from the Alfred Needler surveys conducted in August only since 1990. Critically, this limits our ability to analyze intra-annual variability in fish distribution resulting from behavioural phenomena such as feeding migrations or spawning aggregations and related variations in the marine environment.

A third limitation of research vessel (RV) data stems from the sampling technology used, which limits both the location (depth) and types of fish sampled. Reflecting the trawl technology used (which cannot be used effectively in shallow areas) the ECNASAP database does not contain data records for shallow areas, which have historically been the areas of concentration for the substantial coastal fisheries in the province and include nursery areas for many species.

In the past, scientists often addressed gaps in scientific knowledge by drawing on fish harvesters' local ecological knowledge (LEK). We have tracked the history of interactions between different forms of fisheries knowledge. Each form of knowledge (i.e., science, LEK) about fish and fisheries is a social-ecological product of interactions between knowledge creators, their history and observations, and the social environment that influences what they know, what they observe, and how they interpret that knowledge and those observations (see Neis and Morris 2002 for a discussion). A comparative study of the development of three East Coast fisheries (see Alcock, Ings, and Schneider, forthcoming, for more detailed discussion of this work) shows that the history of the balance between formal scientific knowledge and traditional or local ecological knowledge (TEK/LEK) within marine fisheries in Canada follows a typical pattern over time. In the early stages of knowledge development, catch information is the primary data source, some of it coming from local fishers, little to none of it from science. As fisheries develop, scientific data are collected, and this knowledge becomes more and more accepted as the basis for making decisions in the fishery; TEK/LEK is marginalized within the stock assessment and management process. But when fisheries go into crisis, as in recent years, TEK/LEK tends to reappear as a source of knowledge for science and management, but the relationship between these knowledge systems may well differ in the new contexts of established science and management structures and of resource depletion. That is, the issue of mismatched scale in these forms of knowledge has not yet been seriously addressed.

FISHING EFFICIENCY AND PRACTICE

New Data Sources and Management Options – LEK

That said, if we can solve the problem of scale mismatch and build substantive knowledge of abundance trends, life history, and population structures through RV data, landings data, and LEK, it will help us see more clearly changes in fishing efficiency and their relationship to fish ecology, fisheries science, and fisheries management. Efforts to account for the collapse of the northern cod stocks have pointed not only to the interactive effects of environmental change and the effects of increased fishing efficiency and high levels of capacity relative to stock abundance but also to serious related shortcomings with stock assessment science and management, including a failure to comprehend the dynamism of fisheries and to understand the relationship between such dynamism and stock assessments and the impact of fisheries management initiatives on the quality of the data available to scientists.

Like scientific knowledge, LEK is a complex knowledge system that changes over time, reflecting various different influences, including the social and technical organization of harvesting activities and the ecological relationships associated with the time and place of harvest, as well as the life history of targeted and by-catch species. Awareness of the social-ecological character of the different knowledge systems and of their dynamism needs to be incorporated into the use and interpretation of the information they provide when we seek to improve our understanding of fisheries resources and the effectiveness of fisheries management initiatives (Murray, Neis, and Bavington 2005; Murray, Neis, and Johnsen 2006).

Since the collapse of the northern cod and northern Gulf cod stocks in the 1980s and 1990s, it has become imperative to manage the fish that remain with much greater care and at smaller spatial scales. However, not enough is known about the microspatial and temporal scale dynamics of cod populations in these areas or about the location and health of the habitat upon which cod and other marine species depend. We have been using LEK to reconstruct the migration patterns of cod in the northern Gulf of St Lawrence and southern Labrador. The LEK of retired harvesters indicates that recovering Gulf stocks have begun to renew a migration pattern that is consistent with previous reports of coastal movements by Munn (1922) and with recent tagging data. We have also used LEK to identify potential current and former locations of local cod stocks and to map the presence/absence of many species throughout the East Coast study area, including some understudied species such as sturgeon (*Acipenser spp.*). The continued poor state of cod stocks in Atlantic Canada means that it is imperative that every effort be taken to protect the remaining fish. Until recently, little systematic research has been done on the location of nursery areas for cod and on the location and health of the habitat for juvenile cod, and we have been using fish harvesters' LEK to provide insight into these gaps in science.

We have also done case studies of lobster fisheries in St John Bay (Whalen 2005) and Bonne Bay (Rogers forthcoming) on the west coast of Newfoundland and the Eastport Peninsula (Davis, Whalen, and Neis 2007) on the northeast coast. These three areas have quite different ecological, fishing, and conservation histories. The case studies employ varying combinations of LEK, direct observation, key informant interviews, and scientific research. In Eastport, LEK, demographic, and key-informant interviews were used to reconstruct a thirty-year history of the fisheries, catch rates, and conservation initiatives and to document the history of the Eastport Lobster Conservation Committee. In St John Bay, LEK, demographic interviews, and direct observation produced both an historical reconstruction of the fishery and a basis for studying fishing strategies and inter-harvester communications. Results from this fieldwork component were then used to develop a generic individual-based model (IBM) of harvester-lobster interactions that could be applied to other lobster fisheries. The model has been used to test "what if" statements related to harvester behaviour and the potential impacts of different management initiatives. In the case of Bonne Bay, the historical reconstruction of fisheries in the Bay points to the larger pattern of shifting effort across species over time that we have found throughout the study area. A dive-based survey of lobster in the Bay and other data gave an indication of the health of lobster stocks in different parts of the Bay at the end of the fishing season.

In general, the results of the three case studies, as well as of other LEK research related to lobsters, have shown that dwindling cod landings in all three areas coincided with a tendency to shift effort onto lobster in the 1980s. In the Eastport Peninsula area, harvesters reacted to the perceived threat to their lobster fishery by implementing a multifaceted lobster conservation project that limited access to area lobster grounds to harvesters from the Peninsula and included other conservation initiatives like v-notching and closed areas. These changes, coupled with access to alternative species like snow crab, and retirements have limited the pressure on Eastport Peninsula lobsters and have been associated with some recovery in landings in research years.

In Bonne Bay, the number of lobster licences went down in the 1980s as harvesters from further up the Northern Peninsula came in and bought licences and moved them up to the St John Bay area. Some harvesters in Bonne Bay also have access to other species like crab, which, along with the relatively low number of licensed harvesters in the area, has helped to protect incomes and lobster stocks in Bonne Bay. That said, research by Rogers (forthcoming) suggests that even these lobster stocks are at risk of recruitment overfishing. Rogers was also unable to find any lobster larvae in tows in the area, raising questions about recruitment but also about the relationship between this stock and lobsters from outside of Bonne Bay. It is possible that lobster recruitment depends upon currents carrying larvae into the Bay.

The St John Bay lobster fishery underwent major expansion in the 1980s as licences were transferred in, inactive and underutilized licences were activated, and harvesters concentrated more effort on lobster in the absence of access to other species. Community territories have broken down over time, and conservation initiatives such as limited cuts to the upper trap limit, increased carapace size, and some v-notching do not appear to have prevented serious problems with overfishing and declining catches and incomes from lobster. In addition, a recent analysis of landings, logbook, and onboard monitoring data indicates very low rates of voluntary v-notching in this area and probable removal of undersized lobsters (Ennis and Whalen 2005). Harvesters who moved licences into this now overcrowded lobster fishing area are no longer permitted to move outside the area, and there are few alternative places to go. In addition, these harvesters have had unusually limited access to alternative fisheries, such as those for cod and snow crab. The situation confronting the St John Bay lobster stock and fishery and those harvesters, families, and communities that depend on it is serious, and we have encouraged these harvesters and the DFO to work together to develop a recovery strategy.

Finally, we are carrying out a detailed case study of the snow crab fishery in southern Labrador. Five communities in Labrador have snow crab processing plants as their primary industry. Cod stocks have not recovered in this area, and although a substantial shrimp fishery takes place in 2J and 3K adjacent to this region, harvesters in the region have very limited access to the shrimp fishery, and the vast majority of this shrimp is processed onboard a fleet of factory freezer trawlers or in other regions. Thus, the region is heavily dependent on a snow crab fishery within which, despite dramatic quota cuts in recent years, harvesters are having difficulty reaching their quotas. Harvesters in the area describe increased problems with soft-shell crab in their pots, and it is now known that discarding soft-shell crab (indeed poor handling of any snow crab) can result in high mortality. In addition, landings have declined to the point where small-boat harvesters have contracted their fishing range because it does not pay to travel further afield or to move their pots a lot. Larger-boat harvesters are travelling further and moving pots over larger distances, with the two sectors together displaying the pattern of intensification (small boat) and expansion (larger boat) that has been observed in other fisheries. Fishing costs relative to landings have escalated to the point where many harvesters can no longer afford insurance on their boats. Lower quotas mean that plant labour forces in several communities have been unable to get enough hours of work to become eligible for Employment Insurance

(EI), and in other plants, this has been possible only because of access to crab harvested in some Newfoundland boats.

Our detailed studies of changes in landings, effort, and stock abundance are producing fine-grained reconstructions of changes in coastal marine social-ecological systems with a particular focus on fish populations and fish harvesting over the past several decades. They have already provided important insights into interactions between changes in fish, fisheries, and fisheries policy over time that might, we think, eventually be incorporated into simulation models, thereby bridging the gap between macro modelling and micro data. The St John Bay lobster fishery study used the historical reconstruction of that fishery obtained from LEK and other sources to develop a highly innovative individual-based model for harvester-lobster interactions. "What-if" scenarios for management initiatives point to the depth of the management challenges facing this fishery and to the need for a comprehensive, multipronged approach (Whalen 2005). Thus, the micro and macro methodologies complement one another.

INTERACTIONS BETWEEN FISH, FISHERIES, AND FISHERIES MANAGEMENT AS ILLUSTRATED BY INDIVIDUAL TRANSFERABLE QUOTAS (ITQS)

The restructuring of state policies has been an important component of overall restructuring over the past two decades. Neoliberal reforms within many sectors, including fisheries management, have emphasized privatization, individual responsibility, targeting of programs to the "right" people, and efficiency. This regime shift did not necessarily mean *less* government but rather a shift in what governments did. This shift took different forms in different policy areas. In the area of fisheries management, the policy direction was to impose market principles on what had been a common property resource – i.e., to solve a market dysfunction by creating a regulatory framework to artificially mimic a "normal" free market. The main focus was to limit access to the resource – and create more, not less, regulation in the name of promoting market principles and resource sustainability. In fisheries, responsibilities are being downloaded to fish harvester groups and to lower levels of government such as port authorities (see MacDonald, Neis, and Murray 2005 for a more detailed discussion of these issues).

Issues of environmental sustainability, equity, and economic efficiency continue to plague Canada's fisheries as they play off against one another. Since at least the 1950s, Canadian marine fisheries have been regulated from Ottawa, which has maintained a large-scale approach to resource development, generally favouring large corporate actors and, since the 1970s, a gradual shift from community-based to state-based and, in recent years, market-based allocation of fishing rights. In the wake of the 1992 and 1994 groundfish moratoria and related adjustment program, the federal government has adopted an explicit commitment to rationalizing the fishing industry by reducing the number of fish harvesters, limiting future access and professionalizing those who remain, and by formal constraints on government investment in fishing and fish-processing activities. Most Newfoundland and Labrador fisheries are now operated on the basis of individual boat quotas.

Over the past decade, dramatic restructuring associated with stock collapse and moratoria has made it easier for the DFO to transfer routine management of fisheries over to harvesters (in the process driving up the costs of harvesting) and, increasingly, to the companies that control access to our marine resources. However, such transfers do not guarantee resource recovery and are having very serious consequences for coastal people and the communities in which they live (see chapter 12).

Copes (2003) questions the logic that the now popular policy tools of IQs and ITQs provide economic efficiency and ensure improved conservation. ITQs, he argues, give no private property rights to any specific fish or any specific part of the environment that produces and sustains fish, yet the avowed purpose of ITQs has been to bring the efficiency advantages of private ownership to the fishery. The attempt to do so, essentially, has been a failure. For privatization of the fishery to be substantially complete and to meet the test of economic efficiency would require giving every fishing enterprise exclusive rights to, and exclusive control over, a particular identified set of fish, along with the ecology that produces those fish, in the same way that a farmer owns and controls specific animals and all the productive facilities of the farm necessary to raise and bring those animals to market. This is palpably impossible in the fishery.

Legally, ITQs in Canada are "privileges" and not "rights." Governments can give and take away such privileges. If they are "rights" in a broader sense of the word, they are rights only to catch and land a defined amount of fish of certain species. This is an *access* right to fish, which may be called a property right of sorts, but certainly not a property right to any particular physical asset consisting of fish. "Privatization" by ITQ distinctly is not privatization of a fish stock that could serve as the basis for economically efficient exploitation of that stock. As a result of the continuing collective ownership of marine fish stocks, ITQ systems do not give ITQ holders an unambiguous private-ownership incentive to safeguard the value of fish stocks. On the contrary, as fish stocks under ITQ regulation are shared by access-right holders, they continue to compete for capture of the most valuable parts of the resource. Further, as discussed above, ITQs can sometimes operate in a more wasteful fashion than they do under alternative common-property management systems. The abalone (*Haliotis spp.*) fishery in British Columbia provides a cogent example. The fishery was closed in 1990 after eleven years of ITQ management. An excessive quota, data fouling, and misreporting of catch, poaching, and high-grading are all thought to have contributed to the collapse of abalone stocks, which still do not show signs of recovery today. On the other hand, halibut stocks have responded positively, with robust stocks and excellent quality fish entering the markets.

Most importantly, however, the use of ITQs as a management instrument raises equity concerns, encouraging larger vessels and continuing to concentrate ownership into industrial hands. Small-boat harvesters are suffering accordingly. Both fleet reductions and ITQs have led to further concentration of licence ownership, since rural, community-based fishers often cannot pay the sums involved in acquiring an adequate number of licences to secure a viable fishing livelihood. Today, many licences are held by investors who lease them to fishermen, reducing the share of fishing incomes captured by fishermen and coastal communities (Cruikshank 1991). On the West Coast, examples include halibut, sablefish (*Anoplopoma fimbria*), and groundfish trawl fisheries. These fisheries now have a number of quota holders who do not fish themselves but, rather, lease their quota to others who do so, at rates as high as 70 percent to 80 percent of the revenue from the landed catch (Ecotrust Canada 2004). In other words, ITQs may well be good for fish but still result in inequities in the small-boat harvesting sector.

On the East Coast, IQs are not supposed to be transferable and the DFO's fleet separation policy means IQs are supposed to be held only by fish harvesters and not by processors or others. However, trust agreements between fish harvesters and processors and the growing practice of leasing quota are undermining these fundamental policy goals. The result once again is inequities, owing to removal of self-employment opportunities for current and future small-boat fishers and the consequent erosion of the economic base of coastal communities. The practice of leasing quota is also maladaptive in that it is removing from decision making the real-time knowledge of what is happening in the adjacent marine ecosystem that exists in local coastal communities. This creates a disincentive for compliance, along with the need for more expensive enforcement (Pitcher 2005).

Given the double problem that, on the one hand, fish are scarce and, on the other, people have to eat and their communities have to survive, fishers' insecurity is necessarily heightened when management is quota-based and thus biased more towards large capital interests (the industrial fleets) than small local enterprises. On the coasts, food, human welfare, and local cultures have always depended on some kind of fishery, a situation that existing top-down regulatory regimes are reluctant to deal with. Whether it be salmon, sablefish, or sea urchin (*Strongylocentrotus spp.*), for example, on the West Coast, or cod, shrimp, and crab on the East (also for example), scarce marine commercial resources and serious overcapacity have led the DFO to opt for an industrial fisheries policy with potentially huge social and environmental costs and no guarantee for recovery. In chapters 17 and 18 we look at future options for coastal communities' participation in alternative approaches to resource management and recovery.

SCEPTICISM AND CONSERVATION

Growing awareness in the 1990s of conservation concerns is reflected in a series of major policy documents released in this era:

Conservation of marine resources and habitat, and rebuilding of resources and restoration of habitat where necessary, will remain the highest priority for the management of all fisheries. Within the limits of available knowledge, all fishing activities will be conducted in a manner that leads to sustainable levels of resource use ... The economic viability of resource users and fisheries-dependent communities is precarious, and it will remain so until everyone makes a commitment to conservation and sustainable use ... Fisheries and Oceans Canada recognizes that it must be more effective at enabling resource users to take greater responsibility for sustainable use. (Canada Fisheries and Oceans 2004, 8, 11)

The auditor general, furthermore, has noted that "past management of Pacific salmon and its habitat, together with factors beyond the Department's control, have resulted in serious declines in many stocks." A continuing problem was seen as "a limited information base and poor relations with stakeholders," resulting in the conclusion that the DFO has to figure out how to "conserve stocks while providing opportunities for harvesting. Appropriate action to protect the genetic diversity of existing stocks will likely mean that fisheries will be curtailed in the short term in the interests of rebuilding the resource base to sustain fisheries in the future" (AGC 1999, 27).

Conservation has clearly been more a response to resource depletion than a pro-active policy goal until very recently, with reduced quotas resulting in excess capacity in both the fishing fleet and the processing sector and thus continuing pressure for overharvesting. Despite evidence of other factors contributing to resource depletion, such as changes in water temperature, "over-harvesting has been a sizable – if not the single most important – contributor. Fisheries managers and their political leaders have sanctioned unsustainable harvest levels and encouraged unsustainable technologies ... Capitalization and overcapacity has extended to the processing sector as well," so that today, "Numerous individual stocks, and even entire species, of Pacific salmon are in peril and the Atlantic cod have not shown significant signs of recovery after nearly a decade. Many lesser-known species and their habitats are also threatened and the ecosystem impacts of these changes are evident but poorly understood" (Markey, Pierce, and Vodden 2000, 438). Further, science and management capacity, particularly in coastal regions, has been reduced through cutbacks and reorganization to levels where many argue the fishery can no longer be appropriately managed and safeguarded under the current centralized management regime. On both coasts, government science accepts that stocks are vulnerable and has turned in principle to a consideration of species at risk and ecosystem-based management that seeks to conserve healthy ecosystems. However, how government plans to do this remains unclear and, furthermore, its parallel shift to IQ and ITQ systems potentially runs counter to this kind of approach.

On both coasts, not only policymakers but local people understand the crisis in the stocks and have shifted where they can to alternative fisheries, which also now bid fair to be depleted. One official is quoted on the Labrador coast as saying "we may as well hit 'em (crab) while we can, because they are going to go anyway" (Informant, Port Hope Simpson, September 2004). Moreover, in many rural communities, the costs of getting into new fisheries or purchasing (on the West Coast) additional salmon licences needed to fish the whole coast (as many once did) are prohibitive.

While *Coasts Under Stress* was under way, a study on human resource issues in Canada's commercial fisheries on the Atlantic and Pacific coasts was also being carried out for the Canadian Council of Professional Fish Harvesters (CCPFH) (Praxis 2005). It produced a profile of the fish harvester labour force based on telephone surveys with over 1,200 enterprise heads and 600 crew members on the East Coast, and 300 enterprise heads and 170 crew members in British Columbia. Key informant interviews and financial analyses of market and nonmarket factors influencing the cost of licences and enterprise viability were also undertaken, as were several community case studies that explored how trends in the industry have been affecting wider communities. The results show an emerging labour force crisis in the owner-operator fishery that is linked to an aging labour force, high rates of impending retirements, and poor recruitment of young people into fisheries, all of this consistent with our work on these questions.

The authors also point to the role of policy initiatives in contributing to this impending crisis. Thus they link escalating costs for licences to speculation and involvement by processors and others in markets for licences, as well as to the practice of licence staking by wealthier harvesters. These escalating costs are exacerbating financial barriers for new entrants and economic pressures on some existing harvesters, thus exacerbating in turn the recruitment challenges within the owner-operator sector. Licensing costs interact with other escalating costs for insurance, monitoring, etc., and with market challenges, to erode profits within many fisheries. Significantly for CUS, the CCPFH study also found that "higher licence costs and the erosion of owner-operator control are also having a negative impact on conservation and orderly management. There is credible information that increased cost pressures on some harvesters are fuelling an underground market for several high value species" (Praxis 2005, 3). This finding supports our own work on ongoing problems with data fouling in the Newfoundland industry that appear to be linked to interactive restructuring between environmental, industrial, and policy change in our fisheries.

The CCPFH report, like our own research, highlights the relationship between a viable and vibrant community-based fishery and the resilience of coastal communities. The study argues that erosion of the owner-operator fishery and associated community decline has progressed much further in British Columbia than on the East Coast, owing primarily to policy differences between the two regions. On the East Coast, the fleet separation and owner-operator policies are credited with protecting the owner-operator fishery. That said, the researchers provide considerable evidence of the erosion of these policies through such initiatives as trust agreements, which are particularly well-entrenched in Newfoundland and Labrador. The report finds that "[i]f not constrained by clear public policy objectives, the evidence generated by this study suggests that market forces will rapidly move the industry towards concentration in ownership and geographical location of fishing licences. In countries or regions where such processes are advanced, the consequences for fisheries-dependent communities, and the rural-based fisheries labour force have been far-reaching and predominantly negative" (Praxis 2005, 6). Praxis's international research, along with research on particular initiatives in Canada, suggests, however, that policies designed to develop and "shelter" owner-operator fisheries using such tools as financial support for intergenerational transfer of assets, effective taxation policies, regional development services, and specialized education and training programs (including training in small-business management, stock assessment and other science and "capacity-building for self-management in conservation") could help protect these fisheries and the communities that depend upon them (ibid.). It is very clear now that the future of those coastal communities that depend on fisheries is in danger – that is, the future of both fish stocks and employment remains a concern.

CONCLUSION

The combination of resource degradation and heightened fishing capacity and efficiency means that we have far less latitude for error in fisheries management today than thirty years ago, particularly on Canada's East Coast. Management for recovery will require a more comprehensive, finer spatial-, and temporal-scale approach to studying and managing our fisheries. Management for recovery that seeks to promote equity and the health of coastal communities will be very challenging. We need to address scale mismatches and risk and benefit misalignments and to respect and understand the interactive nature of change in social-ecological marine systems. We will need to employ multiple conservation strategies (a challenge we return to in our concluding chapters) that are inclusive of habitat and rich in knowledge about life history, predator-prey interactions, and larger ecosystem processes and that also reflect a deep understanding of the human dimensions of fisheries. Science and management need to be more socially inclusive of elders and retired community members, as well as the current and future generations of harvesters, processing workers, and other members of their communities. For this to happen, leadership, innovation, and experimentation need to be encouraged within communities as well as in government bureaucracies (federal, provincial, and municipal), within science, and within schools and other venues.

Since the moratorium, the federal government has invested heavily in projects to build tourism infrastructure in rural communities, often through make-work programs. At the same time, our shore-based fisheries infrastructure has been passed over to communities, often without the financial and capacity-building support required to protect that infrastructure. In addition, far too little has been invested in assembling, testing, and deepening existing knowledge of fisheries ecology and the state of our marine ecosystems and in supporting local efforts to experiment with different approaches to management and enhancement. Where innovative conservation initiatives have been undertaken, as in Eastport, they were often achieved despite government resistance, and they are treated by government as pilot projects that are somehow expected to reproduce themselves spontaneously and without financial and scientific support in other areas. Despite the relatively recent requirement for harvesters to generate conservation management plans and with the shift to integrated management, management (and hence capture practices) is still based largely on top-down single-species management, and the science is still running around behind actual fisheries practice.

Illegal fishing is a growing problem in many parts of the world, particularly for high-value species such as lobster, and Newfoundland is not alone in this problem. Indications of persistent problems with data-fouling practices like high-grading, discarding, and misreporting of by-catch point to ongoing problems with quota management and with stewardship practices and incentives in our fisheries. In an age of sweeping budget cuts, few fisheries agencies can field the resources necessary to monitor and prevent illegal fishing or to determine appropriate harvest levels (Hilborn et al. 2003; Davis forthcoming). According to Hilborn et al. local cooperation and control is the appropriate approach. Still, it is clear that greater efforts to incorporate the insights of resource users into decision-making processes will not be sufficient to manage fishery resources effectively. Significant reinvestment in state and academic research will be necessary if we are to improve our very limited understanding of marine ecosystems and increase our likelihood of being able to rehabilitate fish stocks.

Improving DFO-fisher-fishery community relationships is going to take a great deal of time, energy, and resources – all of which now appear to be in short supply. Some structural changes could be made to existing institutions at both the federal and the provincial level. These institutions require credible relationships with all stakeholders, as well as sound advice that reflects the views of a wide variety of industry and community officials. Achieving this goal may involve facilitating communication and improving the stakeholders' capacity to participate in fisheries management processes. Decisions should be science-driven in the management of natural resources such as fisheries, and the science should not always have to be catching up with what is going on in fisheries. The ecological and biological facts about fish, distribution, population structure, and so on are very important and form the core of the Newfoundland fisheries. More effective use of local knowledge from fishers should also be built into the science. These fishers often have knowledge about the spawning and nursery locations and migration patterns of the fish stocks that could be harnessed more effectively to develop a recovery strategy for our fish and shellfish resources based on a co-management structure. Co-management must not mean the abdication of government responsibility for protecting and enhancing marine resources, providing social support, and promoting health in coastal areas. Hard decisions must be made if we are to recuperate ecosystems. The price for recuperation should not be community disintegration, deep social inequities, and related deterioration in the health of fishery people. To date, governments have been unwilling to put the required funding in place to support ongoing recovery of fish and fisheries on both coasts. The DFO needs to come to grips with the poor state of ocean science; if our coastal communities are to survive, it is also vital that they pledge themselves to the long-term survival of a wild fishery and that they work with other government agencies in pursuit of equity as well as efficiency. Aquaculture (see chapter 15) will not suffice. The DFO and coastal communities need to build a co-management structure that will not download all risk and responsibility onto fishing communities.

4

Social-Ecological Health and the History of the Forest Products Industry on Both Coasts

INTRODUCTION

In the last chapter, we saw that accelerating resource degradation in coastal fisheries had damaged social-ecological health. In this chapter,[1] we turn to the history of coastal forestry, to ask the same questions and also to consider social concerns, since our goal is to provide a social-ecological analysis. Confining the discussion in this chapter to forests, there is strong evidence that forest ecosystems across North America have undergone major changes since the advent of industrial forestry and that Canadian coastal communities are experiencing part of the yet-to-be documented, continental-scale downgrading of North American forest resources. Some recovery of previously cleared forested ecosystems in eastern North America has been measured from satellite surveillance (Pimm 2000; FAO 2005) in terms of increasing landmass-supporting trees and in a more extensive canopy cover. Concurrently, however, landscape-scale reduction in the complexity and age of forests logged during the past two decades has been measured through aerial photos and ground surveys, and, more recently, from earth satellites over Haida Gwaii (Queen Charlotte Islands) and other Pacific coastal forests in western North America (Gowgaia Institute 2003). Furthermore, the species composition, structure, and function of forests have been irrevocably altered; the introduction of exotic trees and pathogens to the New World has diminished the extent of native trees (e.g., American chestnut [*Castanea dentata*], American elm [*Ulmus americana*], eastern white pine [*Pinus strobus*], and western white pine [*Pinus monticola*]). Fungal and arthropod populations have declined due to air pollution and other factors, and changes to decomposition and nitrogen cycling rates have been created by urban heat islands (McDonnell et al. 1997).

Figure 4.1   Swanson's Bay Mill, 1909. Photo: Frank
Swannell. BCA 1_33288

Overall impacts on biodiversity and the resilience of forests due to
wide-scale conversion of natural forests to tree plantations are largely
unknown. Initial evidence shows that our remaining North American
forestlands have been inexorably changing from complex, gap-replac-
ing, old-growth forests to young, regulated, even-aged forests dominated by a range of crowded, small-diameter trees, supporting a
simpler food chain, and characterized by altered nutrient cycling compared to pre-European conditions. We carried out two companion

Figure 4.2   Ocean Falls Pulp Mill, ca 1920. Photo: McRae Bros. BCA
G_06558

Fig 4.3   Visit of Sir Wilfrid Laurier to Prince Rupert, 20 August 1910. Photo:
McRae Bros. BCA G_03567

studies to see why this has happened, one a study of the forest industry
and the forest landscape in central western Newfoundland and its
Great Northern Peninsula (GNP) between 1890 and the present and
the other a comparative historical study of West Coast pulp-mill towns

(Swanson Bay, Ocean Falls, and Prince Rupert – figures 4.1, 4.2, and 4.3), which was designed to identify the broad underlying patterns behind individual community histories (Hammond and Griffin 2002).[2] When people (in communities and governments) look at the forests around them, or work in them, they are usually unaware of the long reach of history that makes the present landscape the way it is. This chapter, therefore, is also an explanation of how the overall current situation (with which government, firms, and communities all have to grapple) came to be.

THE EAST COAST

Newfoundland governments were aware from the early nineteenth century that economic exploitation of marine resources could lead to ecological depletion, but successive governments ignored fishers' protests for conservation measures. They instead favoured greater capital investment in and spatial expansion of the fishing industry, apparently feeling that (since it was irretrievably backward) it should not be conserved but should rather be exploited to the hilt in order to gain from it the capital that would then allow for landward industrial diversification. Ordinary people in the outports, by contrast, saw marine resources as the lynchpin of survival, and their seasonal round involved using the coastal forests that surrounded their winter settlements at the "bottom" of the bay as a community source of fuel and material for the buildings, boats, and fish flakes that the fishery required. From the mid-nineteenth century onwards, governments began building first roads and then railroads to open up the Island's interior resources, especially forests, to commercial exploitation. They offered timber rights to large tracts of forest to induce telegraph companies and railway developers to come and start projects. Commercial sawmilling (as opposed to local sawmills for domestic consumption [Coakes 1973]) began to boom, leading opposition leaders and newspaper editors to call for government recognition of the right of fishing people to cut for their own needs on Crown land in coastal areas (Cadigan 2003).

By the late nineteenth century, things had got bad enough for outport people themselves to begin demanding protection of the most accessible coastal forests from depletion by new commercial sawmillers. Governments ignored such demands, just as they had ignored demands for fisheries conservation, and worked hard to promote the expansion of the sawmilling industry. They also worked to gain control over Canadian sawmill operators on the French Treaty shore on the west coast, gaining the right to do so by the Entente Cordiale between Britain and France in 1904 (see chapter 2). Commercial sawmills posed a number of problems for fishers. First, they produced effluent that clogged rivers, polluted estuaries, and damaged breeding grounds for herring and salmon. Second, mill owners had begun to prevent fishers from cutting wood for such things as fence rails and barrel staves, which fishers then exchanged for merchant credit in the winter. In any case, sawmilling expanded so rapidly that some former boosters began (too late) to advise a more cautious approach to industrial expansion, for by the end of the nineteenth century, after a mere thirty years in operation, sawmillers' production of white pine lumber had commercially annihilated pine stands in central Newfoundland (Dyer and Robertson 1984, 302–3; Cadigan 2000).

That did not deter Newfoundland governments. Lumber-quality timber depletion did not worry them, because of the growing importance of the pulp industry after 1890, at which time they had already begun issuing permits to contractors who cut pulp wood for export and who constantly demanded more timber rights. Fishing people, of course, feared that pulpwood cutting would increase the threat to domestic wood supplies, but governments claimed that there were special areas reserved for fishers' needs. However, throughout the late 1890s, while the government allowed the right to cut on Crown land to increasing numbers of commercial sawmills, they refused to allow businesses exclusive rights when fishing people petitioned for recognition of community rights to cut in the same areas. Although anxious to promote the pulp industry, the government refused to allow pulpwood contractors to cut on lands reserved around fishing communities such as Sop's Arm, White Bay, because "the reservation is essential ... in the interests of the Fishermen of the Colony" (Cadigan 2000). Notre Dame Bay fishers' fears that the Island's railway development would allow the developers an exclusive right to cut timber along the line forced the government to establish a three-mile coastal forest reserve as the common property of fishing people (commonly referred to as the "three-mile limit," or the "fisherman's reserve"). The forest reserve was first legislated in the 1898 railway contract between the Newfoundland government and the Reid family, and subsequent governments turned down applications for timber limits whenever they conflicted with the three-mile limit (cited in Cadigan 2000).

When combined with railway developments, the industrialization of the forests posed an even greater threat because of the fire hazards they created. Sparks and hot cinders from locomotives that were improperly disposed of, as well as cinders from the poorly maintained boilers that fuelled steam sawmills, all became regular causes of disastrous wildfires.[3] Despite numerous official investigations that revealed the industrial origins of such fires, the Newfoundland government passed the 1890 Forest Fire Act, which was based on the assumption that forest fires were started by tourists, berry pickers, or careless "trouters," and used the act as one inducement for English capitalists to invest in pulp production, since it allowed authorities to justify alienating forest resources to capitalists by arguing that fishing people would only destroy the woods through carelessness if they were left uncut.

In 1903 the Harmsworth family, British newspaper barons, began to acquire the timber rights of old sawmilling and lumber concessions, and by 1905 they had negotiated with the government to establish a pulp-and-paper operation at Grand Falls, which became known as the Anglo-Newfoundland Development Corporation (AND). The Harmsworths' agent demanded complete control over any timber lots it might lease, ostensibly to control "the danger from forest fires," and government obliged, giving the company the sole right to nominate the colony's chief fire ranger in return for payment of the ranger's salary. In 1905 the government revised its forest fire prevention legislation to accommodate the AND Corporation – with an act that imposed draconian regulations on outport people, empowering rangers to assert the guilt of the first person they discovered close to a fire unless that person could prove their innocence. Although industrial causes continued to be the most important anthropogenic cause of fires, through to the 1930s the woods rangers prosecuted only fishing people (Cadigan 1999, 2000).

With the establishment of the AND, the Newfoundland government banned the cutting of wood for export to pulp markets or for pit props, although such regulations were relaxed during the First World War and the Depression of the 1930s. The Fishermen's Protective Union (FPU), the most important organization to represent fishers in the early twentieth century, came to support the expansion of the pulp-and-paper industry as a way to provide employment to fishers plagued by the long depression that began in the fishery after the First World War. Although it had been a steadfast supporter of the three-mile limit as the common property of fishing people, the FPU accepted small-scale commercial cutting and sawmilling exploitation of coastal timber stands by fishers as a source of supplementary income. Always worried about rural poverty, the FPU was susceptible to pulp-and-paper promoters such as Harry Crowe, a Nova Scotian with a background in the sawmilling industry who had come to Newfoundland in 1900 to promote forestry development with the backing of American financier H.M. Whitney. Crowe wanted government permission to build a mill in White Bay to exploit wood on the three-mile limit around Hampden and Sop's Arm, promising that workers would receive good wages and live in clean company facilities with the best medical attention available. The government agreed to Crowe's plan when it became clear that he had much popular support from the area (Cadigan 2000, 2005).

Crowe had the admiration of the FPU and many local fishers because he built logging camps with amenities far beyond the standards of the day: hot showers, adult education classes, clean bedding, good food, and medical services. Furthermore, he began his projects by promising only to cut deadfall or "blown-down" trees. Crowe used a number of tactics to convince local people and government to back him. First, he hired an American forestry engineer, L.A. Nix, to accompany local officials through the coastal forests in the vicinity of Jackson's Arm, Sop's Arm, and what became Hampden. They determined that the largely unexploited old-growth forest trees were always likely to be "blown-down," effectively establishing Crowe's right to cut any old-growth forest. Second, Crowe immediately moved into Hampden. He began constructing a town site, provided schooling and medical services, and assisted local people who flocked to the area seeking work. The government could hardly refuse him once he had demonstrated that he could alleviate the government's relief burden and once he had acquired the loyalty of local people. Crowe received his timber rights and related grants in 1926, but the Newfoundland government tried to keep him in check by limiting his grants to two years. Crowe responded by dangling the prospect of a sulphite pulp mill in front of the government, in exchange for longer-term rights. The government relented and eventually agreed that Crowe would be entitled to transfer his rights to the Canadian International Paper Company, a member of the family of companies that included the operators of the new pulp-and-paper operation at Corner Brook (Cadigan 2006).[4]

The FPU supported projects such as Crowe's, which ostensibly developed commercial sawmilling on the three-mile limit for relief and humanitarian purposes. The union further supported the development of a new pulp-and-paper development on the Humber River (which later became the mill at Corner Brook) and a possible third project on the Gander River. While the union continued to promote fishers' need for access to forests, its newspaper increasingly boosted Newfoundland's "forest wealth," and the potential for generating employment through the industrialization of the woods.

Throughout the late 1920s the expansion of timber cutting gained greater popular respectability as depressed prices continued to mark the fishing industry, and the FPU increasingly came to defend the three-mile limit as a woodlot reserved for outport people's small sawmills, not as a property held in common for the good of the fishery. Wood cutting for those sawmills had become part of the outport's occupational pluralism – a community strategy that gave flexibility and resilience to the small settlement in difficult times. Intrusion on these reserves by industry was a threat to outport sustainability, and correspondents of the FPU newspaper increasingly attacked cutting on those reserves by large mill owners from St John's or by contractors who supplied the pulp-and-paper industry, and they also criticized speculators who tried to get grants on the reserves. Fishers from Jackson's Arm, Sop's Arm, Brown's Cove, and Hampden, for example, petitioned the government not to allow "foreign" companies to acquire rights to the three-mile limit. They asserted that the "three-mile limit should be exclusively reserved for the use of fishermen," but also "that the royalties on small mills be abolished."

Crowe's transfer of timber rights to the operators of the Corner Brook mill shortly before his death in 1928, however, meant that that particular former three-mile limit would, in future, serve as a timber reserve for the mill. This was the start of a series of transfers that weakened the integrity of the three-mile limit. When Bowater took over the mill in 1938, for example, it also gained control over the former three-mile reserve at the head of White Bay (Cadigan 2005). Another three-mile limit further up the Great Northern Peninsula in Roddickton (Canada Bay) also fell into the hands of Bowater because the need to alleviate desperate poverty and related conditions meant that local people supported the transfer of property rights to a private party, believing that their communities would be best served that way.

This particular transfer exemplifies the way local rights were slowly eroded through a series of property transfers that ultimately gave the timber reserve rights to big industry. In this case, the people of St Anthony and Hare Bay had come to depend on the Grenfell Mission for health services. In 1904 the mission sent people to the site of what became Roddickton to cut timber for a schooner it required, and, realizing the potential of the area, the mission sent people back in 1906 to begin a sawmill that was meant to help support mission activities and provide local employment. In support of the venture, the government agreed to grant timber rights to the mission, including rights to the three-mile reserve in the area. The mission gave up the operation in 1921. By 1935, the Carbonear firm of Saunders and Howell had acquired the mill and expanded production. Between 1938 and 1939, Bowater started cutting pulpwood locally and bought the mill and all the associated timber rights, including the three-mile reserve, from Saunders and Howell in 1953.

The pulp-and-paper industry, although it failed to generate enough employment to counterbalance the impact of the Depression on the fisheries, nonetheless continued to prosper. To many observers, such as the young J.R. Smallwood, it seemed to be a better model for future development. Smallwood thought that the smokestacks, plants, and railway lines of the new mill at Corner Brook were potent symbols of Newfoundland's bright industrial future, and he felt that the mills at Corner Brook and Grand Falls, along with any similar future developments, must be fostered at almost any cost (Smallwood 1931, 22, 48–9, 145–6). The cost was, indeed, significant, and the financial burden of Newfoundland's development policies threatened the Dominion with bankruptcy throughout the 1920s and early 1930s. The ultimate result of its efforts to reschedule debt payments with Canada and Great Britain was Newfoundland's loss of democratic self-government, and administration of the Dominion was put on the shoulders of a British-appointed commission in 1934.

During the 1930s and 1940s, the Commission of Government relaxed restrictions on cutting on the three-mile reserve. Because industrial holders of timber limits had so badly depleted lumber-quality timber stands, most of the large sawmilling operations that produced lumber for export began to close. This left the small-scale enterprises – operated as adjuncts to the fishery and dependent on the forests of the coastal reserve – to supply the domestic market needs of Newfoundland. However, the intensified operations of these small-scale mills soon led to scarcities of timber in coastal areas and the decline of wood-products industries in the outports generally. At the same time, the pulp-and-paper industry was purchasing pulpwood from logging contractors without asking too many questions about where the wood came from and, in the desperate circumstances of the Depression, more and more of it came from the three-mile reserve (Cadigan 1999, 2000).

At Jackson's Arm, Sop's Arm, Brown's Cove, and Hampden local people feared that Bowater would forbid small-scale sawmills from acquiring timber on the former three-mile limit acquired by Crowe, but in fact logging operations for the Corner Brook mill became an important source of local employment. Bowater had constructed a road from Hampden to the headwaters of the Humber River at Alder Lake in 1942, and this made Hampden an important centre of its pulpwood logging operations. Nonetheless, by the early 1950s (after Confederation with Canada), people throughout our study region in White Bay had begun to petition the provincial government to return their three-mile limit rights to them, calling these "God given." They felt that Bowater was not generating enough employment, because it was logging exclusively for the use of the pulp mill. They argued that local sawmilling operations would generate more employment and, with that, related economic development.

RA418 S63 2009

HM861 .O45 2007

HD 1694 .C2 N374 1999.

From time to time, the (now provincial) government explored ways of exchanging timber rights with Bowater to permit more sawmilling in White Bay, but such negotiations usually came to nothing, since Bowater would then hint that any loss of timber rights in White Bay might force it to curtail logging operations there. By 1954, however, local political pressure for the return of the three-mile limit had become so great that F.W. Rowe (one of the MHAs for the area, who was minister of mines and resources) asked H.M.S. Lewin, the vice president and general manager of Bowater, about the possibility of transferring the three-mile limit back to people. Lewin refused to correspond on the matter but instead went to St John's to speak personally with the minister. There is no record of the conversation, but the limit was not returned.

About the same time, the people of Roddickton also began to demand that government return their rights to the three-mile limit from Bowater. Premier Smallwood asked J.D. Roberts, the general superintendent, coastal operations for Bowater, to explain the situation. Roberts argued that Bowater had to manage cutting to make the best use of its current and future pulpwood supplies and felt that people wanted more local cutting because they did not want to leave a good local fishery in exchange for employment in logging camps that were some distance from their homes. Pressure from people at Roddickton continued, and in 1956 officials at the Department of Mines and Resources pointed out that Bowater could probably create more local employment if it integrated lumber, pulpwood, and other wood products operations.

Local people supported Bowater's purchase of timber rights in the area from Saunders and Howell (see earlier) on the understanding that Bowater would cut fifteen thousand cords annually in the area, and also construct a drumbarker, but the company did not live up to its promises. Bowater stopped cutting on the old former Grenfell block when its first cut-over was complete and moved its logging operations to more distant mature stands. Finn Frost, the chief forester, argued that Bowater should not have been using good saw logs for pulp and that local people might have used such good wood for sawmilling, providing tree tops and small trees to Corner Brook as pulpwood. He further noted that Bowater was uninterested in changing its operations but that Roddickton, which would otherwise have been a "ghost town" without the company, should be content (Cadigan 2005).

It is not surprising that the Smallwood administrations let the pulp-and-paper companies have their own way in matters of forestry policy, given the premier's early enthusiasm for the industry. As premier of Canada's newest province, he endeavoured to build a "new Newfoundland" through industrialization and saw a third pulp-and-paper mill as an essential element of that strategy. Provincial government officials often remarked in internal correspondence that they had no independent means of verifying cutting or management practices on the timber limits controlled by the big pulp-and-paper companies but that they felt the companies engaged in sustainable wood harvesting. In 1956, for example, one government report claimed that Bowater and the AND used cutting methods that "are satisfactory and assure the establishment of a new crop." The same report went on to complain about the wasteful cutting by outport people on the remaining three-mile limit and recommended that the government take control of the timber remaining on the limit so that it might be used to encourage the development of a third pulp mill (Tunstell 1956). In effect, industrial greed and local need born of poverty were combining to destroy the resource, with government doing nothing to prevent what was going on.

Indeed, over the next decade, further studies indicated that, at least on the Great Northern Peninsula, a wood supply for the Corner Brook mill was becoming harder to find. One report (commissioned under the federal government's Agricultural Rehabilitation and Development Act) recommended a road-building program on the peninsula to open more remote wood stands and an experimental program to redevelop bog lands for silviculture. It further recommended that the provincial government encourage the consolidation of capital in the cooperative reorganization of the many small sawmilling operations in the area and foster integrated wood utilization with the pulp-and-paper industry (Black 1965). Bowater acknowledged that it was developing a wood supply problem related to its long-term strategy of increasing production, but the company claimed that its own practices were sustainable. It further suggested that the decline in wood supply was the result of insect damage, adding that "we also lose many acres each year resulting from limited alienation for highroads, power developments, parks, and other cultural purposes." They wanted to deal with the shortage by developing a Labrador supply and, if the province would jointly support it, to experiment with the reclamation of bog land (Vardy and Dickson 1967).

In the same year, another report, this one sympathetic to the sawmilling industry, claimed that the provincial government should force the reallocation of pulp-and-paper companies' timber rights to support the development of an integrated industry based on pulp production and sawmilling. This report claimed that, since the pulp-and-paper companies were doing little to ensure that the forests they exploited would regenerate, it was the duty of the province to force the

best use of its woodlands (Ker 1967). However, Bowater and its succes-
sor, Kruger, continued to face no serious regulatory limits, and the
pulp companies continued to possess favourable resource rights
through land grants, long-term leases, and right of first refusal on logs
from certain Crown lands.

Clearly, industrial and government rhetoric notwithstanding,
Bowater's (and subsequently Kruger's) cutting strategies did not
ensure the long-term economic sustainability of communities in our
study area. Quite to the contrary, since the early 1950s, Bowater had
pursued a policy of indirect Fordism, encouraging loggers to increase
productivity through the adoption of power saws. While loggers who
worked for Bowater's contractors in the 1950s felt that these saws
improved their ability to cut more wood, they also noted that the
mechanical efficiency of the saws did not offset completely the fact that
loggers were increasingly cutting in what they called "poor wood," or
else cutting more forest on rougher terrain in more remote areas. The
best forests, in areas closest to the mill and to waterways or roadways,
had been harvested first, of course, thereby keeping transportation
costs low for the companies, although for the workers this usually
meant being far from home for extended periods. As loggers had to
travel to more remote camps, of course, this situation worsened.

In the early days Bowater had logged mainly on its own lands, using
direct employees who lived in company camps. Now they refused to
acknowledge most loggers as direct employees, arguing instead that
they were the employees of logging contractors. Indeed, Bowater's
members on the Newfoundland Woods Labour Board, along with representatives from the AND, persistently refused to increase wage rates
for loggers, arguing instead that the workers could boost their incomes
by cutting more wood. Certainly, few loggers attributed improvements
in their living standards to mechanization. Rather, some loggers
asserted that their livelihoods improved only as a result of industry's
and government's fears about the International Woodworkers of
America, which had waged a bitter and popular strike on behalf of loggers in 1959.[5]

From 1951 to 1971, the total number of loggers employed in Newfoundland declined from 10,333 to 3,085, although only 1,200 to 1,300
of these were union members. Locally, Bowater scaled back and then
closed its pulpwood operations in Roddickton in the 1970s because of
problems in the supply of wood. The development of the Burton's
Cove Logging Sawmill has offset some of the lost employment in pulp
operations. It also constituted an integrated operation, debarking saw
logs and processing the slabs into chips for sale to the pulp mills
(Haynes and White 2002). In White Bay, logging and transportation of

pulpwood continued to be an important, but declining, source of employment for people in Hampden, Sop's Arm, Jackson's Arm, and nearby communities. Hampden's growth as a local service centre somewhat offset the decline in pulp operations, while the development of the Cat Arm hydro project from 1980 to 1985 boosted employment in Sop's Arm, as did a fish plant for Jackson's Arm. However, research suggests that as late as 1998, the forestry industry was harvesting older forests at too fast a rate for younger forests to sustain.

The Corner Brook mill remained under Bowater's ownership until 1984, when the provincial government and Kruger signed a deal that included additional support (Andrews et al. 2000; Norcliffe 1999).[6] When the corporate timber rights of Bowater were transferred, Kruger gradually altered cutting practices and labour relations as resource availability, technology, and markets all created new pressures. The basis of pay reverted from time-based to piece-rate until finally the company ceased being a direct employer altogether and contracted out its logging operations, although Kruger has continued to determine where and when cutting will take place. The union agreement requires contractors for Kruger (working on Kruger lands) to employ loggers under the same conditions that would have applied to Kruger and, over time, the proportion of loggers who are unionized has decreased, and non-union contractors operating on Crown land have grown in importance as a source of labour supply to the company. This outsourcing has transferred more of the costs and risks from the company to the loggers (buying their own saws, providing their own transportation) and contractors. The result, of course, has been increased company flexibility and decreased community flexibility.

The contractors, although relatively few in number, are quite diverse. Four or five out of about twenty on the Great Northern Peninsula operate on a sufficiently large scale to be successful, but the majority have inadequate access to timber. The largest contractors include several who cut exclusively for Kruger, but several can also sell to other buyers because they have their own permits on Crown land. Other differences include whether or not they have employees or subcontractors working in turn for them and whether or not they are unionized. All logging contractors face common pressures around mechanization, increasing scale, and the steady decrease in the number of conventional contractors, and the key to success remains having access to sufficient wood, either by cutting Kruger permits or by obtaining one's own. This pressure on costs, along with declining resource availability, has helped drive technological change, notably a move to mechanical harvesters, which are the only way contractors can reach dramatically higher company production targets. This move is changing the labour process, reducing the demand for labour, and concentrating logging more and more among a handful of large contractors. Smaller operators using traditional methods find they cannot compete, and there is downward pressure on the wages they can pay. This, coupled with Employment Insurance (EI) rules that penalize claimants who have fluctuating earnings, makes it hard to attract workers. Mechanical harvesters have to be kept operating twenty-four hours a day, if possible, to recoup the investment; they obviously intensify the rate of resource exploitation, and they are criticized for damaging the land.

In Newfoundland, a cyclical and competitive industry has continued to become increasingly capital-intensive as surviving companies struggle to increase productivity and thus re-establish threatened profit margins. Local labour is feeling the pinch, which is translating into human stress in forest-dependent places – mill towns like Stephenville and logging towns like Hawke's Bay.[7] Maintaining production, let alone expanding it, also puts pressure on environmental resources and leads to conflict over how forest resources should be used. We found evidence of local conflict, distrust of external powers, complaints about lax enforcement of regulations, and frustration over the limitation of traditional rights of access to wood in places such as Main River.

Through time, labour productivity in the forestry sector has increased, a favourable trend for some, but not all, workers. That increase is the result of the number of forest workers declining faster than the decline in volume of wood cut and processed (figure 4.4). This pattern throws into question the whole issue of the historical sustainability of the forest industry in Newfoundland. While the quality and benefits associated with employment in the forestry sector have improved for individual workers, each job places an increasing demand on the forest. In 1954, 154 cubic metres of wood represented one job in Newfoundland's pulp-and-paper industry. By 1989, technological and labour changes in the forests set each job at 651 cubic metres. This is restructuring with a vengeance.

Associated with these changes over the same period has been a consistent change in the effect of people on the landscape. Cutting before 1950 was concentrated along river valleys and lake edges, which provided productive forest sites and easy waterway transport of wood (figure 4.5). Over time, cutting has become increasingly dispersed, a pattern facilitated by road construction and heavily subsidized by tax dollars, but also determined by insect disturbances and an oldest-first, regulated forest-cutting policy. Annual cut shapes have changed from regular and medium-sized to convoluted and smaller, relative to natural disturbance patterns for the same western Newfoundland forest types. The result of this forest management history has been a movement

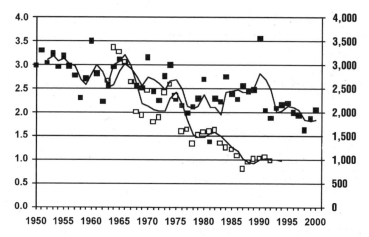

Figure 4.4   Volume of wood cut for all sectors of the forest industry in
insular Newfoundland (solid symbols, variable data sources, including district
forestry reports) and number of loggers employed by the sector (open
symbols, *Newfoundland Statistical Yearbook*), 1950–2000. Three-year moving
averages are applied to each series.

toward an even-age distribution of forests on the landscape when compared to burned and insect-driven ecosystem disturbances elsewhere
(figure 4.6). This change has complemented the increasingly mechanized industry as paper markets expanded especially outside Canada
(figure 4.7). In effect, the landscape impacts of the forest industry in
Newfoundland have rarely mimicked natural disturbances. Ecosystem
health under such practices must be seriously questioned. .

    Policy contexts undoubtedly influence what actually happens in the
forests and mills (through regulations, taxation, and financial support)
but do not fully determine it. While environmental concerns have
resulted in more forest management policies (as noted earlier), many
operators now find themselves squeezed between conflicting, or contradictory, demands of the companies and the regulatory agents. For
example, regulations require that all logs over fifteen inches in diameter be taken for sawlogs or a fine will be incurred; however, the sawmills will reject fifteen-inch logs as too small. Loggers will also be fined
for butt-junking (cutting away too much rot), but logs with rot will be
rejected by the mills. Again, the intensification of harvesting is yet
another instance of the reduction of flexibility at the lowest levels of an
industry, in order to provide flexibility further up the industrial system.

    In short, the future offered to rural people by the forest resource
remains unclear. Present-day corporate strategies are creating path-

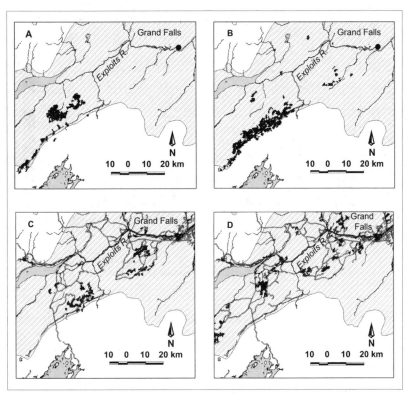

Figure 4.5   Central Newfoundland cumulative cutting patterns during
(a) 1920s and 1930s, (b) 1940s, (c) 1950s, and (d) 1960s. New cutovers
during each period are shown as dark areas over the region of productive
forest (hatched area). Construction of forest access roads beginning in the
1950s is shown as a cumulative network of lines in (c) and (d). The darker
line linking Grand Falls and paralleling the Exploits River is the Trans
Canada Highway, completd in 1966. Figure adapted from Pollard
(2004)

ways for damage to social-ecological health, and government policies
are leaving local contractors, not the large wood processors, bearing
the cost of the environmental regulations, a situation strongly reminiscent of what is happening in the fishery. When one considers the
decreased flexibility now experienced in outport communities as a
result of decline in fishing and forestry – the two main foundations of
their community strategy of occupational pluralism – the resilience of
these East Coast communities can be seen to be in serious jeopardy.

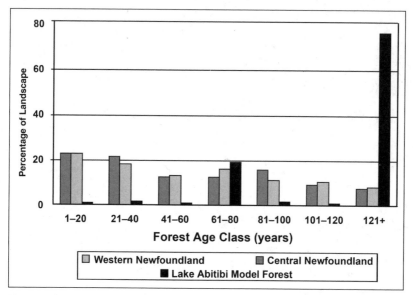

Figure 4.6   Recent landscape age-class distribution of western and central
Newfoundland forests from aerial photography. Adapted from Pollard
(2004). A landscape age-class distribution from a fire-dominated forest in
Ontario, the Lake Abitibi Model Forest (Groot, Gauthier, and Bergeron,
2004), is also shown for comparison.

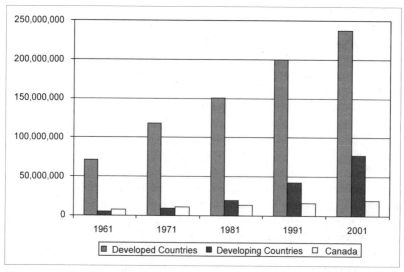

Figure 4.7   Total production of paper and board products, 1961–2001.
Source: FAO Faostat Forestry Data Set, http://faostat.fao.org

THE WEST COAST

In British Columbia we found an equally problematic social-ecological history, and one that unfolded in essentially the same way.[8] On the West Coast, the forests were a major source of wealth. Sawmills, concentrated on the southern coast, dominated the industry, and the northern coastal areas were of only marginal interest, with a few sawmills supplying the numerous salmon canneries dotted along the inlets. Government tenure policies, intended to stimulate the industry, instead created a surge of speculation, with timber interests being acquired as investments rather than for harvesting. In 1903, at crisis point, the issuance of pulp leases was discontinued, and in 1907 the province reserved all unalienated timber lands of the Crown, effectively placing a hold on timber speculation until the royal commission of 1909–10 made its recommendations.

Despite problems related to speculation and policy, the forest industry on the north coast of British Columbia did take some steps forward. Coastal First Nations, already used to capitalist production routines from their work in canneries, cut wood for companies in Haida Gwaii, in Tsimshian territory along the Skeena, and in Hartley Bay. By 1902, an Indian agent for the northwest coast reported that "Salmon-fishing and canning, procuring saw-logs for the three sawmills of the ... coast, the oulachon ((*Thaleichthys pacificus*), also spelled eulachon, ooligan, and other variants) fishing industry, hunting and trapping, fur-seal hunting, procuring and drying herring spawn, catching and drying salmon and halibut for food, boat and canoe making, and cutting firewood for the use of the ... canneries" constituted the seasonal round (Canada 1903). The early speculative pulp leases had also changed hands, and work began on surveying and construction of two pulp mills, Swanson Bay and Ocean Falls. Pulp production did not begin until 1908, and both mills experienced early difficulties of inadequate capitalization. Problems were exacerbated by the relative isolation of the central coast. The extensive Sitka spruce (*Picea sitchensis*) stands on Haida Gwaii also attracted speculators, and by 1910 Queen Charlotte City had four hundred residents. In 1912, H.R. MacMillan was appointed chief forester under the new British Columbia Forest Act, and he was soon organizing provincial forest districts.

World War I set new priorities for forestry in the province. The completion of the Grand Trunk Pacific Railway to Prince Rupert in 1914 also created new opportunities and an industrial forest economy. Sitka spruce wood from Haida Gwaii was in particular demand for wartime airplane construction. By the time of the armistice, "MacMillan had overseen a truly impressive industrialization of north coast lumbering, especially on the Queen Charlotte Islands," although there was also "tremendous waste in the spruce zones, where low-grade spruce and all the hemlock (*Tsuga* spp.) remained on the ground to rot" (Rajala 2006, 53). Nevertheless, as with the fishery, "the most enduring feature of the emerging forest economy, [was] the ongoing marginalization of First Nations at the hands of a timber rights system that denied them access to the resource" (39). The surging forest economy, fed by a demand for airplane spruce, went into rapid decline following World War I; nor were the two new pulp operations as yet successful. The Swanson Bay mill expanded with the demand for pulp by Japanese buyers but closed when that market collapsed after the great Kanto earthquake of 1923. Ocean Falls faced difficulties as well, primarily related to undercapitalization, and it was only with the acquisition of the plant of the American company Crown Willamette and the infusion of new capital that this brought that Ocean Falls was made the success story of the north and central coast. High freight rates, unstable markets, and competition from the large southern mills limited north coast expansion.

At Ocean Falls, Pacific Mills operated a vibrant company town under strict rules of paternalistic authority. Logging contractors there, as on the East Coast, occupied a status akin to southern sharecroppers (Rajala 2006, 74–5), being dependent upon Pacific Mills for annual financing and then a market at the end of the logging season. The firm subordinated conservation to profit, segregated residential areas along occupational and racial lines, and, as landlord, dominated all aspects of community life. As one labour critic of the province's early-twentieth-century company towns put it: "The worker lands on a company wharf, eats at a company hash house, sleeps in a company bunkhouse, buys his booze at a company bar, and his "terbacker" and overalls at a company store. If he dares to make a kick, he is escorted from the premises by a company constable" (*BC Federationist* 1916).[9]

Greater industrialization of the industry in the 1920s meant that new technology was introduced and consolidation of timber limits increased, which left little room for the emergence of an independent logging sector (Rajala 2006, 75–80). British Columbia's future chief forester, E.C. Manning, voiced concern about the way in which Pacific Mills (which had assumed almost total control over northwest logging) sought short-term profit over long-term sustainability, saying that he was "pessimistic about the rate of regeneration on the larger clearcut areas," while the 1924 annual report for Prince Rupert also deplored the "appallingly crude and wasteful" nature of industry practice there (80). Nowhere in the north of British Columbia or on the East Coast in Newfoundland was local timber manufacturing created – there was no diversification around the staple base, no capturing of good linkage structures. "Perhaps," Rajala notes wryly, "the only solace in forest industry circles ... would be that the Great Depression's impact would be felt less sharply in a region already so familiar with hard times" (87).

Until the end of World War II, the entire forestry sector was in difficulties, although people (First Nations and settler) in the British Columbia northwest had their practices of occupational pluralism to fall back on – homesteading, fishing, trapping, local woodcutting, gardening, and wild plant harvesting – a time-tested flexible pattern strongly similar to that in Newfoundland, where much the same kind of community and household maintenance was evident (Thornton 1979; Omohundro 1994). Rajala documents a dreary picture for the northwest of depressed conditions, collapsing mills, and threatened local economies. Barter took over from the cash economy in some areas, and "the new devotion to homestead development gave the region a 'fairly prosperous appearance'" (Rajala 2006, 96) – a condition that parallels the East Coast during downswings in local economies, when improving one's house provided the only way to fill the hours, days, and weeks productively (Ommer 1994).

During the Great Depression, Pacific Mills and other employers "pushed hard to extract every bit of value from workers and the forest ... Wage cuts, incentive payment plans and efficiency drives in the woods and mills drove workers to their limits" (Rajala 2006, 89). Pacific Mills cut Ocean Falls output back by 15 percent and wages by 20 percent as an adjustment to the state of foreign markets. Aggressive forest practices on the mainland and in Haida Gwaii threatened natural regeneration processes, and it is clear that provincial and industry foresters did not have the ecological knowledge to conduct scientific management, being unclear as yet as to the efficacy and long- term sustainability of selective cutting over clear cutting, for example. In any case, Rajala observes, "Production, once again, trumped conservation" (90), and Pacific Mills hired an efficiency expert who created what forester A.E. Parlow described as "a forced-pressure system in which only the younger and fit can maintain the requisite pace to hold their jobs" (96).

Similar productivity strategies were put in place elsewhere on the Coast, and the treatment of the resource itself was even more draconian. Damage was ignored. Large-scale logging, it was explained, left "more extensive cutover lands with a less satisfactory seed supply" (Hall 1937). As the 1930s wore on, markets improved somewhat and technological change also contributed to raised production levels. This was good for the industry and provincial revenues, but it increased the assault on the resource and downsized the labour force. The regulatory role of government became a public issue as "decades of unrestrained clearcutting had left residents of timber-dependent communities worried about their future" (Rajala 2006, 104–5). Chief Forester Manning achieved some slight improvements in management, but promotion, rather than regulation, remained the order of the day, as booster-style reports insisted that production was not damaging the status of the resource.

Matters could only get worse when the onset of war created new and urgent markets. H.R. MacMillan, now federal timber controller with the Ministry of Munitions and Supply, "transformed the fiercely competitive BC lumber industry into a co-operative war effort," says his biographer, Drushka (1995, 191–203). The H.R. MacMillan Export Company, looking at United Kingdom markets, had by 1940 set the Sitka spruce forests of Haida Gwaii once again under assault. Pacific Mills moved into Skidegate (logging spruce for their Ocean Falls sawmill), and other firms of long standing in the area expanded their operations and locations. In 1942 a Crown corporation (Aero Timber Products) was formed and was destined over time to direct all the operations in Haida Gwaii (R. Turner 1990, which has a chapter on this company).

Production continued to run rough-shod over conservation, and a pessimistic district forester, seeing no change in attitude, complained that mature stands were "cut from an economic rather than a silvicultural standpoint" (PRFD 1940, 2–4). United Kingdom and Commonwealth quasi-monopolistic wartime contracts rendered the "need-to-be-competitive" explanation invalid, but it was simply superseded by "the pressures of war," provincial and federal officials both agreeing. A basis for postwar conversion to sustained-yield management was explored (this eventually evolved into the Tree Farm Licence [TFL]) but had to await the end of hostilities. Meanwhile, the forests of Haida Gwaii and elsewhere in the north were ravaged by a variety of cutting techniques. Blow-down and rotting trees did not bode well for regeneration. "Corporate executives, small operators, hand loggers and government officials, all shared a utilitarian view of the forest despite debate about how the wealth generated from its exploitation should be distributed," remarks Rajala, adding that "aboriginal populations shared a quite different view of nature, but a relationship premised on respect and cultural sanction against over-exploitation had no influence in policy-making circles" (Rajala 2006, 113–14).

The end of the war saw the beginning of a new era. Unionization had occurred between 1935 and 1945, strong markets now existed, multinational pulp-and-paper corporations had arrived, and sustained-yield policies were adopted. Resource control continued to go mostly to corporate interests, which could afford to put paper facility infrastructures in place, and while large amounts of new development went to Ocean Falls, almost all the rest went to supply plants in the south of the province. At this time, the central internal government's argument was community stability, and their view was that assured and substantial timber resources were necessary to ensure stable economic development, while at the same time meeting the need for a sustained-yield program. But new technologies decreased labour inputs in British Columbia, as in Newfoundland. First Nations' protests at the cutting on traditional territories was answered by the argument that First Nations were at liberty to compete for Crown timber, with of course no acknowledgment forthcoming that "lack of capital and bureaucratic requirements would preclude such participation" (Rajala 2006, 139). On the West (as on the East) Coast, local communities' well-being was neglected, as was ecological health, with forest science remaining shaky at best.

As Pacific Mills extended its timber source hinterland into Haida Gwaii, annual sales of $30 million were reached in the 1950s. On these islands there was almost no manufacturing, and they "represented in microcosm the classic case of a British Columbia resource hinterland, its timber drained off to create profits and jobs at southern processing centres" – less than 2 percent of production went to local manufacture. "Together, the coastal forests yielded a saw log cut of over 174 million feet. Ocean Falls took almost 51 million feet of this, and only 1.4 million feet went to local sawmills" (PRFD 1949, 6–7). One local district forester, Gormley, pointed to the need for policies that would enhance local development but got no response. "Capital investment, sustainability, economic development and community stability" were held to be under way through the restructuring that was brought about by the new policies and the establishment of new operators such as the Celanese Corporation, who (in exchange for establishing a new pulp plant at Prince Rupert) were the first recipients of a Forest Management Licence, one that handed considerable control over to private interests under Forest Service supervision. Corporate consolidation was also under way on this coast: MacMillan Bloedel, Koerner-Abitibi, Rayonier, Crown Zellerbach (the last two, American), Canadian Forest Products, and BC Forest Products were all seeking their share of what had become, in less than one hundred years, a dwindling resource. The government Sloan Commission of 1957, however, was unperturbed, praising the Forest Management Licences' "contribution to the permanent, sustained, and increasing production of manufactured commodities" (Sloan 1956, 104, 112).[10]

Local resentment grew, both from First Nations and others, and controversy over land title became a serious issue on the coast, as "policies geared to promoting large-scale industrial capitalism severed First Nations from traditional seasonal cycles of resources use and deprived them of entrepreneurial access to the forest" (Rajala 2006, 156). Rigidity in the governance and industrial systems increasingly inhibited the old flexibilities of coastal living, such as had been seen on both coasts in the 1930s when the informal economy and local community enterprises had proved to be a significant safety net in hard times. But by the 1950s, the system was without understanding for the grass roots levels of local community concern for survival. At the top, under government supervision, were the few corporate mills that controlled both log marketing to them and the lumber produced from there. Below them were middlemen, independent (or, more accurately, pseudo-independent) contractors, and then workers. Licensing regulations by 1960 gave established operators "preferential access" to the allowable cut of provincial sustainable yield units (PSYUs) – creating a system that was, since such quotas became transferable, similar to the individual transferable quotas in the fishery. Indeed, in both fishing and forestry the pattern was increasingly one of consolidation of holdings/access by larger companies. Moreover, the British Columbia provincial government held firmly to its belief that First Nations had no special claim to resources, a position it would maintain until 1990, when it finally started negotiations with First Nations on the issue (see later this chapter). Corporate consolidation continued unabated, and by 1962, even though output almost doubled in the previous seven years, there was a miniscule growth in employment of only seventy-six loggers. On the Pacific Coast, as in Newfoundland, the new technologies had reduced labour costs but had made clearcutting standard practice, laying the basis of a future reforestation crisis (Marchak, Aycock, and Herbert 1999; Scientific Panel for Sustainable Forest Practices in Clayoquot Sound 1995a). Headlong growth in both provinces was consistently confused with sustainable development.

Change was, however, at least under way on the West Coast as "an increasingly affluent, urban educated, and mobile population" in North America had begun to "appreciate nature more for its aesthetic value than its commodity potential, generating demands for policies that would provide for ecological integrity, a better quality of life, and greater public participation in the regulatory process" (Rajala 2006, 180) and as environmental non-governmental organizations (NGOs) started to become more active and effective. At the same time, insecurities in the global marketplace after 1970, coupled with a dwindling resource and a slow rate of regeneration in the northerly climate of

Canada, forced mill closures. Unemployment increased, First Nations pressed land claims with increasing fervour, and public discontent rose. Everywhere on the coast resistance to large-scale industrial forestry escalated, as the idea that locally based companies should be fostered also began to gain ground.

In the early 1970s, the Haida formed the Council of the Haida Nation to advance their land claims, and similar movements were afoot among other coastal First Nations. In those years, too, provincial government environmental regulations were changed: Coastal Logging Guidelines were created to promote integrated resource use – patch logging, limited clearcut, restocking of cutblocks, stream bank protection, and better road construction were all involved. Industry at all levels complained about the rigidity of the new guidelines, and in 1975, the Pearse Royal Commission was struck to look at the structure of tenure and forest policy. At the same time, environmental groups and First Nations were making their presence felt in the Nass Valley, the Skeena, Haida Gwaii, and elsewhere (Rajala 2006, 231 ff. for details; see also Pinkerton 1983).

These years saw some sensible local solutions proffered to government by local groups. The Nass Valley Communities Association's Report to the Pearse Royal Commission dealt with small-community issues, which are also part of the concern of our research. They pointed out in 1975 that "the devotion to corporate profit that elevated wood production above all other land and water uses in the valley" had done great damage to environmental sustainability. "Wildlife, agriculture, community development, water quality, and 'the future' were protected only as an afterthought ... The federal Fisheries office felt 'lucky to get concessions on even minor logging practices' ... and 'regional benefits and marginal and control are non-existent.'" The association's solution included settlement of the Nisga'a land claim in the Nass Valley, integrated resource management with a local planning team (whose existence would help to achieve balance in a process dominated by corporate interests), and the return of stumpage revenues to create local-scale environmentally friendly diversification.

Other presentations by other local groups ran along similar lines, underlining their concern for the environment, local economies, and the future and pointing out the inflexibility that the environmental and economic restructuring by and for corporate concerns had inflicted on local and regional economies. Local people needed access to wood also, they said, and the Kispiox Valley Communities Association's brief argued that community involvement "in forest management would consider the local knowledge and social values of people who must live with the consequences of 'environmental damage, depleted forests, or decaying one-company economies'" (Nass Valley Communities Association Report 1975, 1–7; Kispiox Valley Communities Association Submission, 1–12).

The royal commission, reporting the next year, called for multiple-use principles to be adopted and for more timber to be made available to small operators. They pointed out that ten of the province's major companies held rights to 59 percent of the public forests, hardly an equitable share. Unfortunately, the 1978 Forest Act ignored much of what Pearse recommended. Two years later, in 1980, Ocean Falls' declining and increasingly expensive production system forced it to shut down. The mill faced losses of $1 million a month as a result of the doubling of the cost of wood on the regional market, low rainfall (causing reliance on expensive fossil fuels over hydroelectric power), old equipment, pollution problems, and the diversion of logs to more efficient plants to the south. The social impact of the uncertainty before closure had been considerable: a 1977 survey had reported 40 percent of couples experiencing marital problems, 65 percent of those surveyed identified alcoholism as a threat to the community, and the survey also spoke of problems with youth and with depression.[11]

In the event, four hundred employees were laid off permanently at Ocean Falls, and industrial and institutional capacity to adapt to the new circumstances and pressures, in terms of both firm operations and government policies, appeared to be seriously lacking.

Ocean Falls was not alone. For some time now, corporations had been consolidating assets and shutting down inefficient mills, such as Skeena Pulp. In Prince Rupert, the mill had absorbed smaller companies in an effort to increase access to logs for the production of pulp and to chemicals for plastics. The mechanized logging camps that supplied the mill were based one hundred miles inland, drawing wood both locally and from Alaska. TFL (Tree Farm Licence) 1, the TFL granted to the mill owners as part of a new plan for the establishment of sustainable forestry, was the largest in provincial history. But in 1997, it too had collapsed, and the economic and social repercussions were felt as far inland as Prince George, causing the mill to be rescued by government. Even before that, the failure of numerous small contractors and unpaid suppliers had had a devastating effect on the regional economy, especially when combined with declining employment in fishing and canneries, since fishing had also became more capital-intensive and licence limitations there had driven the cost of participation up to the point where ownership was consolidated in a few major companies (see chapters 4 and, especially, 8).

The early 1980s crisis in the world economy rebounded on the Coast, at the same time as reports of exhausted forest resources multiplied, and it became obvious that companies would have to commence logging in less accessible areas if they were to survive. The unsurprising result was capital flight, as "depleted holdings, aging mills, pollution abatement costs, pressure on the land base from environmentalists and First Nations, the softwood lumber dispute, and the appeal of fast-growing forests in the southern hemisphere all contributed to de-investment" (Rajala 2006, chap. 8:194–5). Government faith in the private sector was proving, at best, naïve. Corporate flexible production became the industrial order of the day (as it had on the East Coast), and employment slumped even further as rigidities were shifted downwards in the system to the level of communities, households, and workers. Fisheries and forests also came into conflict: single-sector policies on both coasts failed to link the consequences for households and communities of problems in one sector, thereby exacerbating those in another, as a slipping fisheries sector suffered under globalization, quotas that benefited corporations, and damaged stream environments. Once again local communities offered potential solutions; once again they were ignored.

Then, amid company failures and community distress, the principles of a December 1997 Supreme Court ruling legitimated Aboriginal oral evidence and Aboriginal title, not only to land but also to the resources thereon. On the West Coast at least, patterns of negotiation would now have to look different, and some community voices would have to be heard, but recession continued to take its toll and corporate failure and bankruptcy, along with labour disputes, forced government ultimately to operate Skeena Cellulose itself. To date, the combination of troubled markets, seriously declining resources, and pressure from First Nations, from environmentalists, and from local communities continues. As yet, there is no change of political heart, and the sensible solutions proposed by local communities are yet to be taken seriously.

CONCLUSION

Forest ecosystems across North America have undergone major changes since the advent of industrial forestry and continental-scale downgrading of forest resources across the North American continent seems likely. As in fisheries, little attention has been paid in industrial forestry and forest governance to forest species other than economically important "fibre" species. And, as in fisheries ecosystems, the alteration of natural forest ecosystems proceeds apace without due consideration of possible (even probable) consequences. We show, for example, that the replacement of old-growth forests, with their diverse age structures and multiple successional stages, which result from natural and indigenous disturbance regimes such as fire (e.g., Foster 1983; Lorimer 1977; Hunter 1993; Niemelä 1999), has already had unforeseen consequences. What have we lost in terms of other kinds of forest values through the large-scale export of raw logs and other primary forest products (pine mushrooms and other harvests from mature woodlands)? What is the effect of wholesale exploitation of commercial species on the totality of forest ecosystem services (Costanza et al. 1997)?

Research elsewhere is now showing that in natural situations, gap disturbance determines forest structure and processes more than was previously assumed (McCarthy 2001). We also now know that a higher incidence of fragmentation on the landscape scale is created by logging, compared to wildfire depredations (DeLong and Tanner 1996), and that forestry activities have operated within a much narrower range of variability in emulating fire cycles than the large range of natural variability (Bergeron et al. 2002). Moreover, an overall reduction in diversity due to industrial forestry practices has been noted for tropical forests by Hughes, Daily, and Ehrlich (1997) and Ceballos and Ehrlich (2002), while the risks to temperate and boreal forest populations and habitats are potentially high in terms of lost ecosystem services alone (Balmford, Green, and Jenkins 2003). Worse still, changes in climate or other environmental thresholds can produce unexpected results in mismanaged forests (Post 2003) – "catastrophic regime shifts" can be anticipated in which a system can have alternative attractors on a landscape scale, even if it does not on a local scale (Scheffer and Carpenter 2003). Obviously, such shifts in forest ecosystems will impact aquatic and marine ecosystems as well, as in (for example) the effects of logging on salmonid habitat or the impacts of ocean log dumping and storage on benthic communities and productivity (see chapter 9).

Evidence shows that the forest products industry has become more global in the last few decades (Sinclair 2003). In the global market there is pressure to compete through cheaper sources of raw materials (fast-growing pine plantations in the southern United States, for example); in the context of both coasts, this has translated into local strategies to harvest a dwindling supply of wood more cheaply. While this pressure is not new, the pace of change has escalated in recent decades. The tools used by the companies in their quest for raw material and to protect profits include political pressure to preserve and enhance priority access to forests, technological innovation, and the restructuring of labour relations. However, the outcomes are becoming highly contested, given the competing interests of other forest industry actors (sawmill operators, loggers, contractors), other forest users (ecotourism operators, guides), and environmental groups. In recent years, the competition has intensified as the resource has become depleted and as the employment the pulp-and-paper companies can offer to people living in forest hinterlands has dwindled.

Nonetheless, as the locus of natural-resource production and consumption has shifted from the local to the global, the residents of our coastal study areas have experienced tremendous pressure on local forest resources in a globalizing economy, much as they have in the fishing industry. The legal recognition that First Nations' rights have been alienated may help to redress some of the inequities of history on the West Coast. It is to be hoped that on the East Coast, where settlers also had claims that pre-date those of corporations (albeit without the time depth that applies to First Nations), coastal communities may find government beginning to take more note of community rights. But to this point, in the forestry sector as in the fishery, local communities' needs and rights and the health of the environment have received seriously inadequate attention. The passing down of rigidities to the local level and the subsequent ignoring of recommendations at the community level that might restore flexibility and community resilience do not bode well for the future.

Wilkinson claims (1992) that, in the United States, the legal and policy framework for resource management that was inherited from the nineteenth century has consequences for the existence, or otherwise, of options today. The consequences include, of course, resource degradation, as we show throughout this book, and the policy framework has a serious effect, therefore, on both environmental and (by extension) human health, as we also show (see chapters 13 to 16). We agree that in Canada also, the living are being ruled by the dead (the "lords of yesterday") and that the legal and political frameworks we have inherited have contributed in a major way to the problems and deprivations of today.

In the case of the timber sector, whose social-ecological history we have examined in this chapter, although superficially different "stories" emerged on both coasts, we have shown that in both cases the underlying pattern in the industry was that of the creation of environmental degradation, coupled with market and technological vulnerability, which was global in scale and which has resulted in significant social-ecological distress. The forest products industry on both coasts, that is, has created significant damage, which has had a negative impact over the long term on both the environment and the society in which it has operated. Sadly, government policies on both coasts until very recently (and again now in British Columbia) have demonstrated an inability to think beyond a mono-staple mindset and a very rigid and narrow idea of what constitutes development. Local people in this sector have been seen, time and time again, as "the problem," and industry as "the solution." Resource depletion has passed unnoticed – unless it has been perpetrated by local communities, and then the reasons why this might have happened have been totally ignored. Governments continue to be caught in a mental staple trap, rarely thinking even out of the forest box into diversification or local control, and never into local development and flexibility. The flexibilities that have been approved in this sector on both coasts have been corporate, although the evidence for both coasts is that this has not and does not and cannot benefit regional economies and coastal communities. This generalization is fundamentally important in the light of a global context of decreasing forest ecological health.

5

Social-Ecological Health and the History of Nonrenewable Resources on Both Coasts

INTRODUCTION

Different natural resources have different development implications that depend heavily on how their exploitation is managed, both technologically and structurally. The early staples of Canadian history were renewable resources (fur, fish, timber, wheat). The first nonrenewable staple was gold, to be followed by a suite of minerals as the age of sail gave way to the industrial age of the steam engine. Natural-resource development also has impacts on, and implications for, social-ecological health, depending upon the type of resource, management methods, extraction technologies, and various local circumstances. In the previous two chapters, we considered the fishing and forest staples of the two coasts, showing how cross-scale misalignments and scale asymmetries in governance resulted in unrealized development potential from the point of view of social-ecological and economic health. Temporal scale asymmetries as identified in both state and industrial goals also fostered the wealth of enterprise over the short run, while creating fewer benefits for the region and its citizens over the long run. We turn now to the history of the nonrenewable resource sector to examine the implications of interactive restructuring in this sector for social-ecological processes and options on both coasts.[1] We use as case studies (1) two examples of copper mining – one from each coast – during the late nineteenth and early twentieth century, as well as (2) the more recent example of ocean "mining," which has occurred with the extraction of oil and gas from the seabed on Canada's East Coast. This last case study serves as a prelude to the discussion in chapter 13 of potential oil and gas extraction on Canada's West Coast.

While many of the factors associated with the development of nonrenewable resources are similar to those at play in renewable resource sectors (factors such as ecological impacts, silos in resource management, market vagaries, costs of production, technological change, and human health and safety), nonrenewable resources present particular challenges. First and obviously, regardless of how well or badly they are managed in terms of optimizing benefits and minimizing costs, they will eventually be depleted, and the cross-scale and cross-sector implications of inevitable demise need to be foreseen and planned for from the outset with policy strategies that involve collaboration between industry and several governmental departments, although in the past such collaboration has been more usual at the outset than at closure. Indeed, it has long been recognized that there has been a history in this sector of governments being seduced by large capital investment and major short-term employment potential into promoting such industries as a panacea for economic ills. This is at least in part because such industries pose a significant "entry problem" for small-scale enterprises or local communities, since resource extraction requires heavy capital investments in technology and infrastructure. Consequently, the participation of major corporate interests, along with any elements of dependence and lack of control this entails, is part of the equation. Moreover, the extraction processes have considerable potential to affect human and environmental health adversely, and significant consequences then arise for people, communities, and ecosystems.

Our case studies are designed, not to present more examples of the well-known problems associated with mining, but rather to uncover both the forces of interactive restructuring that drove mineral development in these two regions and the manner in which that development restructured the social-ecological health of the coasts in several ways, both across scales and over the long and short term. Highlighting *similarities* in copper mining between the two coasts allows us to draw out some of the common features of this form of resource development in two regions while recognizing that they were (and remain) very different from each other environmentally, economically, socially, politically, and culturally. At the same time, the *differences* between the two cases underscore how the same basic type of development can be driven by different forces and have different impacts, depending on specific local circumstances. Governments have always talked about this kind of resource extraction as the panacea for economic ills in Canada's coastal areas, as is clearly seen in the early booster-style rhetoric found in both provinces. Only later is there is a progression to more pragmatic assessments of their value to provincial and national economies. What does the social-ecological history of mining/nonrenewable resource extraction and its relevance for social, ecological, and environmental health share with the history of renewable resource extraction? How might it differ? What can it tell us in relation to our central research question? Once again, therefore, we do not present standard history but the view from the coast.

The role of communities varies by time and place; the importance of a region's capacity to transform from one staple to another and the implications of differing technologies over time for social-ecological health also need to be examined. So we also turn in this chapter to the history of the oil and gas industry in Newfoundland and Labrador, which makes it clear that there has, over the long term, been some considerable learning with respect to vital matters of ownership, control, and the achievement of lasting benefits for local peoples and ecosystems. To what extent this is the case and where the difficulties remain will be considered here.

Our overall findings, then, not only tell us about the impacts of past mineral resource development, some of which still linger today, but also raise the question of what we have learned from past developments about the interactions of social, political, and environmental restructuring, what we can still learn, and how that might be applied in contemporary and future contexts.

COPPER MINING: TWO CASE STUDIES

Beginnings

Interest in the mineral possibilities of Newfoundland dates back at least as far as the early seventeenth century, and there were a number of attempts at mining at various sites throughout the eighteenth and early nineteenth centuries. In the 1860s and 1870s mining was thought to be a way to allow Newfoundland to escape from the single-staple (codfish) trap, which was, it was believed at the time, keeping the Dominion "underdeveloped." A newspaper article from 1878 (Harvey 1878) suggested, "Our capitalists are all devoted to the fisheries, and have no eye for any thing else. But for Mr. Ellershausen [who developed the Betts Cove mine] our great mineral treasures would have remained unknown and undeveloped for years to come." In these years, the hook-and-line and seine-and-trap fisheries for cod in many parts of the Island had reached their ecological maximum, and production per capita was declining. The Dominion government considered that this decline, along with a lack of industrial development, had resulted in the Dominion's inability to build a sustainable economy, leaving coastal communities dependent upon and vulnerable to economic forces and corporate decisions outside their influence if they were to secure more immediate needs, such as employment.[2] Even though other ocean-based industries, such as sealing, provided periodic relief in some areas, attempts were made to diversify the economy around agricultural and mineral development, thus providing alternative non-marine sources of employment (Hiller 1967).

Copper mining in particular appeared to be viable, given the 1853 decision to abolish duties on copper ore imported into Britain, and the 1855 Reciprocity Treaty, under which Newfoundland minerals could enter the United States duty-free. Prospectors searched regions of the Island for commercially viable minerals and in 1857, Nova Scotian prospector Smith McKay discovered several promising copper deposits on the Island's northeast coast, including at Terra Nova and at Tilt Cove. In 1860, McKay joined forces with entrepreneur Charles Fox Bennett, one of several determined individuals leading the quest for mineral development at the time, and formed the Terra Nova Copper Mining Company. After having little success at the Terra Nova site, the proponents turned their attention to Tilt Cove and in 1864 began to develop that site under the Union Mining Company.

Around the same time that the Tilt Cove site was being developed, other copper deposits were discovered just south at Bett's Cove, and in 1869 (in an instance of what William Reeves has described as a "lack of any clear distinction between private and public initiatives" [Reeves 1987, 335]), Surveyor General John Warren and several Liberal members of the House of Assembly formed the Notre Dame Mining Company and took ownership of the Bett's Cove site (Martin 1983, 18). In 1874, their property was leased to German-born engineer Baron Francis von Ellerhausen, who formed the Bett's Cove Mining Company, recruited thirty German miners from Nova Scotia along with about seventy Newfoundlanders, erected the surface works, and in 1875 began producing copper ore (Martin 1983, 18–19). The subsequent transformation made Notre Dame Bay into the largest industrial centre outside St John's and turned the colony into the sixth-largest producer of copper in the world from 1870 to the 1880s.[3]

This is a remarkable achievement given that the limited information available on the layout and works of the mines suggests that the operations were very basic and labour-intensive. A visitor to the Bett's Cove mine in 1878 described going down "almost perpendicular ladders, into the gloom" four hundred feet below the surface, where "men worked by the light of candles stuck into the mine walls, using pick-axes, crowbars, and hammers, and hauled the ore by hand along tracks laid from working areas to the hoisting shaft." The ore was loaded into steamers for shipment to Swansea, Wales. Smelters installed when the mine was opened were operating and others were being set up (Harvey

1879, 75–6). The Tilt Cove operation was similar, though smelters were not installed there until 1889.

The use of smelters was a difficult judgment call for East Coast operators. On the one hand, smelters were clearly important in increasing the value of the mines' output. In 1879, for example, the first year for which there is a record of smelted copper being exported from Newfoundland, the price of unprocessed ore was $22.00 per imperial ton,[4] while the price of smelted ore, or "regulus," was $46.00 per ton (*JHA* 1879).[5] On the other hand, smelters added to production costs for equipment, labour, and the coal to run the furnaces. Coal was an especially important consideration in Newfoundland, where no domestic supply existed – coal to run the Tilt Cove and Bett's Cove smelters usually arrived on board the ships that brought the copper back to Swansea. It is therefore not surprising that the search for coal figured prominently in the Geological Survey of Newfoundland, established in 1839. Moreover, coal was also needed for the railway that was being considered in the 1860s although not begun until the 1880s. Coal, it was argued, would fuel the railway; the railway would open up more coal and mineral deposits; coal would smelt the ores that surely lay hidden in the Island's interior, and the railway would transport the value-added mineral products to various shipment points (Wingfield-Bonnyn 1890). Throughout the remainder of the nineteenth century, then, and well into the twentieth, the Geological Survey paid special attention to the possibility of coal deposits, particularly on the west coast of the Island, although few substantial coal deposits were ever confirmed and (apart from attempts to extract some small, low-grade deposits on the west coast) hopes for a coal-mining industry were never realized.[6]

Colonization and development in British Columbia had, of course, taken place later than in Newfoundland and, as we shall see, were at least superficially different in character as a result of the multistaple nature of the resource base there. It was gold that first drew miners to British Columbia and then focused world attention on the region in 1858. Miners spread throughout the colony noting signs of other minerals, but successful exploitation was elusive. Limited capital and difficult transportation meant that even the rich lead ores of the Bluebell (so pure they were used by fur traders and First Nations to cast musket balls) could not defeat these difficulties. The arrival of the Canadian Pacific Railway, stretching across British Columbia, created new opportunities, and a flurry of staking and development followed. Between the 1890s and 1920s, as the province began to open up, there was no overall risk of overdependence on a single staple, but a narrower range of staples had been exploited along the coast, and it was salmon fishing, its associated canning industry, and mining that had provided most employment. Now mining communities sprang to life as miners crept up the coast, until finally the forbidding coastal-range mountains yielded their secrets and several of British Columbia's richest mines – the Premier Gold Mine above Stewart, the immense copper deposit at Britannia, near Howe Sound, and Anyox on Observatory Inlet – came into being.

By the time the Granby Consolidated Mining, Smelting and Power Company began development of its Hidden Creek mines in northern British Columbia, it was already well established as a successful copper mining and smelting company. When, in 1902, the properties and activities of the Miner-Graves Syndicate in south-central British Columbia were consolidated, P.J. Graves of Spokane, who had been unable to obtain backing from his Spokane associates, interested S.H.C. Miner of Granby, Quebec, in the prospect. In 1904 the company moved its head office from Montreal to New York, as New York interests had acquired control, but Granby remained a British Columbia company despite its capital base having initially been in eastern Canada and subsequently in New York. The focus of its production was in British Columbia, where the company town of Anyox rapidly became a regional centre of the northwest coastal mining district. There were only minor incursions into Washington State and Alaska, but the experience Granby had developed in the Phoenix mines was put to use at Hidden Creek,[7] where considerable development work was already in place by the time it purchased the property in 1910.

On the West Coast, the operation was far more complex and mechanized than those on the East Coast. It was also, by contrast with the East Coast, well supplied with coal, and the Granby Company was initially able to purchase coke from Canadian Collieries on Vancouver Island. This was so lucrative for Canadian Collieries that it reopened its coke ovens at Union Bay to supply the smelter, and 5,450 tons were produced in the first year. In 1916 the demand for coke on Vancouver Island exceeded the output, and rising prices and the danger of shortages induced Granby to seek an alternative source (*Engineering and Mining News* 1917). In 1918 Granby opened a colliery at Cassidy on Vancouver Island and shipped coal to Anyox, where a coke plant was built in 1919. A range of by-products was also produced at the coke plant, and while some were used at Anyox, sales of tar, ammonium sulphate, benzol, toluol, solvent naphtha, and naphthalene generated some additional revenue (British Columbia 1919). The by-product plant was the first operation of its type on the West Coast and was viewed by the provincial government as a significant advance, particularly in the reduction of waste during the coking process (Williams 1919, 179–89).

After Granby's 1910 purchase of the Hidden Creek Mine, nearly 24,000 feet of diamond drilling was completed by the end of 1912, and nearly 10,000 feet of tunnelling augmented the earlier underground workings: in fact, even though no ore had yet been shipped, an extensive mine operation was running. Smelter construction commenced in 1913, and it was expected that production would reach 2,000 tons of ore a day once the smelter went into production. At Anyox (L.R. Clapp 1923; Lindsay and Healy 1929) the ore moved by gravity through chutes down to the crusher (230-foot level) and thence to the lowest level (150 feet), where it was loaded into 25-ton cars for haulage to the ore bins at the smelter. The company operated two 42-ton locomotives and 25 hopper-bottom steel ore cars between the mine and smelter.[8] The system was divided into the A and B lines, with the A line covering the distance between the mine and the smelter, while the B line connected to the wharf. Various spur lines connected to ancillary facilities such as the sawmill and later the coke plant.

The success of the operation, coupled to a desire to increase production, resulted in the installation of further crushing capacity. The mine was lighted electrically in all the main tunnels, and in 1916 carbide lamps, which each man had to purchase, replaced candles, while Granby supplied the carbide. Individual electric lighting was introduced several years later. In 1916 the mine began producing some lower-grade ore and as a result began shipping more tonnage. At that time also an auxiliary steam plant was built for use at low water. The company acquired the steam tug *Amur* and six barges for hauling blister copper, coke, and lime rock (British Columbia 1914, 1915, K147). Initial production (March 1914) consisted of 260,809 tons of ore processed, yielding 2,853 ounces of gold, 130,767 ounces of silver, and 1,084,646 pounds of copper. Between 1914 and 1936 the company mined 21,725,524 tons of copper ore, which produced 321,546,2020 kilograms of copper, 206,308,934 grams of silver and 3,772,762 grams of gold. The blister copper the company produced was barged to Seattle and then at first transported by rail to New York for refining; in later years it was shipped to the large smelter refinery complex at Tacoma, Washington.

The smelter was located close to the main mine at Hidden Creek, at a spot where there was also room for a slag dump, for future plant expansion and for relatively cheap waterpower. The furnaces and converters were in one steel building, 80 feet wide and 390 feet long, with an asbestos-covered wood roof. A 40-foot-wide craneway served both the furnaces and converters. Construction on the smelter began in 1913 with more than a thousand men working on the site, and in

March 1914 it was blown in. Its stack was 22 feet in diameter and 153 feet high: situated on the hill above the smelter, the top of the stack stood 300 feet above the furnaces. Slag, with British Columbia government permission, was dumped into Granby Bay in front of the smelter, partly because it was the most convenient site but also in order to increase the ground area in front of the smelter. Power was supplied from a hydroelectric plant, with water moved through a 6-foot wooden-stave pipe from a dam on Falls Creek.

At time of start-up, then, both the East and the West Coast mining operations were the basis of diversification in their area. On the West Coast copper mining was a large-scale industrial venture that included a strong Canadian interest and was supported by local coal reserves. By contrast, the East Coast venture was small, technologically constrained, and without efficient access to coal, and there were significant tie-ins to government, as well as international linkages. In neither case were issues of cross-sector pollution of any concern. Copper was used for such things as the copper sheathing that sailing vessels employed in tropical waters, but the sheathing was manufactured in Britain, not in Newfoundland or in British Columbia, a pattern typical of Canadian staple-extraction thinking, in which the value added of subsequent manufacturing was not developed.

Demographic Impact of Mining Operations

Fluctuations in both coastal mining industries were clearly reflected in demographic trends that included typical "frontier" imbalances in gender and age cohorts resulting from the male-oriented early production processes in the industry.[9] Both areas brought in new settlers, although in Newfoundland new labour supplemented population already in the area. The early populations were made up largely of men who were either unmarried or who came to the mines initially without their wives, to be followed later by young couples with children. Many of the men were recruited for their mining expertise (Census of Newfoundland 1869),[10] and at Bett's Cove miners came not only from Newfoundland and the British Isles but also from Germany, California, France, and Australia.

On the West Coast, the early makeshift construction town of 1912 and 1913 served a transient workforce as people had to be brought to the area. Although it was dominated by British, Canadians, and Americans, considerable ethnic diversity existed amongst the workforce. This was quite typical of the period but was probably more pronounced among the transit population at the mine, the only area for which data exist.[11] In 1931 and 1932 it was found that approximately 50 percent of

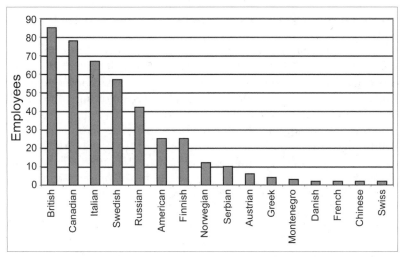

Figure 5.1   Nationalities of miners at Anyox, Hidden Creek, 1917. Source: *Granby News*, 1917

the workforce was considered to be non-English-speaking (*Report Re Strike of Employees of the Granby Consolidated Mining, Smelting and Power Company Limited at Anyox* BC 1933, 5). Senior personnel were usually American or Canadian and the senior managers almost invariably American, but often they had been in Canada for many years working for Granby. Most of the office staff and professional staff were also of Canadian or American origin (figure 5.1).

On the East Coast, however, the demographic impact of start-up was felt by a population already long-settled in the area in small fishing outports. In 1857, Tilt Cove had been home to just 17 people surviving on the inshore cod fishery and subsistence agriculture (Census of Newfoundland 1845, 1857); by 1869, just five years after the start of mining, the population was about 750, 323 being in mining, while 153 were still catching and curing fish. Some subsistence agriculture also continued (Census of Newfoundland 1869). After soaring to 800 during the initial boom, the population of Tilt Cove plummeted to just 155 by 1884 (Census of Newfoundland 1869, 1874, and 1884). The opening of the East Mine resulted in another dramatic increase, to just over 1,000, which then decreased to 808 (Census of Newfoundland 1911). The brief revival during World War I brought the population up to 1,500, but by 1921 it had collapsed to 90 souls (Census of Newfoundland 1921).

As at Anyox, no families of settler origin had lived at Bett's Cove before the start of mining, but by 1878, the population was about 1,800.[12] The Cove underwent its most intense period of industrial expansion and population growth at the same time that Tilt Cove was going through its first major decline. While it is impossible to be certain, given the nature of the sources, it is reasonable to assume that many workers and their families displaced by the decline of Tilt Cove during this period would have moved to Bett's Cove. Then at Bett's Cove the population fell from 1,800 to just 318, following the collapse of the mine in 1883. Once the last shipment of ore left in 1886, the community died completely. The relationship between declines in one community and increases in the other suggest that an internal-migration system existed in which a decline in one community created an increase in the other that absorbed some of the displaced population. Only 12 people remain in Tilt Cove today.

Social Responses to the Mining Industry

Rhetoric. Initial copper mining activity was accompanied on the East Coast by considerable rhetoric regarding the potential of the mines to break the Island's dependence on the fishery and provide much-needed employment. One commentator described the Tilt Cove mine as an example of the "enterprise, capital, and population" that the colony so badly needed to break the grip of those "few individuals" (fish merchants) who had for so long taken advantage of "the ignorance of the mass of our people" in "keeping out capitalists and checking immigration" (Marett 1870). A St John's newspaper described the initial success of the Tilt Cove mine as proof that mining was capable of providing "an employment which would relieve the fisheries from a pressure which at present they cannot support – an employment which is as certain and positive as the other is hazardous and speculative" (*Courier* 1866). Likewise with Bett's Cove: in the winter of 1876, for instance, it was reported that the sealing and the inshore cod fishery on the northeast coast had been failures in the previous year, and while able-bodied relief was on the increase, the mines of Tilt Cove and Bett's Cove were providing some much-needed employment (*Newfoundlander* 1876).

The Communities' Physical Infrastructure. Employers supplied much of the physical infrastructure in both communities. At Tilt Cove, the company had built houses, schools, and churches immediately after the start of mining. Bett's Cove had a school, a hospital, and three churches. The number of churches may have reflected something of the religious and moral sway held by Ellerhausen, who also strictly prohibited liquor in the town. The company also controlled commerce: workers were paid in specially printed company bills redeemable at the company store.

Things were rougher on the West Coast, at least initially. When construction on the Anyox smelter began in mid-1912, one teamster described what he found for accommodation: "the bunks were three high and had straw ticks that were none too clean." In the centre of the bunkhouse stood a large stove, and the workers would hang their socks to dry without washing them, "God what a stink." The floor of the mess house in that first year "looked like it hadn't been washed since it was laid." The plates and cups were all tin, and when they were empty, the men would bang on the platters with their knives until more food appeared. Then they would quickly load their forks and shovel it down; "as to soup, gee there was every kind of sup, slurps and just plains grunts and gulp." The miners' community consisted largely of bunkhouses for the single transients who moved in and out with much greater rapidity than married workers, as did their counterparts on the East Coast. Some houses were supplied: for the doctor, for supervisors and foremen, and for the few miners with families. The bunkhouse at the mine was two stories high with a basement and built with three wings accommodating four hundred men. The showers, drying room, and lockers were all in the basement. The ground floor was divided into separate rooms, while the top floor, called the "bull-pen," was an open space furnished with cots. The change house and recreation hall was a three-story frame building with concrete floors. The first floor housed three hundred sanitary steel lockers, a fifty-person washroom, a water closet, urinal room, twenty-four-person shower bathroom with three porcelain tubs, and a drying closet for clean clothes. On the second floor was a poolroom, a library and reading room, a shift-boss's office, an emergency hospital with four beds, and a doctor's office. The third floor held a dance hall with cloakrooms. A recreation hall, superintendent's office, and storage buildings were all in place before 1914. In 1915 a new bunkhouse was added, and several additional cottages were built, as well as a new electrical substation. In all there were about fifty cottages of modern construction, with porcelain baths, washbasins, water closets, hot and cold water, and electric lights.

As the community developed beyond accommodations only for transient workers, things improved. A large hotel, recreation hall, hospital, store, and one hundred cottages were constructed. The cottages were laid out along a series of wood-planked streets and had three, five, or seven rooms, including a bathroom, and were furnished with electric lights and water and sewer connections.[13] The town was equipped with fire hydrants, and the streets were lighted by electric arc lights. A business core, which was constructed adjacent to the docks, included the company's general store, a telegraph office, a café, the police constable's office and lock-up, and a liquor store that opened in 1921.

Anyox, like Ocean Falls (chapter 4), was very clearly a company town.[14] As with the East Coast, because of the isolated location, the company had to supply housing and other accommodations for its workers. Nonetheless, at Anyox this practice was very much in line with the emergence of "welfare capitalism," which consciously moved away from the "formidable stereotype" engendered particularly by nine-teenth-century coal towns, towns that were rife with oppression and discrimination (Batch 2000, 3).[15] In the early twentieth century, the intent was to create and maintain a happy workforce in which company loyalty was fostered by the comparatively good lifestyle that was possible in such towns, a lifestyle presided over by a benign paternalism. The way the community was configured clearly indicates company intentions (British Columbia 1917, 1918, F47). The smelter workforce was generally accommodated in the flats (worker housing) near the smelter. Middle management was located on Strawboss Street and in the service community on School Street, and upper management was located on Hill Street. Each of these streets was also separated by small topographical features: a creek divided the small-business and man-agement section of the community from the flats and the smelter site.[16]

The distinctions that topography was intended to reinforce did not last long in Anyox: larger families required larger homes, which were located on the business side of the community. As people moved in and out of the community, a more general mixing occurred, with houses being somewhat randomly assigned, depending on availability and need. The vision of the world was changing, as well, as machinists mixed with the police constable and the smelter foreman. Many families took in boarders, though these were usually clerks and office workers rather than labourers. The community across the creek from the flats remained largely separate from the beach community, but this pattern was more a consequence of wage and workplace than of social standing.

An ancillary community developed adjacent to Anyox, just off com-pany property. Initially, a small hotel was built by two miners who had sold a nearby claim to Granby. It was soon followed by a small store and a couple of shacks in which several dance-hall girls lived, working out of the hotel (Hutchings n.d.).[17] Part of the social regime broke down in various ways: at dances, in sports, and at the Elks club. Competition was between beach, smelter, and mine, rather than being based on social class, although senior managers maintained their distance from such events. In short, Anyox was a community with all the amenities, includ-ing churches (the Methodists arrived as early as 1909) for three denominations, schools (including a high school), a hospital (fully equipped with the latest in x-ray equipment and staffed by three physi-cians and nurses), a moving-picture theatre, a dance hall, a pool and billiard hall, and a café run by the community association. Sewer systems and garbage incinerators were in place, and the company maintained street lighting and the plank roads. Even a taxi service was operated, while a regular small-boat service to Alice Arm (a small mining community about sixteen kilometres distant) was also maintained.

As with Bett's Cove, then, social control was part of the reality of Anyox.[18] Union organization had little chance of success, and strikes were also difficult to undertake. Although plenty of space was available off company property, there was little strikers could do to disrupt the flow of commerce for more than a short time, and their position was rendered more difficult by worker dependency on company housing. Indeed, in the isolated conditions that existed in northern British Columbia, as in Newfoundland, and in the light of company control over not only the mine site but also often the town site, most mines were essentially located in company towns that were also communities.

Human Health and Safety

While the historical record on accidents and mortalities among East Coast miners is sparse, there is little doubt that the mines were dangerous places to work. A visitor describes going "up and down ladders, along winding galleries, over planks where a false step would be instantly fatal, and by the edge of yawning chasms, where the faint light from our candles only made the darkness visible" (Harvey 1879, 82–3). A trip to the 150-foot level of what was then a 400-foot shaft left the visitor exhausted from the heat and longing for fresh air, which suggests that ventilation was short of adequate (83). Workers reportedly "especially hated the smelter for its insidious hazards" (Martin 1983, 14). The 1891 census reports 77 deaths in Tilt Cove for the previous twelve months, out of a total population of just over 1,000. The reported death rate for the Electoral District of Twillingate for the same period was 29 per 1,000 (Census of Newfoundland 1891). While the numbers are highly suggestive, only further research would allow for a definitive link to the mining industry.

Statistics are more readily available for the West Coast, and they indicate a relatively high accident rate at the Granby Company in Anyox. In 1914, it was the second-highest in British Columbia with three deaths occurring, although two were in the same accident (figure 5.2). The company began an early program of safety training, and in that year first-aid classes were started at the mine. The resident doctor, Dr Cloud, instructed the class, and work-related accidents were soon in decline. Generally less than half the deaths in the town in any given year were work-related. Falling rock was the single most common

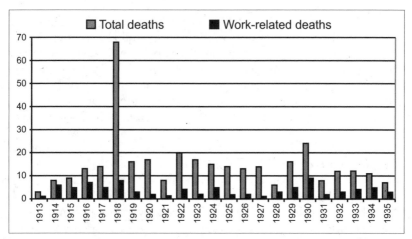

Figure 5.2   Deaths registered at Anyox, 1913–35. Source: compiled by
Robert Griffin and Susan Johnstone from registration of death inquest
records. BC Archives, Vital Statistics Database and GR 1327

explanation for fatalities, but these fatalities were reduced dramatically
after 1918. Influenza took a dramatic toll in 1918, while a jump in
work-related accident numbers in 1930 was caused by the collapse of a
bunkhouse at the nearby Bonanza Mine.[19]

The vast majority of recorded deaths were associated with the mine
rather than the smelter: from 1913 to 1935 four deaths were reported
from the smelter, while seventy-nine men were killed in the mines.
Nevertheless, the smelters also presented serious health hazards. Men
would emerge at the end of the shift coughing and spitting a yellow
substance. Hutchings reports that his mother made masks for the men
to help mitigate the problem: "The sulphur smoke was very bad at
times, and I remember my mother having an order from the company
to make a number of slip-on masks, which the men wore around their
necks and the gas became strong, they pulled it up under their chin
and adjusted a simple copper nose piece, then breathed into the
treated flannelette bag" (Hutchings n.d., 13).

It is also likely that industry-related hazards were not confined to the
workplace and the workforce. On both the East and West Coasts, sulphurous air permeated the entire community. Anyox residents early
recognized the severe sulphur contamination, though some women
took young children into the sulphur-filled atmosphere of the smelter
thinking it would help with congestion problems (Lade n.d.). As with
the workers themselves, in the absence of data we can only speculate
on the health impacts of environmental contaminants on the non-
working population on both the East and the West Coasts.

Short- and Long-term Environmental Impacts

There is little historical evidence regarding the immediate impacts of the copper mining industry at the East Coast sites. The mines and surface works of course altered the landscape, but the smelters apparently had the most pervasive impacts. We know that slag from the smelters was routinely dumped into the ocean and apparently exploded on contact with the water, but we do not know what the effect was on the ocean environment (Martin 1983, 14). There are also accounts of numerous forest fires caused by the smelters, while emissions also undoubtedly led to severe air pollution. Bloxam (1882) reports that in the calcining stage of the smelting process at that time, when the gases and vapours issuing from the calciners were allowed to escape directly into the air, they formed a dense grey cloud of copper smoke, which contained sulphurous acid mixed with a little vapour of sulphuric acid and arsenious acid, which condensed in the air to a fine powder.

Bloxam also notes that because in the 1880s this process was recognized as damaging to "the health and vegetation of the neighbourhood," flues (or what we now commonly call "scrubbers") had been designed to neutralize the sulphuric content of the smelter emissions. Apparently such scrubbers were not used at Tilt Cove, and the air around the community was reportedly filled with sulphurous fumes when the smelters were operating there (1889–92) (Martin 1983, 14).

To assess the long-term environmental impacts of smelting on the region, we sampled slag from the smelter sites and conducted standard major oxide and trace element analyses of the samples by x-ray fluorescence (see Longerich 1995 for analytical techniques). In the vicinity of the Tilt Cove smelter, we found high levels of arsenic, copper, nickel, lead, and zinc in the soils. Elevated chromium levels were directly related to the bedrock geochemistry, but the values for the other metals could not be attributed to that influence. In the absence of other processes or sources, these high values can be viewed as evidence that the environmental impacts of mining and smelting operations that occurred over a three-year period more than a hundred years ago linger in the region today in the form of heavy-metal pollution in the soils. Our geochemical study to examine the potential of using iron-manganese (Fe-Mn) oxide coatings on stream pebbles as an environmental monitor found that conductivity was three to fifteen times higher at the polluted sites, with the highest (1,405 microsiemens per centimetre [μS/cm])[20] in the stream flowing out of the West Mine (this sample also had the highest concentrations of all metals, especially copper, zinc, and nickel) (Riggs 2002; Huelin 2004).[21] At Betts Cove, where the control stream flowed out of a nearby pond, we found polluted streams that had served as drainage from the former copper mine. The polluted sites had pHs in the range of 4.5, whereas the control had pH values in the range of 7; conductivities at the polluted sites were about four times (240 compared to 60) as high, and aluminum, copper, and zinc concentrations were up to ten times higher than at pristine sites.

The Anyox operation also caused considerable damage to the environment. Little vegetation grew around the smelter site; sulfur dioxide emission from the smelting process killed most vegetation. It has been estimated that during the early years of operations nothing grew within a radius of three to five kilometres from the smelter (Errington 1975). Hutchings noted that as early as 1917 timber was dying as far as five miles south and north of the smelter (Field 1993, 22). The forest along Alice Arm was also devastated, and the Granby Company began a logging operation to salvage some of the destroyed timber for wood products. It was also charged $12,000 by the federal government to compensate for timber damaged on First Nations' reservation land. The BC Forest Service watched the destruction, but, as little of the timber destroyed was merchantable, minimal compensation was demanded. The prevailing winds, the heavy rainfall, and the steep sides of the inlet limited the range of the destruction. Parts of the forest along the inlets were destroyed both south and north, and damage occurred in the forests at the heads of both Alice and Hastings Arms. In 1971 Errington (1975) found it difficult to estimate the extent of the damage, for various reasons, including the several forest fires that had raged through the area and the resistance of the dominant species, western hemlock (*Tsuga heterophylla*), to sulfur dioxide poisoning.

For many years skunk cabbage (*Symplocarpus foetidus*) was all that would grow, but a reduction in smelter emissions through increased dust collection meant that by the late 1920s a few individuals with great perseverance were able to grow gardens (Field 1993, 22). Higher copper, zinc, and cadmium levels have also been noted and remain in the surrounding water and land (Odhiambo 1995).[22] The two most pressing problems, however, are acid mine drainage and the two dams left from the smelter and coke plant operations. The Hidden Creek mine has been identified as a severe source of acid mine runoff, but to date little remedial action has been taken to curb this problem. Because the area is heavily mineralized, water tends to have high acid and heavy metal content, regardless of past exploitation. The second serious problem is posed by two dams that could potentially collapse and create severe havoc in the area. The Anyox Hydro Electric Corporation is currently looking at the main dam for potential reactivation as a possible power source. The Provincial Dam Safety Program monitors both dams.

Lessons to be Learned: From Boom to Bust

It did not take long in Newfoundland for it to become obvious that the copper mining industry was anything but a "certain and positive" source of employment. The Tilt Cove Mine operated intermittently until 1917. During the boom period of 1888–1890, smelters were erected next to the mines, the result of the discovery of a new ore body (the East Mine) and a recovery in the price of copper, but by late 1870 it was reported that the difficulty in accessing the high-grade ore demanded by the smelters of Swansea, combined with depressed copper prices, had resulted in a downturn in the industry.[23] In 1889 the Tilt Cove Copper Company installed the blast furnaces to smelt the copper ore at Tilt Cove, but they operated for only three years, after which the operation returned to shipping ore directly to Swansea for smelting. In 1892, the Tilt Cove mines were subleased to the Cape Cod Copper Company, and another brief boom followed. However, the industry subsequently entered another period of decline. The practice of "high-grading" the ore (extracting only the highest grades), along with a steady drop in copper prices, led to the shut-down of the West Mine in 1902 (Symons 1910, 361; Martin 1983, 14).

Throughout the decline that marked the first decade of the twentieth century, the Newfoundland government took various steps to try to stimulate the mining and smelting industries. In 1910, for instance, the government introduced a bounty on copper mined and smelted in Newfoundland, as well as an act to encourage the development of coal deposits (*Newfoundland Gazette* 1910a). Also in 1910, the Discovery of Minerals Act was passed, under which an immediate reward of a thousand dollars would be paid to anyone discovering commercially valuable minerals on private or public land (*Newfoundland Gazette* 1910b). There were other efforts to attract attention to the Island's purported mineral riches. A 1910 trade journal article, for example, pointed out that the settlement of the "French Shore" question in 1904 had opened the west coast of the Island to mineral development, that much of the interior remained "terra incognita," and that Newfoundland's "liberal" mining laws made it an ideal place for exploration and investment (Symons 1910).[24] Such efforts did little, however, to check the decline of the copper mining industry. In 1912, a company apparently established to take advantage of new government legislation, the Newfoundland Smelting Company, went into liquidation (*Newfoundland Gazette* 1912). In 1914, the East Mine at Tilt Cove was shut down. That year the West Mine was revived by two local men, taking advantage of the wartime increase in copper prices, but in 1916 it was shut down again (Martin 1983, 15).

This boom-bust pattern was repeated at Bett's Cove over an even shorter period. In 1880 the death of his business partner and a drop in the price of copper apparently prompted Ellerhausen to sell his holdings to a group of New York capitalists, who incorporated the Newfoundland Consolidated Copper Mining Company in 1880 (*JHA* 1880, 368). By 1883 most of the accessible ore had been removed, and when the company tried to prolong the life of the operation by removing the mine's supporting pillars, the roof collapsed, taking with it surface buildings and machinery. The last shipment of ore left Betts Cove in 1886, and the operation was never revived after that.

In British Columbia, the Anyox smelter closed in 1935–36, primarily because of the low price of copper. The company itself survived only because of the continued improvements in copper prices over 1936, just as the board of directors had begun the process of winding up the company. Instead, the Princeton, British Columbia, operations were reopened. The closure of the Anyox smelter should not have come as a great shock to most residents. Copper prices were down, markets were down, and ore reserves were being rapidly depleted. A recent mine strike had undercut profits and created an atmosphere of discontent. The company had also opened a new mine in southern British Columbia near Princeton.

Still, the isolation and the hopes of Anyox's residents meant that closure was a shock for many of the nearly one thousand inhabitants. Having little money, many residents left furniture stacked in warehouses or even on the dock, and scavengers descended and bought goods at outrageously cheap prices. A few people moved on to other company operations, but most left Anyox for Prince Rupert or Vancouver to seek other work. The hardship, however, could have been worse. If Anyox had not been a company town, the houses would have been privately owned, and they too would have had to be abandoned. Because the community was too isolated to survive once the mine and smelter closed, even if a few hardy souls had held onto the property, it would have had little residential value.[25]

The mineral rights and land were sold to Cominco (owners of the Trail Smelter), who kept a caretaker on the property and undertook some prospecting until 1942, when a fire swept through destroying most of the wooden structures. Since then occasional prospecting forays have been undertaken to the area, and some new ore discoveries have been made, but during the 1990s it was a remnant from the smelter, the slag pile, that attracted people to the site. Mining of the slag was a continuing venture in 2005.

The history of each mine, then, demonstrates how regional development has occurred in the context of changing sources of capital, new technologies, and new community strategies. The mining industries, especially where accompanied by smelting operations, had a detrimental impact on the local environment. Although considerable concern was directed at community well-being, little was directed toward environmental or human health before World War II, unless it affected economic efficiency. Initially, transportation and communications played an important role, while changes in markets and technology later broadened the area of exploitation and the type of material that could be exploited. As panarchy suggests, patterns of government regulation tended to slow or even reverse this trend. The characteristics of the mining activity, the physical terrain, and the geographic location of the community, in conjunction with company and government policy, shaped the social configuration of a company town. Finally, mining contributed little towards long-term community stability, although larger or more productive sites generated short-term wealth for the provinces as a whole – thinking focused more on boom than bust and populations were vulnerable to the industrial and government policies that focused on development at any cost. Have we, then, learned anything since the closure of these mines? We turn now to a modern industry, to see how modern policy treats nonrenewable resource extraction.

THE RECENT HISTORY OF NONRENEWABLE RESOURCES: OIL AND GAS ON BOTH COASTS

The preceding section explored the development of a nonrenewable resource on the East and West Coasts in the late nineteenth and early twentieth century, drawing out the similarities and differences in the two cases and underscoring the influence of local factors on how the resource was exploited. This section retains this spatial comparative dimension by looking at the development of another nonrenewable resource – petroleum – on both coasts, but introduces a temporal dimension by shifting the analysis ahead approximately one hundred years, to the late twentieth and early twenty-first century. More specifically, we explore the development of oil reserves off Newfoundland's east coast, which has been ongoing for some twenty years now, and the absence of comparable development of known oil and gas reserves off the British Columbia coast. At the same time, this account demonstrates what has or has not changed on both coasts since the period of copper mining examined in the preceding section.

Has Canada, then, learned anything from previous Canadian experiences with nonrenewable resource exploitation and their social, ecological, environmental, and human-health impacts? We turn to the "new" staples of offshore oil and gas to answer the question. To compare the history of offshore oil and gas development on the East and West Coasts of Canada is to see that some lessons have been learned, while others have not. It is also to point to the overwhelming importance of government jurisdictional and regulatory powers and to how those apply nationally, regionally, and (by extension) locally. Once again, that is, the political context is important, framing both the manner of resource exploitation (regulations) and its revenue outcomes (jurisdiction).

Starting with that context, we asked why the government of Canada and the government of Newfoundland and Labrador were able to successfully negotiate the Atlantic Accord in 1985, while the government of Canada and the province of British Columbia were not able to enact a comparable Pacific Accord in 1989. In the case of Newfoundland and Labrador, oil and gas development occurred at the same time that the resolution of jurisdictional and regulatory questions was under way. On the West Coast, by contrast, federal and provincial moratoria have been in effect, and they will continue, pending such a resolution. The negotiations that created the regulatory framework on the East Coast – and failed to do so on the West Coast – operated through intergovernmental mechanisms that can play a crucial role in how offshore development affects coastal health, although "development" is a broader issue than the considerations taken into account in federal-provincial agreements like the Atlantic and Pacific Accords. Understanding the nature and substance of the negotiations, however, is important if we are to work towards the establishment of appropriate long-term sustainability strategies for offshore oil and gas development on both coasts.

Debate around the development of oil and gas on the East Coast as it emerged during the 1970s was infused with the rhetoric of economic diversification and industrial development that had changed little since the time of Bett's Cove. During the twentieth century, Newfoundland and Labrador saw the development of a number of land-based resources that were intended to provide a remedy to chronic unemployment and an alternative to using the fishery as the "employer of last resort." For some time, enterprises such as the iron ore mines at Bell Island (1895–1966), copper, lead, and zinc mines at Buchans (1927–84), fluorspar mines at St Lawrence (1933–78), and asbestos mines at Baie Verte (1955–92) went some way toward providing that alternative, as the iron ore mines of west Labrador, as well as pulp-and-paper mills in Grand Falls and Corner Brook, continue to do today. The ill-fated hydroelectric project at Churchill Falls provided many jobs in the construction phase but has failed to deliver much in the way of lasting benefits to date, primarily because of the short-sighted arrangement between Newfoundland and Labrador and Quebec.

While much of the discussion continued to centre on job creation and economic diversification, the debate surrounding the development of offshore oil was marked by two other key features. The first went to the heart of what "development" should mean and what it should bring, and focused on the fundamental continuing belief that economic development had to be built out of the exploitation of a succession (or mix) of natural-resource economic bases. What had changed, as we shall see, were some of the ideas about how to build such development. It was now clear that generating benefits needed to go beyond "jobs" and returns to a primary producer. There was a recognized need for backward and forward linkages such as supplying production equipment and carrying out secondary processing, for developing a skilled labour force to benefit the province in the long term, and for capturing other spin-off benefits such as adequate royalties. These factors were made explicit as a rationale for the development of offshore oil. None of this had happened securely within the renewable-resource sectors of fishing and forestry, the wealth that might have been derived from hydroelectric power had been lost, as mentioned, through a poor political deal made by the Smallwood government with Quebec, and the mining sector had not sustained on-site production (smelters) for very long. The skilled labour forces that had been developed out-migrated when the mines closed, and revenues had not been used for local development.

With the discovery of the Hibernia oilfield in 1979, Newfoundland and Labrador sought to capture benefits that would outlast the period of resource exploitation (Stanley 2004). After years of negotiations and judicial intervention, discussions around sharing revenues from offshore resources emerged between the government of Newfoundland and Labrador and the federal Progressive Conservatives as it became apparent they were likely to form the government under the leadership of Brian Mulroney. While still in Opposition, Mulroney signed a memorandum of understanding (MOU) with Newfoundland premier Brian Peckford. After the Tories formed the federal government, this MOU was introduced as the Atlantic Accord by parallel legislation in both Canada's House of Commons and Newfoundland and Labrador's House of Assembly and was promulgated on 11 February 1985.[26] Among the stated goals of the accord was "to recognize the right of Newfoundland and Labrador to be the principal beneficiary of the oil and gas resources off its shores" and "to recognize the equality of both governments in the management of the resource, and ensure that the pace and manner of development optimize the social and economic benefits of Canada as a whole and Newfoundland and Labrador in particular" (Canada 1985, section 2a, b).

The signing of the Atlantic Accord, therefore, seemed to signal real hope for reaching the kind of economic prosperity that had always eluded the province. Moreover, although the accord was mainly designed to address fiscal issues and was therefore marked by an absence of, among other things, any clauses related to occupational health and safety,[27] nonetheless, under section 61 all federal and provincial social legislation such as that dealing with occupational health and safety was to be extended to the offshore, and under section 63 a joint management board would coordinate this effort with the relevant government departments (Canada 1985, sections 61, 63). In 1987, the Canada-Newfoundland Atlantic Accord Implementation Act became operative, giving effect to the accord, creating the joint management board, and instituting an arm's-length regulatory agency, the Canada-Newfoundland Offshore Petroleum Board (CNOPB). Provisions 188 through 193 of the act provided for safety officers to inspect rigs and enforce health and safety legislation (Canada 1987, sections 188–93). Some effort, then, was made through the CNOPB to enforce health and safety legislation as strongly as any western producing nation, but the arrangement was flawed in that it invested the dictates of production and the dictates for health and safety in a single body.

On taking office in October 2003, Newfoundland premier Danny Williams promised that a "Progressive Conservative government will seek jurisdictional control and ownership over petroleum and other economic resources in the offshore as a means to achieve greater prosperity for our province and more opportunity for our people" (Mandel 2004). Then, in the federal election of June 2004 the Liberal Party, with a minority government, gained seats in Newfoundland and Labrador, including the one held by the natural resource minister. Between the June campaign and the fall of 2004, the price of crude oil rose to over fifty dollars a barrel, meaning that fundamental change of the kind sought by the province would substantially reduce federal revenues. Williams took the issue across the country, arguing that Newfoundland and Labrador wanted "to be a contributing province" (Richer 2004). Benefiting from offshore resources was seen as so important to economic development in Newfoundland and Labrador that not only Williams but also the provincial Liberal Party and several federal Liberal members of Parliament sided with him.

After months of negotiations, Prime Minister Martin and Williams agreed on a new fiscal settlement that does not fundamentally alter the jurisdictional arrangement but ought to provide the province with billions of dollars in additional revenue through the end of the decade, depending on the price of oil. According to the terms of the January 2005 deal, the province gets a $2 billion advance payment against future enhanced oil benefits. If benefits do not reach that level in eight years, the province keeps the difference. If and when benefits exceed $2 billion, the province begins to collect the excess. In return, the province has accepted the federal definition of "100 percent." Newfoundland and Labrador gets a full, 100 percent offset for its oil revenues when it is classed as a "have not" province, but the offset reverts to the declining protections offered by the current Atlantic Accord when Newfoundland crosses the line into "have" status.

The current accord offers less than full protection. The province had wanted the full, 100 percent offset no matter what its financial status. Officials believe the lump-sum advance payment will help ensure the province does not "fall off a cliff" if and when it comes off the equalization program. Williams has said that the additional cash will be used to develop and advance ongoing and new initiatives and "strategic investments." Announcing the deal, he attempted to dampen expectations that the accord – worth a minimum of $2 billion over the first eight years, with a possible extension for eight more – would single-handedly reverse all economic woes in the province (Antle 2005).

When negotiations leading to the Atlantic Accord were going on, the governments of Canada and British Columbia agreed to create the joint, independent West Coast Offshore Exploration Environmental Assessment panel (WCOEEA) to study offshore development on the West Coast. Ninety-two policy recommendations emerged from the WCOEEA, which argued that development could proceed only if these recommendations were met. The two orders of government then sought to negotiate a Pacific Accord. During the negotiations, policy-makers had to consider a range of issues, including exploration in areas that are known to be ecologically and seismically sensitive. A draft agreement was struck in 1987, its objectives (mirroring those of its East Coast counterpart) being "to achieve the early exploitation of resources on and below the seabed on the West Coast offshore area and related Pacific development opportunities for the benefit of Canada as a whole and British Columbia in particular" and "to recognize the right of British Columbia to be the principal beneficiary of the seabed resources of the West Coast Offshore, consistent with the requirement for a strong and united Canada" (Pacific Accord 1987, sections 1.01a, c).

Negotiations were fraught with difficulty over the issue of Aboriginal resource rights, and they became more protracted when an oil carrier – the Exxon Valdez – went aground off the coast of Alaska, causing massive environmental damage. The British Columbia–Canada negotiations disbanded, and a moratorium on exploration was put in place that lasted until 1994.[28] At that time the province's legal basis for its moratorium ceased to exist, and the prohibition on development became a matter of government policy.

The West Coast moratorium continues to be a high-profile issue. An independent review panel from the Royal Society of Canada, commissioned by the government of Canada, tabled a report in 2004 on the potential for future West Coast offshore development (Royal Society of Canada 2004), to be followed the same year by two other reports commissioned by the government of Canada: the Report of the Public Review Panel (the Priddle Report – Canada Natural Resources 2004) on the federal moratorium on offshore oil and gas activity and the report on the First Nations Engagement Process concerning the moratorium. While the Royal Society report identified no discernable scientific rationale for non-development, the Priddle Report indicated that a majority of voices heard in the panel's consultation process opposed lifting the moratorium, and the First Nations Engagement Process revealed unanimous opposition, although many First Nations indicated a willingness to explore the issue more fully provided they were adequately resourced and given sufficient time.

These new findings could alter the federal government's position on the offshore in British Columbia. Successive British Columbia governments have previously maintained a de facto "moratorium" in the province, precisely because there are not as yet in place the regulatory instruments like a British Columbia Offshore Petroleum Board or a comparable body like the Oil and Gas Commission (which oversees onshore development and production), either of which could allow development of the offshore to proceed. However, the battle for and against offshore development in British Columbia is likely to be vocal,[29] as opponents "talk about the ten million sea birds, the rare sponge reefs, the endangered whales and the native peoples who are believed to have lived in the area for 12,000 years" (Hume 2004).

On the West Coast of Canada, environmental concerns, concerns of First Nations, and concerns of local communities have not stood in the way of provincial governments who are determined to proceed with industrial resource development: even in the face of possible negative impacts on the social-ecological health of the coast and real questions about the level of economic development that might be anticipated for local rural communities, as opposed to the urban complexes of southern coastal British Columbia. However, the situation is now more difficult than in the past, given that British Columbia is obliged to address the politically charged issue of Aboriginal resource rights, which may in and of itself place the province's objective of initiating offshore activity by 2010 in question. As British Columbia debates retaining or lifting the moratorium, the experience of Newfoundland and Labrador should prove useful on the issue of impacts.[30] For oil and gas development projects on the East Coast of Canada, an Environmental Impact Statement (EIS), which becomes the basis for an Environmental Protection Plan (EPP), is required.[31]

A key component in moving from the EIS to the EPP has been community input into developing the EPP. The community is consulted with a series of open public meetings for each project (for example: Hibernia, Terra Nova, White Rose, Whiffen Head). Another key component is the use of valued ecosystem components (VECs) as focal points of the EPP. They allow community input by determining which ecosystem components are valued by the public. Now the EIS is taken as a set of predictions that are then tested in a program that is part of the EPP, and the design of the monitoring program is based on knowledge of the project and its interaction with the ecosystem. In short, the regulatory history has been fairly effective in regard to the environment. Discharges are monitored and the environmental impact monitoring program is the best in the world. Post-Hibernia programs (Terra Nova, White Rose, Whiffen Head) for the environment are stronger than those for Hibernia, and local communities have been brought into the process of assessing impacts and monitoring the environment. Of course, this does not mean that environmental accidents will never happen, as the Terra Nova spills of November 2004 illustrate. What it does mean is that the risks are minimized and impacts closely monitored. It also means that the lesson of the value of community knowledge about ecosystems and the importance therefore of their input seems to have been learned on the East Coast: it is certainly central to the development of the EPP from the EIS.

Oil and gas have provided temporary employment to people laid off from the fishery and longer-term employment to a smaller number of people, many of whom were not in the fishery. Early fears that transient workers would bring drugs, crime, and prostitution problems with them have proved unfounded: work camps with the potential to contain social problems were built to house transient workers, and project labour agreements minimized speculative in-migration and prevented "at-the-gate" hiring. The construction and operation activities associated with oil and gas have resulted in high rates of growth in the provincial economy, but of course this growth might be limited to the lifetime of oil production, if royalties were not available to be invested in local infrastructure, because of claw-backs (equalization payments) that existed under previous regulatory arrangements. The regulatory regime must be such that the East Coast will be able to generate the basis for further development over the long run.

The Atlantic Accord resulted in substantial local employment during the construction phase of Hibernia, because the large gravity-based structure took years to complete, but subsequent projects (Terra Nova, White Rose) resulted in less local employment during construction because floating platforms were built. After the construction phase, however, substantial efforts were made to bring local people into long-term operations, including management. By and large these efforts were successful, with important effects on the provincial economy. In the new millennium, Newfoundland and Labrador has one of the fastest growing provincial economies in Canada as a result of post-construction activity generated by oil development.

That said, the offshore oil and gas sector must be able to bring lasting benefits to the economy of Newfoundland and Labrador. When the oil is gone, the jobs will go. Even now that royalties will not be clawed back to offset transfer payments, there will have to be significant investment in local infrastructure by government and business, with support too for local companies that have developed or expanded in response to the industry, if they can show that they are capable of surviving its demise. Beyond this, there are concerns about stress on families associated with shift work on the rigs; and there are the effects of social dislocation associated with imported labour and differential job opportunities across demographic groups. In an investigation of workers in the oil and gas industry in the North Sea, Parkes (2001) concluded that several psychological stressors are related to their work. Work conditions and perceived hazards, as well as separation from family and the community, can result in increased anxiety, stress, and health problems, as well as reduced morale and satisfaction. Disruption to family life because family members are away on the rigs is also stressful: it may result in delayed decision making and generates unease about coping alone. House (2000) notes that adjustment difficulties are felt by both men and women. Men face difficulties when they have to adjust between life on the rig and life at home, while women are cast in the conflicting roles of being at the same time the head of a family and a supportive spouse. Children also suffer anxiety as they get old enough to appreciate the perceived hazards of a parent working offshore. House suggests mitigative strategies that include shorter periods "on" and longer periods "off" shifts, as well as family visits to the worksite, which may be real or virtual. The government also needs to ensure that occupational health and safety concerns are met.

Some observers have also been concerned about the proportional impact on lower-income groups and on women workers. Shrimpton and Storey (2001) stress that appropriate human resource policies are important intervening measures to ensure participation by women, older workers, and others. For example, Husky's Workforce Diversity Policy (Husky Energy 2003) for its White Rose Project "will take special measures to facilitate the full participation of under-represented designated groups." Nicholson (1975) notes that in the Shetlands "disturbance" payments provided annual compensation for disruption caused by the oil industry. Taxes paid per barrel and shipped through Sullom Voe were paid into a trust fund whose revenue was then used to fund social, cultural, and educational projects, as well as to provide loans and equity capital to local businesses. A reserves fund was financed by royalties from oil and was used to support economic and business development.

While success (or failure) in negotiating a regulatory framework for offshore development is important, the existence of a framework does not fully explain why development has occurred on the East Coast and not on the West, since the framework and the beginnings of exploitation in the East were concurrent. The framework is, however, a crucial part of the way in which exploitation will be carried out and helps to explain how (or if) offshore development benefits coastal communities. Joint management through an independent regulatory agency allows questions of ownership to be set aside and allows for exploitation and potential development to proceed – ideally to the satisfaction of both levels of government. As the British Columbia government looks to the future, it will need once again to negotiate with the federal government in order to create a framework to regulate the offshore, should it decide to go forward. Such a process might mirror agreements on the East Coast, or one level of government might take administrative responsibility (Burleson 2003).

CONCLUSION

The story of nonrenewable resources on both coasts is one of ecological damage, some environmental protection, and economic development, underdevelopment, and non-development. What has their history told us from a social-ecological point of view? Along Canada's rugged East and West Coasts, large-scale copper undertakings emerged before World War II, undertakings whose complexity of operations and isolation make them the subject of useful case studies. Industrial activity of this kind has been chaotic and disruptive on both coasts and has proved less than helpful in sustaining social-ecological health. Company towns, owned by the company and completely dependent on the mine and smelter operations, were also ultimately transient phenomena: when the mine and smelter closed, so did the company towns.

Though many things were different on the two coasts and many things changed over time, others remained fairly constant. From the beginning of what might be called the "modern period" in the mid-nineteenth century, the mining industry was marked on both coasts by several distinguishing features that became more deeply ingrained over time: a close relationship between business and political interests, a depiction of mineral potential as a key ingredient in the industrial possibilities of the region, government willingness to grant incentives and concessions to possible developers in order to stimulate industrial development, a tendency to downplay or ignore considerations of safety and environmental impacts in exchange for jobs, a marked lack of security and control that resulted from outside ownership of an unpredictable industry, and the maintenance of economic and technical ties with mining interests from Britain, Canada, and the United States.

The rise and fall of these mining communities was part of a broader debate over the economic direction of the coasts of Canada itself, which in Newfoundland took place in the context of successive crises and policy shifts associated with such factors as fluctuations in the fishery and the thrust towards landward resource development. At the time, since these industries provided much-needed employment, helped to open up land-based resources, and spurred infrastructural initiatives (such as the trans-Island railway in Newfoundland), they were in some ways useful as well as politically and economically attractive. However, they also brought negative ecological impacts on the affected areas, including air pollution. What is more, many of the environmental impacts of the smelting industries are still evident in the soil in the immediate vicinity of the smelter ruins. The rapid growth of these mining centres led to a dramatic increase in population in the affected areas that was ultimately unsustainable when the ore was depleted.

We observe further that the case of the mining industry serves to highlight some of the political and ideological aspects of the "development" question during the nineteenth and early twentieth century and the perceived status of the mining industry within that broader issue. One of the most striking aspects of the cases considered here is the contradiction between the rhetoric surrounding the mining industry and what was actually happening within the communities and the industries. It is ironic, for example, that the mining industry was extolled as one of the keys to eradicating dependence on an unpredictable fishery while at the same time proving to be a wildly erratic and undependable enterprise. Viewed in this light, the Notre Dame Bay mining industry emerges as a striking instance of a misplaced development model, a model that "built on Newfoundland's weaknesses, not her strengths" (Ommer 1985). It also serves as an illustration of how the rhetoric of "development" was used to both political and private ends.

The copper-mining industries that arose and declined in various parts of what is now Canada during the late nineteenth and early twentieth century followed a predictable pattern: rapid industrialization and an accompanying dramatic increase in population, followed by uncertainty and fluctuation, and finally a period of rapid decline in the industry and the community. The detailed stories of the communities considered here reveal much about the perils of the mineral nonrenewable resource sector, the kind of "development" it entails, and its implications for economic sustainability and human and environmental health. The notion of pursuing development through large-scale, resource-extractive industries and of attracting investment and creating employment through concessions, is still very much with us in debates surrounding, for example, Voisey's Bay nickel.

Copper mining, as Richard White (1991) and Kent Alexander Curtis (2001) have pointed out, is not practical for small-scale production: to succeed, it requires capital, complex technology, and insight into the best means of managing the resources available. There have always been social and environmental costs related to this industry, and we have quantified and evaluated these costs where data permitted. In Newfoundland, although no detailed study of the subject has been undertaken, a collusion of political and business interests appears to have formed a closed community of mining interests within the colony. Though the policy of the day ostensibly provided for open access to individuals wishing to explore and stake mineral properties (provided they had the money to pay the fees), in fact a private system appears to have been controlled by a few select individuals from the political and business worlds, in conjunction with outside capitalists and technical staff.

In the end, we determined that the ecological footprints of the mining/smelting operations at Tilt Cove and Betts Cove were actually relatively small and localized. The geochemical signal of the operations can be detected over a century later, but the local environment shows no visible lasting impact, although the actual mines are big holes, due to the "glory-hole" type of mining operation. The mines we have investigated here prompt us to reflect upon the notion of "development" in the nonrenewable resource sector, its benefits and its costs, and the long tradition of using the rhetoric of resource development to political ends. It is, therefore, encouraging to see that some lessons have been learned with respect to the modern oil and gas industry. The process that carries an EIS to an EPP and then to an Environmental Effects Monitoring (EEM) program has evolved on the East Coast over nearly twenty years, resulting in an EEM program with real community input and a strong science basis. The strong science basis rests on the principle that effects are predicted and then tested by the EEM program,[32] and the rigour of this design exceeds that of programs generally used elsewhere in the world. The overall lesson has been that a good environmental program results when community input is embodied in the EPP process (as with the concept of VEC) and when predictions of effects are tested.

We have found that regulation of oil and gas in Newfoundland and Labrador has been effective from a social-ecological point of view in the short term. Environmentally, discharges are monitored, and the environmental-impact monitoring program is the best in the world. Post-Hibernia programs for the environment – Terra Nova, White Rose, Whiffen Head – are stronger than those for their predecessor. The process has succeeded in engaging the community in the assessment and monitoring process. Socially, development resulted in substantial local employment during the construction phase of Hibernia, in part because, as we mentioned, the large, gravity-based structure took years to complete. The skills gained during the construction phase were generally not transferable to the economy, and subsequent projects resulted in less local employment during construction, because they were floating platforms. Beyond the construction phase, however, substantial efforts were made to bring local people into the long-term operation, including into management positions. By and large these initiatives were successful and have had positive effects: Newfoundland and Labrador currently has one of the fastest-growing provincial economies in Canada, as a result of post-construction activity generated by oil development.

We conclude, then, that nonrenewable resources have always been viewed as important opportunities to diversify the economy beyond traditional sectors, even though, like all staple industries, they have always been influenced by economic developments external to the provinces. This influence is even more pronounced today, as the restructuring of national and provincial economies makes them more highly integrated in the national, continental, and global arenas.

With oil and gas, Newfoundland and Labrador sought to avoid the boom and bust cycle that had characterized earlier nonrenewable resource development. Thus, the long fight for ownership ensued, as governmental cross-scale inequities were feared, on top of the other losses of control that transnational corporations and the need for foreign investment bring with them.

The rhetoric of development on the East Coast (though marked by some key differences from the earlier period) is still with us. However, awareness of environmental and (on the West Coast) First Nations' issues is new, while the interplay on both coasts of national and provincial politics remains a crucial aspect of resource development. In addition, with modern oil and gas development, modern fiscal arrangements must be considered, making the deal concluded between the federal and provincial (Newfoundland and Labrador) governments in January 2005 important, because it positions the province for taking responsibility for capturing the full benefit of the resource within its period of exploitation.

6

Cross-Scale, Cross-Sector, and Cross-Purpose Issues: Overlap in the Coastal Zone

We must not assume that human intervention in ecosystems and natural processes will always be only of short duration or have only foreseeable impacts on social-ecological health. The social-ecological history of coastal communities and their staple products on the East and West Coasts of Canada (chapters 2 to 5) makes this abundantly clear. In chapters 3 to 5 we examined crucial social-ecological management issues in the main resource sectors (fishing, forestry, and mining) of the coasts and showed how such issues have developed over the long term: they are partly the legacy of an historical staples mindset. In this chapter[1] – which concludes this first section on interactive restructuring and its historical legacy – we first examine problems that cross sectors, in order to show how institutional regulations fail in these circumstances, since they appear not to be sensitized to the linkages between ecological disruption in one part of an ecosystem and its effects on other parts. That is, they are misaligned, and the economic benefits in one resource sector produce not downstream wealth but downstream degradation. We explore here, then, the connections, interactions, and failures of misalignment between human economic exploitation of resources and parts of coastal ecosystems, particularly with respect to important issues of marine impacts of pollution from coastal natural-resource industries. To get at the fine detail, we once again use case studies, drawn this time from forestry, mining, construction, and hydroelectricity, looking at their impacts on the coastal land- and seascapes.

Second, we then consider alternative ways of thinking that were offered to us by community participants and that could become new models for development. Recognizing that Canadian thinking on the management and exploitation of coastal resources needs adjustment, we explored the cultural (as well as the social-ecological) importance of abundance and biodiversity, mindful always of our insistence on the importance of interactive effects and cross-sector and cross-scale problems in coastal social-ecological health. Over the course of our work, we were struck by the contrast between the thinking behind industrial exploitation of resources and the different mentality that underlay (and sometimes still underlies) community ways of using resources – a mindset that takes into account the social and cultural importance of the coastal zone. In the second half of this chapter, then, we extend the concept of key species to include the idea of *cultural keystone species* and briefly test it using examples of important species on both coasts. Taken together, the two halves of the chapter illustrate the need for new thinking and build a cultural and social-ecological basis for social-ecological stewardship, which we will return to in chapter 15 as one new option for coastal-zone management. We construct our base by identifying some of the components that come together when considering not just economic development or one single sector's development or the activities that occur inside a single time frame or cultural diversity or biodiversity but all of these and how they should work in tandem if we are to achieve social-ecological health.

This chapter then concludes our first (resource-base) section of the book by laying the cultural basis for a mind-change that could promote coastal social-ecological recovery. We argue that to operate at only one scale (institutional, economic, geographic, or temporal) is to fail to deal with the complexities of social-ecological systems; to operate in either sectoral (usually staple) or temporal silos is to fail to identify linkages between them. To do both is to risk the kinds of damage that we have documented in this section and to fail to learn the lessons that damage to social-ecological health has created in the past.

INDUSTRIAL CROSS-SECTOR SOCIAL-ECOLOGICAL IMPACTS: THREE CASE STUDIES

Marine environments suffer not only from inadequate fisheries management, as discussed in chapters 2 and 3, but also from the problems that terrestrial industry can bring in its train. We explore here three coastal case studies that we selected because they involve typical industrial activities: log booming and the consequences of long-standing forest practices for the coast of British Columbia; aggregates mining in British Columbia, which may be important if oil and gas is developed on the West Coast (we also take a very brief look at potential dangers in the interaction between oil and fishing); and hydroelectric power development in Labrador. All three industrial operations have generated problems in other parts of the ecosystem, and all three also demonstrate what can happen when regulations fail to take cross-sectoral impacts into account.

The Ecological Impact of West Coast Forest Industry Practices on the Marine Environment

The temporary storage of harvested logs in estuarine areas has been a common practice in many parts of eastern and western Canada for nearly two centuries. While the practice has been severely curtailed in eastern Canada in recent years, this way of dealing with the handling, storage, and marine transportation of logs is still extremely important in British Columbia, because of the remoteness and rugged character of most of the coast and the absence of land-based transportation routes. Estuaries are preferred sites for temporary storage, both because they are usually adjacent to areas accessible by local logging roads and because they have low salinity. Low salinity is very useful because it inhibits the development of shipworms (teredos, *Bankia setacea*), which are highly destructive of wood.

Most previous studies of the ecological impacts of log handling have concentrated on infaunal (organisms living in the sediment) and intertidal epifaunal (organisms living on or just above the seafloor) communities (Levy et al. 1982; Duval and Slaney 1980). A substantial literature also documents such effects as reduced concentrations of porewater dissolved oxygen, increased elevations of interstitial reducing conditions, and changes to infaunal communities (Pease 1974; Conlan and Ellis 1979; Jackson 1986; Levy et al. 1989; Stanhope et al. 1987; Stanhope and Levings 1985; Sibert 1979; Sibert and Harpham 1979; McGreer, Moore, and Sibert 1984). However, previous work on subtidal epifauna has been qualitative and limited to relatively small areas or small sample sizes (McDaniel 1973; Pease 1974; Conlan 1977; O'Clair and Freese 1988; Williamson et al. 2000). To remedy this lacuna, we developed techniques to identify wood debris accumulations, which allowed us to document the extent of accumulations in areas near log-handling sites. We then characterized the physical and chemical characteristics of sediments in areas of accumulation and compared these areas to places that had not been used in this way. This allowed us to determine the impacts of wood debris on epifaunal taxa in estuarine areas.

Using this methodology, we examined several estuaries on the north coast of British Columbia near Douglas Channel and upstream from the First Nations community of Hartley Bay that are currently used for log handling and/or were used for it during different time periods. Site selection paid attention to exposure and seabed morphology,[2] and

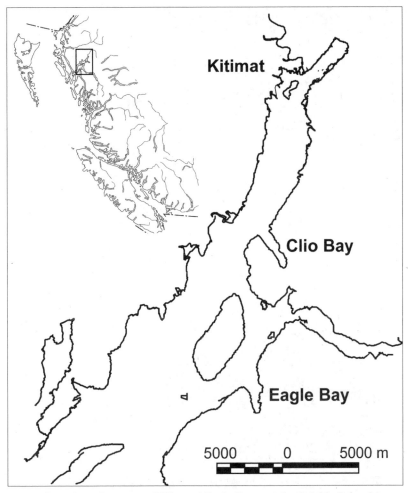

Figure 6.1 Location map of Clio and Eagle Bays on the British Columbia
north coast

as a control, we also examined sites that had never been used. We
report here on Clio Bay and Eagle Bay (near Kitimat – figure 6.1),
which are morphologically similar and also very similar in size (approximately 1.6 km²), shape, depth, and orientation, but Clio Bay has been
actively used as a log-handling site for several decades, whereas Eagle
Bay has never been used for that purpose.

Our sidescan sonographs (figure 6.2) and seafloor video imagery
(figure 6.3) showed that Clio Bay has a significantly higher terrestrial
organic content in the sediments on the seafloor than Eagle Bay.

Figure 6.2   Sidescan sonograph from Clio Bay showing dense concentrations of whole logs on the seafloor

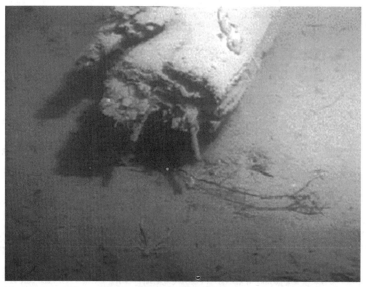

Figure 6.3   Photograph of partly decomposed whole log on the seafloor in Cleo Bay

Indeed, large areas of Clio Bay exhibited very high concentrations of whole logs, in some instances in stacks up to ten metres high above the seafloor. We estimate that this bay contains several thousand partially decomposed logs that have sunk beneath log-booming sites. They are

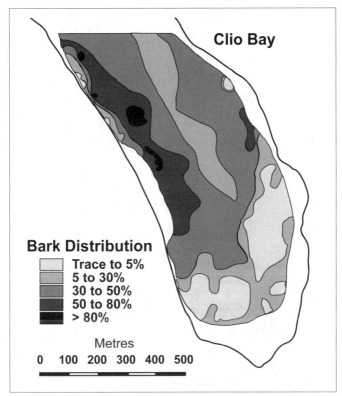

Figure 6.4 Map of wood debris concentrations from towed seabed video imagery in Clio Bay

in this condition because, as the central woody material in the logs deteriorates (in large part through destruction by teredos), a tube of bark often remains for some time (figure 6.3). Eventually this more resistant bark material collapses and forms a layer of coarse debris within the sediments, often up to several tens of centimetres thick. Not only is this easily recognizable in video imagery, but the amounts of it that are in the bay can be quantified. In the Clio Bay seafloor, bark concentrations reach more than 80 percent in some locations and average about 40 percent over the areas surveyed (figure 6.4). Moreover, much of the sediment in Clio Bay consists of fine (about 1 mm) disseminated wood, while in sharp contrast, Eagle Bay (figure 6.5) displays organic contents of less than 10 percent on average, with only small, highly localized occurrences of more than 50 percent.

Carbon/nitrogen (c/n) ratios in recent sediments usually reflect the source of organic matter. c/n ratios of about 7 (the "Redfield"

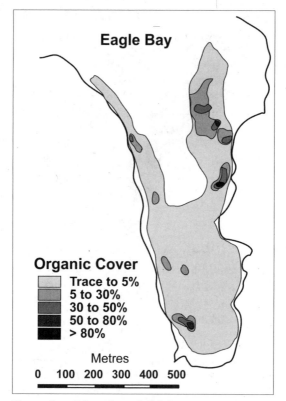

Figure 6.5 Map of wood debris concentrations
from towed seabed video imagery in Eagle Bay

Table 6.1
Mean Carbon to Nitrogen Ratios (c/n) and Percentage of Terrestrial Organic Carbon
(and 95% Confidence Limits) of Sediments from Clio and Eagle Bays, British
Columbia (Collected in May and September 2001)

| | C/N | | % Terrestrial C | |
Bay	Mean	95% CL	Mean	95% CL
Clio	39.5	5.05	67.0	6.55
Eagle	21.5	2.66	31.7	6.04

ratio) in estuarine areas are associated with marine organic matter,
whereas values in excess of this ratio usually indicate terrestrial plant
material.[3] Mean Clio Bay c/n ratios were nearly double, significantly
greater than those from Eagle Bay (table 6.1). Basing our estimates on
the c/n ratios, then, we take the mean terrestrial organic-carbon con-

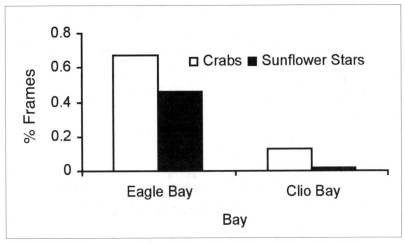

Figure 6.6   Percentage of towed video frames from Clio and Eagle Bays (British Columbia) where Dungeness crab (*Cancer magister*) and sunflower seastars (*Pycnopodia helianthoides*) were observed

tribution in Clio Bay to be 67 percent, as opposed to only 32 percent in Eagle Bay.

There was also a substantial difference in the epifaunal communities of Clio and Eagle Bays. Our examination of towed seabed videos suggests that large benthic predators avoid wood-dominated habitats: we observed Dungeness crabs (*Cancer magister*), for example, five times more often in unimpacted Eagle Bay (figure 6.6) than in Clio Bay. Similarly, sunflower seastars (*Pycnopodia helianthoides*) were twenty-five times more abundant in Eagle Bay than in Clio Bay, and both species were several-fold less abundant in the wood-dominated habitats within Clio Bay than in non-wood habitats in that bay.[4]

Some species, however, displayed a positive correlation with wood debris abundance. Squat lobsters (*Munida quadrispina*), some anemones (*Metridium* spp.), and holothurians (*Parastichopus* spp.) are all more abundant in Clio Bay than in Eagle Bay, and squat lobsters in particular show a high degree of association with sunken logs and coarse wood debris (figure 6.7). The increased abundance of squat lobsters and anemones is probably related to the substrate that coarse wood provides for attachment, and the availability of crevices would explain (especially around whole logs) the evasion of predators, since both these taxa are planktivorous and therefore do not rely on a healthy infaunal community for food. Wood-dominated habitats, however, may support elevated microbe populations by providing an abundant food source for holothurians.

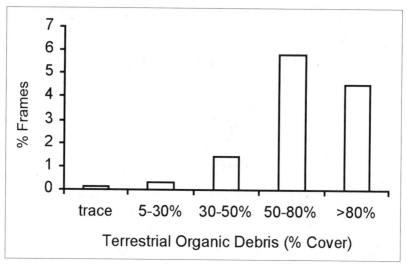

Figure 6.7   Percentage of towed video frames within each terrestrial organic debris surface area coverage category in Clio Bay where squat lobsters (*Munida quadrispina*) were observed

What this means is that we can now say that the wood debris accumulation associated with log handling is extensive in Clio Bay. This is reflected in the significantly higher seabed bark coverage shown by the c/n ratios in Clio Bay relative to Eagle Bay. Importantly, epifaunal communities also appear to be a useful indicator of habitat differences, since large benthic predators (e.g., Dungeness crabs and large sea stars) avoid wood accumulations, while these habitats may be preferred by squat lobsters, some anemones, and holothurians. In other words, the impact of industrial log-storing practices is neither temporary nor trivial. On the contrary, these practices change both local habitat and the ecosystem to one in which less-commercial species replace those that are important for commerce and human diet.

The implications of these findings are obvious. Perhaps less obvious is the need to improve foreshore lease relinquishment regulations for forest practices, since our results show that these regulations are clearly inadequate. We suggest that surveys that adequately assess the status of the foreshore lease and subtidal industrial use (before, during, and after) need to become standard governmental practice. There is also a real need to establish a generic restoration program for heavily impacted sites.

*The Social-Ecological Impact of the East Coast Oil
and Gas Sector on the Fishery*

The relationship between the fishery and the oil and gas industry has been less tumultuous in offshore Newfoundland and Labrador than it has been in comparable regions of the world (such as the North Sea), since all active oil rigs on the Grand Banks are located more than two hundred kilometres from shore, thereby limiting interactions between the two industries.[5] In recent years, however, a number of concerns have been raised about existing or potential conflicts. In offshore areas, there are lingering uncertainties about the impact that seismic blasting may be having on fish communities and marine ecosystems, while other issues have emerged closer to shore.

Placentia Bay is home to both the Newfoundland Transhipment Terminal (NTL) – where "sweet" Grand Banks crude is off-loaded and transferred onto smaller tankers for export to mainland Canada and the United States for refining – and the North Atlantic Refining Ltd (NARL) oil refinery, which ships in "sour" crude bought from the world market for refining. The bay is now the site of about one-quarter of all oil movements in Canadian waters and has been deemed a high-risk site for a major oil spill (Brander-Smith, Therrien, and Tobin 1990). This is naturally of particular concern to fish harvesters living in the area, since Placentia Bay has a lucrative inshore crab fishery and boasts one of the strongest remaining populations of cod in Atlantic Canada. Worries were exacerbated by a spill off Spain's Galician coast, which caused severe ecological damage and has had a lasting impact on markets for Spanish seafood products. Moreover, the main deep-water shipping channel on the east side of the bay sits above excellent crab fishing grounds, where many harvesters fish. This location is risky because tankers are not able to stop quickly, and there are ongoing concerns about potential collisions with fishing vessels or damage to fishing gear.

Fortunately, these concerns are being recognized in an important local cross-sector initiative. In 2002 the organization One Ocean was established through Memorial University in St John's to act as a non-government liaison between the fishing and the oil and gas industries. The organization, which is funded solely by the oil industry, has sought to address issues of mutual concern between the oil and the fishing industry and has helped to coordinate research, workshops, and study tours in support of that goal (One Ocean 2003). In addition, the Department of Fisheries and Oceans has been active in developing an integrated management plan for Placentia Bay (under the Oceans Act) through which the bay and the Grand Banks were identified in

2004 as a key area of focus under the new Oceans Action Plan (Davis forthcoming). This is a welcome initiative and bodes well for co-management thinking, which we discuss in detail in chapter 15.

The Social-Ecological Impact of West Coast Marine Aggregate Extraction on the Littoral

On the West Coast, a similar issue concerning cross-sector environmental impact exists and will become extremely important should oil and gas development come to the Hecate Strait. This is the issue of the extraction of the nonrenewable resource of "aggregates" (sand and gravel) for the construction booms that take place around modern staple development, including oil and gas. Extraction of aggregates and marine minerals from the sea is increasing every year as on-land sources are depleted or access is denied. In northern Europe, for example, marine-aggregate production is extensive, and up to 50 percent of the United Kingdom's sand and gravel consumption is derived from marine sources. In Europe aggregates are extracted under extensive site regulations that allow for monitoring of habitat loss or change. In Canada regulatory regimes and constraints on seafloor dredging for aggregates vary by region. In British Columbia seafloor dredging has in the past been largely unregulated, resulting in significant habitat damage, in terms of both physical loss of near-shore areas and overall change of the habitat. In the British Columbia offshore, if the moratorium on oil and gas exploration is lifted, offshore aggregate deposits may provide the answer to the pending problem of limited local resources for construction aggregate in the Prince Rupert region.

We suspected that offshore aggregate extraction of more than 4 million tonnes of marine sand and gravel in the area in the past had led to significant environmental concerns, including coastal erosion and environmental damage resulting from over thirty years of coastal mining. We therefore used a combination of sub-bottom profiler, sidescan sonar, sediment coring, unmanned submersible systems, aerial surveys, and shore field parties to collect a range of geological and environmental data on the issue. Our findings include, but are not limited to, sediment removal resulting in the loss of the west end of the Tugwell Island tombolo in Metlakatla Bay, tree and property loss due to coastal erosion in Secret Cove, and the unnaturally steep beach found in Welcome Harbour. Biological recovery of these areas appears to have been successful, although because no assessment of the biologic community before or immediately following dredging was conducted, it is unknown if the community that returned was of the same makeup as the original.

The regulatory regimes and constraints on seafloor dredging for aggregates in British Columbia are significantly different from those in Europe. For example, whereas in the United Kingdom dredging is required to occur only after a very thorough environmental review by many government agencies and usually only in water depths greater than eighteen metres, in British Columbia, virtually all dredging for aggregates has been in less than eighteen metres of water and often right to, and including, the intertidal zone. Should offshore aggregate mining resume on the north coast of British Columbia, policies and regulations for mining practices will need to be established. Based on the results of our work, it is advised that Metlakatla Bay, Secret Cove, and Welcome Harbour not be revisited for the purposes of aggregate dredging. Rather, they should be utilized as examples of potential consequences to consider when drafting offshore policies for the province of British Columbia.

Canada and British Columbia have no specific regulations or guidelines regarding marine aggregate extraction and on the Pacific Coast, no institutional experience in dealing with applications for such work. In countries where marine aggregate extraction is common, laws specific to marine sediment extraction have taken many years to be enacted. In all cases, including Canada, an Environmental Impact Assessment (EIA) will be required before a dredge permit is issued, so the most practical focus may be to ensure that the EIA deals with all potential problems and with compliance. Therefore, we recommend guidelines for EIAs, monitoring, and compliance, rather than new laws. General environmental guidelines for marine mineral extraction, including aggregate dredging, are given in the International Marine Minerals Society (IMMA) Code for Environmental Management of Marine Mining. This code is intended for the marine mining industry and contains general principles for industry and more site-specific operating guidelines. Specific guidelines for marine aggregate dredging are provided by the International Council for the Exploration of the Sea (ICES) in their Guidelines for the Management of Marine Sediment Extraction. The ICES guidelines for 2003 cover general principles for mineral resource management and the administrative framework for an EIA, topics to be addressed in a pre-extraction EIA, the authorization of extraction, and monitoring of extraction activities. The more general resource management principles and administrative framework are covered by existing Canadian acts, such the Oceans Act and the Fisheries Act.

One of the most complicated sections of the EIA deals with the evaluation of the impacts, since evaluations should require changes at the extraction site to be put in a larger, regional context. Short-term habitat damage associated with aggregate dredging may be acceptable where affected species are regionally healthy, but their regional health has to be known to allow this determination. This lack of regional knowledge is probably more acute in an area such as northern British Columbia than in Western Europe, especially for less commercially important species. From a practical standpoint this means that, while any EIA-related guidelines that are adopted may be thorough and complete for data collection and may require all the appropriate questions to be asked, nonetheless the data might not be available and hence appropriate questions might be unanswerable. Depending on the number of unknowns, the guidelines may not provide a clear basis for action, and the decision to issue an extraction permit may be based on principles given in other documents, such as the Oceans Act or the Fisheries Act. Thus, the limitations of a set of guidelines should be recognized and anticipated.

In northern Hecate Strait there is a vast supply of aggregate at exploitable water depths. Most of Hecate Strait is a shallow, semi-enclosed tidal channel, very similar to the Bristol Channel or English Channel off the United Kingdom, where mining takes place year round. Within the Queen Charlotte Basin, and particularly Hecate Strait, there is a significant fishing industry that includes bottom trawling and groundfish hook-and-line fisheries. Shipping lanes, marine protected areas, and possible future offshore oil and gas activity and wind power projects are also considerations. It should be noted, however, that the large-scale aggregate extraction in European waters occurs in regions of greater activity in all of the above uses and over smaller geographic areas. Therefore, it is reasonable to anticipate that conflicts between various users of the Hecate Strait region may not prevent extraction from being carried out. It is imperative, however, that guidelines for ocean management of the marine aggregate industry, as with all other natural-resource industries, be established before the development of a regulatory regime. We also suggest that the kind of "One Ocean" initiative that exists between the oil and fishing industries in Newfoundland would be a very useful example to follow in this case. Taken together, these steps should ensure sustainable ocean management into the future and provide a viable new economy for the region.

Furthermore, should offshore aggregate mining resume on the north coast of British Columbia, policies and regulations for mining practices need to be established. Based on the results of our work, we think it advisable that (1) Metlakatla Bay, Secret Cove, and Welcome Harbour not be revisited for the purposes of aggregate dredging – rather, they should be utilized as examples of potential consequences to consider when drafting offshore policies for the province of British

Columbia; (2) potential concerns specific to each proposed location for extraction should be identified before extraction; (3) a complete data set should be established before, during, and after dredging to ensure that the impact on the environment is understood and limited; and (4) ongoing monitoring of dredging activity should be mandatory while it is in progress, as well as for a number of years following the completion of mining.

The Social-Ecological Impact of Labrador Hydroelectricity Development on the Coastal Environment

At the present time on the East Coast, three large-scale development projects are under way or being planned and, as on the West Coast, they will have cross-sector effects on the marine ecosystem and the people of coastal Labrador. There is now nickel mining at Voisey's Bay, offshore oil and gas exploration on the Labrador Shelf, and new hydroelectric development on the Churchill River (figure 6.8). Given time and cost restrictions, we selected the Churchill River development for investigation, in order to cover yet another resource sector. Currently one hydroelectric generating station, which was completed in 1971, exists on the Upper Churchill River. No dam was constructed, but a series of dykes was installed to divert water, and hundreds of existing lakes and bogs in central Labrador were linked to form a reservoir from which water is released through control structures. The upper section of the Churchill River was diverted into a new course along a series of lakes parallel to the original riverbed, in order to supply the power plant (Bobbitt and Akenhead 1982), with excess water being released through a spillway and allowed to flow along the original route.

This power development could change the circulation within Lake Melville and Groswater Bay (figure 6.8) and therefore the coastal ecosystem. Comparisons of data from before and after the hydroelectric development (Bobitt and Akenhead 1982) have shown that the salinity in the upper thirty metres of Lake Melville has increased by one to two parts per thousand (ppt), the result of the lower discharge rate from the Churchill River altering the flow from Lake Melville into Groswater Bay. Fishers have expressed concern ever since the Upper Churchill River was diverted, and some believe a local cod stock in Groswater Bay may have been affected. Overfishing is, however, an alternative explanation for the disappearance of coastal cod in the area, since the decline in the fishery occurred during a period of severe overfishing along the whole Labrador coast, although it is also likely that stock decline here, as elsewhere, is the result of more than one causal factor.

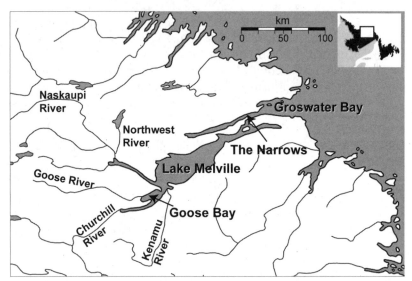

Figure 6.8   Three bodies of water that make up the Hamilton Inlet: Goose Bay, Lake Melville and Groswater Bay. Adapted from http://atlas.nrcan.gc.ca /site/english/maps/reference/outlineprov terr/nfld outline

Nonetheless, recent plans for hydro-power development of the Lower Churchill River have renewed fishers' concerns.

We (Cardoso 2004) therefore constructed a computer model to test the hypothesis that regulation of the discharge of the Churchill River had resulted in significant change in the freshwater-driven component of the estuarine circulation and in the entrainment of nutrients into the marine ecosystem of Groswater Bay. The large volume of runoff in spring may have been important to the seasonal dynamics of the plankton and (through food chain dynamics) production of coastal cod. The model would therefore determine whether fishers' concerns about hydroelectric development on the Churchill River are reasonable or not. From our modelling simulations, we concluded that there are factors in addition to regulation of river flow that can strongly influence plankton dynamics in Groswater Bay. Such factors include annual variation in winter ice cover and far-field ocean climate variation (measured by the North Atlantic Oscillation index). It is therefore difficult to determine whether changes in plankton dynamics are caused by natural environmental variation or regulation of river flow for power generation.

When we were modelling the regional fish production, it became clear that the harvest of fish off Labrador during 1968–92 was not sustainable ecologically. However, since the existing environmental data base for coastal Labrador is quite sparse, assessment of the environmental impact of hydro-power development is very difficult. Given the strong likelihood of new hydroelectric power development on the Lower Churchill River, we recommend establishing a time series of observations on biological production in Lake Melville–Groswater Bay and adjacent coastal areas before such development proceeds.

This, and the other case studies just discussed, make it clear that cross-sector implications of industrial activities have not often been on the radar of regulators. We are encouraged to note that the East Coast oil and gas industry is itself trying to deal with cross-sector impacts, but we would urge federal and provincial regulatory agencies to also start thinking this way and to learn from the co-operation that the One Ocean initiative has instigated. That said, we would further suggest that cross-scale issues may also be important in the interaction between resource exploitation, environmental impact, and community health, and we turn therefore to an exploration of alternative ways of thinking.

THINKING DIFFERENTLY: THE CULTURAL AND MANAGERIAL IMPORTANCE OF ABUNDANCE, BIODIVERSITY, AND KEYSTONE SPECIES

In chapters 3 and 4, we saw that little attention has been paid in industrial forestry to forest species other than economically important "fibre" species and in fisheries to fish species other than those directly targeted by commercial fisheries. In the case of fisheries, we identified large numbers of species that appear to be endangered, only some of which have been direct targets. In the case of forests, the replacement of old-growth forests – with their diverse age structures and multiple successional stages that had resulted from natural and indigenous disturbance regimes such as fire (e.g., Foster 1983; Lorimer 1977; Hunter 1993; Niemelä 1999; Turner 1999) – is having unforeseen consequences across sectors and at different scales.

In view of recent work recognizing many parallels and linkages between ecological and social systems (see chapter 1, and, most notably, Berkes and Folke 1998 and Berkes, Colding, and Folke 2003), it has become important, theoretically, conceptually, and from a social-ecological health perspective, to cross-reference those ecosystem and social features that share commonalities. In the broadest sense, then, just as there is a "biosphere" (the region of the earth's crust and atmosphere occupied by living organisms), there is an "ethnosphere," defined by Davis (2001, 8) as "the sum total of all thoughts, beliefs, myths and institutions made manifest today by the myriad cultures of the world." The term "ecological keystone species" refers to the

significant role played by certain top predators – such as seastars in natural ecosystems – whose presence can influence ecosystem integrity at
levels potentially greater than would be expected from their relative
abundance. As with ethnosphere/biosphere, the ecological keystone
species concept can be broadened to take into account the significant
role that some species play in particular cultures – that is why some
researchers now talk about cultural keystone species. By this they (and
here, we too) mean species that have a fundamental relationship to the
cultural identity of particular groups (Garibaldi and Turner 2004).

In this book, we focus on the restructuring of social-ecological systems. Within such systems, the abundance, distribution, and diversity
of both cultural (extending the metaphor) and ecological keystone
species are understood as the product of interacting processes that can
be triggered by physical, biological, cultural, economic, spiritual, and
other types of events. The profound impact that human societies have
come to exert within social-ecological systems, particularly industrial
societies, supports the contention that humans are the most important
ecological keystone species affecting processes on Canada's coasts
(O'Neill and Kahn 2000).

There are challenges associated with identifying ecological keystone
species. When attempting to do this, it is important to remember that
they tend to be context specific and thus that their importance can vary
across spatial, temporal, and social scales (O'Neill and Kahn). This
means that one individual species may exert quite different influences
at different points in history, during different seasons, or even in different ecosystems and at different geographical scales. We argue here
that, similarly, the cultural salience of an individual species can vary
across cultures, over the life cycle, and over the annual round of activities, as well as over the history of a people. Since in some contexts ecological keystone species may also be cultural keystone species, they may
have particular cultural relevance that stems from their nutritional or
commercial importance, their spiritual importance, or some other factor. This extension of the metaphor is useful, but not intended to imply
that there is exact equivalence between human social and cultural systems and those of the biological world. Both ecological and cultural
keystone species may be the target of intensive exploitation, while, by
contrast, cultural keystone species may be subject to protection and
cultivation. There are additional challenges associated with identifying
both ecological and cultural keystone species that live in marine habitats, since there is less chance to observe behaviour or changes in physi-
ology than there is for terrestrial species, and changes in population
size are even less apparent than on land. For pelagic organisms, including life-cycle stages of many benthic species, it is difficult to estimate

overall numbers or to demonstrate that a species has an impact greater than would be expected from its relative abundance (see chapters 3 and 14).

Cultural keystone species, then, are "culturally salient species that shape in a major way the cultural identity of a people, as reflected in the fundamental roles these species have in diet, materials, medicine, and/or spiritual practices" (Garibaldi and Turner 2004). Some examples of cultural keystone species in the context of the East and West Coasts of Canada are cod and capelin on the Atlantic Coast and cedar (*Thuja plicata*), salmon, and edible seaweed (*Porphyra abbottiae*) on the Pacific Coast. Because of their pivotal roles and cultural centrality in local and/or regional contexts, they can serve as a focus for analyses of the history of human occupation and landscape change, transmission of knowledge and practice, eco-cultural restoration and conservation, and the importance in such endeavours of matching the scale of human practice to an understanding of the various scales at which ecosystem components function. For example, focusing on the knowledge, practices, and changes that affect and are affected by harvesting and the use of edible seaweed by the Gitga'at of Hartley Bay on the West Coast or the use of Atlantic cod in the communities of Newfoundland on the East Coast can provide critical insights into overall change and community resilience in the face of social, economic, and environmental restructuring.

Marine Cultural and Ecological Keystone Species: A West Coast Example

A cultural icon on the West Coast that we choose as one of our two illustrative examples is kelp. The story of kelp (*Nereocystis luetkeana, Macrocystis integrifolia*) on the Pacific Coast is a classic example of a complex web of relationships between humans as keystone species, other keystone species such as sea otter, and a major component of marine vegetation in many ways equivalent to our terrestrial forests: the kelp "forests" of the subtidal coastal zones. The kelp forests provide critical habitat to many other species, both algal, invertebrate, and vertebrate, including salmon and many kinds of rockfish. Kelp beds also protect the shoreline from excessive wave action by serving as "breakwater" zones for beaches and rocky headlands. Kelp is a direct food source for many marine invertebrates, but primarily for sea urchins, which are well known as kelp browsers. The preferred food of sea otters, thick-furred marine mammals, is a variety of invertebrates, like clams, but especially sea urchins (Riedman and Estes 1990). Sea otters have had measurable environmental effects, for example, in creating a "hard bottom" on the ocean floor from clam shell pavement (Kvitek et al. 1992). For many thousands of years, sea otters were hunted by First Peoples all along the Pacific Coast. The European fur trade, however, created an insatiable market for sea otter furs, and as a result, over about 150 years of intensive hunting, sea otters became extirpated (locally extinct) up and down the coast, from California to Alaska (Simenstad, Estes, and Kenyon 1978). The result of this depletion was a burgeoning of sea urchins all along the coast, since one of their major predators had been removed. This, in turn, led to a rapid depletion of kelp because of over-browsing by sea urchins.

Complex as the story already is, it does not stop there. Beaches started to wash away, and intertidal life started to erode because of excessive wave action once the kelp was removed. Furthermore, the kelp forests were no longer there to harbour the young salmon and other fish that lived within this special habitat, and that resulted in further demise of these fish species, which were already suffering from the impact of habitat loss from logging around spawning rivers (see earlier, this chapter) and from other problems such as overfishing (chapters 4 and 8). The seals and sea lions that feed primarily on the salmon started to decline in numbers, and the orcas (killer whales, *Orcinus orca*), whose major food is seals and sea lions, also declined. Efforts to reintroduce sea otters to various places along the coast have met with only limited success, partly because of predation by orcas. Another problem is that sea otters also love to eat abalone (Watson 2000; Estes et al. 1998; Kvitek et al. 1992) and compete directly with humans, whose overharvesting of abalone and subsequent depletion of the stocks have been well documented. People are worried that sea otters will further deplete their abalone stocks, but nevertheless, they are now harvesting sea urchins for the Japanese market. This story is ongoing and the end results of these major ecological restructurings in our marine ecosystems are still uncertain. Among the very many lessons to be learned here are, not least, that keystone species are ecologically vital and also culturally vital (icons) and that damage to them is damage also to the cultural bedrock, which is one of the fundamental pathways along which human societal health is connected to environmental health.

Marine Cultural and Ecological Keystone Species: An East Coast Example

In most areas of Newfoundland and Labrador, since the arrival of Europeans and for much of the history of the province, Atlantic cod has been the most important marine cultural keystone species, and we choose it here as our second example. Historically, however, Atlantic salmon, seals, and whales might have been better candidates among

First Nations, while the Basques fished whale and cod along the Strait of Belle Isle and the Labrador coast. From the beginning of contact on, the main driver behind European interactions with cod was commercial. This is reflected in the cultural terms and understandings associated with the history of the cod fishery.[6]

If we look at the *Dictionary of Newfoundland English*, we find the term "fish" used to refer to cod, and we are told that dried cod acted as a form of currency, in that it was used as credit against the purchase of goods in the merchant stores that operated a truck system in Newfoundland in the period before Confederation. Cod was central to local proverbs like "as cold as a cod's nose," "cod-fish is cod by name and by nature," and "no cod, no cash" (Story, Kirwin, and Widdowson 1990, 104). In the dictionary, there are thirty entries for terms beginning with "cod", including references to the fish itself (cod, codd, codde, cod-fish), the general activity of fishing for cod (cod-fishery, cod-fishing), weather (cod-fish weather), fishing gear used to harvest cod (cod-net, cod trap, cod-line, cod-seine, cod jigger, cod hauler), fishing twine (cod-trap linnet, cod-trap twine), places where cod were harvested (cod-trap berth), time of the year (cod-trap season), infrastructure used to dry cod (cod flake), parts of the cod that were eaten (cod's head, cod sound, cod'[s] tongues), cod blubber, and commercial products produced from cod (cod oil) (104–8). There are also five pages of entries for "fish," most of which relate to cod.

It is culturally significant that almost all the terms for cod-related material culture and cod processing contained in the dictionary relate to the period before the shift from salt fish to fresh fish and from drying cod on stages and flakes to the production of frozen fillets of cod. Earlier terms may, however, have influenced later ones as, for example, in the case of fish-processing factories, the term for which in Newfoundland is "fish plant." The traditional meaning of the term "plant" in Newfoundland related to an area of the foreshore with structures for landing and curing fish (381). It appears to be derived from the term "plantation," which is what the Newfoundland colony was in its early days, when settlers (who had migrated from the mother country to form mercantile plantations for the export of staples back "home") were known as "planters." Many other entries in the dictionary relate to terms for cod of different sizes (tom cod, rounders) and colours (red cod, tally fish), to other codfish and fisheries (bank cod, bank fish, bank fisherman, banks fishery, bank hook, trap skiff, trap buoy, trap fish, trap lead, trap loft, fall fish, winter fishery) and to other activities related to processing and handling cod (fish-beach, fish- beam, fish-book, fish-loft, fish-maker, flake, throater, yaffling, washing tub). Cod was also a very important nutritional staple during the fishery and during the winter as well, although once again, its importance would have varied from place to place in Newfoundland and Labrador. Economically and culturally central to most fishing communities, cod and its various parts (tongues, cheeks, sounds, britches [roe]) were basic foodstuffs for Newfoundland meals.

The cod fishery also features prominently in Newfoundland songs, and there are some indications of a spiritual element in the relationship between cod fishers and the cod themselves. This relationship was sometimes closely mediated by Christianity with, for example, "lady day," and "lady day fish," a day in August that was commonly the last day of the summer cod fishery from which the catch would be given to the church (294). The term "breach" refers to "cod, caplin, etc., to ruffle or break the surface of the water"; "chop" refers to codfish "slapping the surface of the water with their tail and back extremities when swimming in schools or enclosed in a net" (61, 94). These actions were considered to be signs of abundance. A fish forgotten in the boat when unloading was referred to as a "witch" and was seen as a sign of bad luck; a "loader" is a "codfish with a blunt head considered to be a sign of a good catch to follow" (308). Another indicator of the cultural importance of cod within Newfoundland culture is that people and their activities were sometimes described using terms for cod. For example, the term "harbour tom-cod" means a small immature codfish, but it is also used to refer to someone who is foolish.

Are cod an ecological keystone species? Cod, particularly large cod, are top predators in marine ecosystems in Newfoundland and Labrador. There are also strong indications that the removal of cod, including the particularly large old cod known as "mother fish," from Newfoundland and Labrador ecosystems has contributed to other changes in these ecosystems, including increases in the abundance of such invertebrates as shrimp and snow crab. However, such changes by themselves would probably not be sufficient to confer upon cod the status of keystone species.

Atlantic salmon may be another candidate for cultural keystone species in Labrador. Salmon and Atlantic cod were the two most heavily fished species until recent moratoria and the development of crab and shrimp fisheries there. During our interviews with retired fishers, we found local names for two or three annual runs of salmon. These names refer to coloration, for example, blue, green, or gray, but they also signify differences in size of fish, relative abundance, and timing. While different runs of salmon were described for various areas on the island, they were rarely named as in Labrador. As with cod, the candidacy of salmon for ecological keystone status is unclear. In the *Dictionary of Newfoundland English*, there are about two pages of entries starting with salmon. Many are taken from Cartwright's journals from Labrador.

Seals, particularly the harp seal, are another candidate, although the particular type of seal, its role, and relative cultural and economic importance have varied from region to region and over time. As with cod, the centrality of harp seals to the local culture (particularly in the past) is evident in the large number of terms used locally to refer to harp seals (we found seven names for harp seals in our interviews with retired fishers), seal behaviour and habitat, seal hunting, and objects of the material culture. Entries beginning with "seal" comprise seven and a half pages of the *Dictionary of Newfoundland English*. There are terms that describe the colour, condition, and markings of harp seals at different stages in their life cycle. These include white-coat, bed-lamer, ragged-jacket or turner, saddler, or saddle-back. Other terms relate to seal behaviour and related habitat (seal hole, whelp, whelping ice, whelping ground) to sealing activities and technologies (swile, seal patch, seal shot, seal gun, seal twine and seal nets, seal dog, sealing gaff, and capstan), occupational health risks (seal finger) and seal products (seal soap, sealskin boot).

Harp seals are also a candidate for the status of ecological keystone species. Many fish harvesters believe that the large numbers of harp seals and growing abundance of grey seals (*Halichoerus grypus*) in Newfoundland and Labrador waters, and harbour seals in river estuaries, is playing a critical role in the slow recovery of cod stocks and the limits on salmon populations. The total harp seal population was estimated as 5.3 or 5.5 million individuals for 1998, depending on the mortality rates used for pups. Harp seals off the east coast of Newfoundland were estimated to consume 1.2 million tonnes of Arctic cod (*Arctogadus glacialis*), 620,000 tonnes of capelin, and 88,000 tonnes of Atlantic cod in 1994 (Stenson, Hammill, and Lawson 1997). Scientists estimate 445,000 tonnes of capelin, 20,000 tonnes of Arctic cod, and 54,000 tonnes of Atlantic cod were consumed by harp seals in the Gulf.

Similarly, it is commonly believed that capelin play a critical role in marine ecosystems of Newfoundland and Labrador, although whether their influence is unusually great relative to their abundance is unclear. Harvesters, particularly those not involved in the mobile-gear capelin fishery, have consistently expressed concern that the capelin fishery is jeopardizing marine ecosystems in Newfoundland and Labrador. Fisheries managers, recognizing the importance of capelin as prey for a number of important species, set conservative exploitation rates at a maximum of 10 percent of the spawning stock biomass (Carscadden, Frank, and Leggett 2001). In the scientific literature, capelin are often presented as being at the centre of food webs for the North

Atlantic, since many seabirds, fish, whales, and mammals feed extensively on capelin (Jangaard 1974). At historically low population sizes
in the late 1990s, cod were estimated to consume 100,000+ of capelin
annually, down from 2,500,000+ in the 1960s (Carscaden, Frank, and
Leggett 2001). Rose and O'Driscoll (2002) suggested that cod rebuilding is directly dependent on capelin. Bowering and Lilly (1992) estimated that Greenland halibut consumed between 100,000 and
200,000 tonnes of capelin annually during the 1980s. These estimates
may be currently applicable, as stock size in 2000 was comparable to
levels in the 1980s (Carscadden, Frank, and Leggett 2001).

Capelin do not, at first glance, appear to be a good candidate for a
cultural keystone species, at least in a widespread sense. On the other
hand, although respondents from our interviews referred to bay and
sea capelin and used only two other names for them, there are capelin
festivals, capelin songs, and capelin art; families go to gather capelin
each year. So we conclude that in some localities, they would function
to some degree as cultural keystones.

Eelgrass (*Zostera marina*), a slender subtidal flowering plant, is
hypothesized to be an ecological keystone species in some coastal areas
of Newfoundland, as well as in many other regions. Eelgrass meadows,
occurring in offshore sheltered shorelines at depths from one to five
metres, are prime habitat for scallops, juvenile populations of Atlantic
cod, striped bass (*Morone saxatilis*), haddock, and winter flounder
(*Pseudopleuronetes americanus*), providing both food and shelter against
predation for these species. Peterson (1918) presented calculations
that suggested cod and plaice off Denmark depended on eelgrass.
Smaller-scale studies have shown that eelgrass influences the distribution of fish in coastal waters. It is known to reduce predation risk for
juvenile cod (Linehan, Gregory, and Schneider 2001; Laurel, Gregory,
and Brown 2003a). Also, eelgrass promotes high growth rates of Atlantic cod (Tupper and Boutilier 1995). Juvenile Atlantic cod, Greenland
cod (*Gadus ogac*), and hake (*Urophycis tenuis*) were found to be associated with eelgrass (Ings et al. forthcoming) more than any other
habitat type found in Newfoundland coastal waters. In habitat manipu-
lation studies where eelgrass was removed, Laurel, Gregory, and Brown
(2003b) subsequently observed a decline in the number of juvenile
Atlantic cod and Greenland cod.

Ducks and geese also feed on eelgrass, and the dead eelgrass plants
that wash ashore form windrows of decaying wrack on the beaches that
shelter sand fleas and other small crustaceans that are the major food
of many kinds of shorebirds. When the eelgrass microbial "wasting disease" wiped out more than 90 percent of the eelgrass on both sides of
the Atlantic, the effect on waterfowl, fish, and shellfish was drastic. A

number of species declined, although the effect was not as large as Peterson predicted (Milne and Milne 1951). However, the eelgrass limpet (*Lottia alveus*), which depended on eelgrass as habitat and a food source, has not been observed since the mass die-off; it is now considered extinct (Carlton et al. 1991). Brant geese (*Branta bernicla*) populations suffered a drastic crash because they relied on eelgrass roots, stems, and leaves as winter food. After the eelgrass disease of the 1930s, Brant geese were observed to change their foraging habits, with increasing reliance on other food sources, including sea lettuce, culti-vated grass, and cord grass (*Spartina* spp.). Only recently have Atlantic Brant populations started to approach the levels observed before the eelgrass die-off. Scallop populations also dropped precipitously, as did many inshore fish species that lost their protective cover and foraging grounds.

In Newfoundland and Labrador, fish harvesters are often aware of eelgrass (which they sometimes refer to as "goose grass"). However, they do not appear to be particularly aware of the ecological impor-tance of eelgrass, and from a conservation perspective, there are lim-ited mechanisms in place to protect eelgrass beds, and scientific knowledge of their location and of changes in such beds over time in the region is unfortunately quite limited. Similar destruction and lack of attention to the ecological importance of eelgrass is evident on the Pacific Coast, where dredging, log booms, oyster farms, and other activities have been detrimental to eelgrass beds and, ultimately, to the life forms they support (Scientific Panel for Sustainable Forest Prac-tices in Clayoquot Sound 1995b; Wyllie-Echeverria et al. 2003).

CONCLUSION

In this chapter, which concludes this part of the book, we have explored cross-scale and cross-sector impacts of human natural-resource indus-tries on the social-ecological systems of the coasts, as well as exploring a different scale-matching and sector cross-referencing rubric that is deeply rooted in the cultures of the two coasts. In the first part of the chapter we identified yet another lack of environmental concern exhib-ited by governments and industries whose perspectives of purely eco-nomic efficiency contrast sharply with the knowledge, understanding, and appreciation of local species that coastal peoples, First Nations, and white settlers express in their cultures. The human costs for coastal peoples of the changes that we describe in this book are signifi-cant, and we focus on that issue in more detail in the next part of the book (chapters 7 to 11). When we add the evidence presented in the three case studies here to the evidence in the previous chapters, it becomes abundantly clear that not enough attention is paid to the environmental as well as the social (cultural, dietary, and economic) impact of industrial practices in coastal areas (see Parrish, Turner, and Solberg 2007). Higher levels of government and many resource-based industries are usually geographically, culturally, and ethically removed from the remote coastal environment, being for the most part urban-based even when coastal. Perhaps that is why the practices and regulations of government and industry permit the kind of damage we have demonstrated in preceding chapters and in this one as well. They certainly seem unaware of the dangers of their practices until such behaviour downgrades the resources and the ecosystems of which these resources are a part.

We contrast all this with the long-term (but now declining) resilience and viability of the community cultures that are embedded in ecosystems and whose sophisticated understanding of them has resulted in practices that are both scale-appropriate and sector-sensitive.Our discussion of cultural keystone species and abundance and diversity points to the importance of such local cultural-resource practices, which could be an important feature of co-management as we discuss that in chapters 14 and 15.

Taken together, chapters 2 to 6 – the "historical natural resources" chapters – demonstrate that the process of social-ecological degradation has indeed escalated over the past century (and sometimes longer). We have found also that that process of degradation has been masked by discipline-based science, by silo-based policy-making, and by the dynamics of restructuring within and across sectors, which has manifested itself in both the expansion and intensification of exploitation. It seems, moreover, that provincial and federal policies have been more concerned with macro- and meso-level short-term revenue gains than with micro-level long-term security. In other words, the legitimate needs of local environments and communities (which have provided those revenues and have proved surprisingly resilient in the face of consistent neglect) have not been built into the policy calculus of the upper-level orders of government. Interestingly, more concern has been expressed, at least in recent times, around social-ecological issues in the nonrenewable sectors than in the so-called renewable sectors, where the possibility of destruction of natural resources is only now coming to the forefront of policy concerns, as resources actually collapse. It remains to be seen, however, how communities have managed to endure, given that resilience in policy-making has been focused on the sustainability of industrial concerns rather than sustainable communities, stewardship, and resource recovery.

In the following chapter, therefore, we turn to our investigation of the impacts of industrial practices on the health of the people who live in these communities. The links are so close between ecosystem and human health that we need to talk about social-ecological health as we track the impacts of industrial kinds of resource usage and regulations on coastal resources and, through them, on the well-being of coastal peoples.

PART TWO

The Human Impact of Restructuring and Social-Ecological Health

7

The Restructuring of Health Care on Both Coasts since the 1980s

In this chapter, we examine the restructuring of the health care systems of our two provinces in the period since the 1980s.[1] We have chosen to focus on the recent period because it coincides with the phases of the most intense economic, social, and ecological restructuring on both coasts. It would, however, be incorrect to assume that before the 1980s the health care systems of British Columbia and Newfoundland were models of stability and persistence. Nothing could be further from the truth, particularly in Newfoundland, where the early years of the Newfoundland health system were marked by persistent change and frequent transformations.[2] Key turning points were a royal commission in 1929 leading to a significant reorganization in 1931, the challenges and opportunities created by the Second World War and the massive presence of US military forces, entry into Confederation in 1949, a second royal commission led by Lord Brain in 1966, and, shortly thereafter, the formal creation of "medicare." Linda Kealey has found (in work to be published elsewhere) that this turbulent history featured three distinctive themes: a recurrent tension between local approaches and external, largely colonial, influences in shaping the system of health care provision; a shifting balance between centralization and decentralization, between administration on a province-wide basis pivoting around the institutions and expertise of St John's and provision of services based on local initiatives and knowledge; and, finally, a persistent tension between the insistent pressures for increased economic outputs, efficiency, and employment, on the one hand, and the challenges of providing healthy, safe lives for the residents of isolated coastal communities, on the other. As the current chapter makes clear, while the first of these three dialectics has faded from view in recent years, the other two have continued to shape the evolution of health care systems into the late twentieth and early twenty-first centuries, not only in Newfoundland and Labrador but in British Columbia as well.

What are the various factors that have placed the two provincial health care systems under strain and what attempts have been made by the two provincial governments to define and study the stresses faced by their health systems? We looked at these issues, as well as the various attempted policy responses (both financial and administrative), and the impact of the ensuing reforms on health care provision. As in Newfoundland and Labrador, so also in British Columbia (and indeed in Canada as a whole) the health care system has been under considerable stress over the past several decades, and nowhere has this been more the case than at the level of remote coastal regions and communities. Present-day stresses are due partly to the sorts of demographic, economic, and ecological shifts discussed in the rest of this volume and partly to shifts in medical technologies and standards of care. In all places, there has been a dramatic increase in the cost of health services and in public and private health spending. Burgeoning health expenditures have forced provincial governments to undertake or commission repeated studies of the challenges confronting their health care systems and to search for strategies to meet these challenges.

As in other Canadian and international jurisdictions, Newfoundland and Labrador and British Columbia have come up with a wide range of policy responses, ranging from budgetary cuts and the redeployment of staff to full-scale attempts to reconfigure administrative authority over provincial health services. In these two provinces, as almost everywhere else in Canada, the most drastic version of this administrative redesign has been the organizational response labelled "regionalization," which manifests itself in practice in a combination of an *upward shift* of authority from individual local hospitals to regional boards and a *downward shift* of decision making to these boards from the provincial health ministries. In coastal communities, regionalization has been experienced as amalgamations and/or dissolutions of administrative boards, the shifting of administrative boundaries, the closure of smaller clinics and hospitals, and the transfer of personnel and responsibilities. As well, both provincial governments have, through various means, displaced the costs of a number of formerly publicly funded health services onto citizens, community groups, and private health insurance plans. The extent and the impacts of these changes have varied across and within the two provinces.

HEALTH SYSTEMS UNDER STRESS

Health care restructuring in Newfoundland and Labrador and in British Columbia has been driven by two competing and unrelenting sets of pressures, one financial, the other a result of cultural expectations developed over the long term (as Kealey shows for one province). On the financial side, the most important factor has been the repeated effort by the federal government (late 1970s to late 1990s) to curtail its health care and social services transfer payments to the provinces. With the passage of the Established Programs Financing Act in 1977, the federal government stopped matching provincial spending in health care and education. This brought about the decline (from 1979–80 to 1995–96) of cash transfers as a proportion of combined provincial/territorial health care costs from a high of 28.4 percent down to 16.3 percent (Provincial and Territorial Ministers of Health 2000, iii). In 1995, even more drastic cuts were introduced with the passage of the Canada Health and Social Transfer (CHST) program, which replaced the previous system (of fluctuating transfers based on a percentage of overall spending) with a system of fixed and sharply reduced "block" grants. These reductions, particularly after 1995, put the health systems of all Canadian provinces under severe strain.

The impact of the cuts was exacerbated for British Columbia and for Newfoundland and Labrador, because they came at a time when costs within both provincial health care systems were escalating rapidly, as a result primarily of wage pressures and infrastructure maintenance expenses, as well as of new drugs, expensive new diagnostic equipment, and surgical procedures that created additional layers of expense without producing easily recoverable savings elsewhere in the system (Newfoundland and Labrador Health and Community Services 2002; British Columbia Ministry of Health Planning 2002). Rising public demand was a second source of pressure, fuelled in large part by medical advances like magnetic resonance imaging and genetic testing, which are continually reshaping definitions of appropriate care and generating additional demands and expectations for health services. The emergence of newer, more expensive drugs also fuelled citizen demands.[3]

In 1999, the federal government adopted a more generous approach to transfer funding through significant multi-year funding commitments made in the first ministers' "health accords" of 2003 and 2004.[4] Nonetheless, despite the injection of many billions of dollars in new federal transfers (and promises of more) most provinces, including Newfoundland and Labrador and British Columbia, continue to worry that they may still not be getting enough funds to sustain their health care systems over the long run. A 2000 report by provincial and territorial health ministers cites improved communications technologies and the "sense of entitlement" shared by members of the baby boom generation as important factors behind the *culture of expectations* surrounding health care (Provincial and Territorial Health Ministers

2000, 47). These growing expectations, they said, are leading to an explosion in the demand for home care, increased visits to doctors, additional diagnostic testing, demand for shorter waiting lists, and demand for advanced technology and surgical procedures.[5]

Problematic escalating costs and expectations were exacerbated by additional factors that impinge particularly on coastal communities. In rural areas, the distribution of the population over remote, relatively inaccessible locations makes it expensive to deliver services. In British Columbia, for example, some 4 million residents are scattered unevenly over 926,492 square kilometres of territory, and the Northern Health Authority (NHA) – which includes the North West Health Service Delivery Area (NWHSDA) – covers almost two-thirds of British Columbia's land area, yet is populated by only about 300,000 residents (Northern Health Authority 2004). Likewise, the Vancouver Island Health Authority (VIHA), which includes the Central Vancouver Island Health Service Delivery Areas (CVIHSDA) and the North Vancouver Island Health Service Delivery Area (NVIHSDA), covers all of Vancouver Island, the Gulf and Discovery Islands, and a portion of the mainland adjacent to the northern half of Vancouver Island but oversees health delivery to just 706,000 people, the majority of whom reside on the southern tip of Vancouver Island.

Another driver of costs is demographic change. On both coasts, demographic change has always been important, and we discuss it in detail in the next chapter. Here, we point out that overall population decline, out-migration of young families, and the aging of the population (see chapters 13 and 14) have placed new kinds of stress on the health care system and at the same time eliminated many traditional family supports for seniors and others requiring long-term care and supportive services. A final exacerbating factor, which is specific to East Coast rural communities, is the comparatively high level of health risks they confront. Unlike British Columbia – the healthiest province in Canada – Newfoundland and Labrador consistently ranks among the worst in the country for a number of risk factors associated with high levels of chronic illness, including physical inactivity, obesity, and high smoking levels among teens and young adults (Newfoundland and Labrador, Health and Community Services 2002, 2). The figures from our study area are a heightened version of general trends in the province. Twenty-five percent of people in the Grenfell region surveyed for the 2001 Canadian Community Health Survey smoked on a daily basis, compared to 21 percent in Canada as a whole. Fifty-four percent of Grenfell residents were categorized as overweight or obese according to international standards, compared to 45 percent of Canadians. In turn, the province has the highest proportion of deaths from circulatory disease and the highest proportion in the country of people diagnosed with diabetes (Newfoundland and Labrador, Health and Community Services 2002, 2). Indeed, at 8 percent, the percentage of Grenfell residents with chronic diabetes is twice the Canadian average.[6] When added to aging, the relatively poor health of the rural population continues to weigh the system down with additional demands at a time when traditional family supports are in decline.

Overall, then, reduced federal funding, accelerating costs, increased citizen demands, and serious demographic challenges have created an increasingly heavy health care burden that the governments of both provinces have had to shoulder. From 1983–84 to 2001–2 Newfoundland and Labrador's perennially cash-strapped government went from spending 23.6 percent of its total expenditures on health care to spending 31.3 percent. The sharpest and most sustained increases took place after 1996–97, the year the CHST and its associated cuts were first implemented. If calculated as a percentage of provincial GDP, the Newfoundland government's health expenditures were the highest among the provinces virtually every year from 1983–84 to 2001–2, although they have dropped somewhat since then as a result of the recent increase in GDP resulting from oil and gas activities (CIHI *Health Spending*).[7]

British Columbia, for its part, also had consistently high levels of health expenditures during this entire period. Since 1981, total health expenditures as a percentage of the provincial GDP grew steadily from 7.6 percent to 11.4 percent in 2003, despite a minor (3.5 percent) decrease in 2002–3. Even though federal Canadian Health and Social Transfers to British Columbia actually increased overall from 1993 through 2002–3 (BCMA 2004), health spending has come to represent 42.5 percent of West Coast provincial program expenditures, up from 31 percent in 1998. These numbers highlight the financial pressures under which health care on both coasts has been forced to operate. When combined with rising popular expectations and politically generated and media-stimulated debates on "sustainability," what has emerged on the coasts (as elsewhere in the country) is a persistent sense, which began in the 1990s and has intensified ever since, that the health care system, like the coasts themselves, is in crisis.

HEALTH SYSTEMS IN CRISIS: DIAGNOSES AND PRESCRIPTIONS

Both provincial governments have made numerous efforts to diagnose the problem and to propose remedies. Indeed, both provinces have created several commissions on health issues. In April 1983,

Newfoundland and Labrador (one of the first provinces in the country to declare its health care system to be in crisis) commissioned a study of problems and possible responses: the Peckford government's Royal Commission on Hospital and Nursing Home Costs, chaired by lawyer David Orsborn. Orsborn's mandate was to determine why hospital costs were rising faster than the rate of inflation and to recommend methods for achieving greater operating efficiencies. Health Minister Wallace House pointed directly to the declining rate of growth in Established Program Financing transfers from the federal government as one of the biggest problems facing the province (*Evening Telegram* 1983),[8] and later claimed that the federal contribution to health care costs in Newfoundland and Labrador had declined from 36 percent in 1979–80 to 27 percent in 1983–84 (*Evening Telegram* 1984). The royal commission found that the provincial cost of operating hospitals and nursing homes had risen from $65 million in the early 1970s to $370 million in 1983–84, driven by a 44 percent increase in the number of hospital personnel (from 7,335 full-time workers in 1974 to 10,527 in 1983–84), substantial increases in the salaries of nurses and other hospital staff, a 90 percent increase in the number of nursing-home beds (from 1,065 in 1974 to 2,058 in 1983), and a 30 percent increase in the number of specialist and non-specialist physicians.

The royal commission attributed the bulk of these increases not to any expansion in the quantity of services provided but rather to greater intensity of care, as measured by resources per unit of output (Royal Commission on Hospital and Nursing Home Costs 1984, 2–4). In all probability, the greater intensity of care reflected the change in standards of medical practice that required health care workers to spend more time and effort on each patient. New procedures, such as coronary bypass surgery and renal dialysis, also accounted for a substantial portion of spiralling costs. In addition to short-term measures such as freezing hospital budgets and closing beds, the Orsborn Commission recommended that institutions within defined geographic areas should co-ordinate and share administrative services as a means of improving the allocation and management of funds. In general, the commission felt that "the next few years will and should see a decreasing reliance on service delivery through autonomous, free-standing individual institutions in favour of innovative, more interdependent and cooperative organizational structures."

One decade later, this prediction was proved correct. In 1992, the Wells government appointed Lucy Dobbin, former executive director of St Clare's Mercy Hospital and of St John's General Hospital, to review the province's hospital boards. Against the background of a highly public battle with the major public-sector unions, rising demands from the private sector for debt and deficit reduction (*Evening Telegram* 1993), and a general climate of fiscal austerity, the government released Dobbin's *Report on the Reduction of Hospital Boards* in March of 1993. The same day, the premier announced that the province's public debt was $6.1 billion – the fourth-highest per capita debt among all provinces – and that in order to meet its deficit target and protect the province's credit rating, the Liberal government would cut $99 million from the next provincial budget (Loney 1993). The next day, Health Minister Hubert Kitchen explained that the rationale for the hospital board review had been to "look at ways we can best provide health care in regions with a limited amount of money" (Belec 1993). The Dobbin report had recommended that the province's sprawling system of fifty-odd independent hospital boards be consolidated into seven new regional boards, and within a month of its release, two regional health boards were created, in the St John's and central regions, setting the stage for consolidation of all boards throughout the province over the next three years.

Operational efficiency and the need to adopt new approaches to public management continued to be major themes of public health care inquiries into the new century. In November 2001, at the request of the Health Care Corporation of St John's (HCCSJ – far and away the province's largest health board), the Department of Health and Community Services hired the Hay Health Care Consulting Group to perform an external audit of the corporation's financial affairs. The consultants were given the dual task of devising a plan to balance the HCCSJ's operating budget in the short term and of identifying ways to trim costs and enhance revenues over the long term, in the face of a projected 2001–2 operating deficit of $6.6 million, which the corporation had managed to reduce from an original figure of $8.6 million on its own initiative (Cleary 2001).[9] The Hay Group recommended combining departments, eliminating management positions, improving clinical efficiencies by shortening hospital stays, and shifting some services to outpatient delivery.

That same year, the provincial government launched a series of regional health forums to gather advice from citizens, policy-makers, health care providers, and other professionals on issues relating to the structure of the health system and the delivery of services. One of the recommendations emerging from this process was that population densities and the regional availability of the required health care providers should be major considerations in deciding where to locate service delivery centres (Newfoundland and Labrador Health and Community Services 2001a). The forums advanced the notion of a "critical mass" of health care professionals required for effective service delivery in a given region and stipulated that a minimum of four physicians in any region were required to share on-call services. In this manner, they laid much of the conceptual groundwork for primary health care reform and other attempts at service consolidation. Partly on the basis of these consultations, the Department of Health and Community Services released its Strategic Health Plan in September 2002.

The plan set out broad directions for improving health services and population-health outcomes, including the possibility of re-examining the number of health boards with a view to removing artificial organizational barriers and administering the system with the greatest organizational efficiency (Newfoundland and Labrador Health and Community Services 2001a, 26). It is noteworthy that the various studies, commissions, and plans – while finding serious and multiple problems in the province's health system – made recommendations that remained entirely within the scope of the Canada Health Act. In sharp contrast to some other Canadian jurisdictions, including British Columbia, none of the studies conducted in the province over the course of two decades suggested moving away from Canada's traditional reliance on public funding and not-for-profit provision of health care or supported any shift towards greater reliance on the private sector and the market.

The recent history of health administration and policy in British Columbia features an even more formidable array of studies, commissions, policy papers, and reform proposals. The province first adopted comprehensive public health insurance in 1967 (British Columbia Ministry of Health Planning 2002, 5), and five years later, a full-scale evaluation of the British Columbia Health Programme was begun with the report *Health Security for British Columbians* (Clague 1997). Its author, R.G. Foulkes, noted the approximately 70 percent increase in total expenditures for the British Columbia Hospital Insurance Service between 1969–70 and 1973–74 (Foulkes 1974, 25), and suggested that (aside from British Columbia's comparatively high wage rates) spiralling health care costs could be attributed in large part to the structure of the health care system, which he portrayed as excessively bureaucratic, highly politicized, and often featuring poorly considered decision making (Foulkes 1973, II-3–11, IV-A-1). In fact, British Columbia's health services were organized along what was the standard model for health care administration at the time, with the provincial health ministry divided into four departments: the health services branch, mental health branch, hospital insurance service, and medical care program. Foulkes argued that this structure led to fragmentation, inefficiency, overlap, and a lack of coordination, and he recommended that the province move to a regionalized structure of health administration, through the decentralization, co-ordination, and integration of health authorities (II–3–11). In 1975, the newly elected Social Credit government shelved the recommendations of the report and continued to manage the health system as before.

The next attempt to examine and reform the province's health system came in the late 1980s, when the government established a royal commission, chaired by Justice Peter D. Seaton, to undertake a full-scale reconsideration of health care administration, financing, and delivery. Seaton's first report, the 1989 study *Toward a Better Age,* focused in part on long-term care. It recommended shifting the emphasis from facilities (hospitals and physicians) to home-based services, increasing the emphasis on preventive health programs, and allocating some resources to community-based caregivers, but increasing the availability of nursing-home beds in small communities, especially beds for persons with special needs (Brody, Simon, and Stadler 1997). Two years later, the commission's final report, *Closer to Home: The Report of the British Columbia Royal Commission on Health Care and Costs,* was presented to the newly elected NDP government. It endorsed the recommendations of the commission's earlier report with regard to care of the elderly but emphasized the more effective use of existing resources without incurring additional expenses (Brody et al. 1997). It also urged a shift of emphasis from institutional to community-based care and developed a set of principles to guide health care restructuring. People, it said, should feel like residents not patients; individuals residing in facilities should live near their families and communities; and the necessary number of long-term care beds should be available to avoid increased costs for acute care.

The commission's final report also devoted considerable attention to the specific challenges of rural and northern communities, particularly Aboriginal communities. It found that the combination of poverty, lack of education, and underemployment and consequent poor health outcomes was creating a dependence on social support that lowered self-esteem and increased family disorganization, substance abuse, and physical and sexual abuse. The report also found that British Columbia's northern rural/remote populations, whether Aboriginal or white, suffered a number of clear disadvantages and had much greater health care needs than the rest of the province. Northern communities had higher than average death rates (British Columbia Royal Commission on Health Care and Costs 1991, C24–5), higher percentages of children under fifteen, higher proportions of people with less than a grade nine education, below average family incomes, higher rates of underweight births, and higher infant mortality.

The above-average health needs of northern and rural communities were accompanied by a severe shortage of health care workers, thus imposing high costs on people who had to travel at their own expense to receive specialized services. That then influenced the rate at which people could be returned to their homes after hospitalization. The north also had the highest rate of utilization of acute and rehabilitation care: 1,200 days per 1,000 inhabitants, compared to a provincial average of 800 days (Brody, Simon, and Stadler 1997, 29–40). It was also reported that the funding base of rural community hospitals was inadequate, a result of staff shortages and people travelling to major centres for higher-risk procedures. The commission therefore recommended that a rural-hospital transition grant program be established to help these institutions become integrated health care centres (British Columbia Royal Commission on Health Care and Costs 1991, A3–A16). It also recommended enhanced recruitment efforts to attract more rural people into health professional training programs and further advised the ministry to move a portion of medical internships and residencies to rural areas, suggesting the funding of rural specialty training in obstetrics, surgery, anaesthesiology, nursing, and occupational and speech therapy. Health care workers, it urged, should be better prepared for work in rural practices, commenting that measures should be taken to reduce excessive workloads for northern health care workers, along with social and professional isolation. It suggested that this be achieved through increased opportunities for continuing education.[10] Finally, it urged the Ministry of Health (MOH) to copy two programs that had proved successful in Ontario and Newfoundland: Ontario's Underserviced Area Program for recruiting and retaining health professionals in rural areas and Newfoundland's Telemedicine Program, which linked remote areas to larger urban or neighbouring rural centres.

In 1993, the government responded to the broader recommendations of the royal commission with a policy statement entitled *New Directions for a Healthy British Columbia*. This statement proposed shifting the health care system from a physician-centred model with an emphasis on illness toward one that emphasized health for both individuals and communities, and it set in place five "new directions," in the form of broad policy principles that were intended to transform health delivery into a responsive, comprehensive, and integrated system (British Columbia Ministry of Health and Ministry Responsible for Seniors 1993, 2). The government also took up the Royal Commission's challenge to develop explicit health goals for the province, and following extensive consultations in 1994–96, these goals were developed and then approved by the government in 1997. *Health Goals for British Columbia* established six overall population health targets and forty-four related objectives to be adopted and implemented by selected provincial ministries. The ultimate intent was to minimize future financial pressures related to the aging and growing population and the development of new illnesses and to maintain and improve the health of British Columbians by enhancing quality of life and minimizing inequalities in health status (adapted from AGBC 1997–98, 21). These goals and the forty-four population health objectives associated with them were designed to cover all the determinants of health and also to guide the shift away from the bio-medical model towards a greater emphasis on wellness, prevention, and health promotion (British Columbia Ministry of Health and Ministry Responsible for Seniors 1997, 7).

Finally, the *New Directions* document recommended regionalization of the health care system as part of its emphasis on accountability, improved public administration, equity, and democracy (Larson 1997, 1–81; Davidson 1999, s36). To increase accountability to local communities, the commissioners had recommended the creation of twenty regional health boards (RHBs) and eighty-two community health councils (CHCs), and they were created the following year with the passage of the Health Authorities Act. In addition to reconfiguring the regionalized structures, the new policy framework called for a "Better Teamwork" approach, which shifted emphasis away from empowering communities and from making health organizations accountable to citizens and towards a more managerial approach, emphasizing political and fiscal accountability to the ministry and the government.

In 1998, the Provincial Co-ordinating Committee for Remote and Rural Health Services (PCCRRHS) was formed to share information and ideas on health care delivery in remote/rural communities. *Enhancing Health Services in Remote and Rural Communities of British Columbia* (1999) emphasized primary health care services, Aboriginal health care, and mental health services.[11] In spite of recent efforts and expenditures, the report uncovered a widespread belief that rural areas were not receiving equitable treatment. The evidence supported that belief, since the north lacked sophisticated diagnostic equipment, severe problems persisted in Aboriginal health (including a twelve-year disparity in life expectancy between First Nations and the general population of the province), and communication between health workers and First Nations was compromised because of cultural and language barriers. Moreover, medical and physician services suffered because of a lack of co-ordination, and the provision of mental health services was hampered by the lack of physicians trained in psychiatry (Provincial Co-ordinating Committee for Remote and Rural Health Services 1999, 4–5). In its submission to the provincial health department in July

2000, one regional board (the North West Health Authority) provided a detailed analysis of the deficiencies of health status and health services in its largely rural part of the province. It reported severe human-resource gaps, the absence of important programs, and the prevalence of serious health challenges, especially among Aboriginal populations.[12] The authority recommended a number of specific changes and warned against adopting the ministry's plan to introduce a population-based funding formula, since it would be inappropriate for the northwest's sparse population, vast geographical scope, and numerous remote communities (North West Health Authority 2000, 50–2).

In 2001, the new Liberal government mounted yet another round of health care rethinking. They argued that the health system needed change because under the existing system, health care services were fragmented, uncoordinated, and incapable of meeting the expectations or needs of the public. Acute care services were spread too thinly, and there were few care options available for the elderly. In November 2002, *The Picture of Health: How We Are Modernizing British Columbia's Health Care System* announced a new set of "core values" substantially different from those of the previous regime. Rather than emphasizing wellness, prevention, and population health, the health system was to concentrate on "core business areas," to produce health care delivery in the most patient-focused manner possible, with an emphasis on equity, access, effectiveness (through yearly performance contracts for all health authorities), appropriateness of service and location, cost-effectiveness, and safety (British Columbia Ministry of Health Planning 2002).

These core values represented a fundamental shift in orientation from a focus on population health to a corporate approach to health care delivery, emphasizing corporate accountability and fiscal responsibility. The MOH established three goals for the health system: (1) high quality patient care; (2) improved health and wellness for British Columbians; and (3) a sustainable, affordable health care system (adapted from British Columbia Ministry of Health Services 2004, 16). To this end, health policy was focused on developing and managing a network of health services, including acute care and diagnostic and treatment centres. The needs of the chronically ill were to be addressed and effective health promotion programs and policies were to be developed.

POLICY RESPONSES

Financial and Administrative Responses

In the face of persistent financial pressures and an atmosphere of crisis generated by national and provincial media coverage of a string of public hearings and recommendations, the governments and health administrations of both provinces sought to impose financial restraint of various sorts. In Newfoundland and Labrador, efforts at restraint were begun early. Following the recommendation of the Dobbin Commission, the government froze hospital budgets for a period of two to three years and closed roughly half the 303 acute-care beds recommended for closure by the commission. By the end of the decade, these cuts had created critical problems related to aging equipment and inadequate staff levels, which led in turn to low staff morale, high turnover, and high levels of sick leave (NHNA 1993, 4). The government increased its 1990–91 budgetary allotment for health care but announced later in the year that health spending would again be frozen in the 1991–92 budget, with no adjustments for inflation, resulting in a reduction in real funding of approximately $60 million.[13] By 1994, the five main facilities that would eventually comprise the Grenfell health region[14] had lost 46 beds out of a total of 167.[15] This loss included a one-time reduction in 1990–91 of 38 beds at Curtis Memorial Hospital in St Anthony, the major provider of referral services for area residents. At the provincial level, the cuts resulted in a reduction of acute beds per 1,000 population to 3.58 (Botting 2000, 15–16). This approach, in combination with a wage freeze and deferred implementation of a court-imposed pay equity settlement for female employees, did produce the requisite cost savings, but it also produced additional pressure on staff and further travel requirements for health system clients in coastal communities. Work intensification for existing employees, a shift to day surgery and outpatient delivery, delisting of certain medical services, outsourcing of nonmedical services to private companies, and the hollowing-out of middle management followed. In rural areas the impact of these changes was particularly acute.

This extended phase of budget cuts coincided with the implementation of the Dobbin commission's recommendation that the administration of provincial health services be reorganized on a regional basis. During the mid-1990s, the system's fifty-odd independent hospital boards were consolidated into fourteen regionally based boards (see next section). Since regionalization and financial austerity coincided, it is very hard to separate their impacts, so we treat the impacts of regionalization and those of the austerity drive together, as part of a broader process of "restructuring." Restructuring was most pervasive in St John's, where almost half the province's population lived and where specialized services and equipment were concentrated. Eight acute-care facilities were consolidated in the capital region in 1995 under the Health Care Corporation of St John's, and organizational and administrative functions were integrated in the first year, followed by integration of clinical services across sites in 1996. Management costs as a percentage of total employee compensation dropped from 12.8 percent in 1995–96 to 7.0 percent in 2000–1 (Barrett et al. 2003, 16). Administrative consolidation of several acute and long-term care facilities and isolated clinics also occurred in the other regions of the province (1994–96). The result was a major reduction in management staff in each case, such that from 1995–96 to 2000–1, the weighted average of management costs as a proportion of total employee compensation in the five sites declined from 10.3 percent to 8.1 percent. However, rationalization of clinical services did not always occur, largely because of geographic challenges (12). For example, the number of health care facilities in the Grenfell region – one hospital, three community health centres, and five coastal nursing stations – remained unchanged.

These various changes took place as the overall number of acute-care beds in the province declined from 1,893 to 1,601, a 16.4 percent drop in St John's and a 14.5 percent decline in the rest of the province. The Grenfell region lost another 57 beds during this period, including 47 at Curtis Memorial in St Anthony alone.[16] Total acute-care days show a gradual decline outside St John's and a less consistent downward trend within the city. By 2002, the number of acute-care days had dipped by 19.8 percent in St John's and 14.5 percent in the larger hospitals outside the city (Barrett et al. 2003, 17). In part, this change reflects the success of targeted efficiency measures, especially in the St John's region, which reduced the number of unnecessary hospital days due to delays in surgeries, diagnostics, and inefficient medical management (18), but it also reflects a shift toward day surgery and outpatient delivery of services. Moreover, Botting points out that it also entailed a transfer of costs from the government to patients and families, since certain services and drugs are insured only inside the hospital system (Botting 2000, 41). Notwithstanding the cuts, acute-care expenditures across all provincial regions rose during the 1995–2001 period by almost 10 percent per year, a rate far exceeding the growth rate of GDP or of government revenues (Barrett et al. 2003, 15–16).

Although changes in human resource expenditures had a tremendous impact on the overall cost structure of the system, a major part of the hike in spending levels can be attributed to negotiated wage increases and increased overtime. During the period under review, overtime costs as a percentage of regular personnel expenditures increased from 2.9 percent to 10.8 percent for nurses in St John's, while the weighted average for the other five institutional boards went from 1.7 percent to 7.1 percent. For hospital support workers in St

John's, the equivalent figures were 2.4 percent and 7.1 percent, respectively, while overtime costs in all the other sites went from 1.5 percent to 3.9 percent. Except for non-union, non-management employees, this indicator gradually increased at all sites over the study period (Barrett et al. 2003, 15, 20–1).

Interestingly, restructuring did not involve any great decrease in staffing levels across the province. Newfoundland and Labrador is the exception among Canadian provinces in that the number of registered nurses (RNs) employed in its health care system relative to the size of the province's population has grown substantially over the past decade (elsewhere the ratio of employed RNs per 100,000 population has declined since 1993; CIHI 2004a, 196–8),[17] with the biggest increases occurring in the rural parts of the province (CIHI 2002, 31–3), probably mostly as a consequence of the rapid pace of rural depopulation. Indeed, though the ratio of nurses to population has grown by 15 percent since 1993, the total number of nurses employed in the province has increased by only 6 percent and now sits at about 5,442. The province also experienced modest growth in the number of radiation technologists and was one of only three provinces that experienced positive growth in the numbers of laboratory technologists over the ten-year period from 1993 to 2002 (CIHI 2004a, 92, 110). There was no increase in the number of casuals in the nursing workforce: after rising substantially at first, the percentage of casual nurses in the province's health care system ultimately declined from 16 percent to 10 percent over the period 1993–2002, whereas the percentages in Canada as a whole were 14 percent and 12 percent respectively. It would appear, then, that the government has bucked the trend in much of the rest of the country by maintaining or slightly increasing staff levels in the rural parts of the province, even though population there was falling dramatically.

Restructuring has, however, had a less visible impact. Though the Department of Health and Community Services has publicly committed itself to public delivery of health services, a certain amount of what might be called privatization-by-stealth has occurred. Since the late 1980s the government has delisted some services – such as vision care – that were once covered under the provincial medicare plan and has started billing for services for which patients were generally not charged in the past, such as medical examinations for employment or driver's licences, advice over the phone, absent-from-work forms, and some dental surgery, such as wisdom tooth removal (Botting 2000, 31–3).[18] The government has also contracted out a significant proportion of the management and delivery of non-medical services in hospitals, such as dietary, laundry, and housekeeping functions and food services, while outsourcing has entailed a reduction in the number of full-time equivalent positions through attrition, as well as through reduced hours and wages for temporary workers (Botting 2000, 47–8).

In British Columbia, the 1990s witnessed intense financial and organizational change for the health care system. From the mid-1980s to the early 1990s a recession (generating fiscal deficits and high borrowing costs) limited the growth of hospital budgets. This was reflected in declining inpatient capacity, outsourcing of services, closures of beds and institutions, and increased reliance on physician office visits as compared to hospital treatment. Some aspects of care, particularly long-term care, were off-loaded onto the sick, their families, and local care networks in communities (McGrail et al. 2001). Between 1993 and 1997, the number of acute-care beds in the province decreased from 11,265 to 8,847, a cut of 11.5 percent in four years (Penning et al. 2002, 1–39; Sheps et al. 2000). Between 1991–92 and 1996–97 the number of beds in short-stay units, the number of acute-care days per 1,000 population, and the average length of stay declined by 30 percent, 29 percent, and 13 percent respectively (Sheps et al. 2000).

The patterns of change in human resources were more varied. Between 1993 and 1997, the number of physicians and practising registered nurses (RNs) increased, but the number of registered practical nurses and licensed practical nurses (RPNs and LPNs) dropped.[19] By the end of the decade changes to health services budgets and administrative structures had alienated health workers and had plunged the province into yet another round of health care restructuring and crisis.

In 2001, the government introduced legislation that mandated balanced budgets by the 2004–5 fiscal year, leading to proposed cuts in services and a number of reductions in both facilities and acute-care beds. In a new phase of government restructuring, the new health authorities (HAS) were asked to reduce expenditures, collectively, by $550 million over three years. The major thrust of these changes was linked to amalgamation of the old system of regional health boards (RHB) and community health centres (CHC) in an attempt to maximize organizational efficiency. The new round of restructuring was supposed to address operational inefficiencies, with new high-tech diagnostic equipment, specialists, and some acute-care services. These would now be concentrated in major metropolitan areas, chiefly Vancouver, with some smaller hospitals being converted into outpatient or community care centres.

In practice, however, some 1,500–2,000 acute care beds were to be cut provincially (with some being converted to subacute beds) and eleven community hospitals were to be closed or converted into community health centres. Closure of ten to fifteen long-term care facilities had also been proposed, along with a planned reduction of 1,000 long-term care beds. Additional cuts to programs such as maternity and pediatric services, closure of some cancer care beds, and staff cuts for psychiatric and mental health wards had also been recommended (all numbers taken from Caledon Institute of Social Policy 2002, 1–2), while lengths of stay in facilities were to be shortened and some patients were to be moved from inpatient to day surgery (Ireland 2002). On top of these institutional cuts, plans have been unveiled to delist some medical services including physiotherapy, chiropractic, and naturopathy. Administrative budgets are to be cut by 45 percent over three years (delisting and cuts from Moulton 2002), and there is also a planned elimination of some 6,500 full-time health care workers, with the majority of the lost jobs being transferred to contractual service providers (Caledon Institute of Social Policy 2002, 1–2).

Rural communities, including coastal communities, have felt and will continue to feel these cuts most acutely. Health authorities were told to evaluate the delivery of their health programs and make cuts accordingly. The Vancouver Island Health Authority (VIHA) was to eliminate 920 long-term care beds by 2004–5. Changes to the Central Vancouver Island Health Service Delivery Area (CVISDA) included a reconfiguration of acute-care services at its five hospitals and a complete review of the Tofino hospital. The North Vancouver Island Health Service Delivery Area (NVIHSDA), with three diagnostic and treatment centres and five hospitals, endured a series of changes to service delivery that cut $570,000 in expenditures in 2002–3 (Vancouver Island Health Authority 2003). Similarly, the Port Hardy hospital is now undergoing a complete review. Finally, the Northern Health Authority was to revamp its budgeting in order to provide more efficient use of its reduced 2003–4 allocation (Northern Health Authority 2004). This entailed the elimination of six acute-care beds at the Prince Rupert Regional Hospital by 2005 (BCNU 2004). A full-scale review of health services resulting in a services redesign program is also planned for Prince Rupert (Northern Health Authority 2002).

In short, in the period since 1990 both provinces have implemented sweeping financial and organizational changes to their health care systems in an effort to control spiralling costs and to improve efficiency. However, the severity and extent of these reforms, as well as the level of political rhetoric surrounding them, appears to have been considerably greater in British Columbia than in Newfoundland and Labrador, especially after the election of the Campbell government in 2001. Newfoundland and Labrador has not made any significant cuts to staffing levels since 1992 and has closed only one major hospital during the entire course of its restructuring program. In British Columbia, on the other hand, drastic staff cuts and hospital closures are now on the horizon. What the two governments do have very much in common, however, is that their restructuring efforts contributed to severe and persistent labour conflicts and to difficulties for remote coastal communities.

Reorganization and Regionalization

In Newfoundland and Labrador, the concept of regionalization has a long history. The idea of regional health boards can be traced as far back as the 1960s (see chapter 7), when a royal commission under the Smallwood government proposed that the province should be divided into regions, each of which would have a regional hospital (governmental or otherwise) that was to be governed by a board that would include seats for community citizens, with doctors being assigned a prominent role. Regionalization then (despite renewed efforts in the 1970s) fell off the governmental agenda for over a decade, to re-emerge in the 1983 Orsborn Commission. Even so, little happened at the decision-making level until the early 1990s, when the combination of changes in fiscal federalism, the collapse of the fishery, and a nationwide recession created the kind of policy and political context conducive to comprehensive restructuring, and the government introduced regionalization. The system was fairly complicated. A total of fourteen boards were created: six "institutional boards" responsible for hospitals and other acute-care institutions; four "health and community services boards" responsible for an array of social services including home care, mental health, and drug dependency programs; two "integrated boards" combining the two functions in the province's sparsely populated northern regions; and two special boards, one for nursing homes in the St John's region and the other for cancer care across the entire province.

In our study area up to January 2005, there were one institutional board (Western Health Care Corporation based in Corner Brook), one Health and Community Services Board (Health and Community Services–Western, also based in Corner Brook), and two integrated boards (Grenfell Regional Health Services Board based in St Anthony and the Health Labrador Corporation based in Happy Valley–Goose Bay). In addition, the Newfoundland and Labrador Cancer Treatment and Research Foundation, based in St John's, provided services for people living in our study zone. All boards were funded by the provincial government, and members (appointed not elected) were accountable to the minister of health and community services. During its decade of operations, the system was seen by most observers as a significant improvement over the preceding one, with its multitude of independent boards operating under a centralized ministry. However, because they are appointed, the boards have had some difficulty providing effective links to community interests and in generating a sense of citizen engagement and accountability. Some people thought fourteen boards were too many for a small population in a poor province, while many boards, especially in rural areas, had too few resources and capacities to be effective. There are also concerns about the impact on continuity of care of dividing institutional care and community care in many regions into two separate silos.[20]

Beginning in January 2005, the government merged the fourteen boards into four larger ones, each responsible for all types of services in its catchment areas. For our study zone, this involved merging the two boards in the "western area" into a single board and combining the two integrated boards (Grenfell and Health Labrador) into a single organization. Each of these boards will be responsible for the delivery of a broad range of health and community services, as well as all cancer services, to its catchment population. Hospitals in St John's will, however, continue to provide many of the specialized secondary and tertiary services they provided under the previous system.

In British Columbia, the process of regionalization has been considerably more complicated. A regional system was introduced quite early, in the mid–1970s, then overhauled twice in the mid-to-late 1990s, and then again in 2001–2. The 1973 Foulkes report proposed a complete reorganization of the British Columbia health system on a regional basis with seven regional health district boards (RHDBs) and six community human resources and health centres (CHRHCs), which were to have responsibility for the coordination of community health priorities for non-statutory services and for making recommendations to the minister regarding the allocation of funds to the community organizations that administered services. In order to foster community participation in health decision making, local residents were to be elected to boards through municipal electoral systems, and the CHRHCs would have the responsibility for hiring doctors, although on the basis of contracts negotiated provincially by the MOH and the BC Medical Association. Nursing and social services staff were to be transferred to the CHRHCs from the MOH and from the Ministry of Human Resources.

This scheme prevailed until 1994 when a new configuration was introduced, following the recommendations of the 1993 New Directions program. The Health Authorities Act introduced a significant decentralization by establishing twenty regional health boards (RHBs) and eighty-two community health councils (CHCs) and refocusing attention away from acute-care delivery toward community-based health care, with the CHCs addressing community health needs and overseeing health service delivery, while also providing local-level input to the larger RHBs. In turn, the regional boards were to oversee service delivery in larger geographic areas. Membership of the RHBs was to include local members from the CHCs (CHC and RHB information from Frankish et al. 2002, 127). No sooner had the new system been established than the government began to have doubts about its suitability. Under pressure from providers, administrators, and a deteriorating fiscal environment, and concerned about the impact of local participation on policy outcomes, the government became nervous about the original design and began appointing key members to both RHBs and CHCs. This shift made the new regional boards subordinate to government and reduced effective public input into the health care system (Davidson 1999, s37).

By 1995, increasing wait times for some surgeries and in emergency rooms had generated sufficient negative publicity that a decision was made to refocus health policy away from a strict emphasis on improved population health outcomes and towards policies aimed at improving efficiency and effectiveness in conventional health care services (Davidson 1999, s37). Whereas in 1993 RHB results were reported in terms of population health outcomes, from 1995 onwards, RHBs were told to quantify regional results in terms of health care spending (s38). Thus, the system evolved quickly from a focus on population health to one that emphasized managerial accountability, efficiency, and economy. In 1996, the government placed *New Directions* under formal review and the process of creating and amalgamating RHBs and CHCs was put on hold pending an assessment by the Caucus Committee Regional Assessment Team.

In 1997, the report entitled *Better Teamwork, Better Care* proposed improving efficiency by laying out a complicated scheme for the amalgamation and reorganization of RHBs and CHCs, with a sweeping cut in the ranks of management and the streamlining of services (Frankish et al. 2002, 128). The two structures were given jurisdiction over separate geographical areas (there was now no overlap between CHC and RHB service areas) and the number of RHBs was reduced from twenty to eleven. They became responsible for the majority of acute-care services in urban areas and for the control of, and finding allocation rights for, services for continuing-care facilities and community health programs. CHCs were reduced from eighty-two to thirty-four and were to focus primarily on rural and remote areas with responsibility for acute-care hospitals, continuing care, and home support. In addition, the report called for a third class of health board, seven community health services societies (CHSSs), which were to be responsible for the provision of community health programs to a number of communities within geographic areas that overlapped with those of the CHCs. All three levels were to receive funding from the MOH and to be responsible for the management of funds within their area of responsibility. Each had its own corporate services and administrative infrastructure, including chief executive officers and boards of directors, but none had any role in the payment of physicians or nurses, in negotiations with unions, or in the allocation of funds for prescription medications and other related costs (Frankish et al. 2002, 128).

This was a time when new public management theories were popular, and it was assumed that they would enable the health boards to make sound decisions, reflective of the needs of the populations and geographic regions they served: the process created a style of administration that closely mirrored a business model (Davidson 1999, 837). Later in that same year, however, the auditor general's report on the health care system, *A Review of Governance and Accountability* (AGBC 1997–98), found that there were difficulties in establishing the authority and the objectives of RHBs. Roles were not well defined, there was little consistent leadership, and policies were not developed to support decisions made by the boards (Frankish et al. 2002, 144). The complexity of the system resulted, moreover, in contentious overlaps in budgetary allocations and fragmentation of policy development responsibilities among agencies, ministries, and health professionals. There was also some confusion as to who was eligible to serve at each level, and the processes for their election and appointment were found to be excessively complex and detailed (129).

In 2001, steps were taken toward yet another reconfiguration of the British Columbia health system, as the new Liberal government sought to apply its business approach to health services. In December, *A New Era for Patient-Centred Health Care* (British Columbia Ministry of Health Planning 2001) argued that restructuring the network of RHBs was needed to rectify three ongoing problems: historically inadequate planning and management, a growing disparity between public expectations and actual health care delivery, and the failure to ensure the sustainability of health care expenditures. In particular, the new governing party criticized the existing network of boards and councils for inequities in service delivery, administrative overlap and duplication, and failure to meet regional needs. The existing array of fifty-two RHBs, CHCs, and CHSSs was to be replaced with a system of one provincial and five regional health authorities (RHAs), while fifteen health service delivery areas (HSDAs) were created to ensure local participation in health-based decision making. The regional HAs were responsible for the identification and planning of regional needs while ensuring the adequate funding of programs. The provincial HA was responsible for administration of specialized services, such as transplants and cardiac care, as well as for programs provided by provincial agencies such as the BC Cancer Agency, children's and women's health centres, and the BC Centre for Disease Control. It was also to ensure that access and wait-list issues were equitably addressed (British Columbia Ministry of Health Planning 2001, 3–4). The HSDAS, reporting to the HAS, were to manage the delivery of local health services, meet performance objectives as laid out by the HAS, and ensure community participation in health service planning (British Columbia Ministry of Health Planning 2001, 1–8). The composition of the HSDAS was to reflect provincial geography and patient and physician referral patterns (British Columbia Ministry of Health Planning 2001, 3).

It was argued that this streamlined system would better reflect new health objectives, while providing increased financial sustainability and better-quality care (British Columbia Ministry of Health Planning 2002, 12). Just as in 1997, however, it was not long before the Auditor General's Office released a report that was highly critical of the new regional health system, saying that the MOH needed to develop a framework with measurable performance expectations. Budget development and resource allocation procedures in HAS were not informed by rigorous analysis; rather, they were dictated by historical trends and use patterns. Although HAS were to report specific financial and statistical information, the MOH lacked the capacity to analyse the information received (AGBC 2002/2003). Consequently, it was unable to link budget decisions to performance and to strategic priorities. Reporting by the MOH could not provide a comprehensive picture of the performance of the regional health care system.

Despite the different twists and turns the process took in each province, regionalization has been one of the more visible and important measures undertaken as part of the restructuring process in British Columbia and Newfoundland and Labrador. As we have seen, this administrative restructuring in both provinces also played an important role in a distinct but closely related process – financial rationalization. By creating new units of authority with responsibility for implementing the new efficiency-driven approach, governments set the stage for cost control. Regionalization has essentially introduced a new set of structural constraints on the allocation of resources to hospitals; hospitals in need can no longer appeal to their own boards for extra resources but must compete with all other institutions in their steadily expanding regions for the attention and sympathy of the regional boards. The boards are essentially agents of cost control; once the budgetary envelope for each region is set, the provincial government can then step back and allow the boards to make the often politically difficult decisions necessary to achieve the desired restraint.

Also worth noting are various efforts by both governments to enhance the provision of health services and population health outcomes in rural and remote areas. In Newfoundland, these efforts took three forms. The best-known is the Telemedicine Program, an innovative use of computer, telephone, and electronic communications media to bring medical treatment, including specialist care, to remote parts of the province. Another aspect of this effort was the incorporation of a heavy emphasis on rural experience and rurally located training rotations within the curricula of Memorial University's Faculty of Medicine at both the undergraduate and specialist levels, with a particular focus on the Family Medicine Program. Third, the Department of Health and Community Services and the Faculty of Medicine collaborated in the mid–1990s to develop an innovative approach to the organization of rural health care. Three special rural clinics were set up as part of what was called the Primary Health Care Enhancement Project. These were staffed by Memorial clinician/professors and, on a rotational basis, by trainees in the Family Medicine Program. These clinics were all in areas in, or adjacent to, our study zone (Twillingate, Port aux Basques, and Happy Valley–Goose Bay) (Newfoundland and Labrador Health and Community Services 1999).

In British Columbia in the 1990s, the ministry's northern-service delivery support system introduced a number of programs with varying degrees of success, including the Northern and Isolation Travel Assistance Program (NITA), introduced to defray direct travel and accommodation costs for specialists who travel to twenty-nine rural and isolated communities to hold special clinics;[21] the Physicians Outreach Program (POP), introduced in July 1998 and providing honoraria to doctors for time spent travelling to remote sites; the University of British Columbia (UBC) Psychiatric Outreach Program, which funds psychiatrists to travel from Vancouver to outlying areas and to train general practitioners; the outreach programs at the BC Children's Hospital, the BC Cancer Agency, the Arthritis Society, and the Emergency Medical Coverage Program, established in 1998 to provide after-hours emergency medical coverage in communities eligible for the Northern Isolation Allowance (an older program dating from 1978);[22] increasing fees paid to physicians living, working, or taking calls in 103 rural/remote communities; the Travel Assistance Program (TAP) introduced in 1993, offering discount travel fares to the public for non-emergency medical care; and the Northern and Rural Locum Program to assist physicians practising in small communities to secure subsidized vacation relief.[23]

In 1998, an independent central recruitment agency, Health Match BC, was developed. Start-up costs for the agency came from the ministry of health, and today it operates with an annual grant from the ministry and through fees for its services. In addition, various other programs were developed to enhance the preparedness of physicians for rural practice. These programs included UBC's Family Practice Rural and Remote Program, requiring a minimum of one month of rural practice for health care residents; an optional but very popular rural program for all second-year residents; the Special Skills Post MD Training Program, which allows rural and urban general practitioners to upgrade their skills; the Rural Postgraduate Medical Specialty Training program at UBC, offering mandatory rural rotations in various specialties; a rural apprenticeship program for UBC undergraduate medical students; and UBC's Undergraduate Rural Practice Program, offering medical students summer employment as apprentices to rural physicians. In 1997–98 an exposure of four to eight weeks was provided to 108 students in fifty-nine locations in British Columbia and the Yukon. In addition, a two-year residency in family medicine in Prince George was established in 1995, with a total of twelve residency positions, six in each year.

In short, an important focus of the various reports on the troubles of our two provincial health systems involved administrative structures. In both cases, what was recommended was a fundamental overhaul to transfer decision-making authority away from local boards and from centralized provincial ministries towards the regional level. In both provinces, as in most other parts of Canada, governments did, indeed, move in this direction. As we will see, however, the regionalization processes in our two provinces took rather different directions. In Newfoundland and Labrador, the regionalization process emerged in a straightforward manner from the Dobbin report until a partial reversal of direction was introduced by the Williams government in September 2004. British Columbia, by contrast, has undergone multiple rounds of reform, each producing its own highly complex and idiosyncratic model of regionalized health administration.

Although labour relations are not strictly part of the mandate of this book, it is important to note briefly that efforts by government to control costs and contain the growth in numbers of health care personnel and the size of their salaries inevitably took a toll on relations with health care workers. On both coasts, the key issues have been wages and, to a lesser extent, workloads and contracting out, as both governments have attempted to wring as much output as possible from their workforces at the lowest possible cost. The Williams government in Newfoundland and the Campbell government in British Columbia have recently redoubled their efforts in this regard; both have engaged in very unpleasant public battles with their major health sector unions as part of a broader and continuing effort to control costs and balance their respective provincial budgets. In Newfoundland, there has been a striking pattern of issues around wages and staffing levels, stored-up employee resentments towards government intransigence that has been justified on the grounds of extreme fiscal constraints, settlements imposed from above, and grievances left to fester.

The role played by a decade of retrenchment and restructuring is hard to miss. Surveys of health sector workers conducted in 1999–2000 and 2001–2 by Barrett et al. (2003, 22–3) measured health care providers' changing perceptions of the effects of health care reform on the work environment and the impact of system changes on work-related attitudes. Most groups of workers expressed low levels of commitment to their organizations, felt that employers had failed to fulfil original commitments made to them upon hiring, and were unsure about staying with current employers. Most providers had concerns about the negative impact of reforms on the emotional climate of the workplace. Most of the groups surveyed were dissatisfied with managerial support. In both time periods, all health provider groups had concerns about quality of care, safety measures, and standards of care in the workplace. Nurses in particular felt that the staffing situation in their units was inadequate for meeting patient needs and that current availability of registered nursing staff for patient care was much less than before restructuring (Baker 2002). With these sorts of attitudes so prevalent among health professionals, it is not surprising that the system has been plagued by strikes.

In British Columbia labour relations have also been tumultuous. The most significant and protracted disruptions in health services occurred in the spring and summer of 2001, when job actions and strikes, legal and illegal, rotating and total, dominated the entire period and involved nurses, lab technicians, paramedical employees, and other health professionals (Bailey 2001b). The key issues for all these groups were wages and working conditions, with the nurses' demand for parity with Alberta wage rates playing a pivotal role (*Vancouver Sun* 2001). The Liberal government resorted to a legislated cooling-off period, court challenges, and attempts at legislated contract settlements. The result was that in August, with the nurses' union facing a legislated cooling-off period, court injunctions, and a salary offer well below what they were demanding, it threatened resignation of its entire membership (Bailey 2001a). Ultimately, the various disputes were ended, if not resolved, through increased wage offers and legislated back-to-work orders (*Whitehorse Daily Star* 2001).

As we can see, in both provinces labour relations in the health sector have been difficult for the past decade and more. Things have, furthermore, got much worse of late with the election of aggressively conservative governments in both provinces. Both governments are operating under commitments to balance their books and have turned to health sector wages and benefits as primary targets in the effort to control expenditures. Despite two recent rounds of increased federal funding commitments, prospects for improved labour relations in health appear slim. The two provincial governments, like many of their counterparts elsewhere in Canada, continue to regard health expenditures as the key to their budgetary challenges and to see the wages of health employees as the core of the problem.

CONCLUSION

It is clear that both provinces confronted very similar challenges in providing and paying for health services for their populations, especially those located in rural and remote communities. Health officials and professionals faced declining federal funding transfers at the same time that they confronted escalating costs, rising public expectations, and increasingly complex demographics. As we will see in the following chapters, most of the small, isolated, and economically endangered communities on their coasts confronted health status and service challenges that were different from, and generally more intractable than, those in urban areas. In both provinces, governments and health system decision makers scrambled to understand the challenges and to devise appropriate responses. Both governments commissioned repeated studies of health system issues, sometimes with a specific focus on rural questions. Both also devised and implemented a range of approaches to try to make health care more efficient and more effective. Particularly in rural communities, however, the search for greater efficiency (including lower costs) and the search for more effective service provision were often at cross-purposes, as consolidation of services and regionalization of administrative structures threatened small coastal communities with reduced access to health services, increased waiting and travel times, and decreased health-related employment opportunities.

What has been missing in all the efforts governments have made is an appreciation that health responses are themselves strongly related to social-ecological restructuring. More specifically, as we will show throughout this section, restructuring of employment, incomes, diet, educational services, ecosystem benefits, and the like has profoundly affected local communities, increasing health risks among the population. This is occurring at the same time that health services restructuring is affecting their access to services and social restructuring (in the form of out-migration) potentially affecting their access to informal unpaid health services. Occupational risks are also changing as ecosystems and resource-dependent industries are changing. The results, as we shall see in this part of the book, are serious, long-term, and pervasive.

8

The Statistical Face of Restructuring and Human Health

This chapter provides a statistical profile and structural analysis of the impact of restructuring on coastal communities in our study areas on the East and West Coasts, in particular its effect on demographics, income levels, and various aspects of health.[1] We begin by examining demographic and income changes as indicators of social responses primarily to industrial restructuring. In terms of health impacts, the social-ecological model that frames and informs our research (see chapters 1 and 2 and Dolan et al. 2005) posits that the interactions between political, industrial, social, and environmental restructuring have a significant impact on population, community, and personal health through the mediating effects of the multiple determinants of health. It follows that the pathways that describe the restructuring-health relationship are complex and require analysis using multiple methodologies, which include the systemic analysis of the previous chapter, the statistical one discussed in this chapter, and the more qualitative and motivational analyses of the chapters that make up the rest of this section.

Here we examine the health of our communities and their residents on the East and West Coasts from three perspectives. At the broadest level, we consider first the effects of restructuring on *population health* on the West Coast by comparing temporal trends in mortality and morbidity rates in Prince Rupert with those in the province of British Columbia. We then examine *community health* indicators using data from surveys we conducted in selected East and West Coast communities. Finally, we analyze *personal health* status, based on the same survey data. In each case, we are seeking to determine the strength of evidence for the impacts of restructuring on health, recognizing the intrinsic problems of attribution within the limits of our research designs and available data. We then turn, in the following chapters, to the underlying factors for the statistical patterns we identify here.

DEMOGRAPHIC CHANGE

With few exceptions, the communities in our study areas have experienced declines in population that are related to migration but also to changing birth rates. Particularly in rural Newfoundland and Labrador, migration has been a long-term adaptation by local people to perceived opportunity elsewhere, and many individuals on both coasts consider moving away even in relatively good times. However, migration has profound effects on the local society when most of the youth, whole households, and even a majority of local residents take this decision. In Newfoundland and Labrador, young people have always left the province for education or employment opportunities, but as one older woman from the Northern Peninsula observed, "Yes we left in the 50s. We went to Toronto to live and work, but we came back [See also House, White, and Ripley 1989]. Young people have always left. The difference between then and now is that we *could* return if we wanted. People can't do that any more. There are no jobs."

The most marked decline occurred in Prince Rupert, where the population decreased by 11.8 percent from 1991 to 2001; all of the decline came after 1996. Port Hardy, on northern Vancouver Island, has seen similar changes; the population declined by roughly 13.4 percent between 1996 and 2001, following an increase in the previous five years. The sole exception is Tofino; owing to its flourishing tourism-based economy, the population increased by 33 percent from 1991 to 2001. In comparison, the population of British Columbia grew by 16 percent between 1991 and 2001 (all numbers based on Statistics Canada [Canada Statistics 2001]).

The most dramatic change occurred in the years 1996–2001 (figure 8.1), when all five areas on both coasts lost substantial population; only Alberni-Clayoquot was substantially better than any other area.[2] However, in the previous five years, the study areas in Newfoundland and Labrador were already losing significant numbers of people, while the British Columbia districts had held steady or had grown slightly. Moreover, while Newfoundland and Labrador as a whole experienced a 10 percent decline in overall population between 1991 and 2001 and net out-migration of 5.1 percent for the 1991–96 period, the situation in our East Coast study areas was even worse. In the Grenfell health region,[3] which contains many of the communities we studied, the population declined by 18 percent, from 19,345 to 15,805, between 1991 and 2001, with out-migration of young people being a major component of this decline.[4] Between 1991 and 1996 alone, the region lost 2,300 people (12 percent of its population), mostly to net out-migration, including 1,895 persons under the age of 35. Over the same period, the

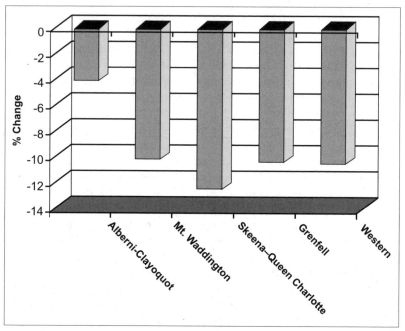

Figure 8.1   Percentage change in population, selected regions, 1996–2001

percentage of the population aged forty and over increased from 31 percent to 37 percent. A similar set of demographic trends has been observed for Newfoundland's Western Health Region, which contains the rest of our study area communities but also includes the southwestern part of the island, an area we did not research.[5] British Columbia's coastal communities faced similar challenges. By 2001, 14 percent of the population was older than sixty-five, and 30.6 percent of the population was older than fifty. In each of our three coastal study regions in that province – Alberni-Clayoquot (which includes Tofino and Ucluelet), Mount Waddington (includes Port Hardy and Alert Bay), and Skeena-Queen Charlotte (includes Prince Rupert and Hartley Bay), the populations of each research area declined after 1996.

If we now consider individual communities, some interesting variation is evident in the trends in total population.[6] Generally, in 1991–96, the populations of towns and villages in our West Coast study areas were stable or expanding. Alert Bay, alone of the five reported in table 8.1, showed a slight decline. Of course, this does not mean that young people did not leave; rather, a combination of in-migration and births provided an effective counter. Indeed for the early 1990s, net migration flows were strongly positive for the twenty to twenty-four age

Table 8.1

Population Change in Selected Locations, Coastal British Columbia and Newfoundland and Labrador, 1991–2001 (Census of Canada 1991, 1996, 2001)

	1991–96	1996–2001	Population 2001	Place Notes
Alert Bay	–2.5	–4.7	583	Excludes reserve, but includes 135 aboriginal residents
Port Hardy	4.0	–13.4	4,574	Fishing, aquaculture, logging and tourism
Prince Rupert	0.3	–12.1	14,643	Regional economic and administrative centre; 4,430 aboriginal residents
Tofino	6.1	25.3	1446	Expanding tourist and fishing centre; Pacific Rim NP
Ucluelet	3.9	–6.0	1559	Tourism; Pacific Rim NP; relatively low earnings
St Paul's	–20.5	–14.1	346	Very small in-shore fishing community
Parson's Pond	–14.1	–19.4	226	Very small in-shore fishing community
Cow Head	–7.7	–23.2	357	Very small in-shore fishing community
Charlottetown	13.0	4.8	346	Revitalized fishery based on shellfish
Cook's Harbour	–27.6	–13.1	226	Isolated and unable to adjust to loss of groundfish
Main Brook	–7.4	–15.8	357	Small logging and sawmill centre
Port au Choix	–11.8	–11.9	1,010	Main Gulf Coast fishing and processing port
Port Saunders	3.6	–7.3	509	Medium-sized community
Port Hope Simpson	–6.0	–11.8	509	Struggling village at north end of new Labrador road
St Anthony	–10.7	–8.9	2,730	Regional administrative, retail, and fishing centre

group in rural areas of Vancouver Island and the West Coast (BC Statistics 2000). In the late 1990s, much of rural British Columbia experienced contraction in its important logging and wood-processing industries, as a result of lumber duties imposed by the United States, and poor markets for many paper products. In addition, depleted stocks of salmon and other commercial species caused problems in the fisheries. Although the coastal areas we have studied in British Columbia often have significant tourist and aquaculture enterprises, they did not fully compensate, and populations fell in all our research areas. Tofino and Ucluelet are primarily tourist and park service centres with secondary participation in logging and fishing. In recent years, Tofino has grown rapidly to rival Ucluelet, whose population had the lowest average earnings among our British Columbia study centres. While some Tofino residents feel that the economy is overly focused on tourism, others see a vibrant and varied economy that has brought people with an assortment of skills and expertise to the area:

[T]here are many opportunities [for work], and I believe at many different levels of society, even for highly, highly educated people where you wouldn't expect opportunities in a small town. This town is very different as I am sure you have picked up. There is a white collar sector in a town this small, that is amazing, it is bizarre ... there are people running private consultant firms; there are people who do bizarre things on stock markets, and lawyers and stuff like that who are able to run their businesses from here and of course these people require ancillary services, people to take care of their computers, and whatever those people do ... so basically this is not in the least a typical small town. (Tofino respondent)

This continuing influx of people to Tofino is one of the key factors that sets it apart from the other study communities. As more and more people choose to make Tofino their home, they bring an assortment of resources to the community while in Port Hardy and Prince Rupert, the valuable contributions of community members decline as residents leave because of the poor economic climate.

On the East Coast, population decline continued between 1996 and 2001, with many communities such as St Paul's, Cow Head, and Parson's Pond on the southern coast of the peninsula struggling to survive. By 2001, Cook's Harbour (located on the edge of an almost barren rock outcrop at the tip of the peninsula and once a thriving village with a local processing plant and an active fishery) was reduced to only 226 inhabitants. In 2004, the provincial government agreed to resettle the three remaining families in the nearby hamlet of Boat Harbour. Thus, some areas appear in desperate condition in the early twenty-first century. An exception is Charlottetown, southern Labrador, where crab fishing and, more recently, shrimp processing were successfully expanded after the cod moratorium and where construction of the Labrador road contributed to population growth. Port Hope Simpson, a forestry-dependent community with a smaller fishing sector demonstrates that the experience of Charlottetown (and Mary's Harbour) was not uniform in southern Labrador. The completion of the trans-Labrador highway and a succession of substantial quota cuts in the crab fishery up to 2005 may accelerate out-migration in these areas if alternative employment opportunities are not found.

INCOME LEVELS

One of the known contributors to human stress is financial insecurity (Wilkinson and Marmot 2003). The Statistics Canada *Survey of Financial Security* shows aspects of the assets, debts, and net worth of Canadians in 1999, while additional information on income has been provided by Statistics Canada for the Canadian Centre for Policy Alternatives (Kerstetter 2001). These data show that, on a basis comparable to 1984, Canadian median net after-tax income of family units was virtually unchanged in 1999, although median net worth had risen by 11 percent. Most importantly, however, there was significant evidence of wealth inequality. For example, the median net worth of the wealthiest 20 percent of family units increased by 39 percent, while the 40 percent with the lowest net worth showed little change. These data show that family units in Newfoundland had the smallest median net worth, and the median worth of young couples across the country fell by 30 percent, although median net worth of all families rose by 10 percent. Finally, British Columbia was shown to "be home to both the highest average wealth in Canada and the largest gap between the richest and poorest households" (Kerstetter 2001, 1). There was no breakdown available for coastal communities specifically in these reports, but the other indicators of community well-being in this chapter put the general Statistics Canada data into sharp relief.[7] While no cause and effect connection can be made, it should be noted that income inequalities between rich and poor groups or areas have been identified in the population health literature as correlates of mortality and morbidity differences (Wilkinson and Marmot 2003).

Table 8.2 shows income and education data for 2001 for the health regions of Newfoundland and Labrador and for the regional districts of British Columbia. Low educational attainment (measured by the percentage of the population with less than high school education) was similar in all regions and much higher than Canadian averages (not shown). In other respects the two East Coast regions are notably worse off than those on the West Coast. They show much higher dependence on government transfer incomes, median personal incomes more than $5,000 lower than the poorest British Columbia region, and much higher unemployment rates. The Grenfell region on Newfoundland's Northern Peninsula experienced the highest out-migration of all regions and still fares worse on four of five measures. Clearly the high rates of out-migration have not left those still living in the areas with adequate employment and independent incomes.

We conclude that to some degree people in coastal communities are finding that they have to "vote with their feet" and, where they cannot

Table 8.2
Selected Indicators of Income and Socio-economic Status, 2001 (Statistics Canada n.d.)

	Grenfell	Western	Alberni-Clayoquot	Mount Waddington	Skeena-Queen Charlotte
Government transfers as percentage of total income	29.6	26.3	15.8	9.7	12.9
Median income of persons 15 and older with income ($)	14,146	14,425	19,913	24,528	20,599
Unemployment rate (%)	44.4	27.1	13.8	10.7	16.4
Less than high school education (%)	25.5	26.0	25.9	28.9	32.7

create new local opportunities for community and household survival, move to other places. The pressures on these coastal peoples are clearly significant. Can we say, however, that their health is suffering as a result of the multiple stressors that restructuring has brought to bear upon their lives? We turn first to the broader aspects of population health.

RESTRUCTURING AND POPULATION HEALTH

On the West Coast, Prince Rupert has experienced major population loss as a result of environmental, economic, and social changes over the last twenty years (table 8.1). Anecdotal evidence suggests that community stress linked to industrial change and economic uncertainty has had negative psychosocial effects, including increased depression, dysfunctional family relations, and alcohol and drug abuse (Gislason, Lam, and Mohan 1996). Scholars looking at international health patterns have suggested that relationships between socio-economic changes and population health can be attributed to changes in intervening social and behavioural factors. For example, Hertzman and Siddiqi (2000) attributed negative changes in health status in rapidly contracting economies in Central and Eastern Europe to conditions created in civil society as a result of increased income inequality, specifically, reduced life control, economic deprivation, and social isolation. Such studies suggest that changes in population health are tied to economic changes through mediating social determinants, including absolute and relative income levels, social capital, social cohesion, and psychosocial stress (Wilkinson and Marmot 2003). However, very little research has explicitly addressed the relationship between restructuring and health in a Canadian context, with the exception of work examining health impacts of restructuring in a cohort of British Columbia sawmill workers (Ostry et al. 2001; Ostry et al. 2002).

We tackled this issue in various ways, using a sample of communities to test our hypothesis that restructuring and health are intertwined. We used hospitalization discharge data from the BC Linked Health Database, an administrative health data repository including hospital admission and discharge data, data on physician services, vital statistics, and mental health and other information (Chamberlayne et al. 1998). The data for our study included a 10 percent cohort sample taken in 1992 of service users from British Columbia for fiscal years 1985–86 through 1999–2000 and a 100 percent cohort for Prince Rupert for the same period. Prince Rupert and provincial cohort results were directly comparable in 1992. Comparison of other years had to take into account slight differences.

We turned first to annual·trends for age-sex adjusted morbidity (disease/illness) rates, estimating the number of individuals hospitalized for a specific condition by using International Classification of Disease 9 (ICD9) data (International Classification of Disease, 9th version). Hospitalization data was used as a proxy for total incidence. On the East Coast the only data obtainable for an examination of hospitalization were aggregated data from the "health accounts" of the Newfoundland and Labrador Community Accounts for 1994–99.[8] We worked with three areas. Area 1 is comprised of the communities of St Paul's, Parson's Pond, and Cow Head. Area 2 includes Port Saunders and Port au Choix, and area 3 is St Anthony. Given the restricted nature of the data, we were unable to carry out the same level of analysis as was possible on the West Coast. However, we examined overall hospital morbidity as well as selected diagnoses for the three research areas and compared these data across areas as well as with the province as a whole. Apart from mental health, for which we did not have data, the same diagnostic categories were used on both coasts.

The highest percentage of hospitalizations for Newfoundland and Labrador was for diseases of the circulatory system, diseases of the digestive system, and injury and poisoning in decreasing order. For British Columbia, selected population health outcomes putatively associated with socio-economic changes were analysed. They included mortality, infectious and parasitic diseases, neoplasms (cancer), mental health diseases, diseases of the circulatory system, digestive system diseases, and injury and poisonings. Two plausible pathways between restructuring and health were explored: (1) cumulative restructuring effects that may have contributed to elevated morbidity rates in Prince

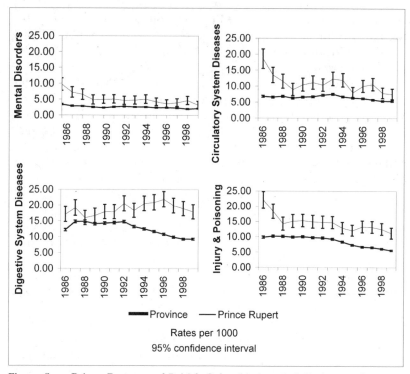

Figure 8.2 Prince Rupert and British Columbia hospital discharges for
mental health, diseases of the circulatory and digestive systems, and injury
and poisonings, 1986–99

Rupert when compared to provincial rates and (2) critical restructuring events in Prince Rupert that may have resulted in significant
changes in mortality or morbidity rates over time.

Age-sex adjusted mortality rates for Prince Rupert were more variable and elevated than those for the provincial sample, although the
rate difference was significantly higher for only 1996. Regression analysis showed no significant slope changes or discontinuities in mortality
rates. We were unable to control for Aboriginal status, a variable likely
to increase mortality rates in both cohorts, because the information
was not available within the mortality data. Age-sex adjusted hospital
morbidity rates for Prince Rupert were higher than provincial rates for
all ICD9 codes selected for analysis, except neoplasms, with the strongest effects being for injuries and poisonings, mental health, and diseases of the circulatory and digestive systems (figure 8.2). In addition,
discontinuities in rates were identified for several morbidity indicators.
Again, these differences may be somewhat inflated by the higher

Table 8.3
Potential Years of Lost Life (PYLL) Rates per 100,000 Population, by Gender, Province, and Health Board, 1997–2001

| Year | Gender | Geographical Region | | |
		Province	Western	Grenfell
1997	Male	6,865.4	5,828.3	6,942.0
	Female	3,929.0	4,635.1	4,137.8
1988	Male	6,401.7	6,691.5	6,065.6
	Female	4,046.5	4,125.9	1,394.3
1999	Male	6,388.0	6,739.7	5,334.0
	Female	4,201.7	4,464.9	4,515.7
2000	Male	7,018.7	7,305.2	7,275.5
	Female	4,109.6	3,738.8	4,152.3
2001	Male	6,275.4	7,522.8	6,696.8
	Female	4,027.7	3,966.7	2,003.1

Aboriginal population in Prince Rupert (approximately 26 percent of the population in 1996) compared with the province as a whole (approximately 4 percent in 1996) (British Columbia Ministry of Labour and Citizen's Services 2001).

While we recognize that the results of an ecological analysis of this sort cannot provide definitive evidence of cause and effect, the question is whether it is plausible to associate the elevated morbidity rates with the cumulative effects of restructuring. In the light of what is known in the population health literature, we need to ask whether or not these higher rates make epidemiological sense, given existing evidence of the impacts of restructuring on the social determinants of health. If we adopt the argument that socio-economic restructuring creates psychosocial stress through its influence on employment and incomes – and there is good reason to think this is true, as we will demonstrate in the next chapter – then we can argue that the changes brought about by restructuring in Prince Rupert have precipitated negative health effects, whether directly through immune system alterations, or indirectly through the adoption of health-damaging lifestyle behaviours.

For Newfoundland and Labrador, we obtained aggregated data to examine mortality and hospital morbidity. Data were available for 1997 to 2001 for potential years of lost life (PYLL), allowing us to examine mortality trends, and for 1994 to 1999 for hospital morbidity. Thus we can provide comparative analysis of mortality and morbidity rates but cannot perform the regression analysis that was done for British Columbia. Table 8.3 shows the rate of PYLL per 100,000 for the

Table 8.4
Hospital Discharge Rates per 1,000 Population, by Selected Conditions, 1994–99

			Geographical Region	
ICD-10 condition	*Year*	*Province*	*Zone 6 (Nordic)*	*Zone 7 (Red Ochre)*
Diseases of the circulatory	1994–95	157	103	130
system	1995–96	164	148	127
	1996–97	168	156	148
	1997–98	174	170	149
	1998–99	176	191	166
Neoplasms	1994–95	68	57	67
	1995–96	67	49	75
	1996–97	68	38	64
	1997–98	68	36	69
	1998–99	68	45	60
Mental disorders	1994–95	51	32	42
	1995–96	59	42	48
	1996–97	60	49	39
	1997–98	60	48	48
	1998–99	66	45	53

province and regions where our communities are located. There are no clear patterns for these rates.

Table 8.4 compares the hospital discharge rates for circulatory diseases, neoplasms (cancers), and mental disorders for the province, for zone 6, where area 3 is located, and for zone 7, where areas 1 and 2 are located. Rates of hospitalizations for diseases of the circulatory system (heart disease is included in this category) increased in the five-year period, as did rates for mental disorders generally. No clear pattern was noted for cancers.

We suggest that our results are significant, and call for additional research using case-control or cohort designs to investigate the relationship and associated pathways in more detail.

RESTRUCTURING AND COMMUNITY HEALTH

The health of communities as entities distinct from their resident populations has come into sharper focus through the healthy communities literature (Hancock 2000, 2002). The thrust of the argument is that the health of communities depends on the quality of their physical and social environments and can be measured in terms of indicators such as pollution levels, crime rates, population dynamics, employment and educational opportunity, family stability, and social capital and social cohesion. Our model identifies community health as vulnerable to the

Table 8.5
Survey Completion by Area and Gender

Newfoundland and Labrador (N=1470)	Area 1	Area 2	Area 3	Total
Males	87 (25%)	87 (19%)	206 (32%)	380 (26%)
Females	264 (75%)	377 (81%)	449 (68%)	1,090 (74%)
Total for areas	351	464	655	1,470

British Columbia (N=1204)	Prince Rupert	Port Hardy	Tofino	Ucluelet	Total
Males	335 (49%)	88 (38.8%)	77 (43.3%)	52 (44.8%)	552
Females	348 (51%)	139 (61.2%)	101 (56.7%)	64 (55.2%)	652
Total for areas	683	227	178	116	1,204

impacts of restructuring, and our research on both coasts has examined community health outcomes using a combination of secondary and primary data. We addressed changes in incomes and demographic characteristics earlier in the chapter and focus now on residents' perceptions of the health of their communities as revealed by their survey responses.

On the West Coast, we conducted a community telephone survey in the summer of 2002 to examine the impacts of restructuring on four coastal communities: Prince Rupert, Port Hardy, Tofino, and Ucluelet. These four communities were selected both because we were doing other case study work in them and could therefore enrich our findings and also because they provided us with variations in size, geographic location, environmental conditions, socio-economic profiles, and restructuring history. Community population sizes range from Prince Rupert at 14,000 to Tofino at approximately 1,450. Households in each community were randomly selected for inclusion in the survey using differential sampling rates (10 to 30 percent) to ensure adequate sample sizes for comparative analysis in the two smaller communities of Tofino and Ucluelet. The distribution of survey completions (n = 1204) by community and gender is shown in table 8.5.

The East Coast survey, while parallel in questions to that of the West Coast, differed in the method of data collection and in timing and sampling. Rather than using a telephone survey, we conducted face-to-face interviews in households using trained research assistants. In community consultations before commencing the research, we promised to visit the communities and respond to local requests, and to do that we could not use exactly the same methodology as was used on the West Coast. Before conducting the survey, we mailed informational brochures to each household in the selected communities telling residents about our research, who we were, and what our survey was about. Our data were collected in the spring of 2004. The communities on the East Coast have smaller populations than those on the West Coast, so we used clusters of communities, identified in this chapter as research areas. We attempted to survey all the households, but our actual response rate was 70 percent of the 2,100 households in the 2001 census (1,471 households). At the household level we approached the first adult who answered the door or who was willing to take part in our survey. While we monitored the gender of respondents, we did not try to have equal male-female representation: 26 percent of respondents were male; 74 percent were female. In many of the communities, men were away from their families in search of work or working in other parts of the country, and so our data are skewed towards women's experience. This needs to be kept in mind when interpreting the results or making any comparisons with the West Coast, where the survey had more balance in gender representation.

The same principles that governed the selection of communities on the West Coast (e.g., size of community and economic conditions and diversity) informed the East Coast selection. We included small-community clusters, medium-sized communities, and one larger one. Table 8.5 shows survey completion by research area and gender for both coasts. Both surveys were designed to assess residents' perceptions of their community environment and their individual health status. In terms of community health, the factors considered included satisfaction with the community as a place to live, satisfaction with a range of public services, and perceptions of social problems. Wording was the same in both surveys, but on the East Coast, in response to community desires, we added questions on care-giving and the elderly, the degree of uncertainty concerning, and confidence in, social institutions, and changing food consumption patterns.

On both coasts, respondents expressed a surprisingly high level of satisfaction with their community, given the extent of population loss experienced in almost all communities, although the counter-argument could be that the residents remaining are the "satisfied survivors." Seventy-four percent of West Coast residents indicated being either satisfied or very satisfied with their communities, while East Coast residents' levels of satisfaction were even higher, at 85 percent. This result is typical for survey questions of this kind, but the general finding conceals significant differences in the ratings for particular communities. On the West Coast, satisfaction percentages ranged from 68 percent for Prince Rupert to 88 percent for Ucluelet. On the East Coast, there was less variability: area 1 had a satisfaction rating of 84 percent, area 2 was 82 percent, and area 3 was 88 percent, almost identical to Ucluelet. By contrast, the percentages of respondents who were either very dissatisfied or dissatisfied with their community were, on the West Coast, 17 percent for Prince Rupert and 3 percent for Ucluelet; on the East Coast, area 1 came in at 9 percent, area 2 at 11 percent, and area 3 at only 4 percent.

When asked about the main things they liked most about their communities, respondents frequently cited the small-town supportive atmosphere and close proximity to nature, but there were differences between communities in regard to things people liked least. In Prince . Rupert, for example, the most prevalent concerns were (in order of frequency) the weather (Prince Rupert has fewer sunshine hours annually than most other communities in Canada), the economic situation, geographic isolation, and the lack of shopping and services. While the issue of the poor economy came up in Ucluelet, it was of less concern in general: issues of geographic isolation, lack of services, inadequate recreation facilities, and the weather were cited more often. The variation in levels of satisfaction was also reflected in the percentage of residents who reported that they had considered moving out of the area during the past year. The figures ranged from 57 percent for Prince Rupert to 41 percent for Ucluelet.

In Newfoundland and Labrador there was a diversity of answers by respondents in all three areas. What people liked the most about their communities was the "quietness" (27 percent) followed by the "people" in the communities (19 percent). Other less frequently given responses were "safety" or a "good place to raise children," characteristics often associated with rural communities. In general, people valued social relationships and qualities associated with the physical environment of their communities. They also associated some negative aspects with living in these places, the biggest drawback being "no work" (25 percent), while lack of services, such as shopping and recreation facilities, were also seen as negative. Not surprisingly, 54 percent of those interviewed in area 1, 52 percent in area 2, and 33 percent in area 3 thought that "more work" would improve the places in which they live. Sadly, 33 percent had considered moving from their communities in the past year.

Additional insight about the possible sources of satisfaction was gained through a series of questions related to levels of satisfaction with public services. In British Columbia, the percentage of people who were satisfied or very satisfied was relatively low for health and educational services but substantially higher for several aspects of physical infrastructure and emergency services (table 8.6). There are, however, significant differences by community in satisfaction levels for almost all categories, so that broad generalizations are not possible.

Table 8.6
Percentage of Respondents in British Columbia Satisfied or Very Satisfied with Various
Public Services

	Port Hardy	Prince Rupert	Tofino	Ucluelet	Total
Health services	29	47	46	41	43
Education system/schools	37	52	25	59	46
Family counselling services	66	33	39	38	41
Road maintenance	62	24	47	23	34
Emergency fire response	87	92	89	86	90
Sewage system	81	64	34	72	63
Water system	82	78	43	38	70
Garbage collection	93	83	87	91	85
Recreation services	64	61	45	39	57
Law enforcement	75	70	71	76	72
Ambulance service	87	78	90	63	80

For example, 35 percent of respondents in Port Hardy were very dissatisfied or dissatisfied with the health services, a substantially higher percentage than in the other three communities. This is an interesting finding considering that hospitals in Port Hardy and Tofino (which also serves Ucluelet) offer similar services and levels of care and may be related to the distance to a larger hospital offering more services: from Port Hardy to Campbell River, the distance is 238 kilometres, whereas from Tofino to Port Alberni it is 126 kilometres. It may also reflect residents' expectations – Tofino is a smaller community than Port Hardy, and people recognized that because of the remote location and small population, there was a limit to the health services they could realistically expect. Port Hardy, however, is a regional centre for northern Vancouver Island and for the central coast. Several residents expressed concern that availability of health care services, especially those offered by specialists, had declined over the past five years, so reduced access to care may also account for the divergence in opinion between Port Hardy and the other communities. In terms of satisfaction with the education system, there are marked differences between Tofino and Ucluelet, which are only forty-two kilometres apart. In Ucluelet, where the regional high school is located, 50 percent were very satisfied or satisfied, whereas the figure was only 25 percent in Tofino, whose high-school students face over a ninety-minute round trip by bus. Some insight into the factors affecting satisfaction with the community overall can be derived from the correlations with service-specific ratings. In British Columbia the strongest correlations were

Table 8.7
Percentage of Respondents in Newfoundland and Labrador Satisfied or Very Satisfied
with Various Public Services

Newfoundland and Labrador	St Paul's, Parson's Pond, and Cow Head	Port Saunders and Port au Choix	St Anthony	Total
Health services	59	62	68	64
Education system/schools	59	74	66	67
Mental health services	28	23	53	37
Road maintenance	14	7	16	13
Emergency fire response	93	94	93	94
Sewage system	72	87	79	80
Water system	42	61	77	64
Garbage collection	95	89	82	87
Recreation services	26	18	49	34
Law enforcement	60	82	86	78
Ambulance service	71	91	90	85

with garbage collection (tau [coefficient of correlation] = 0.19), recreation services (tau = 0.18), law enforcement (tau = 0.18), education system (tau = 0.17) and health services (tau = 0.16).

In the Newfoundland and Labrador communities that were surveyed (table 8.7), the strongest correlations with overall community satisfaction were with recreation services (tau = 0.17), the water systems (tau = 0.17), and the level of health services (tau = 0.16). Levels of satisfaction with services and infrastructure differed between areas within Newfoundland and Labrador and from the findings on the West Coast. Area 1 had the lowest percentage of respondents satisfied with health, education, law, and ambulance services. Area 1 has the smallest communities among the three clusters or combined research areas. As a consequence, area 1 residents are much less likely than those in the larger communities to have good access to these services. One participant was particularly concerned about "the law," which consisted of the RCMP, and about the fact that calls were transferred to St John's on evenings and weekends. Area 2 had the lowest percentage of respondents satisfied with the road infrastructure and mental health services. The latter finding is of particular importance, as this cluster of communities seemed to have high levels of stress related to downturns in the economy, and residents also perceived there to be more social problems in their area.

Perceptions of social problems were determined by respondents rating their degree of agreement that each of six possible problems existed in their community. In British Columbia the percentages indicating agreement or strong agreement (table 8.8) are relatively high in all categories, especially for alcohol abuse, drug abuse, and unemployment

Table 8.8

Percentage of Respondents in British Columbia Reporting Agreement or Strong Agreement about the Existence of Social Problems in Their Area

British Columbia	Port Hardy	Prince Rupert	Tofino	Ucluelet	Total
Alcohol abuse	88	84	77	73	83
Drug abuse	89	87	69	61	82
Family violence	59	62	46	44	57
Unemployment	86	97	27	84	84
Sexual abuse	47	49	37	27	46
Racial discrimination	51	41	47	33	42

Table 8.9

Percentage of Respondents in Newfoundland and Labrador Reporting Agreement or Strong Agreement about the Existence of Social Problems in Their Communities

Newfoundland and Labrador	St Paul's, Parson's Pond, and Cow Head	Port Saunders and Port au Choix	St Anthony	Total
Alcohol abuse	60	77	53	62
Drug abuse	72	80	59	69
Family violence	17	43	26	28
Unemployment	95	96	89	93
Sexual abuse	18	26	21	21
Care for the elderly	35	37	23	30

(except for Tofino). For each problem issue, perceptions differed significantly by community; residents in Port Hardy and Prince Rupert generally reported higher levels of agreement. This result is consistent with the variation in community satisfaction and is confirmed by the significant correlations between community satisfaction and perceptions of social problems, particularly unemployment (tau = 0.14) and drug abuse (tau = 0.13).

We measured perceptions of the same social problems on the East Coast, except for "racial discrimination," because of the expected homogeneity of the local populations, based on the 2001 census statistics (table 8.9). This expectation was confirmed in our question on demographics where we saw little variation in cultural background: less than 2 percent of the respondents said they were ethnically distinct from the dominant England/Irish groups. Instead, we asked about care for the elderly, which was said to be a problem in some communities because of the out-migration of younger people. For all social issues, area 2 had the highest percentage of respondents reporting these problems.

To assess perceptions of change in socio-economic conditions in the communities, residents were asked to indicate whether they thought conditions had improved or worsened over the past year. Again, there are striking differences in the results by community. In terms of economic opportunities, 84 percent of Prince Rupert residents thought that the situation was somewhat or significantly worse than a year before. The corresponding figure for Port Hardy was 61 percent. These high percentages are in contrast to those for Ucluelet and Tofino, which were 18 percent and 11 percent respectively. This striking contrast reflects significant events that have affected local economies. In Prince Rupert, for example, the Skeena Cellulose mill, the largest private-sector employer in the region and the largest taxpayer, closed in July 2001. At the time of the survey, the future of the mill was uncertain, and the economic and social repercussions of the closure were being felt by individuals and businesses throughout the city and region. This change, combined with the downsizing of the fishing industry, reduced employment in fish-processing plants, and reduced port activity, has resulted in a grim economic outlook for the community. In Port Hardy, the closure of the Island Copper Mine in 1996 and concurrent restructuring of the fishing industry because of implementation of the Mifflin plan, has severely impacted the economic base of that region. Although forestry operations continue in the region and there is an expanding, but contentious, aquaculture industry, the survey findings indicate that residents are quite pessimistic about the future of the local economy. In contrast, both Tofino and Ucluelet continue to benefit from expanding tourism industries, as we saw earlier. While the economic rewards are more obvious in Tofino, Ucluelet's reputation as a significant tourist destination is growing, and opportunities in eco-tourism, accommodation, and recreation continue to emerge.

Respondents were also asked to consider how their personal financial and employment situation had changed over the past year. Not surprisingly, the results correspond quite closely with perceived changes in economic opportunities more generally in the community. The percentage reporting that their financial situation had worsened ranged from a high of 38 percent in Prince Rupert to a low of 16 percent in both Ucluelet and Tofino. Similarly, those reporting deterioration in their employment situation ranged from a high of 27 percent in Prince Rupert to a low of 9 percent in Tofino.

On the East Coast, we also assessed perceptions of change in socioeconomic conditions in the three research areas. We found some striking differences. Area 2 respondents reported a marked worsening of economic opportunities in the past year with 74 percent of those surveyed feeling these conditions had worsened. In comparison, in areas 1 and 3 only 27 percent and 34 percent respectively reported worsening economic opportunities.

In addition to these questions assessing economic change, we asked the East Coast residents about uncertainty about the future and their level of confidence in the various levels of government to respond to their perceived problems. Almost half the respondents across the three areas agreed or strongly agreed that they were uncertain about their own future. This proportion rose to 80 percent for the residents of areas 1 and 2 when they were asked about the uncertainty of the future of their community or the province. Area 3 residents differed on these last two dimensions and were generally more positive. Confidence in the ability of the federal and provincial levels of government to solve the economic problems in the province were low across the three research areas (14 percent agreed/strongly agreed). However the level of confidence in the health care system (at 36 percent) and the education system (42 percent), while still low, was somewhat higher than the reported confidence in the government.

The recent population-health research literature has emphasized the importance of social capital and social cohesion as indicators of community health and well-being and as social determinants of personal health. Several questions were included in both surveys to identify residents' type and degree of involvement in their local community through voluntary and political activities and other indicators of social interaction. In addition, respondents were asked to indicate in overall terms both their level of involvement in local community activities and to describe their level of attachment to their community.

On the West Coast, 20 percent reported being either considerably or very involved in community events over the past year, while 22 percent reported being not at all involved. There were only minor variations across the communities and the differences were not significant. Twenty-seven percent reported their community attachment as very strong, compared with 5 percent for whom it was very weak. In this case, community differences were significant. In Tofino and Ucluelet, 73 percent and 71 percent respectively reported very/somewhat strong attachment, whereas in Port Hardy and Prince Rupert, the corresponding figures were 61 percent and 62 percent respectively – a pattern of difference between the four communities that supports already noted trends. It is also noteworthy that community attachment was strongly correlated (tau = 0.33) with community satisfaction.

On the East Coast the overall attachment of people to their communities was fairly high, with around three-quarters of the respondents saying they had a very/somewhat strong attachment. There were some differences in the degrees of involvement of residents in local community events. Respondents in area 1 were more likely to be considerably or very involved (28 percent), followed by area 3 residents (22 percent) and area 2 residents (17 percent). As on the West Coast, the East Coast survey results showed a strong correlation between community satisfaction and attachment (tau = 0.31).

On a more personal level, social cohesion was examined by determining whom respondents felt they could count on to help in a crisis situation. For the majority (73 percent) on the West Coast, they were family members followed by friends (21 percent) and others (3 percent). Overall, only 4 percent reported having no one to call on. Here again, there were only minor variations across the communities, and the differences were not significant. Similar responses were reported to the question "To whom can you talk frankly and be totally yourselves?" For 62 percent the response was family members, followed by friends (29 percent) and others (4 percent). Only 5 percent had no one that they felt they could talk to. Once again, responses varied little by community.

On the East Coast in areas 1 and 2, over 90 percent of the respondents said they could count on family to help them in a crisis, followed by friends and then others. This pattern is similar to the findings on the West Coast. Area 3 differed from this pattern in that a slightly higher percentage (70 percent) would count on friends and 63 percent on family. Area 3 is a major service centre for the peninsula. Since many people from outside the community move there for work, it would more than likely have more residents without family in the immediate area. The same pattern for the three areas held, albeit with lower percentages, when respondents were asked with whom they could talk frankly and not have to watch what they said.

Taken together, our findings on community health on the West Coast are consistent in showing that Tofino and Ucluelet residents were more optimistic about the economic outlook and social conditions in their communities than the residents of Port Hardy and Prince Rupert. While the optimistic outlook of Tofino residents was not surprising given the economic growth this community has experienced, the similar results for Ucluelet were somewhat unexpected since this community continues to face economic challenges. Findings on the East Coast suggest areas 1 and 3 were somewhat more optimistic about conditions, while still being concerned about the future of their communities and the province. Area 2 findings reflect the concern with the aftermath of the fishery crisis, as those communities were more dependent on the fishery in recent years and perhaps felt that they had more to lose. In public meetings and discussions, communities in area 2 confirmed that they felt this to be so: they are feeling the acute effects of restructuring. Area 1 is experiencing more chronic or long-term effects, and area 3, with more economic diversity, has greater buffering potential.

RESTRUCTURING AND
PERSONAL HEALTH

As a process that is extensive over time and space, restructuring is by definition a distal determinant of personal health status as reported at a particular time and in a specific locale. Added to the complexity of determination, and therefore the challenge for inference and attribution, is the fact that personal health is determined by multiple factors, and a self-report is fundamentally a subjective assessment. It follows that the pathways to population health as measured by mortality and morbidity rates through time, or to community health, are somewhat more direct than those to personal health. Nevertheless, there are plausible hypotheses that can be investigated within the limits imposed by cross-sectional surveys and self-report data.

In our 2002 Coasts Under Stress survey of the four West Coast communities, self-rated health status was assessed using the standard question now used in many health surveys: "How would you rate your health, in general, at this time?" on a five-point scale ranging from poor to excellent. Additional questions were also asked to determine self-perceptions of physical and emotional health compared to others of the same age, health satisfaction, and current health status and compared with one year ago. These questions were paralleled on the East Coast. As is typical in similar health surveys, on the West Coast self-reported health status was heavily weighted to the positive end of the rating scale; 50 percent reporting their health as either very good or excellent. The results did, however, show significant differences across the four communities: the results in those two categories ranged from a high of 60 percent in Tofino to a low of 47 percent in Prince Rupert. Correspondingly, the proportion rating their health as either poor or fair ranged from 12 percent in Ucluelet to 23 percent in Port Hardy.

An important question is whether the health status reported by residents in these four communities differs from expected levels based on comparable provincial and national statistics. Data from the Canadian Community Health Survey (2000–01) provided a basis for comparison. Those data indicate that 61 percent and 59 percent of respondents for Canada and British Columbia respectively reported that their health was very good or excellent and that 12 percent for both Canada and British Columbia reported their health as poor or fair. With the exception of Ucluelet, the poor/fair percentage for the British Columbia communities was higher than would have been expected from the Canadian Health Community Survey results and was substantially higher for both Prince Rupert (19 percent) and Port Hardy (23 percent). Likewise, the percentages for very good or excellent were lower than expected, and again this result was most marked for Prince Rupert (47 percent) and Port Hardy (49 percent).

Self-rated health status on the East Coast was lower than on the West Coast: 43 percent rated their health as excellent or very good. There were significant differences across areas from a high of 48 percent in area 3 to a low of 36 percent in area 1. These ratings are lower than the ratings for the province as a whole reported by the Canadian Community Health Survey from 2001, where 67 percent of the respondents in that survey rated their health as excellent or very good. It would be important to monitor health status to see if it is becoming poorer or what is influencing the perception of health among people in these communities.

While these descriptive statistics do not control for possible variation in health-related factors between the community, provincial, and national samples, they are random samples in each case, so the fact that the health status differences observed between the samples and between the four communities are consistent with a possible restructuring effect is an important finding. Overall, all the communities reported poorer health than either the province or the country as a whole. Among the four West Coast communities, the poorest health was reported for Prince Rupert and Port Hardy, the places that have been the more seriously impacted by restructuring. Likewise, on the East Coast the poorest health was reported by area 1, which experienced the longer-term or more chronic effects of restructuring. However, it must be noted that it has been affected more by out-migration and would have a more rapidly aging population.

When asked to compare their physical and emotional health with other people their age, residents in all four British Columbia communities generally reported their health as about the same or better than others, and there were only minor variations in responses between communities. Ratings of health satisfaction were also relatively uniform, and a high percentage (74 percent) reported being either satisfied or very satisfied with their health in general. The majority (60 percent) reported their health status as unchanged over the previous year, and there was a relatively even split between those who rated their condition as worse (16 percent) or better (24 percent). Here again responses were not significantly different between communities.

The findings from the East Coast had some similarities to and some differences with the West Coast findings. All three areas reported their emotional health as about the same or better than others, with only slight variation. On the question of physical health, a high percentage of respondents in area 1 rated their health about the same or better than others of their age (93 percent), while only 67 percent of those in area 3 had the same rating. Area 2 was slightly better: 74 percent indicated this higher rating. Like respondents on the West Coast, most East Coast respondents felt that their health status had not changed over the previous year. Ratings of health satisfaction varied on the East Coast. Area 1, despite the lower rating of health status, had the highest rating of satisfaction with health, at 87 percent. Areas 2 and 3 had lower ratings: 80 percent and 77 percent respectively were satisfied with their health.

Still in the realm of personal health-related factors, residents on both coasts were asked to report on how stressful they considered their lives as a whole, to rate seven specific sources of potential stress, and to compare their overall stress levels to a year ago. The percentage from the West Coast reporting their lives overall as either not at all stressful or slightly stressful was quite high at 55 percent, with little variability by community. The percentage reporting that their lives were either considerably or very stressful was therefore correspondingly low at 12 percent overall, which was also quite uniform across the four communities. On the East Coast, these trends were repeated: a small percentage of respondents across the three research areas rated their lives as considerably/very stressful. The overall percentage (7 percent) was lower than that reported for the West Coast.

The Canadian Community Health Survey data provide a basis for comparison with national and provincial statistics, although based on a three-point (none, some, considerable stress) rather than the five-point scale used in this study. Comparing the percentages in the "considerable" category for the CCHS data with the combined percentage for the "considerably" and "very stressful" categories for our CUS survey, we find that reported stress is considerably higher in Canada as a whole (26 percent) and British Columbia as a whole (24 percent) than in our four communities (12 percent) on the West Coast. The CUS results change markedly, however, if the "moderately stressful" rating is included, in which case the percentage increases to 46 percent. Nevertheless, there is additional evidence to suggest that the stress ratings for the CUS communities are relatively low compared with provincial and national figures. Specifically, 19 percent in these communities on the West Coast reported their lives as not at all stressful, compared with 13 percent for Canada and 12 percent for British Columbia.

While this result on first sight may appear counter-intuitive given the socio-economic conditions and stressors in the communities we studied, it is consistent with findings from other studies that show lower levels of stress for smaller communities relative to urban centres. Given the high percentage of the provincial and national population in urban areas, it is inevitable that the CCHS data reflect urban trends. In the follow-up interviews on the West Coast, we were told by many survey respondents that the local environment was a key factor in mitigating individual stress. Highly developed social networks among residents, a strong sense of safety, and an absence of the typical urban stresses of traffic and pollution were considered reasons for low stress levels. Easy access to wilderness areas and the availability of clean water and air were also cited as environmental features that eased or prevented stress. All these factors would hold for the East Coast findings as well. People in these areas, while admitting that they were concerned with their financial situation and worried over the future of their communities, could somehow put these concerns into perspective, and this translated into lower perceived stress levels. We will provide more detail on these issues in the next chapter.

However, the resilience implied by overall stress levels was countered somewhat by the ratings of specific sources of stress by respondents on the West Coast, particularly those related to employment and finances. In the case of employment, 35 percent reported their situations as either considerably or very stressful, compared with 29 percent rating them as not at all stressful. The corresponding figures for financial situations were 27 percent and 29 percent. Moreover, stress levels for employment and financial circumstances differed significantly by community. In both categories, higher percentages were reported by residents in Port Hardy and Prince Rupert – 35 percent reported employment as considerably/very stressful in both communities, and for financial situations the figures were 30 percent and 27 percent in Prince Rupert and Port Hardy respectively. In contrast, 28 percent in Tofino and 23 percent in Ucluelet reported their employment situation as considerably/very stressful; the corresponding figures for financial situations were 19 percent and 20 percent. On the East Coast, the two leading sources of stress across the three research areas were employment and financial situations: 25 percent and 18 percent of respondents, respectively, rated these situations as considerably or very stressful.

These marked differences in stress related to employment and finances reflect the varying economic conditions in the communities, as discussed above and detailed in chapter 9. Our follow-up interviews with West Coast respondents confirmed the repercussions of these stressors for individuals and families. Many people discussed how financial stress has led to increased spousal abuse and marital breakup in their communities. Unemployment and underemployment have also led to social restructuring, as one spouse (usually the husband) leaves to work elsewhere, while the other spouse and the children remain in the community. Individuals and families experiencing income-related stress may turn to local social-service agencies for assistance. In Prince Rupert, for example, the Salvation Army had seen the demand for its daily lunch program increase by 150 percent in the eight months between July 2002 and March 2003. This rapid increase in need required more resources from the community in the form of food and monetary donations and was putting financial strain on the non-profit groups providing these services. It was therefore having negative repercussions not only for individuals and their families but for the communities as a whole. A similar pattern of community differences emerged for changes in stress over the past year. For Prince Rupert and Port Hardy the percentage rating their lives as somewhat or much more stressful was 32 percent and 30 percent respectively. The corresponding figures for Tofino and Ucluelet were 23 percent and 22 percent.

We conclude, then, that while the ratings of overall stress appear quite positive when compared with provincial and national data, the more detailed results for specific sources of stress and changes over time are consistent with a restructuring effect, particularly in Prince Rupert and Port Hardy, where employment and financial stressors were frequently reported. Supporting evidence comes from examining the correlations between overall stress and specific sources, which showed strong relationships between overall stress and financial (tau = 0.41), employment (tau = 0.41), and social-support (tau = 0.32) stressors. Similarly, there were significant correlations between changes in stress and financial (tau = 0.19) and employment (tau = 0.21) stress levels.

In Newfoundland we also looked at sources of stress. Employment situations were by far the biggest source of stress: 13 percent of respondents rated them as very stressful; financial support was very stressful for 11 percent. Family relationships were not rated as very stressful. As for the correlations between overall stress and specific sources of stress, two areas showed strong relationships: financial situations (tau = 0.48) and employment situations (tau = 0.45).

MODELLING SELF-REPORTED HEALTH AND STRESS

As previously indicated, our work draws on the recent population health literature in identifying social capital and social cohesion as primary mediating factors in the relationship between restructuring and the health of residents in the CUS communities (Dolan et al. 2005). In order to examine their possible effects in more detail and with more precision than the descriptive results reported so far allow, a series of logistic regression models were estimated using self-reported health, emotional health, and stress as the outcome variables. The models also included lifestyle and socio-demographic factors in a dual role: as known determinants of health status, and therefore as control variables, but also as possible mediators of restructuring effects, for example, through the effects of industrial change on population composition, income levels, and health behaviours (e.g., smoking status).

In the model for self-reported health, health status was dichotomized by combining ratings of poor and fair health and ratings of good, very good, and excellent health. Independent variables in each of the four categories (social cohesion, social capital, lifestyle, and socio-demographics) were selected for inclusion in the models based on their statistical association with health status. The variables were entered in two blocks: first, the socio-demographic and lifestyle variables as control variables and then the social-capital and social-cohesion variables. Eight of twenty-one variables were retained in the final model (table 8.10) as significant predictors (p<0.01): age, education, and income (socio-demographic factors); alcohol consumption, exercise frequency, smoking status, and stress level (lifestyle factors); and community satisfaction (social capital). The odds ratios showed that the probability of reporting good, very good or excellent health decreased with age, no post-secondary education, lower income, less frequent exercise, smoking, alcohol consumption, higher stress level, and lower community satisfaction. The model correctly classified 70 percent of the cases.

Nine variables were significant predictors in the final model for emotional health (table 8.10). The probability of reporting the same or better emotional health than others was higher for women, for those married or living with a partner, and for those reporting higher levels of education, an improved personal financial and employment situation, more frequent exercise, and a lower stress level. It also increased for those perceiving an improved economic and employment situation in their community and greater satisfaction with the quality of community services. In addition, those living in Tofino or Ucluelet were more likely to report better emotional health than residents of either Port Hardy or Prince Rupert. The model correctly predicted 84 percent of the cases.

In the stress model (table 8.10), stress level was dichotomized by combining not at all and slightly stressful as one category and moderately, considerably and very stressful as the other. The probability of reporting higher stress increased with no full-time employment, smoking, less frequent exercise, a worsening financial or employment situation, poorer health status, lower community satisfaction, more negative perceptions of community services, and lower levels of social support. Sixty-five percent of cases were correctly classified with more

Table 8.10
Odds Ratios for Reporting Health Status, by Logistic Regression Analyses

Variable	Self-rated Health, N=990 (82.2%)				Emotional Health, N=990 (82.2%)				Stress Level, N=989 (82.1%)			
			95.0% C.I.				95.0% C.I.				95.0% C.I.	
	B	Exp (B)	Lower	Upper	B	Exp (B)	Lower	Upper	B	Exp (B)	Lower	Upper
Gender	.752**	2.121	1.474	3.051	.573*	.564	.324	.981				
Age												
Ethnicity												
Marriage	.538**	1.713	1.199	2.447	.574*	.563	.328	.966	−.201	.818	.603	1.109
Education					.589*	1.803	1.053	3.087				
Employment									.764**	2.146	1.319	3.493
Income	.414**	1.513	1.188	1.928								
Alcohol	.738***	2.091	1.458	2.999								
Smoking	.718***	2.051	1.426	2.951					.475**	1.607	1.189	2.173
Exercise	.644***	1.903	1.331	2.722	.737*	2.090	1.192	3.665	.266	1.304	.990	1.718
Personal, financial, and employment change					.447**	1.564	1.176	2.080	.156*	1.169	1.001	1.367
Emotional health									1.493**	4.448	2.277	8.691
Health status					1.580**	4.855	2.500	9.430	.222*	1.248	1.039	1.500
Stress level	.590**	1.805	1.263	2.579								
Port Hardy					−2.019	.133	.016	1.072				
Prince Rupert					−2.154*	.116	.015	.896				
Tofino					−1.300	.273	.030	2.438				
Community satisfaction	.554**	1.740	1.171	2.587	.788*	2.198	1.171	4.125	.433*	1.542	1.002	2.375
Community attachment												

Table 8.10
(continued)

Variable	Self-rated Health, N=990 (82.2%)				Emotional Health, N=990 (82.2%)				Stress Level, N=989 (82.1%)			
	B	Exp (B)	95.0% C.I. Lower	95.0% C.I. Upper	B	Exp (B)	95.0% C.I. Lower	95.0% C.I. Upper	B	Exp (B)	95.0% C.I. Lower	95.0% C.I. Upper
Community involvement												
Community services									.624**	1.866	1.386	2.512
Community problems												
Community change					-.525*	.592	.383	.914				
Community participation	.429	1.535	.998	2.363								
Community support									-.514**	.598	.425	.843
Constant		-6.715**				.788*				-8.516**		
Model fitting χ^2		129.575				85.630				140.778		
Correct predictions (%)		70.0 (67.6 and 70.5)				80.9 (68.6 and 81.8)				64.9 (47.0 and 79.8)		

Note: ** indicates that the coefficient is statistically significant at the .01 level; * that it is statistically significant at the .05 level. B: coefficient; Exp(B): odds ratios.

accurate prediction for the lower stress level group (77 percent) than for the higher stress group (50 percent).

Taken together, the findings for the three models are very instructive about the relative effects of the four sets of variables included, based on the population health literature, as predictors of health status and stress – socio-demographics, lifestyle factors, social capital, and social cohesion. The contrast between the two health models – general heath status and emotional health – is especially striking. For the first, the predictors are primarily the socio-demographic and lifestyle factors, which are very well established from numerous previous studies as health determinants; in other words, the control variables are dominant, whereas the hypothesized mediating factors for restructuring effects – social capital and social cohesion – are conspicuous by their absence. That said, two important qualifiers bear on the interpretation of these findings: first, the socio-demographic and lifestyle factors are, as previously noted, plausibly playing a dual role, both as control variables and as mediating factors, given the likelihood that restructuring impacts on both and so potentially compounds their effects on health status; and second, the social capital and cohesion indicators included in the analysis are measured as residents' perceptions and not as direct indicators of the quality and quantity of social resources, networks, and relationships. To that extent, their effects may be filtered in ways that are uncertain and unknown within the intrinsic limits of cross-sectional self-report survey data.

Nevertheless, the contrast with the results for the emotional health model is still very noteworthy, because, in that case, the social capital and cohesion variables are stronger predictors, and community of residence per se is a significant predictor. So for emotional health, there is certainly stronger evidence for a restructuring effect, an effect mediated by social capital and cohesion and linked to the particular community in ways that correspond with the recent histories of change in those places; hence the poorer emotional health in Port Hardy and Prince Rupert as compared with Tofino and Ucluelet. The same general conclusion applies to the findings for the stress model in terms of the relatively stronger effects for the social capital and cohesion factors, although in that case community was not a significant predictor, which is consistent with the descriptive statistics reported earlier that show relatively low and uniform stress levels for all four places.

CONCLUSION

A major objective of the CUS project was to examine the effects of restructuring on socio-demographic changes and the health of coastal communities and their residents. The findings reported for demo-graphic and income changes show quite profound effects for both East and West Coast communities that are consistent with, and plausibly attributed to, restructuring processes and events. In terms of popula-tion health, community variations in mortality and morbidity rates, when compared with provincial rates, are suggestive of restructuring effects, even though direct attribution cannot be inferred from the ecological analysis conducted.

Four questions guided the analysis of the health survey data: Do lev-els of self-reported health status and stress differ from expected levels? Are there significant differences by community in health status and stress? What factors explain community variations? And to what extent can the findings be attributed to the effects of restructuring? The results indicate contrasting findings for health status and stress for the first two questions.

When compared with national and provincial statistics from the Canadian Community Health Survey, the results show poorer health status for the four British Columbia communities combined than was reported for British Columbia and Canada in the national survey, while by contrast, stress levels reported for the British Columbia com-munities were lower than those reported for the province and Canada in the CCHS. This latter finding is consistent with a general tendency for residents of smaller communities to report lower stress than those in larger urban centres. In terms of community differences, health sta-tus varies significantly across the four communities on the West Coast, whereas stress levels are relatively uniform. Moreover, the pattern of the health status differences is consistent with the recent history and impacts of restructuring in the four communities, since lower health status is reported in Prince Rupert and Port Hardy than in Tofino and Ucluelet.

In Newfoundland, a different pattern was found for both the current rating of personal health and the level of stress. Areas 1 and 3 were most similar in the percentage of respondents who reported their life as moderately, considerably, or very stressful. Area 2 residents reported higher stress ratings. Likewise, area 2 had the most respondents rating their heath as poor or fair, followed in order by area 1 and area 3. On all measures of health and stress, area 2, the area with the most recent experience with economic restructuring, had poorer ratings of self-reported health and higher stress levels. In essence, this means that the communities on both coasts that had experienced the most recent and severe disruptions as a result of restructuring had both poorer health and greater mental stress. That said, we need to understand more about motivations behind the similarities and differences on both coasts, especially in relation to our last two guiding questions: what factors explain community variations and to what extent can the findings be attributed to the effects of restructuring? The explanation of these differences from provincial and national rates and of community variations in health status required additional investigation at the level of field interviews, and it is to those that we turn our attention in the following chapter.

9

The Human Voice of Social-Ecological Restructuring: Jobs, Incomes, Livelihoods, Ways of Life, and Human Health

In his powerful paper on globalization and sustainability, Rees (2002) speaks of a dominant expansionist view of the world driven by global capitalism and based on structural assumptions that growth and productivity are not tied to the environment and that technology will deal with scarcities as they arise. Across the globe, economies are being restructured to fit this model, while "the rhetorical veil of efficiency actually conceals one of the most wasteful and destructive economic systems imaginable." Rees speaks at a general level of the ills and dangers of this neoliberal remaking of the world, and unfortunately, "restructuring" can take on the flavour of an abstraction when discussed on a large scale – but it is very real. This book, and this chapter in particular, seek to bring the argument down to the local level, to give this global restructuring its human face.[1]

We show here how scale asymmetries in policy – which are created with the nation and province in mind but without adequate understanding of, or concern for, impacts on local communities – result in benefits being directed away from local producers, who are then left to face the uncertainties and risks of a globalizing economy without the ability to generate many choices for survival – a modern staple trap. In earlier chapters we documented the impact of a long history of social-ecological restructuring on the natural resource sectors of Canada's East and West Coasts, in terms both of communities and of ecosystems. Now we listen to the voices of the people who live in those communities. People live the consequences of restructuring at work and at home, in their bodies and in their spirit. If that sounds painful, it usually is. The statistical profile in chapter 8 identified the patterns involved in restructuring for people in our study areas in statistical terms. Here now is the very human voice of restructuring.

For households, these voices tell us about key changes in opportunities in resource sectors as a result of industrial, policy, and environmental restructuring (which affects resource availability). They also speak about the impact of changes in income security policies, especially EI and worker's compensation, as well as factors affecting people's livelihoods, including the kind and amount of unpaid work required of people (such as care of family members in the context of reduced local medical services) and the reduced availability of traditional subsistence resources such as fish or wood. We are told how these apparently disparate bicoastal economies share bitter experiences at the level of the households – a tale that should prompt new thinking about how families cope when one or more of these economic "modes" collapses. There is, of course, also the human face of the health aspect of these changing dynamics, especially the anxiety and stress caused by uncertainty, as well as new occupational hazards (such as result from changing work content, intensity, location, and so forth). Chapters 8 and 9, along with the two chapters that follow, make it clear that the health of communities as resilient places and as networks is negatively affected, although there remains surprising strength in these places and among these people, as we saw in the previous chapter and will now hear.

Statistics (as in the last chapter) are absolutely the best way to tell us what happened when we are looking at large scales and comparative questions. However, we then need people to tell us how it feels to live in coastal communities today and why they have to respond to their restructured social-ecological situations as they do. Our key informant and household interviews speak to what people in our communities – individual workers and households – feel about restructuring, especially in terms of how they have been impacted and how they have responded. Over the past twenty years, the coasts have been seriously affected by resource degradation (the environmental health component in social-ecological health), changing resource management regimes, and industrial restructuring and related changes in employment opportunities within and between the fisheries, forestry, tourism, and service sectors. Even on the West Coast where the diversity of primary resource industries was thought to give much more security than did the single-staple focus on many parts of the East Coast, there have been major problems. In Port Hardy, Prince Rupert, and Ucluelet and throughout the study area, there is a general consensus that the economy is not healthy. Even among individuals whose own jobs remain secure, it is acknowledged that the local economic situation is precarious. The mayor of Port Hardy sees the local economy "probably at its bottom right now, not moving one way or the other," while the mayor of Prince Rupert described the present economic situation as "terrible, absolutely terrible." Many people expressed concern that forestry and fishing, formerly key primary resource industries, no longer have the capacity for employment of ten to twenty years ago. As well, these communities have seen job reduction in related secondary industries (e.g., wood and fish processing). The negative economic situation has been exacerbated by a downturn in several sectors at the same time:

[T]his town was booming, it was doing very well. We had a copper mine, the logging industry was up and running very well, the fishing industry was doing really well. Since then, the mine is shut down ... and it was about at the same time that the fish industry was drastically cut back. And at the same time the logging industry took a bit of a beating, so it all kind of came at once. (Port Hardy key informant)

From a community development officer in Alert Bay in 2003: "That's in Port Hardy, that's in Prince Rupert, that's in Bella Bella, Bella Coola, everywhere that I've been. You go to Port Hardy and that mall right now. There's three stores. You go to those malls in Prince Rupert right now there's probably ten less than there were last year. Everything's closing. All over the place." The situation is clearly grave, and highly stressful for many.

One of the conceptual problems people and their politicians have to face when coming to grips with the human impact of restructuring is that the politico-economic focus on cash incomes, and the prioritizing of per capita or household cash income as an indicator of health and success, misses an important part of the story. That cash-income-based perspective reduces people to atomized individuals – mere inputs into a production process – whose individual incomes could be maximized by moving away from declining places. In this view, gendered and intergenerational effects on households are ignored and communities become disposable nodes in a production network (witness the dismantling of a company town, as we related that event in chapter 5); the human costs are not relevant – or even officially recognized.

Our social-ecological perspective, however – like Rees' ecological economics – recognizes that people are not atomized individuals, that they make decisions based on household and larger needs, that their lives are made up of more than cash income, and that their decisions are based on balancing many factors, such as caring responsibilities, community health, environmental health, cultural resources, familial connections, traditional roots in communities. Taken together, these make up the strategies for decision making that seeks successful and healthy households and communities as its goals. This perspective also situates the cash inputs into a family economy with the other inputs from the informal (hunting, fishing, barter, gift, labour exchange) economy and from state transfer payments like Employment Insurance, the Old Age Security pension, social assistance, and disability insurance. A coastal family gathers these things into household economies (often linked across households) within which it feeds, houses, and clothes itself. Each different kind of input (cash, informal, transfer payments) is balanced against the others in a household economy, which Lutz (forthcoming) calls "moditional," since it melds modern and traditional modes of production and reproduction. Keeping in mind that cash, subsistence, and state payments all come together when they reach the household, we turn now to a more detailed analysis of the problems created by restructuring on both coasts.

THE FISHERY

In both our coastal research areas, people have traditionally depended on fisheries and/or forestry-related work. Both industries have undergone dramatic restructuring, including substantial downsizing in the past fifteen years. Looking first at the East Coast, the fishery was the most important sector for more than two centuries (see chapter 3), but the period between 1985 and 2003 was one of substantial downsizing and reorganization within the industry. This restructuring was triggered by the collapse of the groundfish stocks in the early 1990s and resulting moratoria on groundfish harvesting introduced by the federal government starting in 1992. Fisheries and plants closed, and Newfoundland and Labrador groundfish landings shrank substantially between the late 1980s and late 1990s, with a 90 percent drop from 400,000 metric tonnes in 1988 to less than 30,000 metric tonnes by 1993.

Although there has been a significant recovery in the industry since the groundfish crisis began in the early 1990s, the recovery has been associated with major changes in the shape of the industry related to the species harvested and processed, to the products generated, and to new professionalization legislation affecting individual rights of access of fishermen and women (chapter 3). In the absence of cod, fishers in both the inshore (with vessels under thirty-five feet in length) and midshore (thirty-five to sixty-five feet) sectors are now highly dependent on the harvesting of snow crab, lobster, and, in some cases, shrimp, while shellfish (particularly crab and shrimp fisheries) have expanded since the early 1990s to become the dominant species harvested at present. In 2002, the production value of the provincial fishery reached an all-time high at over $1 billion. The landed value of all fish was $515 million, but $421 million of this (82 percent) came from shellfish. Ironically, a species that was once viewed as worthless by-catch, snow crab, has emerged as the foundation of the most lucrative fishery in the province's history and is now the economic bedrock of approximately forty coastal communities with crab plants and of some others with crab fisheries but no plants. Those without access to snow crab are really struggling as plants have closed in many communities. Snow crab produced $235 million in 2002, compared to $22 million in 1988.

The production of other types of shellfish is also increasing as crab quotas have begun to decline in recent years and the industry faces serious challenges related to competition from China, the increased value of the Canadian dollar, and quota cuts. In 1998, the landed value of shrimp was only $18 million, compared to $143 million in 2002 (Newfoundland and Labrador Fisheries and Aquaculture 2002, 2–3). Shrimp has been less profitable than crab because of problems with low prices and competitive markets, and it is very difficult to sustain a fishing enterprise on either lobster or shrimp alone. Most harvesters who have crab licences have higher incomes than ever before. In some cases, those with licences for crab or shrimp view the notion of promoting recovery of the cod stocks as a threat to these fisheries. Meanwhile, wild cod prices are lower than scarcity would suggest because of marketplace problems linked to lack of sustained supply and competition from other forms of whitefish and from aquaculture products. Under these circumstances, some harvesters and processors appear to be resistant to the idea of a return to the "cod-dominated" ecosystem that came before (others depend heavily on small cod quotas and would welcome an increase).

While there is no question that the snow crab and shrimp fisheries have brought prosperity to some coastal residents, the benefits of the fisheries are much less equitably distributed than was the case with groundfish. Those in the midshore fleet, many of whom got licences in the 1980s, are now entitled to much larger quotas than are those who entered the fishery in the 1990s. Small-boat harvesters were granted temporary permits and small quotas after the moratorium, and it was only through the sustained efforts of inshore crab committees throughout the Island that they were able to have them turned into permanent licences (Newfoundland and Labrador Fisheries and Aquaculture 2002, 2–3). The quotas allocated to these individuals are a mere fraction of those granted to the midshore fleet. While the specific numbers vary from one region to another, depending on the perceived abundance of the resource and the number of active harvesters, it is not uncommon for harvesters in the midshore fleet to receive quotas that are five to ten times those of inshore harvesters living in the same communities. With crab prices in the vicinity of $2.50 per pound in some years since the early 1990s, that could translate into hundreds of thousands of dollars per year.

Workers in the processing sector and other occupations have not benefited nearly as much from the crab boom. Most of them have had a much harder time making ends meet in the post-cod era, and many now face tremendous uncertainty about whether they will be able to remain in their communities in the future. This uncertainty has led to growing tensions in many areas. The problem has been exacerbated by the fact that many crab harvesters are able to catch their quota in a few trips, leaving them free to draw full EI for most of the year. Many people in other fisheries feel that they work much longer hours and just as hard but are not compensated nearly as well, and they often do not get enough hours to qualify for EI. Similar inequities exist in the shrimp fishery.

The local communities have each been affected a little differently by this sectoral restructuring.[2] The Labrador Straits, for example, went from having a plant in each community to having only one remaining plant – the Labrador Shrimp Company, an early player in the shrimp industry. The company took over a local fresh-fish plant in Lance au Loup in 1979 and then decided to build a major multi-species plant in 1988, just as the downturn started. It switched to turbot caught by one company boat and two local big boats and was able to keep going and even expand employment to 200 by 2001. Similarly, the Port au Choix plant had diversified to a mix of cod and shrimp and was able to continue its shrimp production after the moratorium. However, this strategy involved a reduction from over 400 to 150 workers. Furthermore, the remaining workers got much less work, as the operating season decreased from thirty-five weeks in 1995 to just seventeen weeks in 2000. Because White Bay South was never as dependent on groundfish, it was less affected by the moratorium. Its crab plant was able to continue production, and in 1998 a new shrimp plant opened in the area, creating about one hundred fish-processing jobs.

The plants that remain in the case study area all have an aging workforce – the Labrador Shrimp Co. reported many workers have over twenty years seniority, while the average age in the Port au Choix plant is 47.5. At the time of our interviews many key informants raised concerns that the same problems were developing with the newer stocks as had occurred with the cod – shrimp sizes were down, for example, and the market was glutted. The future of the plants, and hence of employment, is therefore still a concern. And workers worry day to day about seniority and hours, hoping to get enough time in each season for an EI claim. They also worry about their health, as new occupational hazards come with the new species, such as the respiratory illnesses associated with shellfish (Howse et al. forthcoming).

On the harvesting side, all areas have seen a great reduction in the number of inshore boats and changes in species. In one Labrador Straits community, for example, the number of inshore boats decreased from fifteen to twenty to five or six. Rather than training sons or hiring helpers, many inshore fishers are fishing with their wives to keep money in the family. As one fisher put it, "Well that's why she's going ... why should I pay some money or something to a young feller, who's only half as good as she is aboard ... and put it all in the one household." Others have taken work on bigger offshore boats. There are also a small number of fifty-to-sixty-foot boats in each area. In the Straits area some displaced cod fishers tried to get licences for other species, such as crab, and some moved into the plant. In the Port au Choix area some inshore fishers turned from cod to lobster, and the mobile fleet changed from a mix of shrimp and groundfish to just shrimp. The recent professionalization of fishers has meant increased annual monitoring to sustain their fishing rights and financial pressures because of additional fees related to certification and training.

While conservation measures at the federal level seek to protect overfished stocks, then, both the industry and the local communities have shifted to new fisheries in an attempt to provide industrial flexibility and community resilience. These new fisheries, however, while generating relatively high revenues, also require fewer people and so reduce employment: a pattern that is in line with much postindustrial restructuring in all natural-resource extraction industries (see chapter 11). Indeed, once again restructuring can be seen to be exacerbating community inequities and reducing access to fish resources for many traditional users. At the same time as employment falls in the wild fisheries, there is increased employment in aquaculture (see chapter 15), although there is growing evidence now that the increase will be short-lived. Closer examination of local fisheries and specific fishing communities demonstrates diversity of circumstance and impact but also illustrates common themes associated with the larger-scale issues discussed above.

On the West Coast, despite the differences in the two marine ecosystems, we found very similar stories. In the 1990s, declining stocks of Pacific salmon, fishing fleet overcapacity, and a 30 to 50 percent decline in prices for all salmon species necessitated substantial restructuring of the West Coast fishery. The Department of Fisheries and

Oceans introduced the Pacific Salmon Revitalization Strategy, referred to as the Mifflin Plan, in 1996 to address these issues. Through a series of changes to commercial fishing regulations, the plan aimed to reduce the West Coast fishing fleet by 50 percent over several years, thus conserving threatened salmon stocks and improving the viability of the fishing industry for remaining fishers. This plan resulted in the buy-back of 798 licences and a loss of fisher-related income for 2,750 individuals in the first year of the program. Associated jobs in the fishery supply sector were also lost (Gislason 1998). By 1999, the number of commercial salmon licences in British Columbia had dropped to 2,557, down from 4,112 in 1996.

All the study communities traditionally had substantial locally based commercial fishing fleets, and Port Hardy, Prince Rupert, and Ucluelet are home to processing facilities for the commercial fishery, while Alert Bay and Hartley Bay have joined the dozens of coastal communities whose processing plants no longer operate. This illustrates a key point: the current restructuring has to be understood as occurring in a time of already severely depleted resources and on the heels of a spiral of downsizing that has been in process since World War II and has accelerated in the last decade.

In 1990, the commercial salmon catch in British Columbia was 96,396 tonnes; by 1996 it had fallen to 34,978 tonnes and in 2002 the annual catch was 33,150 tonnes. As salmon catches declined, some fishers turned to other species, such as hake, the landed catch of which increased from 79,453 tonnes in 1990 to 99,879 tonnes in 1996. However, this species has also proven vulnerable to natural fluctuations and overfishing, seeing a sharp drop in 2000 to 22,347 tonnes harvested. For those who remain employed in the commercial fishing industry, the inherent unpredictability of erratic harvests, variable fish prices, competition from foreign off-shore processing ships, and closure of fish plants is a source of stress and insecurity.

The downsizing of the commercial fishing industry, which was implemented to stabilize the fish resources that were available to communities by limiting extraction, has had economic repercussions for individuals and the communities themselves. Some interviewees recognized that changes were necessary to maintain the viability of the commercial fishery but were concerned that the Mifflin Plan disproportionately impacted small, independent fishers and smaller communities:

And family fishing, they had that big buy-out a few years ago when they started the conservation measures, which were really needed and most people recognize that, you know that if they don't start saving the salmon stocks, that it is going to be bad forever and get worse and worse. But it is the same thing, they kind of forced a lot of guys to sell their boats and so there are very few trawlers, there is more concentration in the big companies that own boats. (Tofino respondent)

When I moved here fifteen years ago, a good deal of the fishing fleet was based in Prince Rupert so it meant that the incomes that were earned remained in the community over the wintertime. Again, regulatory changes have changed that whole mix a lot, so I would say that the bulk of the people who are fishing in the north are actually southern vessels, so the money goes south. (Prince Rupert key informant)

That sense of grievance against "the south" was a repeated theme, emphasizing the problems of a lack of economic diversity and dependence on a limited number of industries that have combined to make communities more vulnerable during economic restructuring. This view was especially prevalent in Port Hardy and Prince Rupert:

I just think it was a lot of things happened at the same time and it made for a really bad combination. You look at it down south, where there is so much diversity in their industry, they can shift between one thing and another whereas up here, while they have been working for a little while, it just hasn't come yet, you know. (Prince Rupert respondent)

Once the third largest port on British Columbia's West Coast in terms of tonnage of fish landed, Ucluelet's fishing fleet has been drastically reduced as a consequence of the Mifflin Plan. The village subsequently faced reduced employment at the three local fish plants and also experienced significant loss of forestry jobs following the 1993 anti-logging protests in Clayoquot Sound. When we spoke with people involved with these sectors, there was marked concern over the viability of Ucluelet as a "traditionally blue-collar high-income community."

The changes hit First Nations communities particularly hard. The fish boats in these communities were typically smaller family boats that were used commercially only in the short salmon season but that provided transportation and access to food resources the rest of the year. It was precisely these smaller, "less efficient" vessels that the Mifflin Plan was trying to get rid of in favour of a high-technology, multipurpose corporate fleet. As Roy Cranmer (2003) attests about Alert Bay: "We had a fleet of 60–70 boats here. Each one of them had 5–6 guys. 300–400 people with them, just on the boats. And that wasn't counting there – we had about six or seven ladies that used to work on gillnets." Today there are only three commercially licensed boats in the community.

FORESTRY

As the traditional industries offer less work as a result of technological changes, resource depletion, and industry concentration, communities and governments have had to adopt a new economic model in which full-time permanent employment is no longer as prevalent as it once was:

> Individuals, businesses and the community outlook as a whole has gone away from that focus on, "I'm a logger," to "Yes, I log, and on the side I do something on my computer, or I go fishing for a month" ... Because so many of the downturns have produced seasonal effects in forestry, fishing and so on, that some people can jump into that concentrated month for fishing, for instance, and then go back to logging for nine months, and they end up with two months of nothing. (Port Hardy key informant)

The forestry industry on the Great Northern Peninsula of Newfoundland has also seen dramatic changes since the early 1980s (Sinclair, Macdonald, and Neis 2006). Pulp companies dominate the industry, the two main actors being Kruger (Corner Brook Pulp and Paper), which bought Bowater in 1984, and Abitibi. Kruger took over Bowater's considerable forest allotments in our study area. Instead of hiring their own loggers, the company changed to working with contractors. Furthermore, mechanical harvesters have come to dominate the logging industry over this period, each doing the work of about twelve to fourteen men. As a result, there have been considerable changes in the conditions of work and the number of contractors and loggers over the past fifteen years (in White Bay South one estimate is that the number of loggers has decreased from around four hundred to barely one hundred). While some older conventional loggers are able to continue to work, younger ones are not being taken on. This period also coincides with the introduction of new forest management policies and regulations, which have further shaped the industry. As part of an effort to improve the yield from the forest, a modern lumber sector has been encouraged, both by regulatory policies and by government support. New integrated sawmills, such as the one in White Bay South, provide employment; however, smaller sawmills, like small logging contractors, have trouble in the new milieu.

Restructuring in the British Columbia forestry sector, which was always more developed than that of the East Coast, is still very similar to the East Coast picture. Large-scale changes in the market economy (e.g., international duties, unstable foreign markets), excess capacity, and degradation of the natural resource base have had serious repercussions for coastal communities that depend in part on forestry for employment and community infrastructure. Primary forestry was traditionally an important employer in Port Hardy, Ucluelet, and Tofino, and wood processing has played an integral economic role in Prince Rupert, while being an important contributor to the economies of Port Hardy and Ucluelet. All four communities have seen employment opportunities in forestry sectors diminish in the past ten years, though Port Hardy has been hit less hard than the other three communities.

In Prince Rupert, the closure of the Skeena Cellulose pulp mill in June 2001 had serious repercussions for the community, since it was the largest private-sector employer in the city. The closure resulted in direct job losses for 750 employees and indirect job losses for 1,331 people (Prince Rupert Economic Development Commission 2002). In a community already experiencing downsizing within the commercial fishery and fish-processing industry, reductions in port activities, and the subsequent closure of another local timber-processing facility in October 2001, the closure of Skeena Cellulose illustrates the economic vulnerability of resource-dependent communities. Elsewhere on the coast, the story was not dissimilar. In Alert Bay, "Oh, it's a lot harder [than ten years ago]. I mean, I know the guys that used to go out logging can't get a logging job now" (Cranmer 2003).

The conservation of coastal temperate rainforest in Clayoquot Sound has significantly influenced community health in Tofino and Ucluelet. In response to mass public protests in the early 1990s over the planned logging of substantial portions of intact old-growth forest in Clayoquot Sound, logging operations in the area were halted and the Scientific Panel for Sustainable Forest Practices in Clayoquot Sound was convened (Ecotrust Canada 1997). The panel's objective was to recommend sustainable forest practices for Clayoquot Sound, integrating the knowledge and values of local indigenous and non-native peoples and incorporating the most current scientific understanding of forest ecosystem functioning within the context of several international conventions to which Canada was a signatory (Clayoquot Sound Scientific Panel 1995). In July 1995, the provincial government adopted the scientific panel's more than 120 recommendations in their entirety, which rendered obsolete former forestry regulations in the area and ended conventional logging practices in Clayoquot Sound (Ecotrust Canada 1997). The land use decisions in this region have had profound, though different, effects in Tofino and Ucluelet, contributing to an expanding tourism industry in Tofino, while causing substantial job losses and business closures in Ucluelet.

The perception among many coastal residents is that restructuring within the forestry sector, precipitated by both economic and political decisions, has not only resulted in layoffs from large forestry companies but has also had negative consequences for small, independently owned operations. In both Port Hardy and Ucluelet, interviewees involved in the forestry industry commented that it is very difficult for small operators to maintain a viable business in the present economic climate. Small forestry businesses that have provided employment and contributed to the local tax base have closed or relocated. There is little incentive for individuals to invest time and money in forestry-based businesses:

INTERVIEWER: Are there a lot of small logging companies in town?
RESPONDENT: Not so much anymore, but there's a number, anywhere from two to twenty men that are working together, there used to be lots and lots of them like anywhere from two to five.
INTERVIEWER: So little companies could make a go of it?
RESPONDENT: Yes, you could keep a half a dozen people going in groceries and stuff and living the lifestyle they enjoyed where that's not quite the case anymore. That's not the way it is anymore, that's the reality of life, I think some of it was an impact directly from government policies but, other things are just the transition of life. (Port Hardy respondent)

If I had known ten years ago what I know today, I would have left this town. In 1994 I would have just sold everything and got out. But I put my trust in the government who was promising all kinds of good things, like value-added industries, jobs, jobs, jobs, I don't know if you remember that or not ... All it means to me nowadays is a photo-op for some bureaucrat or politician, because nothing, absolutely nothing. In fact those [small forestry] businesses, who put up their own money and tried to get started, were broken and basically sent out of town. (Ucluelet respondent)

EMPLOYMENT INSURANCE/INCOME ASSISTANCE

While the fishing and forestry industries have been restructuring and downsizing on both coasts and tourism and aquaculture have been on the rise, some support programs of critical importance to these largely rural, seasonal, and often relatively low-income women and men have also been changing. Changes to unemployment insurance regulations enacted in 1996 made it more difficult for some seasonal workers to qualify for benefits and reduced the level of benefits for many who did qualify. (This rule was eventually abandoned: see chapter 4.)

In a high-unemployment region, where the unemployment rate is above 13.1 percent (as is the case with our East Coast communities), employees needed fourteen weeks of work to get a good benefit (because of the minimum-divisor rule, which will average their earnings over fourteen weeks if they have fewer than fourteen weeks of work). Shorter seasons in fish processing have made this difficult for plant workers, and most tourism operators find it impossible to provide this amount of employment for their workers. Tourist operators have had trouble attracting labour, and fish plant workers are competing with each other for "stamps" (i.e., to get enough work to qualify for UI/EI). An EI regulation requiring new entrants/re-entrants to have 910 hours of work to be eligible for benefits (regardless of the regional unemployment rate), rather than the 420 hours normally needed in a high-unemployment area, also causes difficulty. You are classified as a re-entrant if you had less than 490 hours insurable earnings or benefits in the year before the qualifying year. This limit, which is three times higher than the minimum needed under the old Unemployment Insurance (UI) system, makes it very hard to recover the lost eligibility associated with one bad year. EI changes have also made it more difficult for unemployed people to qualify for retraining and educational support (Payne 2000; MacDonald 1999). Workers go to great lengths, often to the detriment of their health or family life, to try to qualify for EI and avoid the re-entrant trap, leaving their families in order to work in another province for a few weeks to get the hours needed for EI, for example, or working with occupation-related health conditions such as shellfish asthma (Howse et al. forthcoming).

On both coasts there is a strong sense that changes to employment insurance (EI) and income assistance (IA) regulations have made it more difficult for people experiencing employment difficulties to access social support. In the case of EI, individuals must accrue minimum hours or earnings in the year preceding their claim in order to qualify for benefits. Within the fishing industry, shortened seasons at fish plants and reduced commercial-fish openings prevent many processing workers and some fish harvesters from qualifying for EI based on their fisheries work for the duration of the off-season: "What has happened is that because of the short processing season, those who are qualifying for EI, they are working two plants, so it is 24/7 that they are going. The amount of hours that they managed to get only gets them through to maybe December for most of the shoreworkers ... for the average shoreworker, it is a real struggle. There are a lot of people that just don't qualify" (Ucluelet key informant).

The provincial Liberal government of British Columbia, which was elected in 2001, introduced new regulations governing Income Assistance. Meeting the requirements has proved challenging for many individuals in Prince Rupert, for example, who, when we spoke with them, had been laid off from the Skeena Cellulose mill for approximately twenty months. Trying to survive financially until the mill opened again, their EI claims having expired, these people faced the necessity of having to sell off personal assets in order to qualify for IA: "[P]rovincially, all that we are hearing are really bad tales of woe, it is so difficult to go on income assistance now, at a time when people really need it, they find it totally humiliating. It is really bad and if you have any assets at all you are not eligible, so if you own a car worth more than $5,000, sorry. And we are talking again about people that have assets, right. So are you supposed to just sell everything?" (Prince Rupert key informant).

In British Columbia mothers with children under seven qualified for social assistance, but recent reforms require them to go back to work when their children are three. The regulations state that adults are technically able to collect welfare for only two out of every five years. The fear is palpable: "Things are at a really critical point," warns Verna Ambers, Community Development officer for the 'Namgis Tribal Council in Alert Bay. "I don't know what people are going to do if we have to implement those welfare things. What are our people going to do?"

First Nations communities are affected in similar ways by changes to Employment Insurance and Social Assistance but face the additional challenges of being legally assigned to "reserves" that were central locations for the fishing industry of the late nineteenth century but are now suffering the effects of a declining fishing industry and remoteness from most other sources of employment. The high rates of dependence on social assistance in Native communities are relatively recent phenomena, dating from the 1960s and 1970s, but have by now become quite entrenched, bringing the concomitant problems experienced in other dependent populations (Lutz forthcoming). Said Roy Cranmer, former chief at Alert Bay, "Between SA [Social Assistance] and the people working in our office, that's what's keeping this island going." Unfortunately, the provincial model of delivery of social assistance has tended to emphasize individualism and has served to undermine cohesion and enterprise in Aboriginal communities.

IMPACT OF RESTRUCTURING ON LIVELIHOOD

The interrelated effects of environmental, industrial, political, and social restructuring have altered the context in which individuals, families, and communities try to sustain livelihoods and health. The household interviews we conducted on both coasts provide rich accounts of navigating the altered terrain. While the migration figures (in chapter

11) tell part of the story, those still "hanging on" have another part to tell.

Family livelihoods depend on the options available to individual family members for paid employment or self-employment, support from the state, and informal activities and unpaid labour that help maintain the family. Just as the transformation from a merchant economy to industrial capitalism altered the structure of livelihoods (see chapter 2), so has the current round of restructuring. Households are challenged to find new ways of putting together a living. While the new outcomes can be elaborated, it is much harder to understand the processes of "getting by" – how decisions are made, the realm of choice, the gender relations that are at play.

There are certain key differences between the post-Confederation period of industrial capitalism and today's reality on both coasts. First, the former period was associated with expanded wage labour opportunities – whether in the woods working for Bowater or Abitibi Price or MacMillan Bloedel or working in a fish plant or a salmon cannery. The high unionization rates achieved in the resource sectors enabled workers to capture an increased share of the rewards. It is also significant that unionization was achieved in harvesting as well as in processing – in both the forestry and fisheries sectors on both coasts. In the current period, while industrial capitalism is still dominant – in the form of Kruger, Weyerhaeuser, and large integrated sawmills in forestry and Fishery Products International or Patterson and other large interests in the fishery – the combination of technological change, globalization, and the cumulative effects of decades of exploitation of the resources is severing the link between corporate success and local employment opportunities. This result is reflected both in reductions in the number of jobs in the traditional sectors (fishing, fish processing, and logging) and in the nature of the jobs that remain in these sectors (loggers now work on their own, rather than as direct company employees; the structure of fishing crews, vessels, gear, and fishing practices has changed; shellfish processing is more automated and less labour intensive than groundfish processing). With contracting out and the growth of tourism, self-employment and small businesses are on the rise, resulting in a concomitant increase in risk and the loss of some of the benefits that accrue to employees (company-provided benefits, such as health insurance, or state-provided benefits, such as EI entitlements). For example, unionization rates have fallen in East Coast forestry (almost 60 percent of the forestry jobs held most recently by our respondents were non-union, compared to one-third of their previous forestry jobs).

While employment has always been precarious in these communities, the dimensions of that precariousness have changed. Seasonality, for example, has altered. Industrialization of fish processing on the East Coast in the 1970s and 1980s created almost full-year employment for many workers. Now, with reduced resources and new species, the processing season has shortened, and many workers are struggling to get enough work to meet EI eligibility. Approximately one-quarter of respondents reported having had difficulty qualifying for EI. Two-thirds of our respondents reported a decrease in hours in the last five years.

There is also a gender dimension to precariousness. For example, while over three-quarters of our East Coast respondents had seasonal work, women were more likely to have on-call, casual, or temporary jobs; women were also less likely to be unionized. Women's employment showed more volatility; they experienced more job changes over time, perhaps because their jobs are more vulnerable (fish processing, tourism, and retail-sector jobs versus fishing, forestry, construction, and trades) or because they may shift jobs more readily in response to family needs or their partners' work.

As elsewhere in Canada, the national trend of rising labour force participation of women is apparent in our East Coast work. In some cases women are moving from "unpaid" work supporting a family fishing or forestry operation to paid work within family enterprises – for example, as more women now fish with their husbands. In other cases women have gone to work to ensure some stability in household incomes when their husbands' incomes are insecure and volatile. In some areas opportunities have opened up for women more than men (for example, in White Bay South, where men have been hard hit by forestry declines and a new shrimp plant has increased employment for women). The tourism sector also offers employment opportunities for women, albeit at significantly lower wages and with fewer protections than fish processing offers.

We noted different trajectories for East Coast families as they tried to make a living in the context of restructuring. One change in family work patterns we examined is the number of women now fishing with their husbands (Grzetic 2004). This practice is best understood as a household response to deteriorating employment options in fishing communities and deteriorating incomes from fishing alongside escalating costs, which constrain the employment alternatives of these women and their control over where and when they go to work. In some cases it is women's choice to go fishing; in others it is a deliberate decision by both spouses in the face of declining fisheries earnings to keep the money in the family rather than pay a helper. In a third scenario, wives are reluctant participants in a decision made by the men. For example, in one community the men got talking and decided that, as one husband put it, "in order for us to survive, this is what we got to do. We got to get our women. It's not enough into it to get a shareman." This change appears to represent the end of a tradition of training sons to take over the family fishing enterprise. Couples fishing together are able to keep the enterprise going until retirement, while the young people are encouraged to go on to other things.

Whatever the household and community decision-making process that has brought women into the boats, they frequently share some serious vulnerabilities. Women are often poorly prepared for the dangers of the job. Fishing together also has its risks, as the couple is literally in the same boat. This is a source of stress: as one fisher woman told us, "I mean I'm out there ... we're out there together and if anything happens to one of us, it's going to happen to both of us ... So I thinks about, when I'm out there I thinks about, well if anything happened today them two little youngsters home their self. That's what I thinks about, right?" Fish harvesters are not the only couples in the same boat – about half the two-earner couples interviewed are both in the same industry, and thus both are vulnerable to downturns in that sector.

How have households managed in this changing economy, as opportunities and conditions of work change for men and women? As in the rest of Canada, the majority of households now have two earners; 90 percent of the one-earner households we interviewed had household incomes under $30,000, as did one-quarter of two-earner households. On the East Coast, about 60 percent of households interviewed said their incomes had decreased or stayed the same in the past five years, and there was significant variation by local area; as many as 50 percent reported decreases in the Hawke's Bay area, hit by both fishery and forestry declines. Census figures show average individual and household incomes declined from 1996 to 2001 in the Hawke's Bay and White Bay South areas, but not in the Labrador Straits overall.

Income is not the only indicator of how family well-being has changed. For example, one family took the fisheries buy-back and considered it a good financial decision but expressed anxiety and regret at the change in their lives. "Financial wise we made a good decision. But lifestyle ... it ended all my dreams of what I had [en]visioned for us." Temporary seasonal migration for work is another strategy for maintaining income that imposes difficulties on families.

The residents of Tofino described their sense of community in different ways than did the residents of Port Hardy, Prince Rupert, and Ucluelet. Words such as "fragmented," "bitterness," "vague," and "dissension" were used when discussing the local sense of community. Although Tofino has successfully made the transition from a resource-dependent community to a community that is capitalizing on the growing tourism and aquaculture sectors, some residents feel that a paucity of social cohesion in the community is detrimental to continued development.

[P]robably one of the chief restraints on Tofino's development ... is that the council could do it if the people would let them. But there is always the sense of dissension in Tofino, everybody has to have an argument, everybody has to have a different opinion. You say "A"; someone has to say "B." It's Tofino's nature, there is not much to be done about it, it is a contentious town. (Tofino respondent)

While a socially cohesive community is defined as having shared values and hopes at its centre, there is strong evidence that diverse views, beliefs, and cultures within a community do not necessarily result in lower levels of cohesion (Jenson 1998). Jenson argues that "any problem for social cohesion arises only when the recognition of diversity and rejection of difference occurs in ways that reduce feelings of belonging or discourage participation." Anti-logging protests in Tofino in the early 1990s resulted in conflict between community members as environmentalists mobilized to preserve the old-growth forests of Clayoquot Sound, while forestry workers tried to protect their livelihoods. The initiation of a global awareness campaign and the mass protests of 1993 resulted in a halt to conventional logging practices in the area – and in many forestry workers and their families feeling ostracized by their community, while facing unemployment due to the cessation of local logging. Many respondents feel that the tensions that pervaded the community in the early 1990s have had negative effects on social capital by reducing social networks and trust among community members.

THE INFORMAL ECONOMY

What has happened to the role of the informal economy, or subsistence, in household livelihoods? This "unpaid" economy has been critical throughout the history of settlement in rural Newfoundland and Labrador and in remote settler and First Nations communities in British Columbia (Ommer and Turner 2004). In Newfoundland, it originally supported the merchant economy and, in more recent times, subsidized the cost of labour to the modern industrial economy. Indeed, subsistence activities and support networks have remained critical components of rural livelihoods, and subsistence and wage jobs are interdependent – subsistence activities stretch scarce dollars, and cash is needed to fuel the subsistence pump. Many subsistence activities, of course, also have a cultural and recreational value and are not engaged in just for livelihood.

In our research on both coasts, we found that many traditional subsistence activities were alive and well, though some traditional opportunities have been eroded by restructuring: access to the food fishery and wood, for private subsistence use, for example, has been severely circumscribed by resource management policies. On the other hand, other elements of restructuring have encouraged subsistence activities. For example, in the 1990s in Newfoundland, the availability of cash from EI or the Atlantic Groundfish Strategy (TAGS) enabled people to buy some of the start-up material necessary for some cost-saving activities, such as building and repairing homes. Constraints on access to cash and to people with the skills and means to carry out these activities could seriously jeopardize subsistence in the future.

In our household interviews on the East Coast more than half the households reported members who cut wood (mostly for fuel), hunted moose (*Alces alces*) and rabbit (*Lepus spp.*), picked berries, fished for cod, and grew potatoes or other vegetables. There is evidence of some breakdown in traditional gender roles, with more shared activity reported by spouses (hunting, for example). People still value going on the land. There are still people who engage in a fairly traditional seasonal round of work – some paid and some unpaid. One man described it in this way:

It's not the thing of making money all year round, it seems to be something to do all year round. Like I starts, say starts Christmas then I go cut our firewood. Then I go to the sawmill for a couple of months. Then coming on spring you go sealing for a month. Then you get your lobster gear ready. Fish then, fish all summer. Well you fish the lobsters till the last of June. Then you guide for a month. Then you fish eels till October. Then I goes trapping in October till Christmas. Then it's all around again.

Subsistence activities are engaged in with spouses, friends, children, and parents and are not always done for immediate personal gain – though what goes around may come around. Among West Coast First Nations, for example,

[I]nformal exchanges, both resource and cultural, have continued to the present day, both within and among aboriginal families and communities, and between them and their non-aboriginal neighbours. There are numerous examples, especially regarding procurement of traditional food, a key resource in communities with high levels of unemployment and low monetary income ... Virtually every northwest coast family that participates in traditional food harvesting is also a part of an exchange network in which some of their harvest is used to procure other valued products. The informal economy ... is enacted through many different social institutions. Potlatching is a critical one; potlatch gifts accumulated by family members of the host to distribute to the invited guests include many nutritious food items – jars of smoked salmon, wild berry jams, jarred soapberries, dried seaweed, and oulachen grease – as well as basketry and other works of art. These hosts will themselves be recipients of such products when they are invited as guests to subsequent potlatches. (Ommer and Turner 2004, 146)

Environmental changes also threaten subsistence activity. As one hunter in White Bay South said: "I used to go up [the Main River] moose hunting one time. I don't go up there now – there is no moose going to live out in the middle of a open cut-over." And resource management policies further limit traditional access, often to people's consternation and discontent with perceived inequities: "There is companies here catches, you know with their own trawlers and that, goes and catches more fish in one day then all Newfoundland would take probably ... well one boat probably like take more then all Newfoundland would take for to put their fish up for the winter."

Some people complain of a lack of cash to undertake subsistence activities. For example, one man explained that he burned oil now rather than cut wood "because for one thing you got to have a truck; got to have a chainsaw. You got to have a ski-doo. And you got to have gas. And then you got to go then a long way then, you got to go way up there in the County camps right into Tender River and go right up County Cap." On the other hand, some people turn to subsistence activities when other livelihood options decline and for something to do. In the Labrador Straits bakeapple picking has become a source of cash income since the development of a business in the area producing bakeapple (*Rubus chamaemorus*) and partridgeberry (*Mitchella repens*) jams and syrups for sale to tourists and for export.

Other community members, however, feel that challenging economic times and decreased access to some social services have strengthened connections between community members and resulted in a stronger sense of community as people come together to support one another.

When things get tough and people are honest with themselves, you need your neighbours, you need them. We've stayed close with our community here, we're involved with our church, but we are also involved outside of our church ... When things get tough, quite often we'll take the boat and go out get a bunch of crabs and groundfish. We ain't got none in my freezer, it's all shared around. We always share. (Port Hardy respondent)

When people are in dire needs, they do come together, small communities do that ... great support for the community. If they have a special event, sometimes they will hold something in one of the bars ... to help out a family, and people just seem to come out of nowhere and throw money at you. (Prince Rupert respondent)

These statements give the impression that informal (see later) social capital – social networks, trust, and reciprocity – is boosted as the negative repercussions of socio-economic restructuring become apparent.

UNCERTAINTY, STRESS, TENSIONS, AND HUMAN HEALTH

The demographic and survey questionnaire data presented in the previous chapter do not always fully reflect the stress that many people experience when they face pressure for an unwanted move. Nor do they indicate anything of the processes that contribute to stress, which can be multifaceted. Qualitative data suggest the possibility of cumulative traumatic effects as the reaction to the reclosure of a fishery or a job loss in recent years is amplified by memories of the cod moratoria of the early 1990s. Demographic snapshots of out-migration also often miss the complexity of migration patterns and their possible health consequences. While young people may just leave, married couples may leave, return, and leave again. To illustrate, one respondent on the Northern Peninsula reports that she and her husband had spent six months on the mainland some years ago. However, they returned to the peninsula, where they struggled to make ends meet. Ellen worked as a store clerk at the time of the interview, and the week before her husband had headed off again and immediately found employment in construction. This time, Ellen stayed behind with their children.

It's been very hard struggling. All you can get in this place is enough hours for your EI (employment insurance). There's got to be programs in place here. Like us, my husband has to go away for six months to get an income. It's happening to a lot of us here. Not a lot; they usually all pack up and go. I think of my kids and their freedom and the city life is difficult. When we first went there the TAGS (the Atlantic Groundfish Strategy) program paid for half his wages. Then when TAGS ran out he wasn't getting more than ten dollars an hour. The store I worked in was only giving me a few hours and you can't live on that ... Now we realize that it was a mistake coming back, realizing now what we got to go through all the time. At least we were a family there. Now we got to be separated. It's hard to move a whole family.

Ellen had concluded that she could not keep going like this. The local area was in decline, and it no longer seemed possible for her husband to return and find a job. Within a year she expected to join him, although she valued her local support network and was concerned about her children adjusting to the move.

Two interviews from different parts of the West Coast suggest how out-migration decreases the pool of skilled labour in each community and thus has repercussions for individual involvement in community leadership and activities and also reduces local economic viability:

[The closure of Skeena Cellulose has] got huge implications, and that sector, again, you don't see it when it's operating, but that is where the hockey coaches come from, that's where the – so socially it has quite an impact on Prince Rupert as well. Anybody that has skills has gone, they have left. (Prince Rupert key informant)

I think that a lot of people who were leaders in the community were also executives at their companies, and they have left. We have seen the achievement levels in the schools, that's a good indication. We have seen that drop because they are coming from families with less expectation is what we surmise from what we have seen. And some of that sophistication in leadership has gone. (Port Hardy respondent)

When we consider the working-age families reliant on fisheries, forestry, or tourism that have continued to live in these coastal communities, it is clear that life has become increasingly stressful. On the East Coast, when we asked working-age adults in such households, "Would you describe your life as presently very stressful, somewhat stressful, not very stressful, or not at all stressful?" thirty-five percent of respondents said that life is somewhat stressful and 29 percent said that life is very stressful. The levels of reported stress were higher for women than they were for men: 32 percent of women, as opposed to 28 percent of men, said that life is presently very stressful, and 37 percent of women, as opposed to 28 percent of men, said that life is presently somewhat stressful. The level of stress being felt by families varied significantly by area. In the Hawke's Bay/Plum Point area, 83 percent said that life was presently somewhat or very stressful, compared to 75 percent in the Labrador Straits and 48 percent in White Bay South. One reason for these differences may be that, as noted earlier, a new fish plant opened in White Bay South, while the other areas had lost fish-processing jobs.

Respondents were also asked, "Compared to ten years ago would you say that life is much more stressful, somewhat more stressful, about the same, somewhat less stressful, or much less stressful?" Twenty-two percent said that life was somewhat more stressful now, and 35 percent said it was much more stressful. Again, people's responses varied by area: in White Bay South, 38 percent said that life was somewhat or much more stressful, compared to 91 percent in Hawke's Bay/Plum Point and 75 percent in the Labrador Straits.

We also asked these families, "Compared to 1991, do you find it harder or easier to make ends meet?" Eighty-two percent of households said they found it harder to make ends meet. This financial hardship, caused mainly by work uncertainty, was reflected in responses to our further questions about stress. Of the forty-seven households that answered the question, "What is the greatest source of stress in your lives?" twenty-five (53 percent) reported that finances related to lack of work were most stressful. Fifty-nine percent of families in our study area reported that their household incomes had either decreased or stayed the same over the past ten years. The lack of options for work, the rising cost of living, and having to help out with their children's post-secondary education were contributing to the pressure experienced by these working-age adults.

Another aspect of stress is that women and men, even though they may live in the same household, do not necessarily feel the same degree of stress. The pathway between restructuring and stress seems to be mediated by the person's household responsibilities and gender, as well as age and stage of life. In many of the households in our study, women are responsible for paying the bills, and when money is tight or uncertain over time, these women may feel more stress than other family members who do not have a direct responsibility for paying the bills. Harold and Pamela (pseudonyms) have been fishing together on the Northern Peninsula in recent years, and in the following excerpt, they describe how they feel about the levels of stress in their home. They relate their stress levels not just to uncertain incomes from fishing but also to their different household roles.

RESEARCHER: Would you describe your life as presently very stressful, somewhat stressful, not very stressful, or not at all stressful?
HAROLD: I don't know about hers but mine is not a bit.
PAMELA: I finds it stressful. Eh ... very stressful. I would say.

RESEARCHER: Compared to ten years ago would you say it's much more stressful, more stressful, less stressful, or about the same?

PAMELA: More stressful.

RESEARCHER: Can you tell me what that main source of stress is?

PAMELA: Well I guess 'cause you got two kids now. One time it was just the two of us and we didn't care if we were starving to death I guess. And now we got a house and two children and we got lots of bills so you're wondering if you're going to make enough money through the summer to pay it, right, 'cause that would be really stressful. We got two youngsters in school and lots of bills, and a home to keep up, and I finds it stressful. Week to week I'm wondering what we can pay the week, or wondering if we're going to make enough money this week to pay this or that ... Now he ... he's pretty good. He's the same as me but ... I more or less handles the bills right? The cheque comes to me and I pays what got to be paid and then if I got a problem, if I don't think I can pay this today, I mentions it to him and we tries to figures out a way to pay it.

These East Coast household interviews also helped us see that there are other ways to understand the stress and uncertainty associated with restructuring that relate to natural environments as well as to people. Such reflections on the change that has taken place suggest a more holistic approach to health than is normally found in population health research. This holistic approach considers the interactions between humans, animals, and forests and treats them all as actors with varying degrees of agency. The following excerpt is from our interview with Norman (pseudonym). He draws parallels between changes in the fisheries and changes in forests and links animal behaviour and the related availability of moose for food to forestry practices and to the failure to extend the boundaries of Gros Morne National Park across the Northern Peninsula. He is particularly concerned about the ways the actions of government, industry, and local people are increasing stress in both the forests and the moose population in his area, resulting in potentially serious long-term environmental consequences.

In ten years' time the forestry will be like the fishery – they'll be wondering what happened to it ... See, here on the Northern Peninsula we don't have a whole lot of topsoil, right. You take a look around – it's all cliffs and rocks. Now when the fellers, say ten years ago, before the mechanical harvesters, fellers were in cutting with a chainsaw, they'd have a trail and the skidder would haul it out for 'em. That trail was the only ground disturbed. Most of those mechanical harvesters are just built on an excavator train and they tear the little bit of topsoil that is there off the rocks. Now how many years do you think that it is going to be before the timber grows back? First thing you need is the soil has to got to build up again first.

I used to go up [the Main River] moose hunting one time. I don't go up there now – there is no moose going to live out in the middle of a open cut-over. See, really, Gros Morne Park should have been extended from that side of the Northern Peninsula to this side and if the island is not worth that well, hell with it, cut it all anyway. Then that would have given 'em everything because pretty soon we're not going to have any animals left either. We're not going to have any moose left. In the summer they can live anywhere, but in the winter they got to have big timber. You know they yard up in heavy cover winter time. Where are they going to go? A few years ago, they opened up a hunt inside of Harbour Deep in the winter time – a cull. They called it a hunt, eh. There was too many moose in a small area and they were browsing out. The moose went down the Northern Peninsula when they flooded out Cat Arm for the Hydro project. They never had nowhere else to go right? They couldn't come back this way – they'd run into a timberjack. They had to go down the peninsula because they flooded a whole lot of country down there when they put in the Cat Arm station. So the moose moved out. They got down there and the first little bunches of woods they hung up, and then all of a sudden there was too many moose in a small area and people had to go down in the winter and shoot 'em. You know, they called it a hunt but, wasn't much of a hunt. And the same thing is going to happen once the prime timber stands are cut. You're going to see moose yarding up in smaller areas where there is little bits of trees left and then they're going to starve to death. You know, caribou is alright – they're on open country anyway but moose, like I said, moose got to have green timber to be in.

Norman's comments highlight his perceptions of the complex interactions that could exist between such activities as forestry, the construction of hydro dams, and hunting policies and the health of moose populations in his area. When he says, "soon they'll be wondering what happened to it," he is referring to his larger sense that no one is really taking responsibility for protecting the forests or the moose. His holistic perspective on the environment locates moose as actors, embedded in a wider network of other actors such as forests, people, industries, and government.

Our West Coast interviews revealed similar issues linking restructuring and health. In one interview our respondent said, "I hate to tell you how many couples have split up, because financially that is the hugest stress in the whole wide world, if you can't pay your bills. And people have had to leave town to work, so consequently you have wives raising children here on their own" (Prince Rupert key informant). And another talked about stress:

INTERVIEWER: What are the things causing you stress?
RESPONDENT: Anxiety over the future, not knowing is the big one. If you

know what you are going to do next February, not this February, next February – I like to plan things out there, but when you don't know, it makes it very stressful ... Underemployment is a significant factor, and I don't care, anybody that you interview in this town, they may say that things aren't doing too bad, but anybody who is underemployed or unemployed, they are experiencing significant stress. (Port Hardy respondent)

Lack of stability and security in employment affect a person's ability to plan for the future. Many respondents talked about the mental stress associated with financial uncertainty. On the East Coast, when we asked participants, "What if anything causes you stress or worry at work?" issues associated with employment volatility were identified as major stressors. Most of the women talked about issues related to the work environment, such as understaffing, overwork, and work uncertainty, and worry about the children at home. The men described the stress of work uncertainty, as well as threats in their physical environment, and the increased pressures from government rules and regulations in both the forestry and the fishery.

Mental stress is one part of human health, and an important one. But it rarely stands alone. Rather, cumulative health effects are associated with the ongoing struggle to make ends meet, even if the family has managed to keep afloat financially. We heard many accounts of the lengths to which people will go to earn an income. They put up with occupational health hazards or injuries for fear of losing a job or losing time needed to qualify for EI. As one fish-plant worker, who had terribly swollen hands, put it: "There is more then once that I cried my dear. I couldn't get me fingers open ... couldn't get me fingers apart. And you wouldn't dare say, you know, that you couldn't go back. You had to go back." Occupational health risks in shellfish processing differ somewhat from those associated with groundfish processing. Snow crab and shrimp-processing workers on the East Coast are grappling with respiratory problems that may be associated with asthma resulting from allergies to shellfish. By continuing to work they may be putting their long-term health at risk (Howse et al. forthcoming).

Although some fisheries have been converted to individual quotas, many, such as the lobster fishery, remain competitive. As a result, fishing is done in dangerous weather. With a reduced resource and a fixed total allowable catch, there is pressure to get out there and get what you can. "So the person that's fishin' they're gonna try and force it and get out there. It doesn't matter what the dangers are, what the weather is, they're going to get out there and see what they can get out of it for that short amount of time. So they don't take care in what they're doing anymore." In some areas, the location and nature of fishing have shifted as a result of changes in the targeted species and management initiatives. Some fisheries have gone from coastal activities involving day trips to extended trips to sea into areas previously unvisited and in vessels that are not optimally designed for such fisheries.

Fiberglassed vessels have replaced many of the wooden vessels of the past, and vessels are more likely to be purchased than made by the harvesters themselves. Some people have found employment in boat-building enterprises, while others do their own fibreglassing. Working with fiberglass means dealing with potential exposures to chemicals like styrene, which, as research done elsewhere has shown, can be associated with neurofunctional and electrophysiological changes, including colour blindness, as well as neuropsychiatric symptoms (Arlien-Søberg 1992; Cherry et al. 1980; Campagna et al. 1995; Murata, Arski, and Yokoyama 1991). As a result of the industrial restructuring in the province over the past fifteen years the number of boat-building operations has increased. To date, there has been no research in Newfoundland and Labrador on the levels of exposure to styrene and its health consequences for boat-building workers, but anecdotal reports suggest that ventilation is poor in many facilities and exposures are probably high.

Although on the West Coast there were divergent views about how economic restructuring has affected social networks and community involvement in Port Hardy, Prince Rupert, and Ucluelet, there was strong consensus about the effects of the economic recession on local families. Many individuals told us that families have been negatively impacted by the local economic situation, and stories of divorce, spousal abuse, and separation were common.

Their unemployment insurance is running out, families are breaking up, and I think that is fairly consistent in the community, and not that that is totally based on economics, there are probably issues in the family regardless, but this certainly didn't help. (Ucluelet respondent)

INTERVIEWER: What's happened to some of your old fishing buddies?
RESPONDENT: Drugs, divorce, so the typical social [problems] – it makes it really difficult at a social level ... spousal abuse.
INTERVIEWER: You saw it really going on?
RESPONDENT: Oh yes, big time. Again, that's not only physical abuse but it's mental ... I think that financial stress is the leading break-up of the home. And when that happens, then all the mental anguish comes in on top of it, so now you not only have financial trouble, you have mental stress. It's trouble. (Port Hardy respondent)

Verna Ambers, director of community development with the 'Namgis First Nation in Alert Bay observed (2003):

Oh, I'm in the front lines there so there's family violence, there's family breakdown, there's alcohol abuse, there's drug abuse, there's all kinds of things that come from that. The low self-esteem, the inability to work. What happens is you got a man who's been supporting his house forever as a commercial fishermen, as a logger, or as a clam harvester, those types of things, and now you've got there guys that you can't even retrain. They don't have the skills. You can't even retrain them because the majority of these guys can't write, they can't read, so how do you do that? You don't have any pride and then you just get – I think most people don't know what else to do. And we've tried training programs.

There is a strong correlation between healthy families and healthy communities. Families need economic stability in the form of secure employment opportunities and access to supportive community services and social networks in order to be healthy and resilient. Conversely, strong families are essential to promoting and maintaining healthy communities (Weissbourd 2000). The communities of Port Hardy, Prince Rupert, and Ucluelet, already dealing with the serious economic consequences of declining resource-based industries, now face subsequent social issues associated with family vulnerability because of economic hardship. At the same time, some of the social services (e.g., Child and Family Services, women's centres) that are designed to support families are being cut back:

Ministry of Children and Families had the same workers for years and years, and with the changes in government policies, last year it started to change and we basically lost all but one regular worker, and it's my understanding that that worker is now gone as well. These are people who have been in the community who know the families and things like this. And now they've got people that they are bringing in from other communities for three months here, three months there, and you see a huge difference ... They don't have the history. (Port Hardy respondent)

Clearly, unemployment and underemployment resulting from resource sector restructuring, coupled with reductions in social services and changes to employment insurance and income assistance regulations, are proving extremely challenging for many families in these coastal communities.

In Port Hardy, Prince Rupert, and Ucluelet there were conflicting views as to how socio-economic restructuring has affected sense of community. Some residents feel that because of declining populations and shifting personal priorities, people have become more insular and less involved in community activities and governance. This suggests that the social capital created through individual involvement in structured events and organizations is diminished as membership and interest in these social institutions declines.

A lot less people now, less involvement, less people interested and people are keeping more to themselves ... in survival mode you might say ... In the summer, we've got Filomi Days – Fishing, Logging, and Mining. Well, it's just basically logging now, there's no fishing any more really, very little, and the mining's gone. Back then there were organizers and it used to be a lot of fun. We still got it, but it's not the same. (Port Hardy respondent)

INTERVIEWER: Do you feel that people are less invested in their community now than they were, say, ten years ago?
RESPONDENT: Yes. It is those that have and those that have not, is kind of the sentiment. And those that have not, just are like, "why bother, I'm tired of fighting. I have been fighting for this, I have been struggling to keep my job, and I am just watching my life go down the tubes," and they really have a sense of why bother. (Ucluelet key informant)

CONCLUSION

The word "community" is used in several ways that have different implications in this chapter. It may, for example, refer to an occupational group. Taking the example of fishers, it is often difficult to see what unites fishers harvesting different species, using different technologies, and having different ethnic identities or genders. That said, it is unlikely that wild fish resources will be so devastated that the occupation totally disappears in the future. However, the trend to fewer numbers and greater concentration is evident on both coasts. Work is about more than a pay cheque; it is also a key element in a way of life for most. Should the work fail, fishers will be losing not only their jobs but also their homes and community networks. Community in both its everyday and academic sense is a synonym for place, somewhere that people identify as a shared site for residence or employment. These communities can stabilize or gain population only if there is an economic rationale for them to continue to exist or if state policy subsidizes living conditions to a level that holds the population. The latter is unlikely in the political climate of our times. Employment contraction will probably occur in the resource industries, which is likely to put increasing strain on the smaller communities.

Without some alternative they may well disappear or be reduced to hamlets of temporary homes.

Coastal communities have experienced significant changes in the past two decades as a result of social, political, industrial, and environmental restructuring that has occurred both regionally and globally. This restructuring has had mostly negative, but some positive, impacts for many small coastal towns and villages as resource-based industries such as forestry, fishing, and mining have declined, while other industries such as tourism and aquaculture have increased. For the better part of the twentieth century, both primary and secondary industries based on resource extraction and processing were the economic backbone of many coastal communities and provided well-paying employment to residents. These industries contributed to the well-being of communities through wages paid to local employees, payment of municipal taxes, and in some cases, investment in local athletic facilities, student bursaries, and community beautification initiatives.

However, as postindustrial restructuring has taken place in industry, what had been firm rigidities before restructuring are gone – the kinds of protections and restrictions that unions and workforces used to be able to impose on employers have eroded and have been removed. Instead, as we have seen, industry has gained flexibility by imposing constraints on the working population, whether through piece work or through contracting out or other strategies by which firms have been able to remain competitive in a global marketplace where cheap third-world labour forces industry to make cost-saving adjustments wherever possible. This new burden has been exacerbated by government cut-backs and cost-saving adjustments in the institutional (provincial and federal government resource management regimes and income and health care regulations) infrastructures. As a result, the flexibility of coastal communities – a strategy that kept them functioning in the past in hard times – has been narrowed and restricted as a (probably unforeseen) outcome of multiple changes in two higher levels of the national system. We have heard from the people themselves in this chapter that the end result has often been seriously detrimental to community, household, and individual well-being, a problem that has been further compounded by damaged and shrinking resource bases, which have been rendered thus by inadequate and insensitive resource management systems at both the federal and the provincial level.

It is always possible that new discoveries can alter pessimistic projections, but these possibilities cannot obviously be planned. Most threatened rural areas struggle to find new bases of employment. Aquaculture (see chapter 12) has been attempted on both coasts, but especially in British Columbia. However, aquaculture expands in conjunction with concern about its environmental impacts and sometimes in conflict with other marine resource users. Rural areas are more likely to attract industries and businesses that provide lower-paying, less stable jobs. In Newfoundland, Grand Falls–Windsor (outside our study area) is a good example. Then there is tourism, which we also deal with in chapter 12. From 1996 to 2001, the populations of the Bonne Bay villages that are surrounded by Gross Morne National Park all fell significantly (Rocky Harbour by 6.0 percent, Norris Point by 7.5 percent, and Woody Point by 8.5 percent). Clearly, this prime tourist attraction provided inadequate stimulus, even if the population declines were below the regional average. Moreover, population fell in the previous five-year period as well. Close to Pacific Rim National Park, Ucluelet had 6 percent fewer residents, but Tofino grew by 25.3 percent – impressive, but in a regional district that experienced overall decline. Tourism can help, but many jobs in the industry are seasonal and poorly paid. Some small businesses may thrive in niche markets for high-end bed and breakfast or skilled crafts. Overall, barring unexpected political or economic changes, we can expect many small communities on both coasts to decline and disappear over the next twenty to thirty years, while a number of regional service and industrial towns should stabilize or even grow.

Our work on the effects of socio-economic restructuring on the health of people and communities on the East and West Coasts and particularly our in-depth interviews with residents and key informants in each community have provided an insight into, and given a human voice to, the ways in which the restructuring of resource-based industries has affected the health of coastal residents, their families, and the communities themselves. We have heard how interactions between government, industry, and ecosystem changes have failed to promote, and in many cases jeopardized, the health of resources, people, and coastal communities. We have been told how such scale asymmetries in terms of concern in upper levels of the state have failed to nurture lower, municipal levels but instead have abandoned people and their communities to the harsh effects of staples traps that were not of their making. We have seen how work, risks, and benefits are out of alignment.

At the same time, we have discovered a remarkable resilience among coastal people, which is particularly striking in the face of scientific, industrial, and policy initiatives that have failed to nurture and sustain coastal communities. We have shown here, for example, how restructuring of the various resource sectors has affected the availability and quality of local employment, community population numbers, family security and stability, and community social capital, while the concurrent restructuring of employment insurance and social assistance has, in some instances, actually exacerbated already challenging economic situations for individuals and communities. Sadly, it is also the case that the state infrastructure, far from assisting and supporting communities as they seek to restructure themselves in order to act as dynamic contributing partners in the Canadian federation, has in many cases made matters worse by following initiatives that are more likely to challenge and undermine the resilience of these people and the places they call home than to nurture what is a rich Canadian heritage of culture and place.

Restructuring, Nutrition, and Diet
on Both Coasts

One of the most important things to understand about local communities and local cultures is that their maintenance and natural evolution
(not "preservation," which implies stagnancy and even death) speak to
the resilience of the social-ecological system of which communities are
part. One common but usually unrecognized effect of interactive
social-ecological restructuring has been dietary change for communities and households and in school meal services, for example. Such
impacts speak to problems of misalignment of risks and benefits of
restructuring and are yet another example of how local communities
are bearing imbalanced risks as the wider provincial, national, and
international economy restructures and as local diet solutions fall prey
to ecosystem degradation. It is important to understand such changes
because of the potential present and future impacts on human health
(for better or worse) and because of the close relationship between
diet, community satisfaction, culture, and way of life.

Food is a vital cultural expression. Food security, then, speaks to
physical health and cultural and emotional health and well-being –
and, by extension, environmental well-being also, since, in small local
communities, usually a substantial component of their activity is growing (or protecting the growth of), harvesting, and preparing local
foodstuffs. The actions of our societies around food demonstrate that
it represents far more than just nutrition: we identify ourselves and our
societies by the food we eat. Food and human social relationships simply cannot be separated. From the production and harvesting of food
to the processing, storing, exchanging, and serving of food, participation by groups of people is almost always the norm. Likewise, in households all over the world, offering food and drink to visitors is an
essential part of social etiquette. The rituals and conventions around
the serving of food are discussed in many books, both academic and

popular. Margaret Visser's *Much Depends upon Dinner* (1988) and *The Rituals of Dinner* (1992) and Gary Paul Nabhan's book *Coming Home to Eat* (2002) are just three examples. The English word "companion" is derived from the Latin words *cum* (with) and *panis* (bread): a companion is someone with whom you take bread – a basic food of Europeans and many others around the world. Eating in *company* is a fundamental human experience.

We turn first to the West Coast to examine the cultural and economic significance of food for coastal indigenous peoples and others, under environmental conditions that generally provided a wide range of food. We then look at schools on that coast. Next we turn to a brief history of nutrition and diet on the East Coast, and examine food, diet, and nutrition at the level of the household. In that way, we cover the major areas in which diet and nutrition are important – in history, in First Nations and settler communities, among youth (schools), and in households.[1]

DIET, NUTRITION AND FIRST NATIONS ON THE WEST COAST

Food reflects socio-economic status in most, if not all, cultures – indeed, it is one expression of the ordered and intricate nature of society and culture. For indigenous peoples of the northwest coast, for example, some foods or special parts of foods – the longest silverweed roots (*Argentina egedii*), the best parts of the seal or deer, the fattest parts of the mountain goat (*Oreamnos americanus*), cakes of pure salal (*Gaultheria shallon*) berries – are reserved for chiefs and matriarchs. Other foods – perhaps some types of shellfish, the poorer cuts of fish or meat, the shorter silverweed roots, or berry cakes with mixtures of the less tasty types of berries – are the fare of the ordinary people. The order of serving food and seating positions at a feast are also often determined by social role and status; at most communal meals, elders are served first. Depending on one's gender, age, life stage and social status, certain foods may be denied or promoted. For example, in some cultures it was taboo for a menstruating woman to eat fresh salmon; to do so was said to cause offence to the spirits of the salmon and to result in fewer salmon coming to offer themselves to the people. New parents, especially parents of twins, or recently bereaved people also often have special observances relating to food.

When the children of the Nisga'a, Nuu-chah-nulth, and other First Nations of coastal British Columbia sat down at the dinner table and started to eat, recalled several elders, their parents would start to talk to them. Just as they were taking food into their bodies, they were told, they were going to take in knowledge and teachings. A person's body was said to be receptive to learning when they were eating. Feasts, too, were and are a time for teaching. Chief Earl Maquinna George, first hereditary chief of Ahousaht in Clayoquot Sound, remembered when he was a boy how important community feasts were as events when all the people were instructed by the chiefs and elders, the orators of the day, about their history, lineages, and responsibilities to their lands, resources, and communities (George 2003). Whether at a family meal or at a feast for an entire village and beyond, in land-based societies food has provided a venue for passing on important cultural knowledge.

Conservation is a major message in teachings about food in some communities, and there can be strong prohibitions against wasting it. The Haida, for example, have a tradition called *k'aaw k'iihl* in the Skidegate dialect, the practice of taking food home after a feast. Barbara Wilson (Wilson and Turner 2004) explains that in Haida society it is considered an insult to leave food on the table: "For a feast or a dinner, we always make sure that there is extra food so people can take it home. It's a way of sharing." Nowadays, people can freeze some of the stew or other food they bring home; it will not only be convenient at a future meal but will bring back good memories of the feast and of all the people eating together and helping each other out. Pregnant women are often plied with nourishing broths and the best parts of meat or fruits so that they will be able to nourish their babies in the best possible way. Nor are the ancestors forgotten when food is shared. At the time of a feast or special memorial, it is a common practice in Salishan communities for people to place morsels of food in the fire or on a plate out in the forest for a deceased loved one, so that they will be able to enjoy their favourite food in the spirit world.

Many ceremonies, such as the potlatch of the northwest coast, incorporate and relate to food. As Kwakwa̱ka'wakw cultural specialist Kim Recalma-Clutesi explains (1993): "We believe that all life, whether it be animal, plant or marine is sacred and is as important as human life. We are but one small part of the big picture and our food gathering practices and ceremonies remind us of that." The First Foods ceremonies are examples. For many people, when the very first salmon of the season is caught or the first ripe berries are picked, a special rite is held to honour the spirit of that food. In these societies, no one is allowed to harvest salmon or pick berries until the ceremony has been held. As part of the ceremony, the first harvest is often carefully divided up and shared with everyone present; this occasion is both a celebration of the renewed harvesting cycle and a tribute to the Creator and the spirit of the food that sustains the people (Turner, Ignace, and Ignace 2000).

Foods, their origins and influences on people, are the subject of many narratives, ceremonies, songs, and dances. For the Nuu-chah-nulth, bunchberry (*Cornus canadensis*) originated as a transformation of the blood of a young woman who was stranded at the top of a cedar tree by her jealous husband. Soapberries (*Shepherdia canadensis*) were said to have been brought by Raven to the Bella Coola Valley; Raven used his magical powers to snatch some soapberry whip from a feast offered by a mountain of the interior – a mountain who was a great chief with soapberries growing on his slopes (McIlwraith 1948). Raven was also said to have brought highbush cranberries (*Viburnum edule*) to Haida Gwaii, cranberries he had taken from the Beaver People (Turner 2004). For the Hanaksiala of the Kitlope, a man with a sick wife was told about using seaweed (*Porphyra abbottiae*) for food and medicine by a supernatural person on a rock out in the ocean (Davis, Wilson, and Compton 1995). Learning about food from animals was a frequent theme; humans and animals shared many foods. Story traditions, names, and discourse often highlighted these linkages (Turner 1995). The harvesting and processing of food was also socially highly important. For example, groups of women travelling together to harvest sites, picking berries, digging roots, processing fish, or chopping seaweed found such opportunities enjoyable and instructive. Grandmothers and granddaughters, aunts and nieces, groups of friends and siblings would spend valuable time together, teaching and learning, demonstrating and practising, storytelling and simply "passing on the news" (Turner 2003).

All these cultural and environmental aspects of food systems of indigenous peoples in coastal communities are as important today as in the past. However, economic and environmental restructuring has resulted in many changes. A reduction in indigenous peoples' access to, use of, and preference for traditional and local foods has been noted by Kuhnlein and colleagues (Kuhnlein 1984; Kuhnlein and Receveur 1996; Kuhnlein and Turner 1991). In general, members of the oldest generations have retained stronger ties with their traditional foods than the younger adults and youth, and the loss of knowledge of how to gather and process the traditional foods is of great concern to elders, as expressed by Chief Johnny and Helen Clifton of the Gitga'at community of Hartley Bay in the film *Gitga'ata Spring Harvest* (Hood and Fox 2003). Johnny Clifton shared his concerns: "My fear is for the children now, not understanding and not knowing how to prepare the food that we gather here [at the Kiel Seaweed/Halibut camp]. It will be a lost cause if we don't take them down every year, like we do, to try to teach our grandchildren how to prepare the food that we get and we share with everybody that is living in our community." Elders up and down the coast have noted the deterioration in the environments from which their traditional food comes; pollution, the impacts of invasive species, and overexploitation are just some of the problems they have identified (Nuxalk Food and Nutrition Program 1984; Stephenson, Hopkins, and Turner 1995; Turner 2005). Nevertheless, younger people have retained an interest in learning about and keeping their food traditions (Thompson 2004), and we suspect that some of the research undertaken through *Coasts Under Stress* will highlight and reinforce the important contribution of local food and traditional food systems to peoples' health and well-being.

DIET, CULTURE, AND NUTRITION ON THE EAST COAST

Considerably less research on the cultural importance of foods and knowledge related to food production exists for East Coast communities and for non-indigenous communities on both coasts. However, as part of *Coasts Under Stress*, we carried out a pioneering ethnobotanical study of bakeapple (*Rubus chamaemorus*, L.; *Rosaceae*) use and ecological knowledge among Labrador Métis living in Charlottetown, Labrador (Karst 2005). This study looked at both the social and the ecological importance of bakeapples.[2] It found that traditionally, bakeapple had been a vital component of the community diet but that a number of changes (primarily the cod-fishing moratorium) had altered social practices surrounding bakeapple picking. Nonetheless, despite a decreased reliance on wild foods by community members in general, this plant still remains a culturally important species. The ecology of bakeapple was also uncovered using a combination of local ecological knowledge and scientific knowledge. We found three main habitat types in this area that are each associated with different bakeapple densities and fruit sizes. The study brings to light the kind of information that can be captured on the coast for species that have both dietary and tourist potential, since bakeapples are not only eaten on the coast but also sold as a preserve that is part of the coastal local culture during the summer tourist season.

DIET, NUTRITION, AND SCHOOL CHILDREN: TWO WEST COAST SCHOOLS

Non-indigenous people, of course, lack the richness of food culture that belongs to First Nations, but that does not mean that food is unimportant or bereft of cultural expression. Nor does it mean that the two cultures always remain distinct when it comes to food. The meals provided at one school in Port Hardy for and by First Nations students are actually designed with both cultural significance and social inclusion in mind. The story of a special space for First Nations students in this school provides an example of the intersection of food, nutrition, and culture.

In times of economic change, schools in small communities have to be resourceful in funding their programs. As the town of Port Hardy faced the severe decline in its resource industries, its secondary school found itself in a quandary. The school had a long history of providing programs that promoted good nutrition, including Food and Nutrition 9–12, the Young Parents' program, and First Nations food courses, and it also offered various extra-curricular programs. With the economic and social shifts that occurred after 1996 (see chapter 9) however, the capacity of the school to offer youth a range of extra-curricular activities and the ability of families to always provide nutritional meals became compromised. Many of the families who remained in the community experienced social and economic upheaval, which harmed family cohesion and stability. Family connectedness and parental monitoring have been associated with positive adolescent eating attitudes and habits (Fisher, Pastore, and Schneider 1994; Mellin et al. 2002).

The Parent Advisory Committee (PAC) decided to tackle a lack of money for school programs and the nutritional needs of students by two approaches: installing vending machines and providing breakfast to students in need. The proceeds from the sale of pop, chocolates, and chips out of vending machines would benefit students by funding sports programs, school trips, and extra-curricular programs. Unfortunately, this meant that students were spending the little money they had on unhealthy food. Even when healthier choices were available – for instance, water and pure juice – the added cost of these items deterred students from purchasing them. Tension between the perceived need for funds and the desire to provide healthy food choice remains an ongoing issue.

On the nutritional side, the local Parents Advisory Council (PAC) successfully applied for a Canadian Living Grant, which they used to establish a breakfast club open to all students. "In the first two years it was retired seniors who were coming in every morning at 7:30 to help prepare and serve the food, which was a wonderful way to get the generations together" (Umpleby 2003). As residents moved away, there were fewer and fewer volunteers, reducing the capacity of the school to continue the program. The Aboriginal youth counsellor at the school saw the benefits of the program with students who came for counselling. She moved the breakfast club to the First Nations room and continued to provide breakfast to all students who wanted it. "When the kids came in I always had bread. They could make toast and peanut butter sandwiches ... I brought that from home ... and tea. It's sort of like when I was growing up and you went to visit your aunt, you know. That's sort of the way I structured my counselling program too" (Sedgemore, 5 November 2001).

Over the next several months, the counsellor continued to note positive behavioural changes among many of her students, realizing that "when kids act up in school or are angry about something, it's usually because their blood sugar is low" (Sedgemore, 19 February 2004). She negotiated with her colleagues to allow students to take snack breaks when needed during the second half of class. For many First Nations youth, the availability of food made school more appealing and increased the likelihood that they would stay in school. The counsellor continues to expand her program through funding she receives from the three local bands. At 8:30 every school day, the kettle is on, and students gather to make oatmeal, toast, hot chocolate, or tea and to consume a piece of fresh fruit. In the fall of 2003, the counsellor proclaimed a Breakfast for Learning Week and invited members from the town council, the principal of the school, representatives from the three reserves, and the RCMP to cook breakfast for the entire community. Her future goal is to proclaim September as Breakfast for Learning Month and to offer the community one free breakfast a week.

Another aspect of food and schooling is also implicit in this story. We now struggle with problems of overweight people throughout the general public and with diabetes (endemic among First Nations populations), and this has turned attention to issues of locally and organically grown foods and of diet and nutrition. One way to deal with the proper diet for children is to establish school food programs, but opposing theoretical positions surround the efficacy of such programs. One position, which is advocated by the World Declaration on Nutrition, endorsed by the Canadian government, and acted on to varying degrees by provincial governments and volunteer organizations, holds that the provision of food to hungry children is marked by dramatic improvements in their social behaviours and their abilities to concentrate. These programs, according to proponents of food provision, result in improved academic performance.

An alternative position, put forward by several Atlantic Canada researchers, holds that the provision of school meals, which is usually undertaken as a volunteer effort, is conceived of as a "wonderful" program. However, it may in practice contribute to increased dependency (Dayle and McIntyre 2003; Williams et al. 2003), stigmatization (McIntyre, Travers, and Dayle 1999), and a growing bureaucratization of program delivery (McIntyre, Raine, and Dayle 2001). An important point accompanying these problems, clearly articulated by Hay (2002), is that poorly planned and implemented programs – implemented without benefit of rigorous evaluation – can act as smokescreens over the underlying issue of income insecurity and the dismantling of social services.

The theoretical position surrounding school meal programs at a school in Prince Rupert (Harris and Shepherd forthcoming), for reasons that will emerge below, supports the World Health rationale. Hunger is seen within this and other coastal community schools as an urgent problem calling for immediate action. At the same time, we acknowledge and support the warning by Hay and others (Keating and Hertzman 1999) not to lose sight of the network of social and economic issues that underlies Canada's present unwillingness to redistribute wealth and stem the growing gap between rich and poor (Goudzwaard and DeLange 1995; Townson 1999; O'Connor 1998).

In addressing the effect of restructuring on the health and well-being of West Coast community populations, we examined the perceived need for, and delivery of, school meal programs. For our detailed case study, in 2001 we selected a single "community school" in northern British Columbia. At the time, there were eighty-six schools so designated in the province for their high needs populations. The school we studied serves a population consisting of 80 percent First Nations (and off-reserve) people, approximately 10 percent Asian immigrant families, and 10 percent Caucasian families, all located on a hill in or near an inner-city housing estate. This school "on the brow," despite problems brought about by poverty and, in the past decade, by the collapse of resource industries, has gained a reputation for its safe, caring environment and its focus on student literacy.

Funding for the school, and for education in general, comes from two government sources: the Ministry of Education and the Ministry of Children and Families. During the 1990s, the school received $75,000 per year directly, in addition to the usual allocation for education based on student enrolment. We use the past tense, because district funds have now been reduced by $200,000 and are sent directly to the district for reallocation to schools. Nevertheless, the school can still provide for additional child care workers, a teacher to work with children "at extreme risk" of failure, two community outreach workers, some special evening and summer programs, and classes for guardians, parents, and kindergarten children to practise literacy skills together. Although the school district faces additional cuts each year, the school continues to serve breakfasts to children, lunches to all, and a fresh fruit snack at the end of the day. So far, the economies have come through the cancellation of valued evening programs and special meal days.

When teachers were asked what it was like before the meals started, they all agreed that hunger was much in evidence and that it affected the eagerness and ability of many children to concentrate on their work. One recently retired teacher remembers that once the food programs began, "You saw a happier lot of children all around you, more cooperative in the halls, on the playgrounds. The atmosphere was better." She admits that she had "been feeding her [children] for years. I didn't want hungry children in my classroom. And those who provided food often fed just their own class. Then, when we found out that so many of us were feeding our students, we put the sandwiches out in the hall and sort of did it on a broader school level."

In those days, only the hungry children took the food or, as the teacher we interviewed noted, the "brave kids who weren't too embarrassed to ask for food." Her ploy to avoid stigma (McIntyre, Travers, and Dayle 1999) was to prepare a table where all the kids would share their food. "Nevertheless," she recalls, "we didn't get them all, not by a long shot." Today, in comparison, the teachers and support staff are relieved "just to know that they're not hungry. You used to see children, those you had sent out at lunchtime, and they would just sit in the park or under the trees. They would never go home, there was nothing to eat. They didn't have anything to eat."

In contrast, today, the principal confirms, all children receive a meal form. Parents fill this out and sign it, and since 95 percent of the children receive snacks and lunches, "only the lunch lady knows who brings in the dollar per day" charged to those who are able to pay. While it is important that children are not stigmatized by food provision, the major advantages continue to be seen in the children's improved behaviour, increased attention to their work, and "happy faces." Food security, in the eyes of school personnel and parents, "makes all the difference."

Each teacher and support worker, when asked the question about dependency, speaks about the primacy of immediate need (i.e., hungry children) over considerations of long-term outcomes. One Tsimshian teacher conveys the common response: "I don't believe that parents can afford it right now, to put it quite mildly. Even if they are on welfare or EI or any of those things, they are not receiving enough money to cover the basics. So it's not that they are not responsible; it's actually that they just don't have [the money]." When asked if in former times she had seen a difference towards the end of the month, she recalled "a lot of fighting in the classroom, kids not working and sometimes coming in really hungry [because] the kids just had too much on their minds. If you are hungry, you are tired and you don't want to work." Another experienced teacher could not "stress enough" the importance of the lunch program: "[It's] probably the backbone of what is happening here. Because I have been here before and my kids went to the school before the lunch program. The lunch program made the major difference because you can't learn when you're hungry." This teacher identified the difference as "more cooperation in the halls and on the playground; the atmosphere was just better."

At the school nutritional learning in the classroom, in the hallways, and by the example of adults accompanies food provision. All school personnel realize that to be effective, food programs must do more than fill the void of hunger. Cereals and milk are served for breakfast and sandwiches and a fruit for lunch and snack, and after school, "the kids line up outside [the meal lady's] door and she passes out carrots or apples." We observed these orderly line-ups, and although we heard the occasional grumble about treats, most children know what to expect and appear to appreciate whatever is "on" for the day.

Although teachers and support staff continue to provide the extra amenities needed by their student clientele, the principal identifies areas in which monetary cuts are affecting the school. First, class sizes are creeping up, although the school can still afford the child care workers who help individual students. Second, fewer funds mean fewer evening and summer programs for children and parents alike. Third, a reduced meal budget militates against the little extra celebrations, such as pizza day. Perhaps the most disturbing feature of the cuts is the sense of uncertainty experienced by school personnel. Rumour and generalized announcements from the government of future "belt-tightening" policies combine to produce tension, anxiety, and fear. As these school problems multiply, school workers are conscious of the larger socio-economic scene. One community worker put it this way: "You're starting to see families breaking up, kids coming into [foster] care and some parents who say, look, we don't have the funds, we can't *buy* food. And when a parent has to tell you that, you know it's bad. And it's happening more and more now."

Hertzman and Siddiqi (2000) note on an international scale the positive correlation between swift economic/political transformations and economic decline, as well as the growing distrust in civil institutions by the populations of rapidly changing societies. At the same time, these authors point to what they call a "virtuous cycle" in countries that realize increasing redistribution of income. The schools we examined, the community, the province, and Canada as a nation (O'Connor 1998) illustrate the first scenario. The rapidity of restructuring has left this northern town, despite its natural resources, in a state of economic and social crisis. While some people maintain their jobs and professional positions, many sink increasingly into abject poverty. In the province as a whole, a similar pattern is emerging (Fuller, Fuller, and Cohen 2003, 5) as health costs shift from the public purse to the private and individual sector.

Food provision within the school should also be considered in the light of expectations about academic growth, a spirit of safety and trust, and a network of services such as one-on-one counselling for individual students in need, community outreach, and an "opportunity room" for students especially traumatized by poverty, social disruption, and family violence. Adequate meals alone will not correct wider societal problems, but in the views of the people we met, they provide the first and necessary stepping stone of possibility towards social equity. The choices surrounding food security, of course, will be determined by the will of the society at large, for, as educator T.B. Greenfield points out, in the face of conflicting findings about a social issue, "we choose our theories, and then we live our lives by them" (Greenfield 1978).

THE HISTORY OF DIET AND NUTRITION ON THE EAST COAST

During the early years of colonial government a great deal of attention was focused on diet and nutrition, because access to food, and especially to a balanced, nourishing diet, was seen as essential to the health and survival of coastal communities. Although coastal communities in eastern and western Canada had access to plentiful ocean resources in the past, nonetheless there were also years when the fishery failed to produce a living for families. Many coastal communities lacked road connections to major centres, and winter supplies, delivered before freeze-up, often ran out before supply ships could return to harbours in spring or early summer. Typically, outport fishing families would purchase in bulk a supply to last six months or more, including flour, lard, tea, molasses, salted meats, and other foodstuffs that were not obtainable locally. To these supplies were added locally grown potatoes and root crops, as well as berries, wild meat, and fish (see chapter 9), though most of the latter went to the local fish merchant, who provided store-bought goods on credit in return for the catch.[3] Life in many of these outports was hard, although it varied from place to place, and some outports were usually able to feed themselves quite well, while others experienced extreme hardship.

Early on, diet and dietary deficiency diseases drew the attention of doctors working with the Grenfell Mission to the point where Newfoundland and Labrador became, like the Appalachian region of the United States, a laboratory for exploring the effects of deprivation, especially dietary deficiency and its effects. Concern with the diet of the working classes in the United Kingdom and with inadequate diets in far-flung regions of the British Empire (and more broadly during the Depression) also fuelled widespread interest in the subject. The earliest studies coincided with alarm over the poor health of World War I recruits, while later efforts associated poor diet and health with inefficient labour and failed economic development. As Grenfell was quick to discover, many of the health problems encountered along the coast stemmed from the restricted diet available to fishing families whose fortunes were tied to the fishery and to the merchant who supplied them credit. Grenfell stressed poverty as a major obstacle, often blaming it on the truck (barter) system of credit. He promoted a self-help philosophy that he hoped would improve economic and social conditions but that could at times blame the ignorance of mothers for the lack of proper nutrition. This coupling of poverty and ignorance were common to most debates about social problems and social welfare in the early to mid-twentieth century, and, given the widespread poverty of the outports, nutrition and diet were a major cause for concern (Lush 2007; Great Britain 1939).

While significant government intervention in the area of nutrition had to await the Commission of Government era (1934–49), in the early days Grenfell and his staff, stressing more self-reliance and self-sufficiency, encouraged experimental agriculture and the planting of gardens to increase the variety of locally grown food and established cooperative stores and a local craft industry to raise the standard of living. Grenfell also encouraged the importation of reindeer, whose milk he thought might substitute for the more expensively produced cow's milk. Despite several decades of public education and experimentation with alternatives, however, by the second decade of the twentieth century, Grenfell Mission staff had had little success in changing local dietary habits. In the 1920s, Grenfell's continuing concerns around diet and preventative health measures led him to encourage more than thirty female nutrition workers to volunteer for nutrition work along the coast, where they were to study the local diet and set up educational programs for mothers (Lush 2007).

Women were, of course, primarily responsible in the twentieth century for the diet and health of families, whether as mothers or as professional experts in health, and the discourse of scientific motherhood and child welfare increasingly stressed these responsibilities. The responsibility for failure to provide a healthy diet, then, rested with mothers of families. Medical experts, with few exceptions, tended to be male. It was not until the post-World War I era that women professionals, trained in home economics or public health, could claim a niche for themselves in nutrition, and even then, their expertise was not always acknowledged or considered important. While both doctors and nutritionists recognized the limited control outport Newfoundland women had over the food supply and while some recognized the role that poverty played, they also worried about the competence and knowledge of mothers in the preparation of and cooking methods for food.

The Second World War saw a flurry of activity in the area of diet and nutrition. Medical experts conducted numerous studies, many of them in the territory covered by the Grenfell Mission (northern Newfoundland and coastal Labrador). Nine major studies appeared in the 1940s during and immediately after World War II, and these, like earlier studies, found major health problems: beriberi, night blindness, rickets, high rates of infant mortality (sometimes tied to an inability to breastfeed), poor teeth, undersized children, skin problems, deformities, and infantile scurvy. Most of these problems were associated with a monotonous and restricted diet, a lack of fresh foods rich in vitamins and minerals, and a lack of fresh milk. For example, Norris Point's Dr Robert Dove, working with his wife Margaret and Dr Irving Wright and Ellen McDevitt of New York, surveyed over four hundred clinic, hospital, and outpatient visitors in the early 1940s. Results showed that 70 percent of those surveyed exhibited symptoms of vitamin B deficiencies; in addition 70 percent of the population showed normal ranges of blood vitamin C for only two months of the year (McDevitt et al. 1944).

Another study led by Colonel J.D. Adamson involved British, American, and Canadian experts who responded to an invitation from Sir John Puddester, Newfoundland's Commissioner of Public Health and Welfare, to conduct a larger survey (868 individuals) in St John's and several outports. Evidence of nutritional deficiencies was found and blamed on the lack of vitamins A and C and riboflavin, as well as a less frequent lack of niacin and thiamine. This study featured graphic photographs of skin, eye, and dental diseases, as well as physical deformities attributed to malnutrition. The published report also commented on slow mental reactions and apathetic and subdued behaviour among the children studied. The study was published in the *Canadian Medical Association Journal* in 1945.

Another study, also published in 1945, deliberately focused on the most vulnerable people in the Norris Point population – pregnant and lactating women and growing children. Based on medical histories, physical examinations, and hemoglobin and plasma protein determinations, the study reported that it was the women who demonstrated the highest proportion of deficiencies. Like the Adamson study this one also commented on the lack of spontaneity and play among the children, who "appeared older than their years." Apparently stung by the negative publicity surrounding dietary studies, Commissioner

Puddester instructed the assistant director of medical services within the department, Dr James McGrath, to write to the authors requesting them to delete such descriptors as "abject poverty," and "inhospitable depressing climate," and to remove references to undersized children. McGrath noted that newspaper reports often stressed the negative and wrote that "we want to avoid any suggestion that these are exclusively characteristic of Newfoundland" (Adamson et al. 1945; J. Metcoff et al. 1945; PANL 1945).

The immediate postwar period saw a number of significant initiatives with respect to nutrition that were a response to the publication of a significant study conducted in 1945 by Dr D.B. Cuthbertson of the British Medical Research Council (Kealey 2003; Cuthbertson 1947). Cuthbertson's study accorded with the thinking of Newfoundland physicians and the Medical Association's Council on Nutrition and crystallized the findings of previous dietary surveys. As early as 1943 the council had urged Commissioner Puddester to take action to avoid a new outbreak of crippling deficiency diseases. Like Cuthbertson, the council had recommended the enrichment of white flour, a long-range education plan involving teachers and volunteer groups, community provision for refrigeration and the employment of a nutrition specialist in government. Cuthbertson also urged wider milk distribution, more cod liver oil production, more support for breastfeeding, restocking the island with moose and caribou, more consumption of fruit, and the supplementation of the diets of pregnant and lactating women and of children. The Commission of Government rejected his call for price controls on food, as well as his condemnatory remarks on the denominational school system. The overall result, nevertheless, was the adoption of most of these recommendations, many of which informed policy-making right into the 1970s.

While government programs gradually ended or reduced the supplements for vulnerable populations at risk of dietary deficiency diseases in the 1950s and 1960s, some policies remained in place. The enrichment of flour and margarine was continued, and in 1964, as a result of the discovery of infantile scurvy, vitamin c was added to the tinned evaporated milk used by most Newfoundlanders. Educational efforts, particularly in the media, continued in the form of the radio show *Kitchen Corner* (1947–68). Perhaps most significantly, the role of the nutritionist (first appointed in 1947) increased in importance as women with credentials in home economics and public health were hired by government to run nutrition programs and studies. Predictably, in order to obtain professional credibility, nutritionists and dieticians adopted the role of "outside expert" vis-à-vis the women they were trying to educate.

Throughout the period following World War II, there had been considerable concern with diet and nutrition, and it was assumed that changing cooking and eating habits would take a long time. Of basic importance to changing food habits was the role of both the nutrition expert and the women she targeted. The postwar embrace of modernization in the Newfoundland fishery was paralleled by a similar emphasis on modernizing the housewife and mother who was responsible for food planning, purchasing, and preparation. The nutritionist could take centre stage in this process by educating women as modern housewives serving their families, thus ignoring outport women's historical roles as participants in economic activities such as the fishery, where women had long played important roles as part of the shore crew. When they were portrayed as mainly "housewives" who purchased, cooked, and served food to their families, modern women became consumers primarily.[4]

In postwar Newfoundland, nutrition experts assumed a higher profile and couched their advice as wise consumerism based in science. As early as 1945, in her column "Food and Your Health" home economist and Memorial College instructor Edna Baird had stressed the importance of proper planning for cooking. Time could partly substitute for money in planning low-budget meals. Malnutrition, she wrote, resulted not from poverty alone, but also from a lack of knowledge. Five years later Ella Brett, the department's nutritionist, reiterated this message, stressing the responsibility of the housewife for her family's good health and happiness. She also warned that overcooking could destroy food values, a complaint frequently voiced about Newfoundland cooking practices in earlier dietary deficiency studies.

Brett's tenure as the department's nutritionist ended in 1951, and Albertan Olga Anderson, trained in home economics and public health, assumed the position which she retained until her retirement in 1979. During her tenure, Anderson conducted many studies, one on expectant mothers (1965); one on men and women in four Newfoundland communities, as part of a research project on hypertension (1967–68); and innumerable school surveys. Anderson, perhaps more than her predecessors, played an important public relations role in writing scripts for the *Kitchen Corner* series on radio and later, television (all of which had to be approved by the male physicians in the Department of Health). She also wrote press stories and prepared resource kits, manuals, and other promotional material on nutrition and worked with various school boards, parent groups, and the Newfoundland Teachers' Association (NTA) in developing school food policies, the latter in the mid- to late 1970s. During Anderson's tenure the government began to hire other nutrition specialists – a consultant dietician for the eighteen cottage hospitals in 1958, an assistant nutritionist for the St John's schools, a dietary consultant for the Health Insurance Division in 1967, a nutritionist for the west coast in 1968, a dietetic consultant for the Hospital Services Division in 1971, and a nutritionist for the St John's office in 1972 (Baird 1945; Brett 1950; Anderson 1967; Swanson n.d.).[5]

Government programs gradually ended or reduced the supplements that had been given to vulnerable populations at risk of dietary deficiency diseases in the 1950s and 1960s, and by the 1970s, when a national study of nutrition concluded, there had been considerable improvements in diet and a decline in deficiency diseases. Within some age, gender, and geographical categories, however, the results showed higher risks for particular problems. Children needed to consume more milk, for example. Women from rural areas aged twenty to thirty-nine also showed higher risk in winter and spring for anaemia compared with the average for Canadian women of that age. Similarly, rural girls, ten to nineteen, suffered from poor protein status compared with their national counterparts. Vitamin c intake, on the other hand, showed results comparable to or better than the national data. Perhaps one of the most worrying findings was that obesity was a problem.

Thus, despite improvements, concern remained about the effectiveness of nutrition education programs, and in 1975 the Interdepartmental Committee on Nutrition was formed, which included representatives from the Departments of Health, Education, Social Services, and Recreation and Rehabilitation. This group focused on particular areas where intervention was considered to be most efficacious – maternal, infant, and child nutrition, school nutrition, low-income families, and the need for a program in dietetics at Memorial University, which was established in the fall of 1975. The rest of the 1970s contained a flurry of activity aimed at child and maternal health and school nutrition, a trend that continued throughout the 1980s. By the late 1980s, however, changing government priorities and fiscal problems resulted in positions being cut or frozen, so that nutritionists were not replaced. Throughout the 1990s it became increasingly harder to provide community health services such as nutrition, except under the rubric of "health promotion," which included a variety of other initiatives (for example, mental health, continuing care, drug dependency services, and health protection). Generally, restructuring of the health care system, which began in the 1980s with the regionalization of public health services in western Newfoundland, continued elsewhere in the province throughout the 1990s, but nutrition was a diminishing priority (Canada National Health and Welfare,

Nutrition Canada 1975; Levy 1993; Newfoundland and Labrador Health Resource Centre n.d.).

DIET, NUTRITION, AND COMMUNITIES: AN EAST COAST CASE STUDY

We turn now to an examination of changes in diet over the past ten years in selected communities. Historically, a number of health problems in the province are thought to have been either directly (as in the case of beriberi – Aykroyd 1930) or indirectly (as in the case of hypertension – Fodor 1980) linked to nutrition. The nutritional status of the people in the province today continues to be of concern, since Newfoundland and Labrador still has the highest rate of cardiovascular disease in the country. A 1996 nutritional survey of 1,927 adults in the province found that 69 percent of them were either overweight or obese – and thus at risk for cardiovascular disease as well as other diseases (Roebothan 2004). Other key findings suggested low consumption of important micronutrients like vitamins B6, B12, and thiamin and minerals such as zinc and calcium. Documenting some of the changes in nutritional practices and factors affecting these practices contributes to our understanding of dietary change in the province, as well as the availability of food and the choices people are making. Here, therefore, we present the key findings of the Household Food Use Survey done in conjunction with the Health Survey in Cow Head, Parson's Pond, and St Paul.

We were able to build on the work that had been done around thirty years ago (1974–75) in a cluster of small communities on the Great Northern Peninsula (Parson's Pond, St Paul, and Cow Head – Marshall 1975) to investigate the impact of restructuring on the food that people ate, as well as the availability of certain foods. Marshall had surveyed the nutritional changes that had occurred in these communities over the ten years before the study and had primarily documented change in the use of country and home-grown food. We repeated, but expanded on, that survey to determine the access to store-bought food, the use of the fast-food outlets that have been introduced into these communities, changes in food consumption at the level of the household and, within households, the eating and activity levels of children over the past five years. We carried out this modified form of the survey interviews in 350 households within the same communities where the interviews were first undertaken in the 1970s.[6] In our 2004 study, in addition to the new questions we added, we administered this survey in other communities in our study area: Port au Choix, Port Saunders, and St Anthony.

The survey on changing food consumption patterns specifically addressed the categories in the original survey: milk, fats, fish, and other seafood that would have been locally available; meat, both commercial meat and game; home-raised chicken and eggs, common garden vegetables, and fruit. In the present survey, 2004, we added home-baked bread and "fast" foods. The latter category was not included in the original survey thirty years ago, as it would not have been commonplace, whereas now it is. The 2004 survey also had some additional questions about food availability and where residents obtained their food, as well as about the problems the respondents had in obtaining the food they would like, since we wanted to understand some of the factors that were influencing changing food consumption patterns.

Milk consumption patterns have changed a great deal since the survey of 1974–75. Three types of milk for consumption were studied then and also in the present survey. The first was "tinned," or canned, evaporated milk. Thirty years ago it was the milk of choice in households, and over 90 percent reported they used it "often," while only one household said they never used it. It was used in tea or coffee, for cooking, and even as an infant formula. The advantage of evaporated milk was that it did not require refrigeration before opening, and even when opened it could last much longer than homogenized milk. The consumption of tinned milk has declined a great deal; only 50 percent of households report frequent consumption of evaporated milk presently. The second type of milk included in both surveys was homogenized, or "fresh," milk, as it is referred to by many people in the province. Greater availability and widespread refrigeration for storage in both stores and homes makes this milk readily available. Its use has greatly increased: 68 percent of households say they use it frequently, compared with 34 percent of those responding to the same question in the survey three decades ago. Although powdered milk was promoted by nutritionists as an alternative to homogenized milk and instead of evaporated milk for children, it was used by around 14 percent of the population ten years ago, but only 4 percent of households say they use it "often" now.

The second category we enquired about was fat consumption. Heart disease is common in the province, and one health-promoting and disease-preventing strategy is to get people to reduce fat consumption. During the period between the two surveys, Canada's food guide was changed, in 1982, to reflect a recommended decrease in fat intake (Canada Health 2002), and this information would have gradually filtered down to the local level. In particular, we looked at the frequency of the consumption of lard, butter, margarine, and cooking oil. Over the past ten years the household consumption of lard and salt pork has decreased with a subsequent increase in the use of vegetable oil for cooking. Fifty-four percent of households use vegetable oil "often" for cooking, compared to the 5 percent who "often" use lard and the 13 percent who often use salt pork. In contrast, thirty years ago only 12.4 percent of households reported using vegetable oil for cooking. Margarine has always been used more frequently than butter in these households. Their use of margarine remained fairly constant over the years surveyed; about 90 percent reported using it often. Now only 6 percent of households use butter frequently, whereas in 1974, 31 percent did so.

There have been some noticeable trends in the consumption patterns for fish and seafood over the past thirty years. Seasonal consumption of lobster has decreased from frequent use by 28 percent of households in 1974–75 to frequent use by 12 percent in the most recent survey. The opposite is true for fresh or frozen crab in season, in that three decades ago only two households reported frequent consumption and slightly more reported consuming it "now and then," but the majority did not eat it at all. Now 9.4 percent of households eat crab frequently, and 76.6 percent eat it now and then. The consumption of both fresh and frozen herring has markedly decreased from 28 percent of households using it frequently thirty years ago, to only 6 percent using it now with the same frequency. Around the same percentage of households reported using salt herring frequently now, as well as in the two earlier periods. Fresh or frozen salmon consumption is higher than herring consumption: 14 percent of households "often" use it, and a further 79 percent say they use it "now and again." What has changed is that farmed salmon is more common on the dining table, and fewer households are bottling salmon as a means of producing food for the household. Consumption of cod (often simply called "fish" in Newfoundland and Labrador) shows some changes. Only 26.5 percent of households say they consume it "often" nowadays, compared with 46 percent three decades ago. The use of salt cod, one of the staples of the provincial diet, has changed dramatically over the past thirty years. In 1974–75 consumption was high: 55 percent of households said they ate it often, compared with 20 percent reporting the same frequency of usage at present.

Meat consumption patterns have changed as well in the decades covered by the surveys. One of the more marked changes has been in the consumption of moose. In the most recent survey 71.5 percent of households said they ate fresh or frozen moose frequently, and only 2 percent said they never ate it. By comparison, in 1974–75, 36 percent reported having moose for their meals often. Additionally, 5 percent at that time said they never ate moose. Moose is also bottled and preserved, and 46 percent often consume moose in that form now, versus the 18 percent engaged in this activity with any frequency thirty years ago. The communities are not far from a national park where all forms of wildlife are protected, contributing to abundant moose and caribou (*Rangifer tarandus*) resources in adjacent areas. Hunters from these areas take full advantage of this and supply their families with game throughout the year. Recipes for moose meat are commonly exchanged, and the research assistant doing the survey was treated to curried moose after the completion of a survey. Fresh or frozen beef is, in fact, reported to be consumed less frequently than game. Only 16 percent of the households surveyed said they ate it often, and almost one-quarter (23 percent) said they never ate it. The second staple of the provincial diet after salt cod is salt beef, and here we see dramatic changes. Thirty years ago 74 percent of respondents reported eating it often, whereas now only 35 percent report the same frequency of consumption.

Raising livestock for food consumption is not uncommon in rural areas of the province, and it varies from region to region. Hardly any of the households (four, or 1.1 percent) in the communities studied raise chicken for consumption or to produce eggs for the household. This is a change from the survey of 1974–75, when 23 percent of the households engaged in that activity. Even thirty years ago people were reporting a decline in raising livestock from ten years before. Omohundro (1994) noted that in general, animal husbandry of any kind was diminishing when he did his field work on the Great Northern Peninsula in the 1980s.

Garden production of root vegetables has provided households in this area with fresh vegetables, but they are now less likely to produce potatoes for frequent consumption: 34 percent of households produce potatoes, followed by beets and turnips at 21 percent. These percentages have been decreasing since 1974–75, when 81 percent of households said they grew their own potatoes for frequent consumption and around 60 percent said they grew the other root vegetables for their household. We also asked about cabbage, since it too can be stored over the winter and provides a source of fresh vegetables. While three-quarters of our respondents said they did not store cabbage, 18 percent reported frequent consumption of it from their own gardens. The frequency of household consumption of this crop was down from 60 percent thirty years ago.

Along with vegetables, the consumption of fresh fruit is encouraged as a strategy for disease prevention. It is thought to protect against heart disease, and while the evidence is conflicting, some advocate the consumption from these food groups to prevent cancer. In national surveys the consumption of fruit is lower than that recommended by Canada Food Guide. In the original survey, households were asked about the frequency of their consumption of apples and oranges, two fruits readily available in rural parts of the province. We included these two food items in our 2004 survey. While the frequent consumption of both apples and oranges remained stable over the ten-year period we covered, at 64 percent and 59 percent respectively, what is of concern is that consumption is down in these communities from thirty years ago, when 75 percent said they ate apples frequently and 67 percent said they ate oranges frequently. Consumption of other fruit may be a possible cause of this change, but that possibility needs to be examined. Locally gathered berries (such as bakeapples – see above and Karst 2005) continue to be a source of nutrition in these communities. They are not used as often in the household today – 52 percent of households consume them often and 4 percent never use them, whereas thirty years ago 63 percent reported frequent consumption, and only in two households were they not eaten at all.

We also found that many households bake and consume their own bread. In fact only 28 percent of respondents said they did not do so at all. A decade ago 19 percent of families did not do so. This question was not asked in the 1974–75 survey, so we do not know how consumption patterns have changed from thirty years ago.

The other area that was not covered in the original survey was the use of "fast foods," since the research assistant from that time said it was pretty well a non-issue. Few fast foods were available. However, with improved road connections and some tourism activity in these communities, "take-outs" have been established. Food from these commercial outlets does not make up a large portion of household food consumption: only 2 percent said they ate pizza frequently, and only 1 percent consumed fried chicken and other take-out foods. More households said they used these foods now and then, but around one-quarter never used them at all. We asked about "household" consumption, and no doubt if we measured adolescent use of these foods we would have higher consumption patterns.

As for the food supply in the communities we studied, over 50 percent of the residents were satisfied or very satisfied with it, 3 percent were neither satisfied nor dissatisfied, and 43 percent were dissatisfied or very dissatisfied. Many residents travelled long distances to access larger supermarkets, but 62 percent could shop within a ten-kilometre radius. None of the communities have a large supermarket. Reasons for dissatisfaction varied among the respondents. Limited fresh produce (31 percent) or fresh meat (14 percent) and not enough variety in foods available locally were the main causes of dissatisfaction, although some respondents cited the high price of food (5 percent) as another problem. For a few, the lack of access to fresh fish and other food items contributed to not being satisfied. Over half the respondents suggested that "more fresh produce" would improve the food supply.

In conclusion, there have been some marked changes in the diet in this area over the thirty years since the 1974–75 health survey was conducted. Homogenized milk is at least to some extent replacing canned evaporated milk, and fat consumption patterns have changed as cooking oil has replaced lard and salt pork. The latter two are more closely associated with what we think of as the "traditional Newfoundland" diet, although as Amilien (2001) indicated, this concept can have different meanings. The consumption of other traditional foods in the province, salt fish and salt meat, has also been declining in frequency. Game in the form of moose shows high consumption rates, and this is not surprising given the high availability of moose in the area, the importance of hunting as an activity in rural Newfoundland and Labrador, and the fact that the meat has been promoted by the Department of Health in the province as a good low-fat source of protein. The consumption of home-grown vegetables has also declined as a smaller percentage of people now grow their own vegetables, probably because vegetables are readily available in local stores. What does need more thorough investigation is the low, and indeed lower, consumption of fresh fruits like oranges and apples.

Despite the promotion of the consumption of fresh fruit and vegetables both in the United States and in Canada, our results are in keeping with recent nutritional studies (Morland, Wing, and Diez Roux 2002; Roebothan 2004; Krebs-Smith et al. 1995). Fruit and vegetable consumption is important, as it is believed to indicate healthy eating (CIHI 2004). At the individual level, decision making around food choice is complex (Birch 1999; Connors et al. 2001; Dibsdall, Lambert, and Frewer 2002; Drewnowski 1997). There are, as well, a host of factors beyond the individual level that will affect food choice, and it is important to focus on some of them (Beaudry, Hamelin, and Delisle 2004).

CONCLUSION

It is clear that a considerable amount of work has to be done to improve diets in rural communities. Some factors contributing to poorer nutritional practices are inadequate dietary education, inadequate quality and selection of food by food suppliers, and inadequate nutrition programs in the school. The alleviation of all of these factors is feasible and not particularly costly to achieve. It is, however, also important not to ignore larger issues that influence food purchases and thus consumption: household incomes, the relative cost of different types of food, legislation of food quality, and the advertising of certain products. In this respect, broader pressures are beginning to affect the quality of "junk" food, as the pressure to remove trans-fats in French fries, for example, becomes fashionable. What we know as a result of our work is that social-ecological restructuring has penetrated coastal communities right down to the level of food consumption and food security, having had a significant effect on traditional local diets and thus the health of local communities, as we have shown here. This is a fine example of the kind of unwitting cascade of effects that social-ecological restructuring has imposed upon coastal communities over time, and it will take raised awareness by families, schools, municipalities, and provincial and federal departments to restore security to these places.

Finally, it is important to remember that food is more than physical sustenance. It is also culturally important, as we discussed at the outset of this chapter. To lose local food production and local knowledge about foods is to endanger cultural as well as physical health, while producing unnecessary stress on household food budgets.

11

The Human Voice of Social-Ecological Restructuring, II: Youth, Education, and Health

INTRODUCTION

While restructuring was a prominent feature of the late twentieth century at the global, national, and local levels, not all communities have been affected in the same way, since the characteristics of any given community and its inhabitants influence how people handle their circumstances. By the same token, the impact of such changes also varies with people's age, gender, and/or developmental stage (Bronfenbrenner and Evans 2000; Bronfenbrenner and Morris 1998; Elder 1998). We might, for example, expect that older members of coastal communities (who have known a very different life) are likely to be affected differently than youth, who in their turn may experience different effects from children growing up after the major changes began to take hold.

In chapter 8 we looked at the statistical face of restructuring, in chapter 9 we gave voice to the experience, and in chapter 10 we focused in to consider the effect of restructuring on health as it has been affected by changes in food and diet. Now we focus in again, this time to consider the response of youth to restructuring and the implications of their responses for the health and education of young people, and hence the future resilience of coastal communities.[1] We start by setting the wider context through an examination of the attitudes of adolescents in urban and rural settings on each coast, and then we turn to a bi-coastal analysis of their feelings about what is happening to them and their communities and of what they think they need to do about that in their own lives in order to prepare themselves for the future.

POLICY AND PRACTICE: EDUCATION AND YOUTH
ON THE EAST COAST

The imperatives of efficiency and competitiveness have triggered profound changes within primary and secondary education as Canadian society develops a "knowledge economy" with which to maintain its place in the developing world economy. Between 1969 and 1995 in Newfoundland and Labrador, for example, boards and boundaries in the education system remained fairly constant. Before the mid-1990s, the province's twenty-seven school boards oversaw roughly five hundred schools (Dunning 1995, 1). In the 1990s, however, one part of the cascade of effects from federal efficiency measures was the emptying of provincial coffers throughout the country. In tandem with public demands for reduced costs, this created strong pressures for school board consolidation (Dunning 1995, v). A key goal of school reform in Newfoundland and Labrador has therefore been to reorient primary and secondary education, in order to meet the needs of urban-based new-economy firms for technologically skilled, flexible, and, above all, mobile workers. According to this vision of reform, however, only larger, urban-style schools are capable of providing the comprehensive range of instruction required to prepare students for the new economy. In a marked departure from the government's traditional approach, therefore, reformers have targeted the small rural school as an obstacle to the generation of a mobile, high-achievement work force.

On this basis, the government of Newfoundland and Labrador has proceeded over the past decade to overhaul governance structures within the province's school system and accelerate the pace of rural school closures. The province's traditional system of confessional schools has been scrapped, not without considerable controversy, thus cutting the number of schools, both urban and rural, sharply. The number of school boards has been cut from twenty-seven to five; this major reform was introduced by the Williams government in 2004. The Department of Education website reveals that the number of schools declined by 36.7 percent between 1993 and 2002, and currently sits at about three hundred. The website gives no indication of how these closures were distributed over the different areas of the province, but it does report that all the forty-two schools that have been closed or consolidated since September 1998 were located outside St John's, and seventeen were located in the three school districts within our study area (Newfoundland and Labrador Education 2004b).[2]

Moreover, the schools that have survived these changes must operate within an increasingly strict regime of cost control. Section 77 of the 1997 Schools Act stipulates that only those schools that "must be maintained and operated because of isolation or because the students cannot reasonably be accommodated in another school" will receive an additional allocation of resources above and beyond what is already provided on a per-pupil basis. Previous government policy had provided all small schools with supplemental human and material resources, not merely those that could claim isolation status (Mulcahy 1999). Responsibility for offsetting this declining financial support has fallen to the new parent-teacher volunteer school councils created by the act. Section 26 of the act authorizes the councils to develop practices and policies for enhancing programs and student achievement at individual schools (Newfoundland and Labrador Education 2004a). In practice, however, the councils devote much of their energies to securing grants and arranging community-based funding. The teachers and principals who serve on these councils must increasingly take time away from pedagogical activities within their schools to engage in fundraising and management tasks.

In addition to widening the disparity between town and outport schools, this quasi-privatization of education has resulted in work intensification for the remaining school staff (C.E. Harris 1998, 119). Although the specifics of reform vary across the country, the general thrust of reduced cash, an increased extra-curricular workload for teachers and principals, and a strong sense of frustration with aging facilities and inadequate support is shared by schools throughout the coastal communities we have studied. Rigidities caused by lack of resources are making it very difficult for schools outside main urban areas to respond creatively to the training requirements of the new knowledge economy, and in practice they are placing increasingly serious constraints on education in coastal communities.

Because we wished to see how this might operate on the ground in circumstances that are very different from those of previous generations, particularly in the coastal communities, we designed a survey that would let us compare rural school students with their urban contemporaries and that would enable us to determine what might be general to adolescents and what might be unique to students in the rural and coastal communities we are studying. We also wished to see if more students were staying on at school in Newfoundland and Labrador than had done so until the 1986 report on education by the provincial Royal Commission on Employment and Unemployment (Newfoundland and Labrador 1986a, b). We found that most students in both geographical settings intended to continue their education after high school. Both rural and urban students thought that a "good life" would be one with a good job, with lots of money, and with family and friends.

To that end, 96 percent of the students in the rural community and 92 percent in urban schools planned to attend a post-secondary institution, and a wide variety of professions were listed as possible career choices by all students regardless of location. When asked whether they thought they would be able to have the education and career to which they aspired, more rural school students than urban (85 percent versus 73 percent) thought they would (which was somewhat surprising given the higher costs associated with rural students' post-secondary education, including the need to live away from home), but the majority of them wanted more help from school personnel with career planning.

As might be expected, there were differences between the rural and urban students' responses to a number of questions relating to their communities. Rural students reported more changes in their world, such as fewer employment and recreational opportunities and more family members and other people moving away. In spite of this, fewer than one-quarter of them thought that these changes had affected their health, the health of their families, or the health of others in the community. This perception is similar to that of most adults in the province, who, in spite of having among the highest provincial rates of a number of diseases (e.g., heart disease, stroke, diabetes) (Canada Health 1999; Heart and Stroke Foundation of Canada 2003), report that they experience good health (Canada Statistics 2004; chapter 8). More students in the rural community (92 percent, versus 79 percent for urban students) expected to be living outside their home community after they completed their schooling, and even outside the province (58 percent, versus 50 percent for urban students). Although slightly more rural (23 percent) than urban (19 percent) students would prefer to stay in their home communities to work, more urban students were satisfied with their life (79 percent) than their rural counterparts (62 percent).

Rural students reported that one of the changes they perceived was that there were fewer teachers in their schools, as a result of fewer students in the schools rather than cutbacks in funding. One rural high school, for example, had a full-time pupil-teacher ratio of 13.4 to 1, which is the same as the provincial average and somewhat lower than urban districts, where it was 15.4 to 1 (2003–4). That said, having fewer teachers overall might be expected to affect the atmosphere and opportunities in the school, particularly with respect to after-school programs and activities, and there were indeed fewer reports of such opportunities in recent years in the rural school. We will return to this point later.

In addition to students' perceptions and attitudes, we examined a number of performance indicators (including school attendance and completion rates, provincial exam grades and participation in post-secondary education) to get an overall picture of the well-being of adolescents in the study communities. We found that in the past fifteen years Newfoundland and Labrador has experienced rapid change in the percentage of adolescents who stay in school beyond the mandatory age and who complete high school. Provincial high school graduation rates increased by 31 percent from 62.7 percent to 82.8 percent between 1987 and 1996, and similar increases were seen in the rural district we studied. By 2002–3, rural and urban provincial graduation rates reached 85 percent, and the overall local study district figure was only slightly less at 83 percent. These increases appear to be a result of a number of factors. Economic conditions, the decrease in the availability of low-skill occupations, and the failure of the cod fishery have contributed to the relatively rapid increase in the percentages of students completing high school, but at the same time, schools have also received federal and provincial government-funded programs aimed at keeping students in schools. However, although more students are finishing school, average rural-school grades on provincial exams (54.2 percent) continued to be below the provincial average (59.6 percent) in 2002–3.

In spite of significant increases in the percentages of both males and females graduating from high school since the 1980s, males continue to lag behind females. In the 1990s and through to 2002, approximately 10 to 15 percent fewer eligible males than females graduated each year in the province. More females continue to graduate than eligible males, but the gap has decreased in recent years (92.2 percent versus 86.1 percent) (Newfoundland and Labrador Education 2004b). Whether this indicates a trend toward increasing numbers of males completing high school is yet to be seen. Overall the differences appear greater in rural areas than in the more urban districts. The lower level of male achievement corresponds to their lower levels of school satisfaction and higher dissatisfaction than their female peers. Therefore, it was somewhat surprising that there were no sex differences in attendance at the district (92.9 percent and 92.8 percent) or provincial levels (91.1 percent and 91.4 percent).[3] Like the relatively high percentages of students now completing high school in comparison to earlier decades, the percentages attending post-secondary institutions have risen dramatically in the past two decades. In the rural community school we studied, in 2001–2 (the latest year information was available) 79 percent of those who graduated from high school went on to post-secondary studies: eight percentage points above the provincial average (K. Harrison 2005). In short, rural youth were no longer seriously behind their urban counterparts, although there was still room for improvement.

THE EDUCATIONAL EXPERIENCE OF
WEST COAST YOUTH

Recognizing that rather similar cost-cutting restrictions were also operative on the West Coast, we sought to ascertain from youth their views of what they face in the present and in planning for the future. We developed a survey in which a total of 930 youth from secondary schools in five communities on northern Vancouver Island and the Prince Rupert area on the northern coast participated. They were aged 13–19 and were in grades 8–12; we surveyed 418 males and 510 females, of whom 29.3 percent identified themselves as Aboriginal and 70.7 percent as non-Aboriginal. Of students who completed the survey, 85 youth are currently in grade 8, 203 in grade 9, 225 in grade 10, 243 in grade 11, and 177 in grade 12. The self-report survey was divided into sections that included "Tell Us about Yourself," "Family," "School," "Work," "Behaviour and Activities," "Youth Issues," and "Neighbourhood and Community."

West Coast youth told us that, for the most part, their families were doing well when it came to meeting basic needs: 84.8 percent said they never had a problem meeting basic needs or that the question did not apply to them, but 14.3 percent indicated they experienced financial difficulties "monthly" or "weekly," while 18.5 percent experienced employment problems "less than monthly," "monthly" or "weekly." We found that a large percentage of students work for pay: 10.8 percent worked fifteen hours or more per week, and the majority of youth worked for pay as babysitters and/or had paper routes (45 percent), while 26.7 percent worked in the restaurant and sales industries. Given the time children and youth spend in school, there is an obvious pathway to their health and well-being from the school environment. Only 19.1 percent indicated they were unsatisfied with school, which is a positive reflection of the "school environment," reinforced by the perception of 55 percent of participants who believed that teachers cared about how they were doing and of 60.8 percent who felt that most teachers liked them. In addition, 63.3 percent had grades that were "mostly as" or "mostly bs," 31.9 percent "mostly cs," and only 3.7 percent said they received "mostly ds or fs." School, then, can be taken to provide a positive, not negative, pathway to the health and well-being of youth.

In contrast to this mostly positive picture, however, a surprising number of West Coast youth (56 percent) said that often or sometimes they "feel hopeless," and 53.3 percent said they felt "unhappy, sad or depressed" sometimes or often. This was reinforced by the 47.4 percent who indicated that they had "trouble enjoying themselves."

Consequently, when asked questions pertaining to "efficacy" or to expectations regarding their future success, only 20.2 percent agreed that they "can change important things in my life." Nonetheless, and on a brighter note, 73.9 percent of youth said they "were looking forward to the years ahead" and 72.2 percent believed that "what happens in the future depends on me." Sadly, a number of coastal youth felt that racism was a problem; almost one-quarter of youth (23.2 percent) saw problems of acceptance of people from different ethnic groups. As one participant said, "I like living in this community. It's just that there is so much racism here between Aboriginal people and other races."

We also sought crucial information on youth and the problem of tobacco and drug use. While 79.6 percent "somewhat agreed" or "strongly agreed" that there was "too much tobacco among ... youth," 79.8 percent "somewhat" or "strongly agreed" that there was a "problem with youth using drugs." There may be a link between this result and the lack of varied activities for youth: 76.9 percent agreed that their communities "need a wider variety of recreational activities for youth." As one participant commented, "there isn't enough things for youth to do. In the future I can see the population here being low, and what population there is, would be that of seniors. The youth get bored here easily and turn to things like drugs and alcohol. We need better things to do here and more job opportunities. Most youth want to get out of this town ASAP, including myself." Another youth stated, "I don't really like it here because it is too boring. A lot of youth drink, have drugs and party all the time and they talk about it. I think it's because they have nothing better to do and their home life isn't that great." In addition to lack of activities, many youth did not believe that there were enough opportunities to be involved in the community to motivate young people to stay in the area, nor did they feel that their contributions to the community were valued. Only 14 percent "strongly agreed" that there were good opportunities for youth to be involved in the community. In contrast, 55.8 percent either "strongly disagreed" or "somewhat disagreed" with the statement that there were "enough opportunities for youth to stay in the area," while less than half of coastal youth (47.2 percent) felt that "contributions by youth to the community are valued."

Overall, these West Coast youth were not strongly optimistic about their futures. Less than 40 percent rated the economic outlook in their communities as "bright," and only 37.4 percent were satisfied with employment opportunities. Furthermore, only 38 percent agreed that their communities did a good job of planning for the future, and only 13.6 percent strongly agreed that their quality of life was good. When we asked youth whether they would encourage their children to live in

their communities, only 8.4 percent "strongly agreed" with this statement. One person said, "I would never ever encourage my kids to live
here," while another responded that "it is a good place to raise a child
but once they get older they don't have as many opportunities as people that are living down island."

We were encouraged to find that, in spite of the major changes in
the economic and social life of this community over the past decade,
there are signs of adaptation and resilience among the students. As on
the East Coast, educational achievement has improved substantially.
Most youth plan to leave the community after completing high school,
and the vast majority intend to pursue post-secondary education.
These young people are adapting to the reality of few vocational or
professional career opportunities in their communities and to the
need to pursue further education and training in order to have a good
life. Their optimism is reflected in their belief that they will indeed be
able to get the kind of education to which they aspire. Although these
advances in school achievement are seen in both males and females,
females continue to get greater satisfaction from, and do better in,
school than males. This gap is not unique to this community and, in
fact, is seen at the provincial and national levels, but it is larger in the
coastal study area. Since educational attainment can be considered a
marker of "healthfulness" (Power and Hertzman 1999) and is strongly
associated with adult health status, attention must focus on why males
continue to perform at lower levels than females and what schools and
communities might do to close this gap. That said, the increased educational attainment of coastal youth, despite cuts in education budgets
at the provincial level on both coasts, is an encouraging sign for these
young people themselves and may be so for their communities if
enough of them decide to stay.

THE STRENGTHS, POTENTIAL, AND PROBLEMS OF COASTAL EDUCATIONAL SERVICES

Over the years of our study we observed many fine education programs
on both coasts. Education can either operate as a means of developing
future skills that will enrich the employment potential of coastal communities or function in such a way as to force out-migration. Such
things as education in small-scale entrepreneurship, art and craft skills,
hotel management, and the like, build on cultural and environmental
strengths and would contribute to keeping communities alive. We will
look more closely at these kinds of options in the next chapter – here
we look at any programs that engage youth in their own communities
(Harris and Umpleby forthcoming; Marshall et al. forthcoming;

Thompson 2004b). We also examine programs that prepare youth for study and work elsewhere – what Corbett (2007), from Canada's East Coast, refers to as "education for leaving" – including some programs that were well-established in 2001 but were discontinued by 2004 because of declining budget allocations. In one West Coast in-depth case study, we noted the engagement of high school students in their community as reflected in art displays mounted in the local mall, in liaisons between students and local business projects, and in the journalism program through which students produced newspapers on a regular basis for the public, practised research skills, and fine-tuned their journalistic writing. Some programs that the schools developed drew on already existent skills and interests of the First Nations student population. A principal from one West Coast secondary school said of a locally developed silversmith program:

We have many First Nations students [approximately 50 percent of the student population] who come with skills already in place. If you ask them how they know how to make an eagle or thunderbird, they would say it is in their blood. These often are not academic students, but ones with wonderful sculpting or painting skills. We had an information technology person [on staff, five years ago] who actually did silversmithing on his own. So now you have a teacher with the skills and the interest.

The customized program, which is "still running and involving non-native as well as native kids, brings disparate groups together," is "one of those programs that kept kids in school." It is also one that might help to keep young people in their communities in the future, and its success invites a closer examination of the idea of "academic success," which is understood all too often only in reference to the dominant culture.

We also found that two other West Coast programs for youth groups in the school were particularly vulnerable to a combination of socio-economic restructuring and government policies of financial restraint reminiscent of what Newfoundland and Labrador has been facing since the 1980s (Newfoundland and Labrador 1986b). One program involves "alternate" learning, and the other involves special needs. Many towns in British Columbia have what are called "alternate schools": the title implies that a student can alternate between the regular school and one specially designed for students who, for one reason or another, are experiencing difficulty with their academic work and are at risk of dropping out or of failing their grades. Often these programs are situated in separate schools, but the school we examined had had an alternate learning program in place since 1978. With a low teacher/student ratio, teachers in this school-within-a-school have been able to provide students with the attention, coaching, and positive reinforcement they would not get in the regular classes and which we were assured they need in order to succeed. As places are limited, students must demonstrate their determination to succeed before they gain admittance and they then have to attend regularly. They occasionally fall off the tracks, and if they do, the teacher and support worker are there to give them a hand so long as this behaviour does not become habitual. "The consistent theme in the educational lives of all these kids," we were told by the teacher of the alternate program, is that their lack of "success in school stems mainly from the home." He contends, "it's easier to look at all those kids and trace improvements in their social-emotional well-being and their skills in that regard, as opposed to their academic abilities." Now, "most of them ... would be average, or better than average in the regular system."

A family resource worker provides liaison with the community, calling in on the families at least "three times a week for an hour or two, to discuss skills and help them with things" and cooperating with social workers and other agency representatives who cannot themselves visit families on a daily basis. Students respond positively to this treatment, and each year several graduate to find jobs and to live much more productive lives than they envisaged for themselves when they first entered the alternate program.

Unfortunately the school does not have the resources to provide for the needs of the whole town now that many people are caught in the web of layoffs and chronic underemployment. One successful student had been in the alternate class for two years after having spent his first two years at the high school skipping classes, neglecting home assignments, and behaving in such a manner that he was perceived to be uncooperative by teachers and classmates. He and ten of his friends had come to the high school from a village several miles out of town, and they were all behind the town kids academically and even in their own group. When asked if all eleven from his village skipped school, he explained that while "a few of them hung out with me, the rest stayed in school all the time and listened to the teachers. But I didn't like any of it." He did not understand what was going on, he said, and so, realizing that he was going to fail if he did not go to the alternate school, he was going to have to drop out, or else not graduate (he joked) until he was twenty!

Things changed when he entered the alternate program. From the first, he "felt right at home," because the teachers actually helped. "Say you don't understand something," he explained, "you can ask a question without them flipping out on you. In my other classes, I'd just ask a question and they wouldn't explain it right, because I'm a bit slower and I don't know half the stuff they are explaining. When I was younger, I never grew up reading books or anything, so I never caught on until about now. So I was a bit slower than everybody." Now he reads, and he actually took four courses – social studies, science, metalwork, and math – in the regular classrooms and passed, while remaining in the alternate learning program to do "peer-tutoring." He now tutors the whole class and does "marking and stuff – but most of the time it's just for my homework and my outside classes." Now he says that his "outside classes ... they're not too bad because, now, I've caught onto everything, and I basically know quite a bit more." This spring this student will be the first person in his family to finish high school. He has gained some self-confidence – although he considers high school the pinnacle of his education – and for now looks forward to finding work and is confident that he will do so.

It is through programs like this that many youth have gained confidence from people who expect them to succeed; they have achieved an awareness of their earlier problems and have now obtained the personal satisfaction that makes productive citizenship possible. The programs will not promote an active involvement in citizenship or create community leaders, but they demonstrate the way ahead and show that the youth who benefit from them can help others on a one-to-one level. They provide a direct pathway to mental well-being.

As the economic situation of coastal communities worsens and increasing numbers of students and their families fall into poverty, the effort to focus attention on student needs becomes increasingly urgent (Maxwell 2003). Unfortunately, with increasing class numbers and fewer support workers assigned to classes and community liaison, the opposite scenario is developing in some areas, and mental health and well-being, not just for youth but also for those who care for them in and outside the home, are suffering in the process. Take, for example, another vulnerable group – those with moderate and severe physical, emotional, and/or learning needs. In chapter 8 we discussed the changing demographics of coastal communities. At one school we worked in, the student population has decreased from 550 ten years ago to approximately 400 in 2003. Of these 400 students, 134 need specialized attention (this number includes the 17 students from the alternate learning program). Many, suffering from a combination of moderate learning disabilities, Asperger's syndrome, and foetal alcohol syndrome (FAS), by definition are categorized as "high incidence." These students tend to move back and forth between regular classrooms and classes with extra support staff dedicated to more individualized attention. Approximately a dozen students are "low incidence" (some are in wheelchairs and severely physically disabled, others intellectually disabled), and all need constant assistance. One autistic youth of advanced intellectual capability until recently could function only on a one-to-one basis with his support worker. While all these young people have shown marked improvements in their development over the past few years, often the signs are meaningful primarily to the special education coordinator and her support workers. In the last two years, however, the number of support workers throughout the school has declined dramatically. For example, the lunch monitors have been discontinued, rendering the duties of the team, who have no break from their tasks, "almost intolerable." Formerly, as well, support workers would circulate around the regular classroom, taking notes about assignments, and later informing the resource teachers about work covered in class. As support workers are withdrawn, teachers face a greatly increased workload; students, in turn, receive far less attention and help than in previous times.

It became obvious to us that the most serious health issues accrue to special-education teachers and workers responsible for the school's dozen or so low-incidence youth. What used to be a team of six is now reduced to three members who assist with the many physically demanding tasks such as toilet visits or managing the portable lift needed to lower a youth into the swimming pool. Cooking lessons, which used to be offered on a regular basis, are now a rare occurrence: "The safety issues," we were told, "make this simply prohibitive." As the special educators attempt to execute these operations with fewer helpers or are forced to cancel trips to the outer community, they endanger their own health and decrease the opportunities for youth in their charge.

Teachers, and thus youth, are also affected by the recent emphasis on accountability. In the case of the understaffed special-education team, excessive accountability contributes to stress. This conflict between bureaucratic demands, and responses deemed feasible by school staff, also applies to high-incidence youth, those with less severe needs. The progress of low-incidence youth, like the autistic student, is meticulously recorded by the support worker who has – and will continue to have – a case load of one. A dwindling budget, passed from government to district and then to schools, does little to lessen the stress on high-incidence teachers. In 2003, the operating budget of our case school was under $200 per student, whereas the year before, "it was $1,000, and we barely made it. This year," the special-education coordinator pointed out, "we did our cooking program with donations from the food bank. Isn't that pathetic? The public school system has to get its food from the food bank, which is meant to serve the poor of the community."

In short, throughout our work, we saw many innovative school programs for youth, in which issues of personnel, workloads, and budgets combine to present serious challenges to schools and to special services in particular. Indeed, a number of students on both coasts expressed concerns about their academic preparedness, and many suggested that the lack of school resources resulting from reduced enrolments was a key reason why they were not receiving the education they felt was necessary. A number of the Newfoundland youth commented that they do not believe that they are academically, financially, or emotionally prepared to leave school. Some believe that the cutbacks to the local high school (a product of restructuring) have affected the curriculum and that they are not as prepared for post-secondary education as youth from other places. Some youth commented on the difficulties of moving to larger centres, not only because of the economic costs involved but also because of the emotional attachments to the community and the feelings of loneliness that arise when one lives in a larger centre away from family and friends. These concerns were echoed on the West Coast, where some students suggested that the variety of courses needed to ensure a quality education (especially in science and technology areas) was not being offered because the number of students attending the high school was declining as more and more families left the community in search of work and other opportunities. In addition, some students observed that young teachers were not coming to and staying at the school and that the quality of their education was suffering as a result. They feared that this process was just going to continue and become worse over time as the community became a less and less attractive option for new teachers:

New teachers aren't staying. They come for a year, they leave, they come for a year and they leave.

When our teachers retires, like, we're not going to get any new teachers to come in because a lot of new, of younger teachers that's coming in now they are not going to want to come to [name of school] to teach.

Well there's nothing here really for them to do when they could probably go to the mainland and get more money than they can to come here.

So we're going to lose courses as our teachers retires and we're going to get less and less.

The West Coast youth noted that courses and training in trades were a particular lack. Shop equipment was not being replaced, trades courses were diminished if not actually cancelled, and work experiences were a challenge: "We have no equipment left. It all got sent to another school because there weren't enough here who wanted to take the courses. Now there are more, but the stuff is all gone."

BOTH COASTS

Given the similarity of educational issues on both coasts, we sought to understand the attitudes, desires, and actions that fuelled what we consider to be two related pathways to the health of youth: (1) the impacts of social and environmental restructuring on health, life, and work, and on development and planning for coastal youth (as covered statistically in the previous section), concerning which we interviewed people in five communities on the West Coast; and (2) the impacts of that restructuring on the health of youth, concerning which we used a northern Newfoundland community for our case study.[4] Our results from both coasts have proved to be sufficiently consistent that we are confident that we are addressing national (systemic) problems and potentials. Where there are coastal differences, we indicate them.

In coastal communities, health is related to a number of complex and interacting factors. For young people on both coasts the major concern is social and emotional health, although "place" (landscape, nature) is also important. Peer and family relationships, as well as schooling and education, are major contributors to the well-being of youth and are closely linked to overall family and community health, especially in smaller, remote communities where access to services is limited. It is clear that restructuring of all kinds has affected the health of youth in a number of different ways and through various linkages and relationships, as our survey made clear. Most importantly, restructuring has influenced the emotional and mental health of youth, often in conjunction with their social health or their relationships with family and friends. These aspects of health are a key issue of concern because during the adolescent period, feelings of "belonging" and the development of identity are mediated largely through social relations, particularly with peers (Harter 1999). Negative social relations are often closely connected to poor emotional and mental health in youth and can jeopardize both their short-term and long-term health. Moreover, the nature of social relationships can influence risk practices, which affect their health and are often connected to injuries and diseases.

The pathways between these health issues and young people are direct. Many of the young people we spoke to described the effects of restructuring in terms of stresses in their own and friends' families. On the West Coast, they saw economic difficulties linked to marital separations, divorce, estrangement from (or fights with) parents, illness, and substance abuse. All of these links we had confirmed in other work (see chapters 8 and 9; Dai 2006). On both coasts, youth spoke of lack of services and youth-centred resources further affecting their daily lives. They expressed great uncertainty about their future, because decisions about what they do and where they might live in the years ahead are closely tied to the social and economic conditions in their communities. These themes were repeated on the East Coast, where the community we studied has historically been economically, politically, and socially organized around the fishery, and especially cod fishing, although sealing and other types of fishing have also been part of the culture. Since the cod moratorium in the early 1990s, various types of economic ventures have developed and attracted the attention of community members seeking to find new employment opportunities. Shrimp processing has been established, as we saw in earlier chapters, but employment opportunities in this industry are limited because of the vagaries of the market, because there are few such factories in the region, and because the local shrimp factory is highly automated: industrial efficiencies and market uncertainty are minimizing this option. Although the West Coast communities have traditionally been more resource-diverse, relying on fishing (salmon, halibut, herring) and logging and some mining in Port Hardy and Prince Rupert, the buy-back of fishing licences and the decline in the forestry sector through the 1980s and 1990s have hit them just as hard, as we also saw earlier.

However, one theme that came through strongly in all our interviews was that attachment to place among youth was strong on both coasts. Attachment to community, rooted in part in love of the rural landscape, affords a sense of security and freedom that many youth feel they would not have in an urban setting. Nonetheless, they clearly recognize that urban settings present more educational and employment opportunities and, indeed, in Newfoundland many of the youth we spoke to feel that they are being "pushed" out of the community because of the uncertainty about the future of the community and lack of employment options. As one older youth commented, "I would love to be able to live here. I loves it here, but there's nothing here for me to do." Young people with a strong attachment to the place and the culture of the Northern Peninsula appear to be under tremendous pressure to reformulate a new sense of self and identity as they attempt to start a new life elsewhere.

POST-SECONDARY EDUCATION, FUTURE CAREERS, AND EMPLOYMENT

Not only did some of the students express concerns about the quality of their current education, given the changing context within their community, but some suggested that even though they are preparing for college or university it was uncertain how many would actually complete a degree. The reasons for non-completion centred not only on not being prepared – academically, socially, and emotionally – for university life but on the economic costs of attending university, as has been the case in Newfoundland (Canning and Strong 2002). We found that a "lot of people are going to university but not finishing it." One person explained that this is because it is really expensive there: "I think one of our teachers said ... four people graduated from university on that year. Four people that's it. Like of the ten, fifteen or twenty that went." Moreover, "[T]hey can't afford it. You can only get a student loan for five years now or something. So it's hard. A lot of people's parents can't afford to put them through it."

A few of the students were quite confident that they knew what they would work at in the future. In one focus group with young women, for example, a couple were certain that they were going to become teachers or nurses. A few of the women indicated that they would like to return to work in the community, but some were uncertain about whether or not the community would be a viable option for them in the future: "I want to do education next year and I'm thinking if I could get back here teaching I would. I mean it's a beautiful area but with regards to having come back to rely on the place to support you and your family it's not ... You don't have a lot of those choices, not here."

Some of the young men suggested that the future for most young men who were not university-bound involved moving elsewhere in search of manual, non-professional work. As one young man explained, the process involves a great deal of mobility – back and forth and in and out of the community. "I've got a buddy, he's like he's a bit older than what I am, he's twenty, twenty-one something like that right, he's never got no education other than grade twelve. And he went out to Ontario sandblastin' and he came home and draws his unemployment for as long as he could and when he ran out of unemployment he went out west like way out to Alberta or somewhere looking for a job and now he's in a bush, working on a pipeline with the rigs." One older youth spoke about how he already had to come and go from the community in search of work. He had returned to the community for a short period to take some college courses, and now he is preparing to move again: "It's ridiculous and I have to move all over again. Because I just came back from Toronto for a couple of years right, and I got to move again." Others were clear that they would be returning to the community only for holidays: "I never thought about, about coming back here ... the economy is in a downfall and there's nothing available that interests me a lot. But mainly just come back for vacations and stuff."

OUT-MIGRATION

Two interrelated themes have emerged from our interviews with youth on both coasts – "uncertainty" and "mobility." The young people's discussions of their current educational and work experiences centred on issues of uncertainty about whether or not they were prepared – academically, socially, emotionally, and financially – for post-secondary education. They were unsure of the future and what it would hold for them. The issue of mobility was a second major theme. The mobility of teachers and others within the community was one element that was affecting the quality of their education, as we have seen. The students' discussions about their future job and career opportunities also centred on the mobility that was certain to be a part of their own futures – mobility related to pursuing their education in order to obtain future employment and mobility related to obtaining employment, whether in professional or in non-professional work.

In Newfoundland, the uncertainty expressed was situated in a context of a changing social, economic, and political environment, and of the major changes in the resource industries that have dominated the communities in the past few years. Although youth acknowledged that the "fishery is gone" or the "community is dying," some did suggest that there was still some hope of renewal and the possibility of industrial development in the near future. On the East Coast, the uncertainty about the future of the fishing industry and the community in general is aptly revealed in the following two exchanges between some of the older youths.

RESPONDENT: Well we just had one of the plants shut down here as well and that took a lot of workers away.

RESPONDENT: But it was bought wasn't it? Are they going to keep it going?

INTERVIEWER: So what do you think the community is going to be like in another five years? What is the future going to be like here?
RESPONDENT: It's going to be a ghost town. I think.
RESPONDENT: I don't know, there is supposed to be a plant opening up within the next couple of years.

RESPONDENT: Water-bottling plant.
RESPONDENT: So I think if that opens, you will see people coming back.

Newfoundland youth's descriptions of supports, issues, and challenges in life and career planning yielded several themes and subthemes. Consistently recurring major themes include (1) staying in the community versus leaving it, (2) the implementation of "possible selves," and (3) cultural identity and self-awareness. A number of youth and adults noted that some youth try to move away and start a new life but return to the community because their strong ties to the community and Newfoundland culture, family, and friends pull them home. One older youth remarked that he had tried to work in different places outside the community but found it was difficult to "fit in." "Me, I've been to quite a few places out west, and you know, all around Atlantic Canada, but I always keep coming back though for some reason. I just can't fit in anywhere else." Another older youth pointed out that although he made a lot of money while working in Ontario, it was very expensive to live in a major metropolitan centre and it was also costly to return to Newfoundland to visit. Other youth who had spent time away from the community thought that it is easier to adjust if you lived in a centre where there were other Newfoundlanders, and they argued that many youth were seeking places of employment and schooling in centres where they knew there were other Newfoundlanders.

For at least some of those wanting to remain in the community, attempts are being made to develop a career within what are perceived to be the more stable fields of education or the health professions. It is assumed that as long as there is a community, there will be an educational system, so teaching will remain a profession with secure employment, although it is recognized that out-migration is even affecting the number of schools in the area (many outlying smaller schools are closing) and the number of teachers within the educational system. Likewise, the long-standing health centre that serves the surrounding area is assumed to be relatively stable, but it is recognized that restructuring is affecting the health care sector as the number of beds in the local hospital have been reduced. Others feel a sense of resignation that the community cannot survive the massive changes that are occurring and that out-migration will spell the death of the community. For others, there is still hope for a future in the community, although this hope also carries an emotional toll resulting from the uncertainty about what the future might look like in a community that is struggling to establish a new economic structure.

Local jobs and careers in both the educational system and the health care system were mainly discussed as options for young Newfoundland women. Most of the male Newfoundland youth spoke of either leaving the community permanently or working for periods of time and returning to the community to collect EI and then continuing this cycle of employment away from the community and unemployment while living in the community. One youth commented that he will probably have to follow in the footsteps of what he sees happening in his family. "Well, Dad has been going away now for the past seven years in Ontario working. Now, that's where my brother's gone; and most likely when I'm finished school that's where I'll be going."

On both coasts, there is also an attachment to people and family that some youth feel has been fractured because of out-migration. This was expressed very clearly on the East Coast where, as more and more individuals and families leave in search of employment and a secure future for themselves and their children, sources of support and connection are lost. For youth, key networks of friendship are then either broken or transformed into long-distance relationships. Youth noted that the changing social relations within the community are having a serious impact on the mood of the community. "Cause I mean back years ago when the fishing was good, or the employment rate was up I mean everybody was happy, they had no worries, they had all their families around, but now it seems like everyone is scattering, so you are losing your family, a lot of jobs are gone, and it don't seem like anyone is happy." Some youth also commented that there was a lot of "negative gossip" in the community, which contributes to negative feelings and relations, and that restructuring has added to the problem of "small-town gossip." As one youth noted, "If everyone was working and had their own lives to worry about, they wouldn't be concerned about everyone else's."

The health and well-being of these youth are also affected by what is presently available to them, and their assessment of how current opportunities stack up against opportunities for youth in other places. Many of the youth commented on the "great outdoors," and by comparison to youth in urban centres, they feel privileged to have the wonder and freedom afforded by easy and free access to the physical environment. A few also commented that they are lucky to be able to be able to grow fruits and vegetables and "kill their own meat" (such as moose) rather than having to buy food that has been "pumped full of chemicals." Winters are considered to provide great outdoor recreational opportunities. Snowmobiling and associated parties are key pastimes for many youth, and winter is, for this reason, the preferred season, although some spoke of the summer months as also a time of highly active recreational activities, including swimming in local lakes.

SUBSTANCE ABUSE AND OTHER ISSUES

The fact that most young people are strongly tied to their communities is both problematic and positive. Some participants in our study are aware that these very ties limit the choices available to them. They are struggling with whether to focus on staying or on leaving. The role of parents, teachers, and other adult mentors is particularly important in this process, to help them explore specific life role and career possibilities. The focus needs to be on generating alternatives, keeping options open, and offering practical support with educational planning, work experience, and economic management.

Although they benefit from their close relationship to the physical environment, many youth commented on the lack of options provided in communities that are under transformation. Many youth noted that since the declines in fishing and logging extra-curricular activities have been reduced and that some indoor places or spaces for youth to "hang out" or socialize have been lost. Also, because of the decreasing number of younger people in the communities, friendships may no longer be chosen – you just "got to go with whoever's there." This appears to have some healthy implications, as some youth argued that there is not as much fighting because the person you fight with today might be the only person to spend time with tomorrow. At the same time, there are negative implications, because younger youth are spending recreational time with older youth, and a few younger youth have become involved in unhealthy activities (such as drinking and smoking) at a very early age. Many of the organized activities extra-curricula that do exist are centred on sports, and youth with other interests have few, if any, structured community or school opportunities to develop their skills and talents. This relative void represents a health-related issue insofar as it creates boredom and fails to provide options for young people's intellectual and social development.

A number of youth, as well as key adult informants, commented on the use of soft drugs. The excessive use of alcohol and the increasing tendency for youth – and younger and younger youth – to drink alcohol was discussed. For many, drinking is linked to a lack of structured and variable activities, to relieving the stress associated with restructuring, and to a culture of drinking within the community, which is associated with socializing and partying. One older youth attending a local community college remarked that for many youth bars are one of the few places where they can hang out.

There was no general agreement on whether alcohol is used more or less than in previous generations, although, given the number of youth who commented on its excessive and continuous use throughout the weekends, it appears that it is currently a health issue of some concern for many, whether or not it is used more than in the recent past. Consistently with the literature on youth and underage drinking, the youth reported that much of the drinking occurs away from home and involves groups of youth without older adults present (Lindsay 2003), although most youth argued that the drinking is acknowledged by adults and is not a secret activity. The use of alcohol in conjunction with recreational activities, such as swimming in the summer and snow-mobiling in the winter (on the East Coast),[5] suggests that there is a combination of healthy and unhealthy activities in these communities that is similar to what has been identified elsewhere by other research-ers (Lindsay 2003). There is no doubt that excessive alcohol and drug use is a major health risk for young people, in part because it is associ-ated with other high-risk activities, including drinking and driving.

Youth are generally very aware of the risks and benefits attached to health and life-style choices. They know the importance of nutrition and fitness and the risks that accompany alcohol, tobacco, and sub-stance abuse. Many spoke of friends who had made "stupid choices" (sometimes fatal) about drinking and driving or doing drugs; some admitted to their own excesses. However, there did not seem to be a high level of concern about these issues – they were often seen as part of natural experimentation, or "rites of passage," into adulthood. Many youth do not seem to see a clear connection between certain behav-ioural choices (such as those of the peer group) and health outcomes (substance use).

Some of the youth are very involved in church activities, most with their families. A few spoke of spirituality in the broader sense, some-times related to moral values, or "doing what's right for others." Some of the First Nations participants acknowledged the role of the Creator in their future.

Some youth are very aware of and interested in environmental and community "health." Most, however, do not use the term "health" when referring to the environment. The specific form in which the interplay of influences within the two spheres (social and natural) impinges upon youth's health or shapes risk practices and feelings about their future within the community is affected by many different and specific factors within their "place" or location. These include nat-ural and material resources, cultural and social contextual factors, the nature and length of the seasons (which shape the youth's relationship to the physical or natural environment and affect how they interact with it), as well as the youth's relationship to others within their com-munity, especially the adult population. By highlighting the impor-tance of the natural or physical environment and the interplay between the social and the natural worlds, this work has opened up and pointed to a new direction for much of the research on youth and health, and sets the stage for more integrated and inclusive frameworks for future health research.

CONCLUSION

Throughout all our work we found it significant that the wider range of staple options (not just marine options) in our West Coast communities has not really helped them to survive, because all resource bases are now in serious decline. This consideration is particularly poignant in the case of youth, because if they leave in large numbers, reluctantly or not, coastal communities will not survive. In other words, the old argument that to be caught in a mono-staple economy is to be vulnerable to the "staple trap" now needs to be recast. These days, to be caught in resource-based economies with a wide range of staple options that are vulnerable to global market and competition forces is also to be caught in a staple trap, since the capacity to move from one resource to another (fish, timber, mining, for example) does not solve the fundamental problem of declining opportunities in a restructured economy.

This is where the force of the argument for diversification around the staple base really takes hold: in a global economy, local diversification, undertaken some time ago and now established, would have been a better pathway to stability. Once more we see the impact of policy misalignments where the youth of coastal communities are carrying the risks (without the benefits) of poorly thought-through resource sector policies, both economic and environmental. For youth, who are the future, the old options are in peril, and the new ones (see part 3) may prove inadequate. In the light of the decline of resource-based communities and the tentative nature of other economic development plans, youth find themselves uncertain about the future in a way that has implications for their emotional and mental health. Discussions about what they think their future may be reveal a huge tension: a strong sense of attachment to their community and its location and, for many, a sense of grief and loss because they feel relatively powerless to steer the direction their communities might take in the future.

In the current climate of economic uncertainty and socio-political restructuring, it is vitally important that young people participate actively in life-career planning. The situation is particularly urgent in communities on both coasts that have been devastated by recent fishing, forestry, and mining closures. There, youth face the challenges of living in areas where there are dramatically changed economic bases, limited work experience options, high unemployment, isolation, and other factors that limit their knowledge of and exposure to the world of work. In addition, young people in families and communities that are experiencing stress related to social, ecological, and economic restructuring are at high risk for injuries and health problems. Substance abuse, peer violence, depression, and high-risk sexual practices are behaviours associated with the effects of societal restructuring and the resultant family stress, economic hardship, and reduced community services. Communities and families are deeply concerned about the diminishing educational and work opportunities available for their children.

Our findings have several implications for policy and practice in mental health services, education, and community development. From the individual and focus group interviews and the descriptions of supports, issues, and challenges in life and career planning we were offered, several themes and subthemes emerged. There are limited and decreasing opportunities, information, and contacts related to work or career development and planning – most youth expect they will have to leave to pursue work or further education. Not surprisingly, then, youth feel disregarded or "disenfranchised" in the decision-making process. We need to be aware of the cultural and gender role differences that exist among youth with respect to expectations for life and work choices. We also need to be aware of the strong relationships and feelings of attachment to the community and the people in it that most youth (First Nations and others) feel, but most First Nations youth demonstrate a strong cultural identity that goes beyond affection for place and is rooted in a sense of heritage stretching back thousands of years. Mentoring and role models are vital in this age group as important sources of help and support, but the consequent stress on families, teachers, and community professionals is evident under conditions of a diminishing population and diminishing resources.

Although the discussions with the youth revealed that they felt very uncertain about their academic readiness and about their futures – where they might work and live and for how long – the students were certain that they needed to work. Many indicated that they were not going to "sit around and do nothing" or that they were not going to go on welfare. Some were somewhat hopeful that forestry and fishing would be renewed and that other industries – including the tourist industry – would develop into a bigger business and that there would be spin-off effects in terms of younger people working and living in the community. Other students were less than optimistic and suggested that in the future it would simply be "a community of old people."

The adolescent years may be, by definition, a period of uncertainty, as youth attempt to grapple with developing their independence and their identities. Yet it appears that in some contexts, this developmental period of uncertainty may be exacerbated by the particular setting within which the youth are living and the restructuring changes occurring within their communities. Thus, youth have not only developmental issues with which to deal, but also the uncertainty of what their community is offering them now, and what it will offer them in the future. This uncertainty has added another layer of complexity to their lives as they think about mobility not only as a fairly immediate issue but also as a potentially long-term solution to the process of restructuring.

The implications of mobility for the youth are numerous. As many of them noted, it is expensive to constantly have to move and to travel back and forth to the community. These economic issues may affect the youth's ability to get ahead economically in the future. In addition, there are potentially some very serious emotional and health-related issues connected to this mobility, especially issues related to attachments to family and friends and the social supports that are important for health and well-being. As we have seen, these challenges to family connections appear to have put First Nations youth at particular risk. It is vital that we address these issues of uncertainty and mobility when generating solutions to the effects of restructuring in coastal Canada.

Finally, youth themselves are aware that they live in uncertain times, and they are ambivalent about how to deal with this problem. Some simply have decided to leave for a better chance elsewhere; others hope against hope that things will improve and that they will be able to live in the places they deeply love. First Nations youth, despite long-standing problems, have an enormous commitment to place, and some government resources now being targeted at First Nations may help them, but they are struggling, not knowing how to cope and whether to go or stay. For those who want to leave, we need to understand that dislocation will have dangerous effects if the transition is not well managed. As for the young people who wish to stay – and they are the majority – they, and the places they live in and are devoted to, will constitute an enormous strength for the future if the wherewithal to help them and the communities they love can be found. We do not think Canada can afford to write off its coastal communities (see the conclusion to this volume – Chapter 16). It is time, therefore, to address the challenges that the youth of coastal communities face and that Canadians need to face with them.

PART THREE

Towards Social-Ecological Health: Coastal Problems and Potentials

12

Future Options I: Aquaculture, Hatcheries, Tourism, Transportation, and Local Initiatives

In this last part of the book we consider possible future solutions to the problems facing coastal communities, that have been identified by people on the coasts but that are often fraught with difficulties and dangers.¹ Some of these are already under way, others are merely potential, and many are problematic. In all cases, we seek to assess their viability for the local people and their ecosystems. In line with recent thinking about adaptive social-ecological systems, but giving it much wider and deeper application than has been possible heretofore, we consider interactive effects, taking note of cross-scale interactions and asymmetries and considering knowledge requirements, economic structures, and modes of governance that could perhaps in future prevent some of the worst problems of the past.

In part 1 we saw that restructuring is an interactive process that has been ongoing for a very long time indeed: it has operated across spatial and organizational scales and across time, and all too often, it has left a legacy of social-ecological degradation. In part 2 we looked at the fallout from interactive restructuring on human health and well-being, showing how entrenched policy and scientific maladjustments have interacted with environmental degradation to produce serious hardship for the residents of communities on both coasts. We also documented their voices as they spoke of multiple stressors: degraded resources, inappropriate social policies, lost nutrition, and other difficulties that people, young and old, have suffered as a consequence of misguided long-term practices, both scientific and political. Finally, we highlighted the resilience of these coastal people and itemized some of their strategies for survival.

The most pressing worry in coastal communities is the decline in employment opportunities triggered by a combination of resource degradation, industrial restructuring, policy change, and associated out-migration. In the face of such concerns, many households and communities on both coasts have shown remarkable resilience and a determination to seek ways to survive and grow. The relative costs and benefits of interactive restructuring have not been evenly distributed across communities, households, genders, generations, and First Nation and non–First Nation communities. Those interested in staying include those who have benefited from restructuring (at least in the short term), such as wealthier fishermen. Many have been encouraged by community entrepreneurs and (sometimes, sometimes not) government agencies to turn their attention to the development of new opportunities, although some of these are not real solutions so much as interim stop-gap measures.

As jobs decrease in the traditional resource sectors, shifting patterns of employment allow individuals to take advantage of self-employment in some sectors and short-term or part-time work opportunities. There is often, however, little protection for the worker and even for many of the self-employed in most restructured business organizations, since governments and larger industrial corporations have passed down to these groups many of the costs and risks and few of the benefits associated with interactive restructuring – we detailed this pattern in our discussion of the logging industry, for example, in chapter 4 and of the fishing industry in chapter 3. In addition, some local initiatives appear to be contributing to deeper ecological degradation, while others demonstrate creative ways to resist degradation and promote the recovery of ecosystems, communities, and individuals, and there is a danger of policy frameworks and directions promoting the former and handicapping the latter. Deeper inequities, ongoing degradation, and serious barriers often result in bitterness and a further loss of security when much full-time permanent employment ceases and this change is coupled to low wages, reduced benefits, a lack of job security, and a lack of advancement. The accommodations families and communities are making in the face of long- and short-term changes suggest that they expect to have to deal with the continuing impacts of resource-based industrial declines and that they are trying to protect themselves by seeking ways of developing more local control (see chapters 14 and 15 on governance).

People in coastal communities are pragmatic, not romantic, about their options, and they regard the future with the wisdom that comes from generations of experience "on the margin." Here, we join them in examining the risks and potentials inherent in some of the major possibilities that they are exploring, either by themselves or in partnership with government. These include aquaculture and hatcheries, tourism, transportation (which provides basic infrastructure), and other local initiatives. We consider them all both in the light of their interactive role as pathways between environmental and human health, and as routes to potential long-term community survival, based on the principle that if these options cannot sustain both roles, then coastal communities should not place any hope in them.

AQUACULTURE AND ENHANCEMENT

In the postindustrial twenty-first century, new opportunities tend to be sought either in the service sector or (in the Canadian "staple state") in transformations of old resource-based industries into putatively more advanced forms of those industries involving some kind of manufacturing. The production of hatchery fish for release in the wild, farmed fish, or the protection of very young wild fish are major initiatives of this kind and are seen by government as being more like agriculture. They are thought to have other strengths, such as producing more stable jobs with good opportunities for advancement in dying coastal communities and substituting in the marketplace for now-endangered or commercially extinct stocks. On both coasts, then, enhancement and aquaculture have become very popular with federal and provincial governments.

We ourselves carried out some research that sought to improve the success of hatchery-based efforts to promote Pacific coho salmon (*Oncorhynchus kisutch*) production. Salmon hatcheries in British Columbia have been attempting to counteract the declines in wild stocks associated with increased fishing effort and habitat destruction, but a more complete understanding of the biology of each of the species is required to make these attempts successful. In the case of coho salmon on the Pacific Coast of North America, the historic catch records tell a tale of an industry that rapidly outgrew its resource base as new and improved fishing boats and equipment were made available (Lee 2005). By the 1970s it had become apparent that salmonid stocks were in peril, and in 1977 the Department of Fisheries and Oceans initiated the Salmonid Enhancement Program (SEP). Hatcheries were built to attempt to strengthen stocks by effectively short-circuiting the natural losses that occur between when the eggs are laid (spawning) and when the fish head out to sea (smolt migration). The hatcheries collect the eggs from returning salmon and then rear the young in tanks and outdoor ponds, feeding them well and protecting them from predators and thereby greatly increasing their survival during the freshwater phase of the life cycle. Then the salmon are released to the rivers or estuaries about the time when they would naturally migrate to sea, completing the remainder of the life cycle as wild salmon do.

However, the survival rates from smolt release to spawner return are consistently lower than those of wild salmon. Evidence has pointed towards high mortality at the time of smoltification, possibly due to inappropriate timing of their release from the hatcheries.

Research into the biology of smoltification has led to improvements in the timing of release, but there is still no reliable, efficient, quantitative indicator of preparedness for release from hatcheries. Smoltification in salmon is a metamorphic event that prepares young fish (parr) to migrate to the open seas. During the parr-smolt transition, coho salmon go through various morphological and behavioural changes, some of which have been used to determine readiness for release. The most commonly used indicators have been silvering and saltwater tolerance. "Silvering" refers to the change in body colour as the parr slowly lose the dark bands on their sides that act as camouflage in the streams and rivers and replace them with a silvery finish that reflects the blue light of the open ocean. However, judging this change of colours has been a very subjective measure of preparedness for release. Conversely, while saltwater tolerance can be measured quantitatively as the amount of sodium taken up in the blood, the procedure may be inaccurate because of stress and other factors affecting the ability of fish to regulate their salt tolerance.

One of the other physiological features that is thought to change when salmon go from fresh water to salt water is their visual pigment chromophore. A long-standing theory, based on the research of Nobel laureate George Wald (1960), that fishes migrating from fresh to saltwater shift their visual sensitivity to match their new photic environment and visual tasks has been supported with research into the visual pigment chromophore shifting in coho salmon. Chromophores are the molecules involved in the first step in converting light that enters the eye into a neural signal that is sent to the brain and interpreted as seeing. Recently Alexander, Sweeting, and McKeown (1994) found a correlation between the timing of the shift in chromophores and the ability of coho to tolerate salt water. These authors proposed that the shift in chromophores could be used as an indicator of smolt status.

In order to test the efficacy of using this indicator of preparedness for release from hatcheries for coho salmon, we followed the chromophore ratio in fishes taken from two hatcheries and the wild populations associated with each hatchery as well. We used a technique called microspectrophotometry to measure the absorbance of individual photoreceptors (Hawryshyn et al. 2001). We collaborated with a Department of Fisheries and Oceans hatchery in Port Alberni, British Columbia (Robertson Creek hatchery), a First Nations hatchery near Hazelton, British Columbia (Kispiox River hatchery), run by the

Gitxsan First Nations, and a private aquaculture company near Sechelt, British Columbia (Target Marine). These different locations provided us with fish of varying age and from different environmental conditions.

After following the chromophore ratio in coho from different locations and life stages for two years, we discovered that not only did the fish that were about to migrate (smolts) shift their chromophore ratio, but so did the parr, which would remain in the fresh water for another year, and so did the adults from the open ocean, which would not be returning to the freshwater environment for another two years. Furthermore, the shift in chromophore ratio followed a seasonal pattern that was more closely correlated to the seasonal shift in temperature than to a change in photoperiod. However, in the Kispiox River hatchery, where the temperature was kept constant, the chromophore ratio still shifted. The continued existence of the seasonal pattern at the Kispiox River hatchery suggests that either the coho are responding to changes in photoperiod or they have an underlying rhythm that may exist in the absence of external stimuli. Such rhythms in various physiological systems and behavioural patterns have been recorded in other vertebrate species and are commonly synchronized with photoperiod (Gwinner 2003). Our results of a seasonal pattern in chromophore ratio at all life history stages suggest that the chromophore ratio is not directly linked to the parr-smolt transition or the migration from fresh water to salt water.

Our findings may appear to contradict the proposal put forth by Alexander, Sweeting, and McKeown (1994) and the earlier work of Wald. However, the coincidence of the timing of migration with the underlying seasonal cycle means that earlier studies may have misinterpreted their findings as a result of having sampled only during the short period around the time of smoltification and only from one age class. Unfortunately, our results suggest that the shift in chromophore ratio in coho salmon is not an appropriate means to predict the timing of release from hatcheries for this species. While the outcome of our research does not have the practical implications we had anticipated, it has added to our growing understanding of the functional significance of the labile pigment pair system, which permits some species to shift their visual sensitivity by altering their chromophore ratio, a system that has remained enigmatic despite over half a century of investigation.

We also undertook research related to aquaculture on both coasts. On the West Coast, aquaculture is not new, even though the sector may be. In the past, coastal First Peoples intensified their production of salmon, various kinds of shellfish, and seaweed through enhancing the reproduction and productivity of these resources, as well as through innovations in processing and storing them for year-round use and trade. Productivity was enhanced through maintenance and improvements in the flow and quality of salmon-spawning streams and beds, through transplanting salmon eggs from one stream system to another, and through use of fish weirs that helped not only to harvest fish but also to monitor them and ensure that enough passed upriver to reproduce (Deur and Turner 2005). Clam "gardening," another practice developed in some areas, probably thousands of years ago, increased the numbers and productivity of butter clams (*Saxidomus gigantea*) and other species (Harper 2004). Intensification of plant products included selective harvesting, transplanting, weeding, clearing, pruning, and in some cases fertilizing (Deur and Turner 2005). In effect, First Nations "farmed" the sea – a process of manufacturing, as is agriculture, not just primary production, which is purely extractive. Overall management was linked to ownership and stewardship patterns that enabled individual families, clans, and lineages to maintain constant control over production and to reap the benefits of their labours directly (Turner et al., forthcoming). It also focused on local indigenous species, not on exotics, and the managed areas were *in situ*.

In Port Hardy and Tofino today, salmon aquaculture plays an important role in the local economy, providing employment to residents both on fish farms and in processing facilities owned by aquaculture companies. The west coast of Vancouver Island is also becoming an increasingly important production area for shellfish aquaculture. At present, Pacific oysters (*Crassostrea gigas*) are the predominant species grown, while Manila clams (*Venerupis philippinarum*) are a distant second. Several other species are emerging. One farm now operates a successful scallop operation; another is experimenting with the culture of abalone, and growers have expressed interest in expanding into both mussels and scallops. There are currently eighty-one shellfish tenures on the WCVI, where leases occupy approximately 480 hectares of the marine and foreshore area (Clayton and Smith 2003). An additional 380 hectares has been set aside for shellfish aquaculture development under a memorandum of understanding between the province of British Columbia and Nuu-chah-nulth First Nations. Farm gate values are projected to increase from $2.5 million in 2003 to over $5.5 million by 2007, and growth is encouraged and supported by a number of groups in the region, such as the Clayoquot Sound Oyster Growers Association, the Working Sound Shellfish Committee, the Nuu-chah-nulth Shellfish Development Corp., Ecotrust Canada, the Regional District Economic Development Commission, and the Regional Shellfish Taskforce. Barriers to further growth include, ironically (given excess fish-processing capacity in the region), a lack of a local shellfish processing facility and of brand/location recognition (Vodden 2004). People are hopeful:

I would say that a lot of us came here at a $26,000 a year type income, very difficult to survive. And a lot of us have doubled our incomes in a ten-year period and people are buying houses now, people are having families, people are buying new vehicles. We have matured, everybody within the [fish farming] industry ... so I think everybody's socio-economic status that has stuck with ... this single industry has benefited. (Tofino respondent)

We now turn to an analysis of how modern aquaculture has markedly changed the "ecological footprint" of the industry in coastal British Columbia. In parallel with the DFO's Salmon Enhancement Program, no less than five federal government departments and eight provincial ministries have been involved in regulating and guiding the development of British Columbia's aquaculture industry since 1972. Despite seemingly obsessive government attention, the bureaucratic environment in which the industry has existed and continues to exist remains a jury-rigged collection of modified capture fishery and terrestrial agricultural policies. The result has been largely ineffective, leaving farmers to learn by their mistakes. In spite of some rather spectacular failures, aquaculture grew rapidly in British Columbia from 1972 to 1985, when salmon farming expanded from zero to 100 companies operating 185 coastal farm sites (Keller and Leslie 1996). At first, the typical farm was a small family business that was undercapitalized and lacking adequate government technical support.

Norwegians pioneered salmon aquaculture and remain a leading salmon-farming nation. A near collapse in the early 1970s, because of disease and a saturated world salmon market, prompted the Norwegian government to overhaul its regulatory policies and institute an agenda designed "to lead and control the industry" (Keller and Leslie 1996). Rigorous controls on production levels and expansion limits were instituted. In Canada the 1984 election of the Progressive Conservative government led to the replacement of the Foreign Investment Review Act with the Investment Canada Act, which aimed to encourage foreign investment. This legislation set the stage for a new era of aquaculture in British Columbia. While producers had demonstrated the potential of salmon farming, Canadian banks remained skeptical. If the British Columbia industry was to realize its perceived potential, financial and technological backing would have to be imported. Norwegian companies, eager to expand but restricted by a new regulatory climate at home, quickly took advantage of the newfound political support for their industry in western Canada. A Norwegian parliamentarian explained why to the Canadian Parliamentary Committee on the Environment: "We are very strict about the quality and the environmental questions [in Norway]. Therefore, some of the fish farmers went to Canada. They said, 'We want bigger fish farms; we can do anything; we can do as we like.'" (Canada House of Commons Standing Committee on the Environment 1990). Vigorous Norwegian-financed expansion and consolidation then exploded in Canada.

In 1985, the British Columbia industry was made up of one hundred small businesses; today, the BC Salmon Farmers Association represents eleven producers, of which five multinationals control 81 percent of production. Before the Norwegian-sponsored restructuring, British Columbia salmon farmers were in the business of raising and selling Pacific stocks, specifically chinook (*Oncorhynchus tshawytscha*) and coho. Thanks to technological innovation and market domination – often through novel but relentless product marketing – Norway became the world leader in farm salmon production (using Atlantic salmon) and often created markets that did not previously exist. Today, Atlantic salmon is the favoured species for culture in British Columbia for a number of reasons. First, the Norwegian-dominated industry had decades of experience culturing Atlantic salmon but was unacquainted with the culture of Pacific salmon species. Second, Norwegian companies had invested heavily in developing international markets for Atlantic salmon, and chinook and coho products could not easily fit into this marketing strategy. Third, on average Atlantic salmon convert feed to meat more efficiently and are less aggressive (leading to greater growth and lower mortality) under culture conditions than chinook or coho salmon. Thus, from both a production and a marketing perspective, it became clear that Atlantic salmon would be a more profitable product than native chinook or coho.

Atlantic salmon were introduced to British Columbia in 1984, foreshadowing the industry's Norwegian-influenced restructuring that was soon to come. Today, Atlantic salmon dominate the British Columbia aquaculture industry (82 percent of production), whose wholesale returns are three times those of the entire Pacific salmon capture fishery, as well as being British Columbia's most-valued legal agricultural export crop. In 2002 nearly four times more farm salmon was brought to market than wild salmon. Note that a tonne of landed wild salmon is worth only one-quarter of what the same amount of salmon sells for at the farm gate. However, the relationship is reversed at the wholesale level, where the value of wild salmon is one and one-half times the value of farm salmon. Product demand, supply consistency, and value-added processing drive much of the price differences. One of the most significant points to be taken from these data is that farms earn 89 percent of the wholesale value of their product, while fishers realize only 15 percent.

While the private economic benefits of aquaculture are obvious, the industry's externalities or hidden costs are rarely discussed. Externalities include government subsidies (job training, infrastructure grants) and nature's subsidies. Ecological costs include, but are not limited to, the genetic implications of escaped Atlantic salmon in British Columbia waters, the potential for disease transfer to and from wild salmon stocks, the development of antibiotic-resistant pathogens, organic pollution from uneaten food and from faeces (which can lead to diminished ecosystem functioning), questionable methods of predator (marine mammals and birds) deterrence, and unsustainable extraction of marine protein raw materials, leading to depletion of the fish stocks that are used as feed. Since 2002, the Broughton has received ever-increasing media attention due to the collapse of pink salmon (*Oncorhynchus gorbuscha*) stocks as a result of sea lice (*Caligus clemensi* and *Lepeophtheirus* spp.) infestations from surrounding Atlantic salmon farms. The ultimate "take-home message" here is that the ecological issues facing industrial salmon farming (as in most extractive industries) are in fact physical manifestations of underlying social imbalances. Hence, to focus solely on the ecological issues is to treat the symptoms and ignore the disease.

On the East Coast, since the 1980s Atlantic salmon aquaculture has been practised in coastal Newfoundland and New Brunswick. Atlantic salmon continue to be the major product of the aquaculture industry on both coasts, although bivalve aquaculture has been growing rapidly, and in Newfoundland there are now considerable efforts towards commercialization of cod aquaculture. Blue mussel (*Mytilus edulis*) aquaculture has been the most successful of the cultured shellfish, but since the 1980s the Newfoundland government has encouraged scallop aquaculture development in Newfoundland because of the ready market for scallops in the United States.

The techniques that are currently used for salmon and cod aquaculture grew out of methods designed to optimize the use of precious resources or attempts to reverse their depletion. Cod farming started in Newfoundland as a means of fattening up undersized wild cod caught by fishermen using traditional cod-trap gear and of using male capelin, mackerel, herring, and "squid-pips," for which there was no market, as feed. Subsequently it was shown that this farming approach was biologically viable not only in Newfoundland bays but also in the colder waters of Gilbert Bay, southern Labrador (Wroblewski, Bailey, and Russell 1998), and in Williams Harbour.

The development of co-managed aquaculture of cod on the Labrador coast is now under way. Gilbert Bay has attracted the attention of biologists because it has a resident population of cod recognizable by their golden-brown colour (Gosse and Wroblewski 2004). This is a genetically distinct subpopulation (Ruzzante et al. 2000) of the endangered northern cod population (Canada Committee on the Status of Endangered Wildlife in Canada 2003). Nutritionally the Gilbert Bay cod is similar to the shelf cod (Van Biesen 2003) and has traditionally been used by the Labrador Métis people for sustenance (food fishery) and recreation (Powell 1987). However, these cod grow more slowly because they live at subzero temperatures for six months of the year, while offshore cod over-winter in deeper, warmer ($3°C$) waters (Ruzzante et al. 2000). More generally, since the early 1990s limited supplies of wild cod have limited the expansion of this form of cod aquaculture.

The residents of Port Hope Simpson and Williams Harbour initiated the establishment of a Marine Protected Area (MPA) to protect the Gilbert Bay cod from overfishing, and it is likely the federal minister of fisheries and oceans will declare Gilbert Bay as the first MPA in Eastern Canada. Without the MPA, this slowly growing subpopulation could be extirpated by overfishing, as almost happened when the northern cod moratorium was ended in 1998. The quota was based on faster-growing offshore cod in NAFO Convention Management Areas 2J3KL. The management plan for the Gilbert Bay MPA will allow local fishers to pursue aquaculture in the outer part of the bay. Development will not be restricted any more than elsewhere, so that the local people will not be penalized for their stewardship of Gilbert Bay, providing that the aquaculture development does not pollute the bay. Cod and scallop aquaculture will be permitted near the headlands of the bay, which are open to the Labrador Sea.

Both cod and scallop in Gilbert Bay contain high levels of healthful omega–3 fatty acids (50–60 percent of total fatty acids: Copeman and Parrish 2004), which have beneficial effects on human plasma lipids and lipoproteins, cardiovascular disease, cancer, adipose tissue mass, and inflammatory diseases. In addition, the sterols in Gilbert Bay scallops are composed of over 60 percent phytosterols, which also have beneficial effects on human plasma lipids and lipoproteins. Thus the MPA status, combined with the development of local aquaculture, may help secure these highly nutritional cold-water seafoods for the coastal communities.

Herring or herring-based feeds are commonly used in aquaculture as a source of omega–3 fatty acids for fish. At first, the feeding of omega–3 fatty acids to fish may seem paradoxical, but fish are themselves unable to synthesize omega–3 fatty acids. They have to acquire them from preceding trophic levels within the food web, ultimately starting with microalgae or phytoplankton. Phytoplankton make fatty acids with certain "omega" designations (omega–3, omega–6, etc.) which cannot be modified by animals, yet some of these omega–3 and omega–6 fatty acids are essential to animal growth. In addition to their nutritional importance, these fatty acids can be used as food-web tracers delineating the pathway between phytoplankton and fish. Zooplankton make some other long-carbon-chain fatty acids with designations such as omega–9 and omega–11. These get transferred to zooplankton-eating fish such as herring. When cod or salmon consume the herring, they preferentially retain the omega–3 fatty acids over the less important and less easily digested zooplankton fatty acids. These then become useful biomarkers of the dispersal of organic waste from fish farms. The form in which these molecules appear in the environment permit us to distinguish between uneaten food and faecal material. If they appear in the free form rather than combined into other larger molecules (e.g., triglycerides) then they represent faecal material. This was the case in our small cod enclosure study in Gilbert Bay, Labrador. Free fatty acid levels and the proportions of the long-chain omega–9 and omega–11 mono-unsaturated fatty acids in settling particles close to the pens (5 metres) were significantly higher than before the fish enclosure was operating and higher than further away from the pens (Van Biesen 2003).

The same organic compounds that we established to be useful biomarkers for tracking organic waste from the cod enclosure in Gilbert Bay were then investigated off the coast of Vancouver Island. We sampled around current and former Atlantic salmon aquaculture sites in the Broughton Archipelago and in Clayoquot Sound and found that free fatty acids and omega–9 biomarkers were higher in British Columbia mussels than in Labrador mussels, suggesting bioaccumulation from large salmon farms. This footprint extended over a kilometre from the farm. In addition we found that bacterial fatty acid levels were very high in the water column within 150 metres of the farm edge: higher than the highest level measured in Labrador.

While we were readily able to detect a chemical signal, the biological signal resulting from the release of particulate and dissolved wastes directly to the surrounding environment was harder to discern. Salmon farms are anchored close to shore, potentially increasing nutrients available to primary producers and suspension feeders in adjacent intertidal areas. Neither the possibility nor the consequences had been investigated in British Columbia. To address this question, using a nested sampling design, we measured intertidal community structure, primary productivity, stable isotope composition (c and n), and biomass of mussels and algae at five salmon farms and four reference sites in Clayoquot Sound from May to September in 2001 and 2002. Our results will interest conservationists, policy-makers, and First Nations, whose traditional territories support salmon aquaculture in Clayoquot Sound. No differences were found between farm and reference locations for any of the variables measured. In most cases, location (large scale, kilometres) and sampling error (small scale, tens of metres) explained most of the variation. Locations were largely homogenous at the medium scale (hundreds of metres), and most of the variables measured indicated no measurable effect of salmon farm wastes. However, weak but non-significant predicted trends were observed for *Fucus* biomass and mussel dry biomass. It is possible that farms are far enough away from the intertidal zone to ensure that their wastes are too dilute to produce strong measurable effects or that natural variation prevents the detection of effects. The fact that location was usually the most significant variance component suggests to policy-makers that each farm must be examined within the context of its immediate surroundings, i.e., within the channel or inlet it occupies.

Invariably, contemporary coastal communities have welcomed aquaculture development initially, but in recent years it has had a very bad press on the West Coast as the industry profile changed from large numbers of small-scale, horizontally integrated independent local producers to fewer vertically integrated multinational companies. Unlike the sustainable aquaculture of First Nations or that of earlier small-scale producers, as in the case of cod farming based on the traditional East Coast trap fishery, the current sector is being developed along the lines of international agribusiness, with efforts intensifying to reach major economies of scale. Farm salmon mainly feeds the export market, while most wild salmon is consumed domestically. Over 82 percent of aquaculture production is exported, bringing new money into provincial and federal coffers, while domestically consumed salmon simply redistributes revenue and thus does not deliver the same economic benefits. Therefore, the aquaculture industry is now a powerful agribusiness with formidable lobbying power, commanding a significant presence in government policy development.

There is real fear on the West Coast that aquaculture development will lead to loss of access by local people to their usual fishing grounds; as well, there are concerns over pollution. In consequence, in the Broughton Archipelago and in Clayoquot Sound, where many of British Columbia's salmon farms are located, there is now strong opposition to current operations and any further expansion. On the East Coast also, aquaculture is now getting a decidedly mixed reception, although in Gilbert Bay the new MPA approach may provide an alternative, more acceptable model for other areas. Issues of commercial viability of aquaculture in remote coastal Labrador, however, remain to be addressed, as do issues of pollution and the appropriate size of farms on both coasts.

Finally, we consider potential paths forward that would permit the growth of the aquaculture sector in British Columbia while not jeopardizing the coastal heritage of all citizens. While salmon, and fish meat in general, offers an excellent dietary source of protein and essential polyunsaturated fatty acids, it is important to consider the ecological costs of intensive production of a carnivorous species. For many years promoters of the "blue revolution" (aquaculture expansion in the last few decades) have proclaimed that salmon is a more environmentally conscientious choice of protein than beef, pork, and chicken; however, their reasoning is flawed. The comparison is made between how many kilograms of food are required to produce one kilogram of meat. Beef requires approximately 8 kilograms (kg), pork about 4 kg, and chicken around 2.4 kg, but salmon is boasted to require as little as 1.4 kg of food to produce 1 kg of salmon meat. The argument breaks down when you consider the source of food.

Cattle, pigs, and chickens can be reared on grains and grasses, which obtain their energy directly from the sun at a conversion (Lindeman) efficiency of around 1.5 to 2.0 percent (Colinvaux 1986). Even in the worst-case scenario, where the production efficiency between grain and meat is only 10 percent (equivalent to a 10:1 food conversion ratio), the energy of the sun is still converted at about 0.15 percent (figure 12.1B). However, when we consider the food chain between the sun and salmon we see that the production of salmon meat is not nearly as efficient. Aquatic food chains tend to be much longer (figure 12.1A). Adult salmon eat small fish, which eat zooplankton, which eat algae, which, like grasses and grains, obtain their energy directly from the sun. While the path from sun to beef has only two steps, that from sun to salmon has a minimum of four, and although the production efficiency is greater for poikilotherms (which do not waste as much energy on heat production and metabolism as homeothermic mammals) the net result is still a much lower efficiency of conversion of the sun's energy (2.5% x 20% x 30% x $35\% = 0.05\%$: values are conservative averages from similar trophic levels from table 4.4 in Colinvaux 1986).

Because salmon are so much farther up the food chain than cattle, pigs, and chickens, raising salmon is more like raising lions and tigers. Instead of eating the salmon we could increase our ecological efficiency by eating further down the food chain and consuming the small fish that are presently being caught and turned into salmon food;

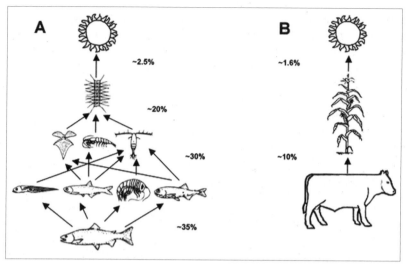

Figure 12.1   Food chains for both marine (A) and terrestrial (B) food sources, showing the approximate conversion efficiencies at each trophic level (Colinvaux 1986)

better still, we could focus more on raising bivalve molluscs that eat micro-algae right at the bottom of the food chain. As things stand, we fear scale asymmetries in current aquaculture practices and a serious risk of local communities and ecosystems bearing the brunt of the risks involved while not receiving many of the benefits.

TOURISM

One of the biggest hopes for the future on both coasts has been the development of tourism. We point to the example of Tofino on the west coast of Vancouver Island where the economic outlook is relatively positive. In addition to the growing tourism industry, aquaculture is expanding (as we saw), and two companies provide direct year-round employment for approximately 245 people, not all of whom are Tofino residents. Interestingly, residents vary in their satisfaction with the new look of Tofino. Some residents describe the economy as "booming," "diverse," and "improving," while others see it as "seasonal," "terrible," and overly dependent on tourism. There is, however, no doubt that Tofino is experiencing substantial growth in tourism-related employment in the accommodation, food, and beverage sector, which has gone from employing 190 residents in 1996 to 305 in 2001, an increase of over 60 percent (British Columbia Ministry of Labour and Citizen's Services 2001).

While tourism is now a fundamental sector of Tofino's economy, some residents feel that the community is overly reliant on tourism as an economic driver: "There has been a lot of energy focused on tourism, and a lot of people are saying, whoa, enough tourism, let's put our energies somewhere else, on year-round sustainable employment for people" (Tofino respondent). Others however, feel that Tofino is a "money generating place [which] makes it really stand out on Vancouver Island; it makes it very different" from other places such as Port Hardy. There is a sense that this economic success is rooted in Tofino's early economic diversification:

[I]n most small coastal communities we are only reliant on primary industries, fishing, logging, that sort of thing and long before that fell out, this community was already diversifying ... This place became an alternative place to stay, it became the bastion for the environmental community and then for the artist's community and everything else that goes along with that. We were well diversified long before the fallout of those primary industries. (Tofino respondent)

In Ucluelet, too, there is a strong sense that successful diversification is under way. The people there think in terms of a range of opportunities that, taken together, provide some hope. They argue that through new internet-based businesses and tourism ventures, in combination with forestry- and fishery-related work, their economic outlook is bright:

I find one thing with this community, I think it is one of the reasons that I am really proud to work here, it never, I never saw the community roll over and give up. And god, it has reasons and opportunities and justifications to do that at so many levels and they just don't. And we keep saying, okay how can we reinvent ourselves this time? And that's what's exciting, that's exciting to be involved. (Ucluelet respondent)

I am getting the sense that a lot more people are getting less stuck on wanting to hold on to industries that are obviously no longer sustainable at the level that they were a decade ago. People are beginning to diversify ... the focus is starting to shift, I find, and there is a lot more diversification in people finding different opportunities. (Ucluelet respondent)

This sense of economic opportunity in Ucluelet is probably influenced by the town's proximity to Tofino (a forty-two kilometre distance), a community that has seen tourism-related business expand rapidly in the past ten years. There is no doubt that Ucluelet has benefited economically from the huge number of people who visit Tofino and the Pacific Rim National Park (one million each summer), seeing tourism-related jobs increase substantially:

In the last couple of years, I would say it's a turn to the tourism industry for sure. It seems to be one of our only viable, sustainable industries that are apparent right now. Forestry used to probably, I would say it likely employed ... at least 30% of the population here, now it is maybe a handful, so things are turning over. And the fish plants actually employ quite a few people in the summer, but again, it is seasonal, so fishing isn't necessarily one of the larger employers either. So yes, I would say that tourism is. (Ucluelet key informant)

Many of the jobs associated with traditional resource-based industries did not provide stable work, although they often paid better than tourism and used to guarantee seasonal employment and a sense of security and stability to workers before restructuring in the 1980s. "[M]y husband was a commercial fisherman up 'til 2001. We sold out. It is not a life like we used to have ... I think the quality of life was much better when we fished years ago. We talked about it just the other day. We knew where we stood with fishing" (Port Hardy respondent).

First Nations communities are also turning to tourism and, indeed, Canadian and international tourists are attracted to their rich cultural heritage. Because of the relatively remote location of many communities, such as Skidegate on the Queen Charlotte Islands, tourists gain access to wilderness areas, wildlife viewing, and sport fishing. As well, the ongoing use and knowledge of traditional territories means that local people are in an excellent position to guide and teach visitors, as can happen in, for example, the traditional territory of the Gitga'at of Hartley Bay, which includes Princess Royal Island and the home of the rare white black bear, also known as the Kermode or Spirit Bear (*Ursus americanus kermodei*).

Such opportunities are something of a double-edged sword, however. The remoteness of many communities has served to protect their culture, while at the same time making it difficult for them to get a significant volume of tourist traffic. Moreover, the crisis in the fishing industry has adversely affected the sport fishery, where commercial fly-in floating tourist operations have been set up in prime locations, including old village sites, by highly capitalized tourist operators out of the United States, Vancouver, and Victoria catering to the high-end market. Although these operations sometimes employ local Aboriginal guides or cultural ambassadors, the local communities capture little of the profit and have no control. Worse still, since traditional Aboriginal territories are not recognized in law, industrial logging has destroyed a vast proportion of the coastal forests and river valleys as sites for scenic and cultural tourism. Finally, the persistence of limited access to trade and professional training has limited the capacity to diversify into the hospitality and cultural industries.

That said, the manner in which the Haida and Parks Canada operate the National Park Reserve of Gwaii Haanas in Haida Gwaii (Queen Charlotte Islands) is an object lesson in what can be done. There, Haida "watchmen" (often local Haida youth together with older people) reside in the abandoned First Nation villages in the area and teach visitors about the local culture. The park limits the impact of tourism on the fragile ecosystems within it by limiting the number of visitors at any one time, and requiring them to take wilderness instruction from park officials before entering the area. The park is run by an effective partnership between the Haida Nation and Parks Canada and could well serve as a model of what can be achieved, given goodwill and intelligent forethought.

On the East Coast, where we interviewed tourism operators in three areas outside the main tourist centres, as well as people on the Great Northern Peninsula (the Viking Trail and Gros Morne Park) and southern Labrador, we found similar kinds of development and similar concerns. Positively speaking, in Hawke's Bay, for example, residents have derived benefits from some long-standing outfitting operations in the Hawke's Bay area that serve hunters and anglers, and they have also been able to take some advantage of people visiting the Viking Trail, though a lot of tourists bypass the area on their way from the park to L'anse aux Meadows. The Labrador Straits have also seen growth in tourism, which was expected to increase with the extension of the Labrador road and with changes in the Labrador ferry service. We spoke to three tourism operators there who moved from the fishery to tourism in the early 1990s. In southern Labrador in recent years, road construction (see later) has made parts of the coast more accessible, as well as providing employment, particularly for men, although such work tends to be short-lived.

White Bay South has the least-developed tourism sector of these three areas. That said, in each area we were told that "tourism can't replace the fishery" as owners and employees struggle with a short season and low earnings. As one person said, it is "OK for a second income or students" but won't keep young people from moving away. Others emphasized that tourism can succeed only if the other sectors do also, as it will be local business that will maintain the restaurants and B & Bs in the off-season.

Tourism, then, provides service-based employment, which is a weak way to diversify local economies, since tourism is tied to short-term seasonal projects that depend upon market conditions and the weather and since its desirability can be questionable, given ongoing problems with environmental and wildlife protection. Tourism-related employment is also problematic on both coasts, at least in isolation, since the pay is generally very low and the season is short. Indeed, on both coasts tourism operators often have trouble finding workers, even while unemployment rates remain high, and labour shortages continue to be exacerbated by EI rules (see chapter 9).

As tourism takes hold on both coasts, there are concerted efforts to "sell" the beauty and history of these places on both the national and international scale. This reminds us of the argument of Whitson (2001), who, writing about the rise of "consumerism" in rural and remote locations, notes that business groups from urban Canada and other countries have turned their eyes to rural communities that can provide world class recreation and tourism destinations. The resulting gentrification of the rural countryside is often seen as leading to a lack of affordable housing for local residents, since municipal services are aimed at supporting affluent newcomers, while local young people struggle to make ends meet in low-paying service jobs. Finally, unless it is very carefully handled, the tourism industry can self-destruct. There is a troubling lack of ecological concern about protecting just exactly the beauty that has brought tourists to these areas in the first place. As things stand on both coasts, tourism is seen as a necessary but not sufficient condition of ongoing community resilience, one that has the advantages of significant degrees of local control but the disadvantages of short-term employment and the instability consequent upon seasonality and consumer tastes.

Coastal community residents on both coasts are aware that tourism can only be a complement or a supplement, not an alternative, to traditional resource employment, since it cannot sustain families and communities on its own:

Tourism is now growing, but that is a whole different thing, and it doesn't provide full-time employment so it is not anything that many people can raise a family on. The way I see it, tourism makes a handful of millionaires richer, and it provides part-time work at $10 or $12 an hour for three months a year. (Ucluelet respondent)

Well there is no work for people, and the work is, if you do want to work, it's pretty menial jobs and there doesn't seem to be a lot of money coming in, and people want to hold on to their money and they don't know what is coming next. So things are pretty bleak right now. (Prince Rupert respondent)

Many people voice concern about the quality of the employment opportunities that are available, especially those jobs associated with tourism, and several West Coast interviewees explained that tourism-dependent jobs, being seasonal and mostly low-paying, are not ideal for people raising a family or trying to buy a house. On both coasts it is becoming very clear that some services that tourists use in the summer (hotels, restaurants) depend for their long-term viability on year-round business traffic. It is also clear that visitors wish to see living communities and cultures, not places that have become theme parks or have been returned to "wilderness."

TRANSPORTATION: A CASE STUDY FROM LABRADOR

For industrial diversification, including tourism development, to succeed, appropriate infrastructure is required. On the West Coast, road and ferry services connect most of the communities we studied, but on the East Coast there had been, until very recently, no road linking the Labrador communities from Red Bay to the regional centre in Goose Bay. Dreams of a railway line or road through Labrador westward to Quebec go back over a century (from the 1870s onwards) but have always been abandoned, first until France renounced its treaty rights to the French Shore in 1904 (see chapter 2) and thereafter because of the immense cost of any such initiative, including either a tunnel under, or a causeway over, the Strait of Belle Isle to Newfoundland, which (the rhetoric claimed) would have shortened communication and transportation time between North America and Europe (Morton 1975). Other proposals and applications – for a rail line, a port (near Cape Charles or Battle Harbour), and a causeway across the Strait – were filed during the first three decades of the twentieth century (Kennedy 1995, 128–9), but these too were abandoned. However, after Newfoundland joined the Canadian federation and after the development of the Upper Churchill hydro project in the late 1960s, the idea of a fixed link between Labrador and Newfoundland and a road north to central Labrador was born again.

Serious thoughts of such a road started in the early 1990s (Russo Garrido and Stanley 2002, 6) and finally, in November 1996, representatives of the government of Newfoundland and Labrador travelled to five isolated communities along Labrador's southeastern coast to discuss the possibility of constructing such a road. Not surprisingly, the Labrador people strongly supported a road in and out of their isolated winter settlements, since (although some fishers had turned to the lucrative new snow crab fishery) a road linking their communities suggested a new beginning. Their hope was that federal monies would be used to build roads as essential infrastructure for the development of a new, but as yet unclear, future economy. That construction became inevitable once the province had acquired responsibility for coastal ferry services from Ottawa, which had developed the Labrador Transportation Initiative Agreement (LTIA) but had found it economically unfeasible to keep the coastal services running, given the size of the LTIA (Russo Garrido and Stanley 2002, 6–7). That change, coupled with the need to jump-start new initiatives in the wake of the groundfish moratorium, led to a 1999 provincial announcement awarding six road contracts totalling over $36 million, which were estimated to create approximately 5,500 seasonal jobs for construction between Red Bay and Goose Bay. It was hoped that the roads would be a "symbol of freedom, linking Labrador to national and global networks, and fostering economic development in mining, forestry, fishing, and tourism" (Russo Garrido and Stanley 2002).[2] The next year, the province appointed a transportation committee composed of local people in southern Labrador to canvas local opinions on transportation priorities; a similar committee met the following year in northern Labrador. The province also commissioned an independent assessment (Abbott 2002) to examine changes in coastal marine services and how their potential replacement (at least in part) by roads, would affect the people of coastal Labrador, with the result that vessel services to Labrador were divided between Lewisporte and Cartwright (www.gov.nl.ca/releases). The phase 2 section of the road between Red Bay and Cartwright was officially opened on 13 September 2003, and the (revised) environmental assessment of the final (phase 3) section, from Cartwright junction to Goose Bay, was approved by the federal and provincial governments in July 2004.

Anthropological studies by Southard (1982), Schneider (1984), and Kennedy (1996) – conducted before today's road – describe a seasonally ordered way of life in which people moved annually between summer fishing stations on outer headlands or islands and winter settlements normally located in forested coves or bays near sources of wood, food, and water (Dyke 1969). These seasonal moves to and from places where essential, seasonally available resources could be harvested were typical of the historical pastoral societies of the northern North Atlantic, as they were also for First Nations on the West Coast. Until the 1960s, when snowmobiles replaced dog traction, transportation in winter (before motor vehicles on today's new road) had been by dog team. In summer, boats were used, water transportation technologies having evolved from the masted bully and row boats of the nineteenth century to the motorized trap boats of the twentieth century, supplemented (after World War II) by faster speedboats propelled by outboard motors.

Such technological change, of course, affected the people adopting it: women (who had depended on men to harness and drive dog teams) were, for example, suddenly free to drive their own snowmobiles. The settlements and fishing stations were also affected by such restructuring of transportation networks: the summer stations gradually lost importance as the twentieth century wore on, and government eventually required that people reside in permanent settlements, most of which had been their seasonal winter villages (Kennedy 1995). It is these winter villages, now occupied year-round, that the new road connects.

In the fall of 2000, one of the team returned to Labrador to see what restructuring had occurred and how it had affected the region since he first started doing research there in 1971 (Kennedy 2001). In the communities of southeastern Labrador, it was clear that most locals welcomed the new corridor through their lives, even though many accepted that there would be negative as well as the positive consequences of employment in an area badly hit by the groundfish moratoria.[3] Many agreed with government's hope that after the construction phase was over,[4] the road would bring new opportunities for diversifying the local economy in tourism, forestry, and the shellfish fishery. There was a widespread suspicion that government had a hidden agenda in building the road: people said that they anticipated a government-blessed corporate deforestation of Labrador (chapter 5), and many people expected an increase in the outside corporate exploitation of local resources, chiefly fish and forests. One man from St Lewis commented that the road was being built as a "woods road for sixteen wheelers." People also worried that rivers and lakes contiguous to the road would be fished out. Still others predicted that out-migration would be facilitated and that strangers would change their communities – they complained that they now had to lock their homes. Few, however, seemed concerned about how the new road might encourage a greater centralization of services such as schools, medical clinics, or airports.

Wanting to quantify the human impact of the new road, we conducted random household interviews in communities between Lodge Bay and Cartwright, talking to fifty-eight households and conducting thirteen key-informant interviews. We found, among other things, that 62 percent of locals believe that the road was constructed for wood extraction, that locals are travelling and/or visiting people in other communities more often, that there has been a shift from ATVs, snowmobiles, and boats to trucks, that increased numbers of locals are travelling to stores in the Straits to shop, and that travel by road has improved access to social and medical services.[5] We also found that only 21 percent of those interviewed have benefited economically directly from construction, a percentage lower than either government or others might have predicted. Moreover, while 80 percent of those interviewed in Port Hope Simpson are content with the centralization of airplane or school services in their community, others (from communities such as Lodge Bay and St Lewis) are very concerned that their future access to these services may not be adequate, given that in winter it will now be necessary to drive on a road that is dangerous and often closed as a result of major snowfalls.[6] Although it is too early to provide a full critical assessment of the impact of the road, we note certain trends. Today's shrimp and crab fisheries occur one hundred or more kilometres offshore, implying that whether "home ports" are located in interior bay communities or closer to coastal headlands is less important than it was during the old inshore cod fishery. Shellfish products are now largely trucked from the area, while plants can now truck in their essential inputs into the production process.

Whether the road will be the hoped-for catalyst for development, then, remains uncertain, depending to some degree on what is expected of future "development." For some, including government, development is synonymous with economic growth and the exploitation of new resources. For others, real development means the enhancement of people's capacity, within the bounds of a given geography, to have greater control over their lives. So far, the hurried construction of the new road has squandered the opportunity to develop the kind of infrastructure and expertise in the region that will be needed in the long-term maintenance of the road. Just where the new road will take the people of the region is not yet clear. What is clear is that there are still many problems associated with it. In the late fall of 2004, Air Labrador said it would cancel its service on the south coast of Labrador, at Cartwright, Mary's Harbour, and all communities in between in March 2005, citing competition from the road and saying that the air service was no longer viable. The problem is, of course, that the road is closed between Lodge Bay and Red Bay in the winter, and residents do not know how they will get in adequate food supplies (CBC 2004).

There is no doubt that the environmental impact of road construction has been severe and negative. In 1992, local people so wanted a road that many appeared ready to deal with any potential problems accompanying it (Kennedy 1996, 114), and many worried that the Labrador Métis Nation's (LMN) criticisms in the media about potential negative impacts of the road might slow construction. The LMN continues to oppose what they characterize as a road through Métis lands, on the grounds that it destroys salmon and trout habitat, and our work (Gibson, Luther, and Haedrich 2002) confirms this (see Kennedy 1997 for details). We examined stream crossings (where conduits transmit water beneath the road) along the new road, in particular those between Mary's Harbour and the intersection to Charlottetown, a distance of 210 kilometres, in order to determine whether construction was conducted according to DFO guidelines, which are explicit and stipulate that (for all year classes of fish) fish passage must not be impeded. The Fisheries Act also states that there is to be no net loss of productive fish habitat,[7] while DFO guidelines further suggest that open-bottom crossings (steel archways) are preferable to cylindrical culverts, because the former do a better job of retaining the integrity of the natural stream configuration and conserve the natural stream substrate.

However, we found that only 2 of the 47 culverts were open-bottom crossings; the rest were cylindrical culverts, none of which were embedded with natural stream substrate with the minimum of 30 centimetres, as stipulated in the guidelines, resulting in loss of stream habitat of 3,000 square metres due to the culverts themselves. In the 210-kilometre section of road that we surveyed there were 4 bridges and 47 culverts, which we considered representative of the road as a whole: the 325 kilometres between Red Bay and Cartwright (including access roads to Charlottetown and St Lewis) has 11,000 steam crossings in all. Of the culverts we studied, 25 (53 percent) presented barriers to fish movements, resulting in fragmentation of stream habitat and the loss of spawning and rearing habitat. The variables (single or in combination) creating the obstructions at the culverts were insufficient depth (36.1 percent), high velocity (25.0 percent), outlet drops (perched outlet; 16.7 percent), dryness (with water going under or around the culvert; 13.9 percent); and outlet barriers (8.3 percent).

Of course, the problem of improper culvert installation is not unique to the Trans-Labrador Highway, but wherever it occurs it involves significant costs to correct fish habitat and restore hydraulic capacity. In the Labrador case and elsewhere, culverts are often installed improperly because contractors are unaware of the ecological role small streams play in the rearing habitat of juvenile fish, and, in any case, they usually place low priority on ecological consequences. We should add that open-bottom crossings cost much more than cylindrical culverts, and limited economic resources prevent sufficient enforcement of guidelines by environmental officials, both in Labrador and elsewhere. We explored the economic obstacles to proper culvert installation and concluded that the most common and cost-effective conduits in road construction are the round 800–900-millimetre culverts, which cost about $200 per metre (or about $4,000 for the average stream crossing). The province, which funded the culverts with its limited budget, understandably requires bottomless arches or bridges only when the DFO insists.[8] As a result of our work on culverts, a partnership with the Labrador Métis Nation (LMN) was formed, which led to high-level meetings with the DFO, who undertook further studies. Five problematic culverts were reinstalled in 2003, the designs for stream crossings of phase 3 of the Trans-Labrador Highway (TLH) were changed in some instances, and LMN fishery guardians have been trained to ensure correct installations.

We conclude that while initiatives like the road are essential in this day and age, it is important to consider their human and environmental costs. Misalignment in the form of environmental risks to fish habitat need not have happened if due caution had been observed, and it is worth noting that such damage is likely to have negative impacts on tourism. It is also important to maintain consistent policies towards all transportation services in this, as in any, remote area.

LOCAL INITIATIVES

It is imperative that governments break the old staples mindset of single-industry growth as a solution to the survival of coastal communities. Coastal communities need to get away from dependence on just one sector. Diversification is essential if flexibility is to be achieved, and coastal residents certainly understand this, and they understand that integrated planning between developing sectors is essential to long-term social-ecological health. We select a few of these local initiatives to illustrate the richness and creativity that exists in these places and to support them as they build new strategies that work at an appropriate scale and out of a deep understanding of local social and ecological conditions.

The community economic development (CED) strategies being used in Alert Bay provide a good example of the kinds of things that can be done and the barriers that are met along the way. Alert Bay has sought to diversify, both outside and within the fishery, and residents would like greater involvement in the management of the fishery and other local resources (Vodden 2002). Community adjustment strategies have included attempts to get more involved in fisheries management and stewardship, an historic role for the Kwakwaka'wakw. Efforts have had some success but are limited by a number of barriers, particularly relations with government (e.g., the DFO) but also limitations in community capacity, finances, and the ability to work together within the community/region. "A demonstrated sense of resource stewardship, commitment to an identified geographic area, a core of dedicated individuals with invaluable local knowledge, and organizational experience in fisheries projects, however, represent a foundation to build upon" (Vodden 1999, 162).

Unfortunately, key leaders have left for various reasons (health, business closures, community conflict), and as a result early efforts have had limited success. Vodden suggests that it took several years after the Mifflin Plan for the community to accept the need for change and that "Alert Bay organizations could be using their resources much more effectively, increasing their likelihood of success, through better coordination and more strategic planning and action" (1999, 116). Now a number of initiatives are being undertaken, particularly by First Nations with a long-term commitment to selected strategies. Development is under way in tourism, energy, and mining. A gravel extraction project is being considered, along with a review of offshore oil risks and opportunities, battles over finfish aquaculture (which has expanded in the region despite opposition from First Nations and others), and developments in shellfish aquaculture, independent power, value-added wood products, and non-timber forest products (e.g., mushrooms, ferns, salves, traditional plant use in tourism interpretation). New and growing sectors such as tourism, resource management, silviculture, watershed restoration, and aquaculture, along with employment by First Nations governments, have provided work for some who have been displaced and have helped to supplement declining fishing incomes. Women have also become significantly more important within the local economy, surpassing the number of men employed for the first time in 1996. Employment has become less important as an income source (relative to other sources), while self-employment and government transfer payments have increased.

On the west coast of Vancouver Island, the wcvi Aquatic Management Board Structure was set up over more than seven years to provide more local input/control of fisheries decision making in this region. The board has reopened the gooseneck barnacle (*Pollicipes polymerus*) fishery (thirty-one part-time jobs, four months per year) and created a trust to hold and distribute such things as licences and tenures, which could be transferred within the community, based on criteria beyond capital alone (Aquatic Trust). Since shellfish opportunities are threatened, the board has started addressing issues in water quality monitoring and working on local sales and a brokerage firm for investment, as well as a consulting firm (Day 2004). There is also a local environmental organization partnered with other local agencies to pursue environmentally and socially responsible economic development options (Clayoquot Green Economic Opportunities Project). Moreover, several different local organizations are involved in encouraging development in the following sectors: shellfish aquaculture, fisheries, value-added wood production, community forestry, tourism, non-timber forest products, seaweed cultivation and harvest, research and education, e-business (facilitated through broadband access), arts and culture, alternative energy, and waste management and recycling.

Other structures, as well, have been set up to increase local involvement in and control over decision making in their area, typically structures involving both First Nations and non–First Nations participation. These include, particularly, the Central Region Board (which reviews forestry plans, for example) and also CLARET (Clayoquot Alliance for Research, Education and Training), Isaak Forest Products (a joint venture between the Nuu-chah-nulth and a major forest company), the Ucluelet community forest, and others. There is a clear change. "Coastal communities are not only fishing communities anymore," says WCVI Aquatic Management Board executive director Andrew Day (2004). But, he asks, "Will new opportunities be subject to the same dynamics? Will there be an ability to influence policy so community benefits are ensured? There is potential for change through intelligent action, but we need to get to the point of empowerment, not be caught in cynicism." We deal with these questions in more detail in chapters 14 and 15.

On the East Coast there are many local initiatives under way, including the natural heritage site at Point Amour.[9] However, we focus here on one unusual initiative, a remarkable local effort in individual and community health and well-being. The Moulder of Dreams pottery workshop in Port Hope Simpson builds strong pathways between economic and human health. The pottery provides a wonderful vocational opportunity to many of the people in the community who suffer from myotonic dystrophy. Centrally located in the town hall, the workshop helps to ensure that encounters and interactions between local townspeople and those working there are routine and positive. It provides opportunities for social interaction and helps to integrate potters into the larger community of Port Hope Simpson. Moulder of Dreams pottery workshop is a source of pride for the client workers, their support staff, and the volunteers behind the initiative. We were unable to carry out a focused assessment of this initiative as part of our research or to compare it to other initiatives attempting to provide support and employment to the disabled across Canada, but it is our perception that Moulder of Dreams represents a highly creative and effective vehicle for integrating the disabled into a vocational and larger community environment and for promoting the health of both the clients and the larger community of Port Hope Simpson. It must also provide some significant support to the families of its clients.

With sustained and adequate government support to supplement the enormous volunteer contribution from the larger community, we think Moulder of Dreams has the potential to become a leading initiative – a model for other communities, particularly those in rural and remote areas, that are grappling not only with major structural changes in their economies and with out-migration of youth and young families but also with aging populations and high rates of disability. Sadly, in 2005 the funding for Moulder of Dreams was – at least temporarily – discontinued, and we hope to hear soon that enhanced and sustained funding and support will be forthcoming. It is well-recognized that intersectoral collaboration is essential if we are to promote the social and economic health of our rural and remote communities. Such collaboration appears to be alive and well at the level of the community in Port Hope Simpson, as many local agencies have worked to support Moulder of Dreams over the past seven years. If such collaboration is lacking at the level of government, however, projects of this kind face enormous obstacles. Small businesses in relatively remote areas are hard to get going; those that also carry a labour force that suffers from a disability are even more difficult to develop to the point of self-sustenance. Primary health care and health promotion cannot be separated from employment generation and rural development initiatives if we are actually to achieve greater equity and to promote the resilience of our rural and remote communities. Moulder of Dreams is a perfect example of the kind of project that has the capacity to promote health *and* economy in this and other communities.

Finally, projects like Moulder of Dreams are, we think, important models that should be supported and talked about with pride at the national level, because they have the potential to play a central role in Canada's efforts to reduce health and employment inequalities between rural and urban areas and to enhance primary health care and health promotion across the country. They are worth supporting, and they are worth boasting about.

CONCLUSION

What the examples of new opportunities detailed above tell us is that there are many creative thinkers out there in coastal communities and many good examples of resilience that governing bodies at all levels could draw upon to promote reconstruction and stewardship of resources, local environments, ecosystems, and social systems: in short, social-ecological health. We have, however, also seen a real danger that governance structures will do more to hinder than help. There are serious misalignments of approach: government thinking conforms more to traditional, resource-based, industrial, multinational-capital thinking and hopes for quick fixes and spectacular statistics, but this kind of thinking demonstrably does not work. There has to be joint understanding and agreement by coastal communities and their government on what development paths are appropriate, what supports have to be in place, and what pitfalls can be expected. So long as opportunities are heading in the right direction, it is wise for communities and their governments to move slowly.

Restructuring in the past led to the destruction of an old functioning social-ecological relationship on both coasts, a relationship between people, their local environments, and their economies. In the future, the only certainty is more change. It is therefore incumbent upon us, as stewards of our environments and our communities, to make sure that change, when it comes, is for the better.

There are lessons in the history of coastal communities (see the section on tourism, this chapter) that have not yet been learned. It is time that we pay attention to the long-term implications of our solutions to problems. Seasonality can, as it used to, be a strength – any future interactive restructuring should recognize the potential for positive links that could be established between seasonal pursuits and economic and ecological diversity. This kind of interaction was once the way of the past, and it needs to be recaptured in modern terms as we seek recovery of ecosystems, communities, and the interactions between them that can promote the health of society and the environment. There are examples out there of co-management, such as we saw for Gwaii Haanas, that are worth implementing elsewhere.

Transportation infrastructures can enhance or damage (perhaps even destroy) the social-ecological balance of a region: it is up to us. Tourism may, with infrastructural support and residentiary discrimination, provide a decent livelihood for some and pleasure for countless visitors – if its footprint is not too great and too damaging. Aquaculture and enhancement of the current damaged marine ecosystem can be made to serve all of us well, rather than damaging ecosystems and generating further economic inequalities in an area. New opportunities in coastal communities are needed and welcome, but they must be wise. Developing them takes time and respect for knowledge across the scales at which a nation functions, from the local to the federal level, and includes scholarship, which may send unwelcome but necessary warnings.

13

Future Options II: The Oil and Gas Potential of the Queen Charlotte and Tofino Basins

INTRODUCTION

Oil and gas development on the West Coast would involve major restructuring of the economy; it could pose threats to the environment; and it would involve First Nations land/sea claims. Some potential could also be developed for employment, income, and the economic diversification of coastal communities. However, the expectations of risks and benefits associated with oil and gas development need to be better understood, and the legal situation needs to be clarified. The issue is so complex that since 1972, a federal moratorium and subsequent British Columbia provincial moratoria have prevented all petroleum exploration and development activities off the West Coast of British Columbia (see chapter 5). However, recent economic and political factors, the experience on the East Coast, and improved technologies and safety in the industry have revived an interest in this region, and it must now be examined as one potential new opportunity for coastal community revitalization. The compelling question, of course, has to be: will the development of oil and gas in the British Columbia offshore be beneficial or detrimental to social-ecological health? In this chapter, therefore, we update the existing estimates of reserves to get a better handle on whether or not it will make economic sense to develop them. We then ask what environmental dangers lie ahead should such development occur and how can we protect against them, using the East Coast experience to provide information on recent environmental regulations. Finally we ask, again using the East Coast for comparison, what are the likely benefits and drawbacks of such development for coastal community economies?[1]

The British Columbia offshore area includes four sedimentary basins that have potential hydrocarbon reserves: the Queen Charlotte,

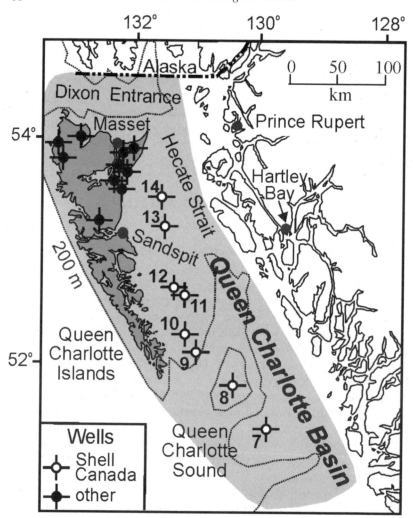

Figure 13.1   Location of the Shell Canada Limited offshore wells, other wells and cores, and the Tofino Basin Carmanah Group strata that outcrop on western Vancouver Island

Tofino, Georgia, and Winona Basins (figure 13.1). The petroleum potential of the Queen Charlotte Basin (QCB) is the primary focus of the current political interest in oil and gas development, because it is estimated to contain significant oil and gas reserves, whereas the other basins are likely to contain predominantly gas (Hannigan et al. 2001). Hydrocarbon shows in the Sockeye B–10 well (Shell Canada Report 1968c; Shouldice 1971) and oil seeps on the Queen Charlotte Islands (e.g., Hamilton and Cameron 1989) indicate that hydrocarbons are present in the region, but the existence, volume, and grade of any reservoirs is unknown. A single gas show is known in the Tofino Basin (TB) within the Pluto 1–87 well (Shell Canada Report 1967a; Shouldice 1971), and there are other hydrocarbon shows in nearby wells on the Olympic Peninsula in the United States (figure 13.1).

We initiated new studies in the Queen Charlotte and Tofino basins to update existing data sets, collect new data and information, provide new interpretations on basin evolution, and develop new ocean floor survey technologies and models to assess basin geology and hence petroleum potential. This let us update baseline geoscience and hydrocarbon resource data, information, and knowledge that could be used for economic and community planning, regulation, and policy development. We have integrated our results into socio-economic models designed to assess the impact of offshore development at the local, regional, and provincial levels.

The British Columbia coastal region is environmentally diverse, geologically and oceanographically complex, and climatically variable. Much of it has rugged wave- and wind-swept shorelines interspersed with inlets and fjords, many of which are within parks. There is an abundance of diverse wildlife, flora, and marine life, and small communities and the economies and cultures of First Nations' villages are linked to fisheries, forestry, and tourism, as we have seen. Off Vancouver Island, the continental shelf extends for about fifty kilometres to a water depth of two hundred metres, but off Haida Gwaii (the Queen Charlotte Islands) it is narrow. Inboard waters of Queen Charlotte Sound are characterized by shelf, bank, and channel deposits with depths that shallow significantly into Hecate Strait, where there are areas of high erosion and sediment transport (e.g., Barrie, Lutnauer, and Conway 1991). Large earthquakes, oceanic currents and waves, and storms known in the region may be of concern for industrial development, raising serious questions about protection from industrial damage of environmentally and culturally vulnerable regions. There are also issues surrounding First Nations' land and marine claims and the maintenance of local fisheries and tourism.

Understanding the geology and structure of any basin is required before any hydrocarbons can be located and developed. The Queen Charlotte Basin (QCB) and Tofino Basin (TB) have related but distinct Mesozoic (~245 to 65 million years ago) and Cenozoic (65 million years ago to present) geological histories. They both share an active tectonic margin with displaced (faulted, sometimes over great distances) and emplaced Mesozoic and Cenozoic strata that commonly have marine sedimentary and volcanic origins. During the late Mesozoic and early Cenozoic, a change in offshore tectonics resulted in the collision of the Eocene Crescent Terrane (an area over which a particular rock or group of rocks is prevalent) with the North American Plate, both within the TB and further south. This collision was followed by further activity that continues to the present day, because the Juan de Fuca Plate is plunging beneath the North American Plate off central and southern Vancouver Island, and the Pacific Northwest of the United States. Just to the north of the Juan de Fuca Plate, the Explorer Plate is moving obliquely, creating a complex triple junction region between the Explorer, Juan de Fuca, Pacific, and North America plates off Vancouver Island and Queen Charlotte Sound. To the north, the Pacific and North American plates are sliding by one another. The petroleum-rich basins in California and Alaska were formed at the same time as the QCB. From north to south, Dixon Entrance, Hecate Strait, and Queen Charlotte Sound (figure 13.1) have experienced a slightly different sedimentary and tectonic evolution. The Cenozoic Skonun Formation strata are composed of sandstone, siltstone, and conglomerate, with some coal deposited in both non-marine and marine alluvial fan, fan delta, detail plain, shelf, and slope settings (Higgs 1991; Dietrich et al. 1993).

We deal first with the Queen Charlotte region, where most of the currently known oil occurrences appear as surface seeps or as oil stains in subsurface samples encountered in sandstones within the Sockeye B–10 well (Bustin and Mastalerz 1995). However, poor hydrocarbon potential is proposed for Upper Jurassic/Cretaceous successions (Dietrich 1995), and we conclude that Jurassic rocks appear to have the best potential (Fowler et al. 1988; Snowdon, Fowler, and Hamilton 1988). In some places, organic-rich mud-rocks and coals of highly variable hydrocarbon potential with total organic carbon (TOC) values up to 30 percent are present, although an average value is closer to 1.2 percent (Bustin 1997). These strata have good gas and fair oil source potential (Dietrich 1995; Bustin 1997). Jurassic formations bear significant organic-rich potential petroleum source rocks in the region, which reach up to six hundred metres in thickness with TOC up to 6.1 percent, comprising oil-and-gas prone rocks (Bustin and Mastalerz 1995). Organic-rich shales with 5 to 10 percent TOC occur in beds up to ten metres thick (Dietrich 1995). Proximity of Cenozoic strata to Mesozoic source rocks, faulting, thermal maturation, reservoir, and cap rocks all occur within the QCB. Of interest is the recurring interbedded nature of the strata that can offer reservoir and seal conditions. This is in addition to source materials (such as coals and carbonaceous materials).

Based on previous studies and exploration data, it is thought that the QCB may have substantial petroleum accumulations. Current estimates of oil are about 1.5 billion cubic metres (m³), or 9.8 billion barrels (bbl) (e.g., Hannigan et al. 1998). The projections of natural gas in place are around 730 billion m³ or 26 trillion cubic feet (tcf). Potential *recoverable* reserves are lower, possibly 400 million m³, or 2.5 billion bbl oil and 550 billion m³ or 20 tcf gas. According to Hannigan et al. (2001) the QCB region is predicted to have 103 oil fields and 120 gas fields. The largest oil field size is predicted to be about 440 million barrels, with six fields over 100 million barrels, i.e., 1.3 billion bbl in total for the top six fields. The largest gas field is predicted to be 2.7 tcf, with fourteen fields over 500 bcf, i.e., 12 tcf in the top twelve fields.

While the estimated British Columbia onshore reserves are 8.2 billion bbl oil and 73 tcf gas, the potential recoverable reserves for the QCB of 2.5 billion bbl oil and 20 tcf gas are considerably lower. Using current direct economic valuation, the offshore oil and gas reserves translate into $49B and $83B, respectively, compared to $41B and $146B for onshore oil and gas, which sums to a direct British Columbia total (on- and offshore) of $320B and, over a fifty-year production period, $6.4B per year. British Columbia petroleum revenues have been increasing steadily over the past decades. In 2001 the onshore revenue generated from oil and gas was $1.7B, which could be substantially increased (up to $6.4B per year) with offshore potential. Estimates of oil and gas in the QCB compared with the East Coast Jeanne d'Arc Basin are difficult to make because of the differences in the stages of exploration between the two regions: for example, offshore well coverage in QCB is approximately 8 wells/80,000 km², or 1:10,000, whereas for the Jeanne d'Arc Basin the well/area ratio is approximately 1:700. Other differences include greatly different geologic situations and differences in available supporting information. First estimates indicate that the British Columbia offshore has greater oil (9.8 vs 4.6 billion bbl) and gas (42 vs 18 tcf) potential than the Jeanne d'Arc Basin, but recoverable reserves should not be confused with potential resource estimates, which are generally much lower.

We also employed Petroleum Systems Modelling in the QCB. This comprises numerical simulations of geologic structures through time based on physical and chemical reactions. Numerical modelling of petroleum systems that were developed in the early 1980s has improved greatly in recent years due to advances in organic geochemistry, multiphase fluid flow models, numerical methods, and computer performance. Basin modelling has now become an essential tool in the exploratory strategy of petroleum companies because it provides a dynamic, objective, and integrated view of processes such as sedimentation, compaction, fluid flow, heat transfer, source-rock maturation, petroleum expulsion, migration, and accumulation. It can also be applied to remote and unknown areas (such as QCB) where only sparse information is available. The modelling is based on a number of geological concepts.[2]

Basin modelling allows simulation of basin evolution and petroleum generation, expulsion, and migration in a physically and geochemically consistent way. Its main purpose is to reconstruct the history of a sedimentary basin and to simulate the main phenomena leading to petroleum occurrence and entrapment, phenomena such as source-rock maturation, hydrocarbon expulsion, migration (pathways) and accumulation, and petroleum phase and composition evolution. Such modelling also provides significant insights into fundamental questions about the effective source rock kitchen areas; the timing of the processes of petroleum generation, expulsion, and migration for each source rock; the possible migration pathways from source rocks to reservoir; the role of faults as migration pathways; and the problem of how effective the drains and seals must be in order to cause a commercial accumulation along with the expected oil and gas compositions in a petroleum trap. Results from such modelling studies have been used to understand petroleum systems better and, most important of all, to identify the possibilities and risks concerning new exploration targets. The modelling requires the input of hundreds of variables. Some are well known; others (such as the distribution of source rocks and mantle properties) are conventionally accepted to lie within a restricted range but are largely unknown. To verify the simulation, calculations must match known vitrinite reflectance values[3] and formation temperatures, as well as heat flow measured in the region.[4]

Judging from vitrinite reflectance measurements in eight wells (Bustin 1997), on average the Cenozoic sediments in the QCB just deeper than 2,100 m (0.5 %R_o)[5] are in the "mature oil window" (figure 13.2). Restricted, if any, production will result from a shallower marginally mature zone (1,200 to 2,100 m). The mature oil window today extends to depths of approximately 2,800 m (0.8 %R_o), while the mature oil and gas window currently lies between 2,800 and 4,100 m. The bottom of this zone has been projected to lie deeper than most of the wells drilled in the QCB. Most of the offshore QCB wells encountered Cenozoic sediments that have a maturity of less than 0.8 %R_o; i.e., they are in the formation stage of the early petroleum window or younger. The Sockeye B–10 well is an exception in that it was drilled much deeper than the rest (TD 4,773 m), and the measured maturity of the sediments at the base is approximately 2 %R_o (late mature).

Most of the rich Jurassic source rocks that are assumed to lie under Hecate Strait are modelled to be presently over-mature, even if significant heating was restricted to Cenozoic rifting. Generated hydrocar-

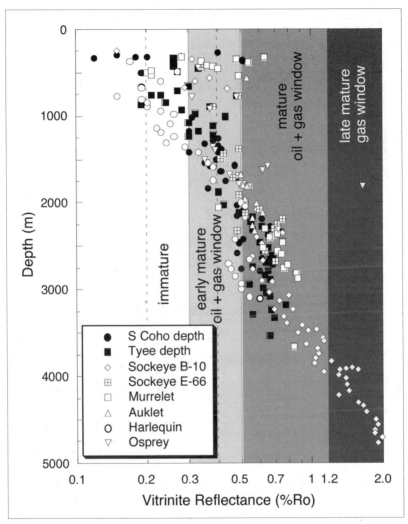

Figure 13.2   Combined depth plots of vitrinite reflectance (VR) of eight offshore Queen Charlotte basin wells showing maturation zones (data from Bustin 1997)

bons are predicted to have been mostly expelled from their original beds, even though some shales were assumed in the model to be thick and impermeable. Extensional faults are predicted to have provided migration pathways, and some of the hydrocarbons might have leaked to the surface. If marine Jurassic source rocks (type II) underlie Hecate Strait, then more than 80 percent of the kerogen is predicted to have been transformed into hydrocarbons and to have produced more than

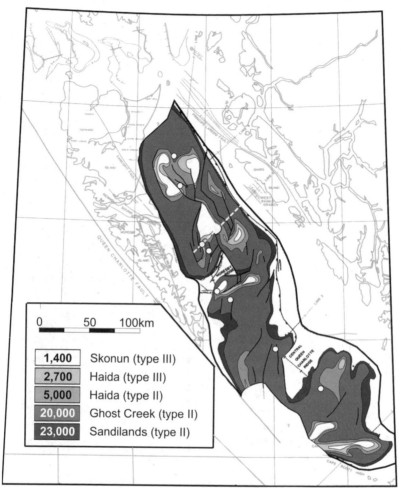

Figure 13.3   Areas where kerogen in the investigated source rocks is predicted to be more than 50 percent transformed into hydrocarbon. For the Cretaceous source rocks of the Haida Formation, scenarios for both types of kerogen are presented. The line marks the location of line 88–05. Note that the location of the Mesozoic rocks under Hecate Strait has not been mapped; we ran the model assuming a uniform thickness of this source layer underneath Neogene basin fill mapped by Rohr and Dietrich (1992)

twenty-seven kilograms of hydrocarbon per tonne of rock as of today. There are indications that Triassic/Jurassic beds lie below Hecate Strait, but any assumption of their aerial extent is speculative. Faulting, uplift, and erosion over the last 150 million years have probably reduced the presence of these units.

The Cretaceous source rocks that are assumed to underlie Hecate Strait are not predicted to generate significant amounts of hydrocarbons. Cretaceous beds on the Queen Charlotte Islands have variable but generally low TOC levels, the Haida Formation being the exception. If time-equivalent sediments exist under Hecate Strait, then they are likely to be mostly terrestrial (type III) in origin. Approximately 50 percent of terrestrial kerogen is predicted to have been transformed into hydrocarbons where these sediments are buried deeper than 4.1 km. No economic accumulations of hydrocarbon are predicted to be generated from Cenozoic Skonun Formation source rocks. These rocks, where sampled, are mostly low in TOC and contain gas-prone terrestrial (type III) kerogen ($HI<100$ mg/g C_{org}), which is concentrated in thin carbonaceous layers. Most of the kerogen probably was not heated enough to be transformed into hydrocarbons. Only below 3.7 km depth, more than 50 percent of the kerogen is predicted to be transformed. Not more than approximately 1,400 km² of the Skonun Formation strata meet this criterion (figure 13.3). Minor expulsion of hydrocarbons from the Skonun is predicted to occur in Hecate Strait.

The Queen Charlotte Basin provides interesting challenges to oil and gas assessments because of the lack of knowledge about the stratigraphy, timing of rifting, and occurrence and maturity of significant source rocks. Small perturbations on dates of rifting and stratigraphy did not yield significantly different hydrocarbon generation scenarios in the models. Clearly, large changes would yield very different results. Uncertainty regarding the nature and occurrence of source rocks, migration of hydrocarbon along faults, and possible trapping in finely layered sediments still pose major difficulties for assessing hydrocarbon prospects in the QCB.

We initially re-examined the most recent (1988) multichannel seismic reflection data for the Queen Charlotte Basin. These data were processed to locate and analyze features that are associated with the presence of hydrocarbons. Although conventional seismic processing techniques revealed many anomalies in all the data sections, the overall complexity of the geological structures did not allow us to make a straightforward interpretation of the features as hydrocarbons (Rohr and Sutherland 2002). In parallel with this study, we also examined the seismic data from northern Hecate Strait to assess geohazards due to earthquakes in the basin. The multichannel data revealed the presence of broad anticlines cut by vertical faults and networks of smaller branching faults in northern Hecate Strait, structures that are consistent with strike-slip and thrust events (Bird 1997). However, the data did not provide sufficient resolution of the shallow subseafloor structures to show the kind of evidence of seafloor disruption that would indicate active faults (Rohr 2002). We concluded that, while other types of marine geohazards might constrain petroleum resource development in the southern portion of Hecate Strait (hazards such as sediment gas seeps, buried channels, sediment slumps, and scours are known to exist from previous surveys [Barrie and Bornhold 1989]), the relationship of the seafloor hazards with the deeper structure in the sediment is not well enough understood, nor are the regional distributions of any of the types of geohazards.

Consequently, we designed a new multisystem seismic survey concept that resolved the structure in the upper portion of the sediment column (600–800 m below the seafloor) (figure 13.4). We used three different sound sources and receivers that can be operated simultaneously along the survey track lines to provide information over a wide range of spatial scales. Although the acoustic power radiated from each type of sound source was well below accepted thresholds for damage to marine ecosystems, marine mammal observers became part of the experiment, since we regulated the use of the sound sources when mammals were observed. The integrated data set provides a comprehensive geological picture of the survey site for analysis of marine geohazards.

We carried out an experimental multisystem survey in May 2002 at two sites in southern Hecate Strait where evidence of seafloor geohazards was suggested from previous data from the Geological Survey of Canada (Barrie and Bornhold 1989), collecting over 400 km of high-quality, new data from each system on the thirty-five grid lines that were surveyed at the sites. The low-frequency seismic images revealed a multitude of small faults and deep buried channels below the seafloor. The combined images over the survey grid show the alignment of channels across the basin. The shallow high-resolution Huntec data provide a detailed picture of more recent buried channels that intersected the seafloor and indicate the presence of gas seeps and associated mud mounds throughout the region.

We found that the shallow subsurface is composed of three units.

1 The deepest unit (30–40 m) is highly folded, and often incised by channels.
2 Overlying these sediments is a unit whose upper surface commonly outcrops at the seafloor, forming prominent moraine-like features. It forms a strong reflector, though the continuity is often disrupted by gas pockets and likely represents glacial till.
3 The uppermost unit is likely composed of sand/gravel. Seafloor classification of the echo-sounder data using feature extraction and cluster analysis indicates that this unit is acoustically distinct from

Figure 13.4   Geohazard survey in Hecate Strait showing the
deck layout at the stern of the CCGS John P. Tully for the
multisystem seismic sound sources. The sound sources are
suspended from the orange float behind the ship, and the
acoustic receivers are towed from winches on the stern.

the glacial-till unit, suggesting different sediment properties for
these two units.

Combined with the seismic data, the presence of gas near the
seafloor can be related to the fault structure in the deeper sediments

that provides pathways for upward migration of the fluids containing
the gas. However, the origin of the shallow gas is unknown: it may be a
result of microbial degradation of local organic matter or of gas diffus-
ing upward along faults or bedding planes from the Cenozoic sediments, as suggested by the multichannel seismic sections. We also used
seabed classification data to assess the seabed habitat for groundfish
productivity in the basin, because our acoustic data provides a means
to determine bottom type remotely by processing the echo signals to
classify the acoustic diversity. We extended the echo-sounder survey to
other sites in the basin where different types of groundfish were known
to predominate.[6] The results of the survey are important. They demonstrate a new multisystem seismic technique that reveals the presence of
geohazards near the seafloor and also shows linkages that connect the
seafloor features with the subsurface structure. Also, the technique
provides the means to assess seabed habitat. Although we had to limit
the survey to two sites, the technique can be applied at other sites
where oil and gas development activity is focused.

Turning to the Tofino Basin, we know that during 1967–69, when
the offshore Shell Canada exploratory wells were drilled, the nature of
faulted, emplaced, and accreted terranes (Hyndman et al. 1990) and
strata along the shelf was not understood. Most of the wells in the TB
targeted anticlinal structures that were thought to have the best hydrocarbon prospects or provided a stratigraphic framework (Shell Canada
Reports 1967a–d, 1968a–c; Shouldice 1971). No significant accumulations of hydrocarbons were located, but a gas show in the Pluto 1–87
well was reported (Shell Canada Reports 1967a; Shouldice 1971).
Shortly after these wells were drilled, a 1972 federal moratorium and
1982 and 1989 provincial moratoria on hydrocarbon exploration and
development within Canada's West Coast basins were put in place, and
about the same time the Pacific Rim National Park was established.
More geoscience knowledge in the Tofino and Queen Charlotte
regions was needed, and funding for new surveys and programs was led
by the Geological Survey of Canada (GSC) (e.g., International Boundary Studies, the Frontier Geoscience Program, the Lithoprobe Project)
and the international Ocean Drilling Program (ODP). These programs
assessed resources and hydrocarbon potential, provided geoscience
information for policy-makers and the Canadian public, looked at and
interpreted deep structure and the tectonic history of the margin,
and/or monitored earthquakes to assess hazards (e.g., Hyndman et al.
1990; Woodsworth 1991; Hyndman 1995).

An exploration potential of 266 x 10^9 m^3 of in-place gas for forty-one
fields within the Tofino Basin, Strait of Juan de Fuca, and northwest
Olympic Peninsula was estimated by Hannigan et al. (2001). The Juan

de Fuca Strait area was considered the most prospective, with greater volumes of coarse, clastic sediments having reservoir potential and being near the source rocks with known thermogenic gas seeps, and with gas shows in four onshore drilled wells (figure 13.1, nos. 27–30) (Snavely and Kvenvolden 1989; Niem and Snavely 1991; Johnson et al. 1997; Hannigan et al. 2001). Although estimated to be about 85 percent of the play, TB hydrocarbon potential is uncertain and considered to be poorly explored (Johnson et al. 1997; Hannigan et al. 2001). For the TB, Hannigan et al. (2001) found

1 mainly outer shelf to bathyal mudstone and siltstone with rare submarine channel sandstone and conglomerate;
2 variable porosity and permeability in coarser units (higher porosities were from 20 to 46 percent);
3 regions of anomalously high geopressure gradients (Shouldice 1971);
4 isolated lenticular sandstone within thick mudstone and common folds and anticlinal structures that could provide reservoirs and traps; and
5 offshore Cenozoic strata that generally had poor source-rock potential of type III terrestrial organic matter with a low organic content (average 0.8 %TOC) and a low hydrogen index (Bustin 1995).

Cenozoic stratigraphic data and results for the Tofino Basin were typically reported over twenty years ago (e.g., Cameron 1980), in some cases over thirty years ago (e.g., Shell Canada Well Reports), and in other cases much farther back (e.g. Jeletzky 1954; Clark and Arnold 1923; Clapp and Cooke 1917): updates are needed. Our approach was to re-examine archival geological data, re-interpret the data to bring it up to current geological standards and understandings, and conduct new investigations and analyses that would lead to a new synthesis of the geological framework of the Tofino Basin. The results are relevant for coastal community and government planning and policy development should offshore hydrocarbon resources be exploited commercially. This section summarizes briefly new biostratigraphic, strontium isotope stratigraphic, and thermal maturation data and results that are integrated with previous biostratigraphic, geologic, offshore multi-channel seismic, and magnetic data and results to provide new information on structures, strata, and the Cenozoic evolution of the Tofino Basin.

Our sample database includes cuttings from the six exploration wells that were drilled by Shell Canada Ltd from 1967 to 1969 in the offshore Tofino Basin (figure 13.1). Microfossils (e.g., foraminifers, ichthyoliths) were identified and their distributions analyzed in addition to associated lithologies and other sample contents. Well reports and data (available for viewing at the British Columbia Ministry of Energy and Mines, Victoria) were analyzed and reinterpreted. The reports include lithological descriptions of sidewall core and cuttings, hydrocarbon mud analyses, well logs, microfossil reports, and other data (Shell Canada Reports). The wells were drilled mainly into or near anticlinal structures. Only one gas show was encountered in a thin sandstone bed in the Pluto 1–87 well (unit 21). Over 2,145 drill cuttings were processed for microfossil recovery. Sidewall core subsamples were unavailable.

Other study materials include Carmanah Group surface samples from outcrops on the west coast of Vancouver Island (Nootka Sound area, 92E), and one GSC vibrocore from offshore Flores Island (END–76B–6) (figure 13.1). About 1,025 Cenozoic surface samples are included from the Hesquiat Peninsula, Nootka Island, Flores Island, and Tatchu Point. They were collected during 1969–74 by B.E.B. Cameron (GSC, retired) mainly for foraminifer studies. Samples were predominantly from shale, siltstone, and mudstone and rarely from sandstone and conglomerate.

Strontium isotope ages were determined to obtain better chronostratigraphic resolution in our established biostratigraphies. Archival GSC seismic data (mainly 1985 and 1989 surveys; GSC Pacific, Sidney, BC) were reviewed and interpreted to identify specific TB structures relevant to interpreting our stratigraphic results. These data are released in GSC Open File (Yorath et al. 1987; Spence et al. 1991) and selected lines in papers (e.g., Hyndman et al. 1990; Hyndman 1995; Clowes, Baird, and Dehler 1997).

Magnetic data from the TB was available from the GSC National Geophysical Data Base and prepared as a histogram equalized colour image by C. Lowe (GSC, Pacific). Magnetic anomalies were identified and interpreted and with our other results, a broader interpretation of TB structures and positions emerged.

Ichthyolith thermal alteration data was obtained by documenting ichthyolith colour, which correlates with thermal alteration of organic material within teeth and is calibrated to ichthyolith/conodont indices established through conodont heating experiments (e.g., Epstein, Epstein, and Harris 1977; Rejebian, Harris, and Huebner 1987). Maximum thermal alteration was estimated for specimens from both onshore and offshore Cenozoic strata intervals. An estimated maximum temperature is important for assessing a basin's potential for oil and gas generation within a 50°C–300°C window (e.g., Legall, Barnes, and MacQueen 1981).

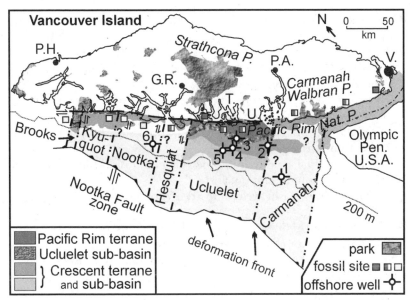

Figure 13.5   Tofino Basin categories of potential offshore hydrocarbons, sub-basins, and regions. Nearby western Vancouver Island parks, communities, and important fossils sites are indicated.

We have reviewed current geoscience knowledge and used archival data and new analyses. Foraminifer, ichthyolith, and strontium isotope stratigraphic results were integrated with geophysical and Shell Canada well data to provide a new interpretation of the geologic and tectonic evolution of the Tofino Basin. This contributes toward Tofino Basin modelling and the assessment of oil and gas potential, which is important for the development of government policies before lifting of the over thirty-year federal and provincial moratoria on British Columbia offshore exploration activities. The results show that the wells penetrated different Tofino Basin strata and structures and that offshore strata are in part correlative with the Carmanah Group on western Vancouver Island and provide new information on the inboard displacement of segments of the Crescent Terrane (CT) and its interactions with the Pacific Rim Terrane (PRT) and the Cascadia Accretionary Complex (CAC) (figure 13.5).

Using the published geophysical results that give a more current understanding of the tectonic regime in the Tofino Basin area, magnetic anomaly data from the GSC National Geophysical Data Base, and our stratigraphic results from a review and analysis of archival GSC Cenozoic onshore samples and Shell Canada well-cutting samples and data released to the GSC and the provincial government of British Columbia, we have been able to

1 provide an updated interpretation and correlation of Tofino Basin Cenozoic stratigraphy with new integrated foraminifer and ichthyolith biostratigraphies and strontium isotope age controls;
2 identify regions, sub-basins, and structures (figure 13.5, no. 5);
3 provide a new interpretation on the positions of the Crescent and Pacific Rim terranes;
4 provide an initial assessment of thermal maturation from analysis of ichthyolith colour alteration; and
5 interpret Tofino Basin areas of hydrocarbon potential.

Detailed new and refined biostratigraphic results from our study are reported by Narayan (2003), Narayan, Barnes, and Johns (2005), and Johns, Barnes, and Narayan (2005; and forthcoming). A new interpretation (Johns et al. forthcoming) on the position of Pacific Rim Terrane (PRT), Crescent Terrane (CT), and Cascadia Accretionary Complex (CAC) segments is mapped with five Tofino Basin regions, three sub-basins, and other structural features (figure 13.5). The location of potential offshore Tofino Basin hydrocarbons is considered in the context of

1 proximity to potential Mesozoic or Cenozoic source rocks (PRT-Ucluelet Sub-basin);
2 potential fluid flow associated with thrust faults and accreted and uplifted strata;
3 significantly faulted or segmented regions that may have smaller field targets (e.g., Hesquiat or Carmanah Region); and
4 greater thermal maturity (e.g., the Nootka-Kyuquot regions).

The Ucluelet Region contains considerable strata and structure diversity and may offer the best hydrocarbon prospects.

Accreted and displaced terranes and strata, Cenozoic structural highs and sub-basins, and thrusted and faulted strata segments are the fabric of the Tofino Basin – a complex basin architecture at an active tectonic margin. Results from our study emphasize the value of a multidisciplinary approach for basin analysis that incorporates paleontological, geological, geophysical, and geochemical techniques, using mainly archival data to provide a new geological interpretation of the region.

From our results (Johns et al. forthcoming), we define five Tofino Basin regions that contain different terrane segments, structures, faults, and strata configurations (figure 13.5). We interpret, map, and rate categories for hydrocarbon potential, placing particular importance on potential Mesozoic and Paleogene source rocks, coarse- grained reservoir facies and associated structural highs, faults to facilitate fluid flow, and thermal regimes.

Of the five Tofino Basin regions, the Ucluelet Region (figure 13.5) shows the least faulting and deformation; the largest segments of PRT and CT strata; the thickest Cenozoic strata with the greatest potential for hydrocarbon source, reservoir, and trap facies; and the only Tofino Basin Shell Canada well hydrocarbon (gas) show (unit 21, Pluto 1–87). Smaller plays may be located in the significantly faulted Hesquiat Region and the faulted strata of the Carmanah Region. Deep thrust faults between accreted strata (e.g., the Tofino Fault, the Crescent Thrust Fault, and others) may provide pathways for fluid movement outboard (oceanward) from potential source rocks (e.g., PRT and Paleogene strata) to the Ucluelet Sub-basin, the Crescent Sub-basin, and the CAC. Thermal anomalies (determined from ichthyolith colour and thermal alteration) are associated with faults. Overmaturity (>300°C – too cooked) of organic materials from Cenozoic strata is indicated in the northern Nootka Island area.

To conclude this section, we note that the petroleum system model we used for our QCB hydrocarbon assessments incorporated geophysical, geological, and geochemical data and is typical of what is used by industry today. Results indicate a petroleum window, where hydrocarbons could exist, at or just deeper than 2,100 m in offshore Cenozoic strata. The oil window may extend to 2,800 m and the combined oil and gas window to 4,100 m. Jurassic hydrocarbon source rocks are assumed to underlie Hecate Strait but are interpreted to be over-mature with generated hydrocarbons having been expelled and migrated upwards along faults. For example, Cretaceous Haida Formation strata are interpreted to have expelled hydrocarbons at depths greater than 4,100 m. Only small amounts of hydrocarbons are predicted to have been generated from the Cenozoic Skonun Formation to date, which has typically been buried to shallower depths. Results are sensitive to the stratigraphy and age of the rocks, timing of rifting, and occurrence and maturity of source rocks. Much more of this data could be collected to improve regional assessments (Whiticar et al. 2003; Whiticar, Schuemann, and Rohr 2004).

For the Tofino Basin, we accessed and reinterpreted multiple data sets (seismic, magnetic, and Shell Canada well samples, logs and reports) in addition to completing new studies (biostratigraphic, strontium isotopic, thermal maturation). Our results helped us to define complex basin structures better, including emplaced and accreted

terranes and faulted, thrusted, and folded strata. Our new interpretations focus on the position of different blocks of the Crescent Terrane
and outboard (oceanward) strata of the Cascadia Accretionary Complex (figure 13.5). Of the five regions of the Tofino Basin that we
defined, our results indicate that the Ucluelet Region (offshore
Ucluelet to Tofino) may have the best hydrocarbon prospects, because
it shows the least faulting and deformation; the largest segments of
Pacific Rim Terrane and Crescent Terrane strata with possible source
rocks; the thickest Cenozoic strata with the greatest potential for
hydrocarbon source, reservoir, and trap facies; and the only Tofino
Basin Shell Canada hydrocarbon (gas) show in the Pluto 1–87 well.
The Carmanah and Hesquiat Regions to the south and north are considerably faulted but may have small hydrocarbon plays. Data from the
Nootka and Kyuquot regions are limited, but the few studies indicate
some areas of thermal overmaturity that are not favourable for hydrocarbon generation. Deep thrust faults (e.g., Tofino Fault, Crescent
Thrust Fault, and others) may have provided pathways for fluid movement outboard from potential source rocks to the Ucluelet and Crescent sub-basins and the Cascadia Accretionary Complex (figure 13.5).
The most complete offshore shelf (<200 m depth) Tofino Basin data is
within the Ucluelet Region. Geological interpretation and hydrocarbon assessment of the other offshore and nearby onshore regions
could be significantly improved with new coring, stratigraphic, and
geochemistry studies, seismic surveys, and Mesozoic-Cenozoic outcrop
sampling and stratigraphy.

IMPLICATIONS FOR SOCIAL-ECOLOGICAL HEALTH

Whether or not development of these reserves takes place will depend
to a considerable degree on resolution of the concerns that exist
among coastal communities, First Nations, and the three orders of government involved. All recognize considerable implications for the
social-ecological health of the region. As we would expect, much of the
current interest, as well as the public hearings, centres on lifting the
moratorium on offshore exploration and development in the Queen
Charlotte Basin. Indeed, the federal public hearings have excluded the
other basins from current consideration, perhaps to make the process
more manageable. If offshore hydrocarbon exploration and development should occur in the near to intermediate future (e.g., in the next
two decades), then any such activity in the Tofino Basin will have
rather different implications for the nearby coastal communities than
will any development in the Queen Charlotte and Georgia Basins.

We turn first to a consideration of the Tofino Basin. Large segments of that coast are now developed as parks (the federal Pacific Rim National Park between Tofino and Port Renfrew, and the provincial Juan de Fuca Park continuing to the southeast between Port Renfrew and French Beach) (figure 13.5). The communities along this coast have always depended on logging and fishing for their existence, but as these industries declined, tourism became a new focus (see chapter 12). We may therefore anticipate substantial opposition to offshore hydrocarbon exploration and development if a perception were to develop that such activities could result in environmental damage or deter tourism. By contrast, since we have argued in the previous chapter that more-diversified economic activities would support healthier coastal communities, at least in the short run, the development of supply industries, accommodation, shore facilities, and the like, probably has its attractions for communities such as Gold River, Port Alberni, and possibly Ucluelet, which were industrial when initially created.

Of course, if the offshore development was only for gas, not oil, then the risk of oil spills and the possible impact on beaches would be minimal. There are other locations in the world where offshore hydrocarbon exploration and development are occurring close to major coastal parks and tourism centres, most notably off the southwest corner of the North Island of New Zealand (near Palmerston). The gas could be piped onshore at some appropriate location, or it could be taken by tanker out of the region as liquid natural gas (LNG). The options available have not been fully explored by the different communities, which can be expected to take different stances, depending on the balance of their anticipated future economic development options, especially tourism.

Development of offshore oil and gas involves highly contentious issues. It is seen by some as a possible lifeline for depressed coastal communities, since it will provide employment and increased commercial activity over the lifetime of the resource. Others are deeply concerned that such development may put marine life and "wildernesses" at risk, affect global temperatures, and change the livelihood, heritage, and ways of life of First Nations' peoples, as well as possibly affecting regional fisheries, aquaculture, and tourism. Clearly, environmental-impact regulations will be crucial under any development scenario, and, as we saw in chapter 5, state-of-the-art regulations are becoming very effective at preventing environmental damage, although they are still not perfect, as was shown by the November 2004 Terra Nova spills. It will also, of course, be necessary to negotiate the kind of political accord that will permit the creation of long-lasting regional infrastructures capable of sustaining regional development when the resource no longer exists. We discuss these issues in more detail below.

We turn now to the Queen Charlotte region to introduce our economic model for the area and to consider oil and gas development issues from the perspective of what needs to be done both to maximize benefits for coastal communities and to minimize the various risks that might occur should offshore energy activity proceed. We also discuss economic impacts from the viewpoints of both regional and provincial economies, as well as environmental and socio-cultural issues.

For evaluating economic impacts at the provincial level, both the BC Macro-economic Model and the BC Input-Output Model are available. The former is a large multi-equation system describing the behaviour of households, firms, and government. The latter is a description of the expenditures and receipts that occur between sectors of the economy at a given time. Corresponding models have been used to estimate the impact of East Coast energy activity on the economy of Newfoundland and Labrador (Community Resources Services Ltd 2003). At the level of individual communities in British Columbia, the provincial government has calculated impact ratios based on 2001 data, updating earlier data (Horne 2004) that could be used to give approximate indications of the employment impact in selected centres. Of course, any simulations using economic models will have to be based on hypothetical data until expenditures incurred by the industry and the public sector to develop the resource, as well as details of the management and regulatory regime (including the fiscal arrangement), are known.

As part of our work, we are developing a prototype computable general equilibrium (CGE) model at the regional level for the regional districts (RDs) of Mount Waddington (RD43), Central Coast (RD45), Skeena-Queen Charlotte (RD47), and Kitimat-Stikine (RD49), which surround the Queen Charlotte Basin.[7] Our new model will have a provincial dimension capable of facilitating comparisons with results generated from the British Columbia models described above. For an overview of the model showing its role in policy simulation, see figure 13.6.

Our model simulates the interaction of different economic agents (households, firms, governments, and the rest of the world) across markets. Optimizing behaviour is assumed and incorporated in equations describing the behaviour of each agent. For example, firms (producers of goods and services) are assumed to maximize profits, subject to technological constraints, while households (consumers) are assumed to maximize satisfaction, subject to income constraints. These are standard assumptions in economic theory. Also, the model defines the "rules of the game" (i.e., the structure of the markets in which agents interact with one another and receive the price signals that influence their

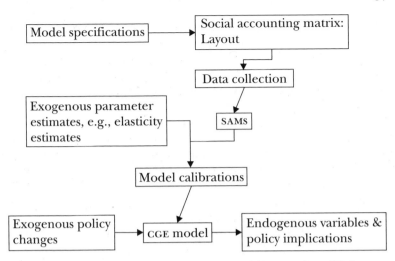

Figure 13.6   Overview of the prototype computable general equilibrium (CGE) model

behaviour). Finally, the "equilibrium conditions" for the economic system are specified. These are the system constraints that must be satisfied. In competitive markets, for example, all excess demands amount to zero, through price adjustments, so that markets are cleared. There are blocks of equations that describe the assumed behaviour of agents in the context of "the rules of the game," the system constraints and the accounting identities embedded in any system of income-expenditure accounts. The blocks define the conditions for equilibrium in (1) each of the production sectors of the economy; (2) the markets for factors of production (land, labour, capital); (3) the markets for goods and services; and (4) various identities defining economic relationships.

The model is calibrated to a dataset (a "social accounting matrix" – SAM) using data from the years 1996–99. This SAM is a matrix of receipts and expenditure accounts for the different components of the economic system and combines data from the British Columbia input-output table, the annual economic accounts of the province, and other sources. The SAM distinguishes between different levels of government and households of different income levels. The calibration process computes parameters for the model's mathematical functions to reproduce the SAM data as an equilibrium solution of the model. The model will be used to simulate the impact on regional and provincial output, incomes, employment (by sector), and government finances (by level of government) of various "shocks" to the regional economy. Such a shock would be the lifting of the moratoria on oil and gas activity in the Queen Charlotte Basin.

Evidence from elsewhere clearly indicates that there will be economic benefits for a region from oil and gas activity.[8] In addition to increased revenues for government from royalties (assuming a reasonable accord with the federal government – see chapter 5), licence fees, and other taxes related both directly and indirectly to the industry, the benefits occur partly in the form of the increased capacity created in communities (e.g., improved infrastructures, health and social services, training and education) from investment by industry and spending by governments and from oil and gas revenues that can contribute to current well-being and future economic development. Also, benefits occur in the form of job opportunities and incomes that arise (1) directly from spending by the industry on exploration, development, and the production of oil and gas and from the public spending of royalty and other public revenues; and (2) indirectly through the so-called multiplier process as initial spending generates demand for output in other sectors. However, a proportion of spending at all stages in this process leaks away in taxes paid to outside jurisdictions, savings, imported goods and services, and remittances by employees in the industry whose domicile is elsewhere. These leakages are greater – and the benefits therefore smaller – the smaller the region. The extent to which spending leaks into imports and remittances depends on the capacity of local communities to provide the labour, services, and other commodities required by the spending.

In the region surrounding the Queen Charlotte Basin, there will be limited capacity to supply many of the heavy industrial or sophisticated needs of the industry, so that there are likely to be substantial leakages into imported goods and services. Likewise, there will be some reliance on imported labour, given the skills required. On the other hand, the region is expected to be able to supply some of the needs for labour and other inputs (Campbell and Community Resource Services Ltd 2002), while supply deficiencies may be mitigated by establishment of new enterprises in the region in response to new demand. Within the region it is the larger places with the stronger industrial-commercial structures – such as Prince Rupert and Port Hardy – that can be expected to reap the main rewards. Either or both of these centres could become supply bases if the industry develops. Smaller, more remote communities will lack the capacity to benefit greatly. Their opportunities will arise chiefly from work on the rigs and at the supply bases, rather than at home, and from the possible expenditure of government proceeds on local infrastructure, health, social and business services, and training. Such opportunities are not without drawbacks, as we have seen in the case of Hibernia (chapter 5).

A key factor in determining the extent to which the communities of the region will benefit from the proceeds of energy sales will be the fiscal arrangements for sharing of royalties between the federal and provincial governments, First Nations, and local jurisdictions. The assumption of various commentators is that a Pacific Accord will be developed along the lines of the Atlantic Accords negotiated between the federal government and the governments of Newfoundland and Labrador and Nova Scotia. Those accords outline the East Coast regulatory regimes, including revenue-sharing arrangements, but have proved less beneficial to the East Coast than was anticipated (chapter 5). Some adjustments may be made to those accords; however, ongoing discussions are not available to the public at this time.

Beyond federal/provincial revenue-sharing arrangements, other institutional arrangements in place elsewhere influence the extent of local benefits, and the precedents might be considered for possible implementation regarding offshore resources in British Columbia. Unless waivers are approved, the Atlantic Accords also require operators to submit, for approval by the offshore boards that administer the accords, a benefits plan that outlines arrangements for local procurement, employment of local labour, and targets for education, training, and research support. In other jurisdictions, such as the Northwest Territories and the Yukon, "impact benefit agreements" have been negotiated between companies, various levels of government, and communities to enhance local benefits from projects. In the Peace River Regional District of British Columbia, a so-called "fair share" agreement between the provincial government and the regional district provides a share of annual provincial oil and gas revenues for infrastructure renewal in the region over a ten-year period. In still other settings, protocols and guidelines developed by local communities govern resource extraction (one such example has been developed by the Innu Nation of Voisey's Bay and the Inuit of Nunavik). Also, resource development committees are used to provide entry points into communities for companies wishing to engage in a consultation process or solicit community participation. Such committees have been used by the Tahltan of British Columbia and the Innu of Voisey's Bay. Savings funds from oil and gas revenues have been set up elsewhere to finance infrastructure improvements and to support local economic development and other initiatives. Examples include the Alberta Heritage Fund and a Reserves Fund set up in the Shetlands (Nicholson 1975). Finally, a stringent regulatory regime minimizes not only risk to the environment itself (see below) but also any negative socio-economic implications of environmental damage.

While it is doubtful that sufficient community support for lifting the moratoria on offshore oil and gas activity could ever be marshalled without the satisfactory provision of local benefits for coastal communities, including First Nations, there will nonetheless be significant challenges in putting effective arrangements in place. These challenges are likely to include differential power relations in negotiations, skill limitations in the local labour force, lack of local business capacity, the global nature of the industry (the use of technologies for seismic testing, pipe laying and other operations provided by specialized companies operating internationally, as well as the use of rigs and supply vessels brought in from abroad), possible constraints on regional development measures imposed by NAFTA, and the monitoring and enforcing of local benefit agreements that have been a problem on the East Coast (Hudec 2003). Notwithstanding the difficulties, there are myriad lessons from elsewhere to help guide developments.

Other sectors of the regional economy will benefit through the multiplier process initiated by spending in the oil and gas industry. However, there are concerns about the possibility of negative impacts on selected sectors. The greatest concerns relate to the impact of oil spills and blow-outs on commercial, recreational, and subsistence fisheries and on tourism and aquaculture. Since accidents can happen, particularly in an area of powerful storms, prevalent onshore winds, and periodic earthquakes and tsunamis, the risks here need to be minimized through tight regulation of oil and gas activity, including controls both on areas of permissible operations and on operating practices. In order to address other negative impacts on the fishing industry, the Hibernia project in Newfoundland established safety zones around oil rigs to minimize the risk of accidents and damage to fishing vessels. Block-by-block development has been used there, allowing fishing in areas not actively developed for oil and gas purposes. Instead of cutting off a whole area for oil and gas development, certain sections are restricted to energy development, while others are open for fishing. This arrangement requires strict management control to avoid debris and spills, and oil and gas operators are required to keep the fishing industry fully informed about the timing and sequence of development. We think that in the event of oil and gas development, further advantages might eventually be gained from these "no take" fishing zones, since they could be made to function somewhat like marine protected areas (see chapter 14) should the development of oil and gas (especially beyond the early stages) be relatively free of environmental impacts. The hospitality and retail components of the tourism sector will presumably benefit from increased business visits in the so-called "business tourism" sector. But the eco-tourism industry in British

Columbia depends in part on unspoiled coastal waters, and there is concern about environmental damage from development activities and unsightly rigs in Hecate Strait and Queen Charlotte Sound.

Although the issues are not identical, Johnston and Hildebrand (2001) suggest that experience from Atlantic Canada indicates that good communications and strong partnerships between industry, government, and other stakeholders are the key to resolving negative impacts, as well as enhancing positive opportunities. Recognition of the mutual benefits of cooperation and respect for the ways each stakeholder attempts to exploit and manage the marine environment are important mitigation strategies. A "process design group" selected from different groups, including government, First Nations, business, fisheries, environmentalists, and the general public, was recommended to work on consensus-building, following community consultations by the Northern Development Commission created under the BC Northern Development Act of 1998.

The environmental issues have been extensively reviewed elsewhere (e.g., British Columbia 2002; Royal Society of Canada 2004; Living Oceans Society n.d.). Clearly, oil and gas activity can affect marine and shoreline systems through seismic surveying (noise damage to finfish, shellfish, and marine mammals), exploratory drilling (waste discharges), and production and transportation (waste discharges, spills, and blow-outs). A rigorous regulatory regime will need to be in place if the moratoria are to be lifted, and preparedness in terms of response capabilities in crises will need to be enhanced. A stringent regulatory regime will be essential to maximize safety and ensure environmentally sensitive development using best practices. Examples of elements of the regime would be tanker exclusion zones, marine protected areas, social-ecological protected areas (for rare sponge reefs, for example, and other cultural icons of First Nations, such as ancient burial sites – see below), ecological windows (activity controls at sensitive periods in marine life such as periods of migration, foraging, and nursing), designs for tankers, pipelines and other facilities, seismic and production technologies, environmental impact assessment requirements, and effective monitoring.

There is no doubt that the industry is seen by the First Nations as a threat to the ecosystem that sustains some Aboriginal communities. They fear that traditional fishing and gathering will be affected, because serious operational accidents could damage vulnerable stocks, and it is also argued that development could encroach on spiritual and burial sites. Land claims are also a complex issue: the Haida have already initiated a lawsuit against the provincial and federal governments claiming title not only to all the lands contained within Haida Gwaii but also to the resources in and under the sea, including oil and gas reserves. It is clear that if oil and gas development is to occur, one vital precondition will be negotiated agreements with First Nations.

Matthew (2001) notes that, regarding First Nations' title agreements and petroleum exploration licences in Cooper Basin, South Australia, the Aboriginal Lands Trust Act leases land to Aboriginal groups for ninety-nine years. The government and the trust have signed an agreement to the effect that the government will pay the trust an amount equal to all royalties it receives from mineral or petroleum development on that land and Aboriginal peoples who live there will receive 50 percent of the royalties under the act. Trust policy requires consent by the Aboriginal community to whom the land has been leased before development. In another example, from the United States, the Alaska First Nations formed the North Slope Borough, which collects taxes from oil and gas facilities and reinvests them in the community economy (Vodden and Pierce 2002).

Whether or not either of these arrangements could be a model for securing acceptance of proposed developments by First Nations' communities remains to be seen and will depend to some significant degree on what agreements can be reached on land/sea title (see chapter 4 for similar concerns over forests). The process design group referred to above has been recommended to address issues of concern like these, with a view to developing a – perhaps professionally mediated – consensus. Given the conflicting interests, perspectives, and ideologies involved, significant obstacles will face any community-based consensus-building process. One important step could be to work with local communities to develop greater understanding of the industry and the implications of offshore development, a matter addressed through the Northern Coastal Information and Research Program at the University of Northern British Columbia. A reciprocal development of understanding by the industry about local conditions, concerns, and cultures would also be invaluable.

In chapter 5 we discussed the concerns in Newfoundland about stresses on families that are associated with shift work on the rigs, the social dislocation associated with imported labour, and the differential job opportunities that have arisen across different demographic groups. Here, as there, the health effects of working in the oil and gas industry will have to be considered, including the creation of satisfactory work conditions, ways of dealing with perceived hazards, and the health effects that come with separation from family and the community, which are significant for men, women, and children. Here, as there, we consider that government monitoring, to ensure that occupational health and safety concerns are met, is essential.

CONCLUSION

Our research in the Queen Charlotte and Tofino Basins has produced new interpretations of old issues and new results that have been made possible by new technologies. In the QCB, when archival seismic data were reviewed and basin structures and faults interpreted, deficiencies were found in the existing data sets. This led to development and testing of new combined seismic survey methods and technologies to improve the resolution of seafloor imaging in the upper part of the sediment column, 600–800 m below the seafloor. Different sediment facies have been classified based on differences in acoustic properties, in addition to structures associated with faults and escaping gas. Seafloor geohazards have been identified and linked to subsurface structures. Also, seafloor habitats have been identified and assessed for biological productivity and possible marine protected areas. The new technology and the surveys we carried out can be applied in any of the coastal basins of British Columbia.

We confirm that the QCB has the best prospects for oil and gas development, being both the largest of the four basins and having what appear to be the most reserves. The Tofino Basin appears to be promising with respect to gas, but the prospects for oil are less clear. The Winona Basin, lying to the northwest of the Tofino Basin, is the least likely prospect and would probably be the most difficult to develop. That said, it is not yet clear how much oil and gas are out there, but it appears to be less than was originally projected.

The socio-economic outcomes of potential offshore oil and gas development and production are very much on the public agenda as issues are brought to the forefront and new studies and research are initiated. Primarily, issues and concerns have been raised about the potential environmental effects but there are also economic and social implications: the overall issue is one of social-ecological health. Before development can proceed, appropriate fiscal and regulatory regimes will be necessary to address environmental and other concerns. Agreements will need to be negotiated over revenue sharing from oil and gas at the federal, provincial, and municipal levels; assurance will have to be given to First Nations peoples about revenue, land, and marine usage, about a fair resolution of ownership rights, and about best-practice environmental protection; and assurances will also need to be given to other coastal communities about best social-ecological health practices. Following all this, industry will need to judge whether or not the investment is worthwhile.

For the province of British Columbia, the economic benefits of offshore oil and gas production could be significant, although how such benefits might be distributed and whether or not coastal communities would partake significantly in such benefits remains to be seen. Benefits are likely to occur in the form of direct and indirect increases in incomes and employment at the exploration, development, and production stages of energy activity, and the greatest annual impact will likely be at the development stage. In addition, the province will benefit from increased capacity in the form of enhanced infrastructure, trained labour, the expansion of knowledge, and new knowledge acquisition in a range of service sectors. If these benefits are handled well, they will underpin future economic development. The provincial treasury will also benefit from royalty and other revenues derived from oil and gas activity (assuming that the new accord, should it occur, would avoid problems over claw-back), as well as from taxes levied on labour income, company profits, and expenditures on commodities generated through initial- and subsequent-round spending in the multiplier process.

At the regional level, larger coastal communities (e.g., Prince Rupert, Port Alberni) may receive economic benefits if they support supply centres or other service industries, although some communities may be too small to supply what industry needs. If coastal people are to benefit, it is likely to be mostly through jobs on the rigs or at the supply bases, rather than at home, and through possible revenue sharing with the provincial and federal governments, local spending from any "legacy" or savings fund that might be set up, or local benefit agreements. At the sectoral level, it will be necessary to guard against negative impacts on the fisheries, tourism, and aquaculture. Increased opportunities associated with the industry will attract transient workers and, if communities are not careful, attendant social problems. In short, any development of coastal oil and gas reserves needs to be cognizant of the importance of guarding against many forms of damage to social-ecological health: environmental risks and negative socio-cultural impacts, including effects on aboriginal livelihoods, heritage, and ways of life, as family stresses associated with shift work and differential job opportunities will undoubtedly arise. The good news is that we have examples of how to guard against these ills.

14

New Options for Governance I: Marine and Coastal Waters

Throughout this book, we have identified the interactive effects of social and ecological restructuring on Canada's coastlines, and we have shown how social-ecological degradation has built up on Canada's East and West Coasts to the point where social-ecological health has been damaged and the people and ecosystems involved have been subjected to serious stress. In the previous two chapters, we looked at some of the new options that exist for coastal communities, either potentially or actually, and we have shown how it will be fundamentally necessary to encourage and support these efforts in a deliberate and focused manner at the institutional level if they are to have any hope of reversing, rather than perpetuating, the problems with degradation already in place. Now, therefore, we turn to the institutional structures involved, to consider, in this and the following chapter, what governance structures and changes are needed if Canada is to improve the fate of its remote East and West Coast resource-based communities. The first of these two chapters is a companion to those parts of this book that are concerned with the marine aspects of social-ecological health.[1] The second looks at the coastal marine/terrestrial interface. In it, we consider new options for the landward side of governance, as well as integrated coastal-zone management as a potential tool for recognizing the marine/terrestrial interface in coastal social-ecological systems, which we explored in chapter 6.

Together, these chapters address the policy issues that need to be considered by focusing on the kinds of new options for governance that are aimed at providing a cross-scale institutional framework that would allow various orders of government to have as their goal the social-ecological health of the communities making up their constituencies. To do this, we must draw on details of both potential and actual forms of governance that take into account environmental and human well-being in the marine-based resources and environments on which our coastal communities depend. We also need to offer, along with a brief explanation of institutional frameworks, a detailed analysis of, and potential solutions to, the social-ecological problem of depleted marine ecosystems, which we know have serious consequences for human community health and well-being. In short, these last two evidence-based chapters of the book provide details of potential and actual forms of governance that consider environmental and human well-being in the land and marine-based resources on which our coastal communities depend.

We recognize that management of resources needs to be genuinely social-ecological, which means that to manage a resource we must manage its top predators – humans – and, by extension, we must also seek to match governance practice to ecosystem function and the lives and life cycles of people, and also to communities. Thus, for example, to target only individuals (as the TAGS policy did), rather than households and communities, was to produce serious unintended consequences and to fail to address the fundamental issue of community survival. We also hold that, given the problem of resource degradation currently confronting both coasts of the country, resource management must be management for recovery; it must not merely sustain the present misery.

It is unclear to what extent actual recovery is possible, but, given the global state of many resources, taking recovery as our goal is appropriate not only on Canada's coasts but globally. Throughout this book we have seen that because global competition has increased the need for flexibility in resource-extraction firms, such firms have responded by making themselves able to respond efficiently to changing conditions over which they have little control. They have achieved this, unfortunately, by effectively reducing options for people at lower levels of the state system – one such example is the practice of contracting out, which leaves firms flexible but workers insecure. This is particularly problematic for communities with employment that is often seasonal, because over a long history and until very recently, they could always rely on having flexible responses that allowed them both to survive and to make a strength of seasonality.

It seems that similar problems have existed in government management of marine resources. Holling (1973), for example, has shown elsewhere that the hierarchical and rigid procedures of the former DFO regime resulted in a series of governance actions that, paradoxically, drove the groundfish fisheries to commercial and nearly to biological extinction. With this example before us, we examined new oceans management policies, focusing on the issue of species at risk, and the problems and potentials of marine protected areas (MPAs) as one possible approach to conservation. We also offer here a new approach, which has been developed as part of "Back to the Future," whose estimates of earlier ecosystem structures and abundances were discussed in chapter 2. These different approaches are all important tools in ecosystem conservation and fishery management for the future and should be thought of as complementary to one another.

CANADA'S OCEANS: THE INSTITUTIONAL FRAMEWORK

We start with the institutional framework within which Canada intends to manage its marine ecosystems in the future.[2] This is encapsulated in the Oceans Act (Canada Fisheries and Oceans 1996) and its companion policy documents "Canada's Oceans Strategy" (Canada Fisheries and Oceans 2002a) and the "Policy and Operational Framework for Integrated Management of Estuarine, Coastal and Marine Waters in Canada" (Canada Fisheries and Oceans 2002b), along with the recently released Oceans Action Plan. In 2002, "Canada's Oceans Strategy" and the "Policy and Operational Framework for Integrated Management of Estuarine, Coastal and Marine Waters in Canada" were released. Together, these documents outline the approach that will be used to bring the broad goals of the Oceans Act into action. They represent a significant and promising transformation in the philosophical direction of Canadian oceans policy, replacing top-down regulatory approaches to managing ocean industries with a more participatory approach to governance, one grounded in the idea of integrated management. The Oceans Act, as well as providing the legislative foundation for Canada's much anticipated ratification of the United Nations Convention on the Law of the Sea (UNCLOS) – which eventually occurred in November of 2003 – also gave the DFO the capacity to establish and regulate MPAs and to develop regulations pertaining to marine environmental quality. Most significantly, it identified three core principles that were to be used to guide all future management of ocean activities: sustainable development, the precautionary approach, and integrated management.

The act was received enthusiastically by many organizations with an interest in ocean conservation, but progress on its implementation has been very slow. Serious questions remain about how it will relate to existing legislative regimes and what role other government departments will be expected to play in making integrated ocean management a reality. There are also lingering uncertainties about how, exactly, the act will change the ongoing operations of existing ocean users, although the 2005 budget promised substantial funding for its implementation, for fisheries conservation, for coastguard activities related to the act, and for other programs (P. Harrison 2005).

The Oceans Strategy identified three interrelated goals: understanding and protecting the marine environment, supporting sustainable economic opportunities, and providing international leadership, and it reaffirmed the Oceans Act's commitment to integrated management, as well as calling for the promotion of stewardship and public awareness as a key vehicle through which ocean conservation is to be achieved (Canada Fisheries and Oceans 2002a). The Policy and Operational Framework called for the designation of two types of integrated management plans: Large Ocean Management Areas (LOMAs) and smaller Coastal Management Areas (CMAs), which would be contained within LOMAs. Each of these new entities would be steered through collaborative decision-making processes that involved multiple stakeholders.

A similar approach is advocated for the management of MPAs (Canada Fisheries and Oceans 2002b). At the time of writing, two offshore MPAs have been formally designated: one for the Endeavour Hot Vents off the British Columbia coast and another for the Sable Gully on the Scotian Shelf. In Newfoundland and Labrador, there is a new MPA in southeastern Labrador, and Areas of Interest (AOIs) for MPAs have been identified in Leading Tickles and the Eastport Peninsula on Newfoundland's northeast coast. A major integrated management plan is also being developed for Placentia Bay. In British Columbia, AOIs have been identified for the Bowie Seamount, the Gabriola Passage, and the Race Rocks, and a large integrated management plan has been proposed for the central coast area. We deal with MPAs in more detail later in this chapter. In addition, numerous pilot projects are in various stages of development throughout the country.

The Oceans Action Plan, as identified in the Speech from the Throne of 4 February 2004, has four pillars: international leadership, sovereignty, and security; integrated oceans management; the health of the oceans; and oceans science and technology (Government of Canada 2004). Key areas of interest will include overfishing, maritime security and terrorism, Canada's continental shelf claim under UNCLOS, maritime boundary issues, coastal planning, MPAs, pollution prevention, conservation, ecosystem science, seabed mapping, and sustainable development of ocean resources (Canada 2004). There will also be a number of large management areas, which will include the Queen Charlotte Basin (British Columbia), the Beaufort Sea (north- west Arctic), Placentia Bay/south coast/Grand Banks (off Newfoundland), and the Scotian Shelf (off Nova Scotia) (P. Harrison 2004).

Although these developments are promising, their potential to protect the marine environment and empower coastal communities remained largely unfulfilled until funding became available. Now, with funding secured, questions of process become compelling. These include: What does "integrated management" mean and how should it be put into practice in concrete settings? How effective will integrated management frameworks be in protecting and/or restoring damaged environments? How acceptable will integrated management decisions be in particular contexts? Who has the authority to speak for whom, and what consequences will this have? Who has the right to participate, and in what capacity? How do integrated management processes help to overcome, create, and/or exacerbate power relationships between particular individuals and/or groups?

The Oceans Action Plan's business-friendly language is likely to attract greater attention and additional funds for ocean research. However, the ability of rural communities in coastal areas to benefit significantly under this new regime remains in doubt, unless these communities can become self-sustaining. The "sustainability of economic livelihoods for coastal communities" is mentioned in Canada's Oceans Strategy, but there is no clear vision offered of how this might be achieved. Instead, both in Canada's Oceans Strategy and documents like the *Atlantic Fisheries Policy Review*, the emphasis seems to be mostly on economic efficiency and self-sufficiency, along with "new economy" industries that may be more amenable to market-based regulation. As a result, some local people argue that, unless MPAs can produce the kind of economic self-sufficiency required by the neoliberal policy agenda, there may well be no local people left to benefit from stock recovery in MPA areas.

Moreover, while the documents all insist on the importance of making it possible for citizens to take part in decision-making processes, there is little mention of system equity. Nor are there plans to assist fishery-dependent coastal communities so that they may weather ecological crises, although many coastal communities remain highly dependent on fishing and fish processing, both of which face considerable uncertainty, as we have shown throughout this book. There seems to be little recognition that, as the globalized fishing industry becomes more and more footloose, we must protect, not just our fish stocks, but also our fishing communities, something that we have been arguing for some time now (Ommer 2000, 134–5). We strongly recommend, once again, that the principle of adjacency needs to be tightly tied to the precautionary principle of the UN, because only those who have a place-based commitment to the ocean can act as responsible stewards of our coastal waters.

With the collapse of cod, we are now "fishing down the food web" (chapters 2 and 3; Pauly et al. 2001; Pauly and Maclean 2002). The additional strain on the remaining target species has also created problems, as in the case of the lobster fishery on the Northern Peninsula and the snow crab fishery in Labrador and some other areas (chapter 3). We need better research to provide the basis for better management, and we have discussed some of this in chapter 3. The biological cycles of the marine ecosystem are "set" – things grow only so fast – and management structures have to take this into account. Microscale historical reconstruction (for example) might make it possible to determine more precisely the nature and timing of major changes in marine food webs, which would then allow us to generate appropriate management frameworks to handle such changes wisely and to set appropriate targets for recovery by informing us more fully of what ecosystems are capable of producing (see, for example, Rosenberg et al. 2005). This is urgent work, because it is now clear that we must manage our damaged marine ecosystems with much greater care and at smaller spatial scales. There is mounting evidence that gaps in scientific knowledge, as well as the industrialization of fisheries and interactive effects between fisheries, forestry, tourism, and transportation, have been eroding critical habitats and thus reducing the potential for recovery in many systems and further degrading others (see the discussions earlier in this book of aggregates, the Labrador road, and eel grass and also the website http://www.mcbi.org/DSC_statement/sign.htm, which has a scientists' statement on deep-sea corals).

To the extent that environmental degradation interacts with industrial, social, and policy change to amplify risks to environmental and human health – as suggested here and by Dolan et al. 2005 – it is also placing fishing communities at risk, as quotas are cut and control over quotas and licences is concentrated in the hands of fewer and fewer organizations, and as younger generations leave and communities lose their population and skills base for sustainable development. Most people living in rural coastal communities cannot be expected to achieve self-sufficiency without sustained support from policy-makers: some are of an advanced age or lack formal qualifications and are in locations unlikely to benefit from new industrial opportunities as they occur (see chapter 9). We return to this issue in the next chapter.

How marine ecosystems will fare under the Oceans Act and associated strategies remains unclear. The relevant policies offer little guidance about when, if at all, economic growth will be curtailed in order to protect and rebuild already degraded marine ecosystems. The Oceans Strategy adopts the definition of sustainable development made famous by the 1987 World Commission on Environment and

Development ("development that meets the needs of the present without compromising the ability of future generations to meet their own needs" – Bruntland 1987) but does so unreflectively, without addressing a number of critical questions.

Are we sustaining environments or simply sustaining development? Should we be sustaining those environments or supporting rebuilding? If our focus is recovery, recovery for whom? Who will still be around when environments and stocks recover? How are we to differentiate between needs and wants? How well are the needs of the world being satisfied now? How are we to know what the needs of the future will be? How many future generations need to be considered? Is "integrated management" an adequate vehicle through which to protect the marine environment, since the short-term interests of industries and communities are not necessarily the same as those of ecosystems? The task of speaking on behalf of the ecosystem falls to the DFO scientists, but can they do it, given the severe financial constraint to which they have been subjected over the last fifteen years and given the fact that rapid policy and industry shifts mean too often that scientists are following fisheries around (Neis and Kean 2003; Haedrich, Merrett, and O'Dea 1998)?

Integrated management and MPA development appear to direct resources to areas with the capacity to mobilize – such as Eastport. While intelligent at one level, this approach leaves aside the issue of the social and ecological consequences of spotty and opportunistic allocations of federal and other resources – perhaps these are not the key areas or issues from a social, economic, or ecological point of view. What is our commitment to the marine ecosystem as a whole and to each of its constituent parts (Murray, Neis, and Bavington 2005)? This type of approach is similar to the shift in social programs towards targeted funding for particular groups – it runs the risk of the sort of piecemeal programming that dedicates more and more resources to those with power and that marginalizes those most in need of support (MacDonald, Neis, and Murray 2005).

The move toward research partnerships with industry and academia is plagued by conflict of interest issues and currently seems to be unable to complete the task. In Canadian ocean waters we now carry out seismic testing, large-scale aquaculture operations, and a variety of new fisheries that have no precedent here. Without any longitudinal data on the ways in which these activities affect marine ecosystems, it is unclear how we can assess whether the precautionary approach is being followed. Indeed, Fisheries and Oceans managers now face a serious legitimacy crisis in many coastal areas as a result of the ever-widening gap between the rhetoric of ocean management and the achievements of ocean science.

SPECIES AT RISK

One way to address the overall problem of fishery declines is to show, through an examination of potential species at risk in the Newfoundland area, how the whole ecosystem is affected. The fish community in an ecosystem is composed of many different species, but it is those with commercial importance that are often of the most interest and therefore the best known. However, governments and others have increasingly recognized the concept of species at risk and the need for their protection. IUCN (the World Conservation Union) is the international body that keeps tabs on the global situation (Hilton-Taylor 2000). Canada has only recently passed its own Species at Risk Act (SARA), but the Committee on the Status of Endangered Wildlife in Canada (COSEWIC) has been recognized for over twenty years as the lead organization in the country for assessing species (Shank 1999).

The assessment of just one species by COSEWIC takes at least two years, and the committee does its work one species at a time. But even after COSEWIC has made its recommendation, the final decision about whether to list is the responsibility of the minister of the environment. Additional consultations at that stage can take another two years, so the full process moves at a very slow pace. We have addressed the issues of data collection and scientific assessment in chapter 3. Here we note that as of November 2003, only five Atlantic marine fishes had been classified as at risk by COSEWIC (COSEWIC 2003) – Atlantic wolffish, (special concern), and northern wolffish (*Anarhichas denticulatus*), spotted wolffish, and cusk (*Brosme brosme*) (all threatened). Four populations of Atlantic cod had been recognized and given individual status designations – Arctic (special concern), Northern Labrador (endangered), North Laurentian (threatened), and Maritimes (special concern).

The process of listing a species under SARA includes a socio-economic impact component. According to Environment Canada, the term "socio-economic" within the realm of SARA is based on the *Concise Oxford Dictionary* definition of "socio-economic" as "relating to or concerned with the interaction of social and economic factors." The term "socio-economic evaluation" is taken to mean a process by which both the positive and negative impacts on social and economic factors of a planned activity in a given area are identified and assessed using both qualitative and quantitative assessment techniques. The results provide the means of conveying to decision makers the nature and magnitude of the proposed activity's socio-economic impacts. Decision makers can then use this input to make informed decisions when planning and implementing recovery measures.

There are four references to socio-economic evaluation in the SARA:

- Preamble: "community knowledge and interests, including socio-economic interests, should be considered in developing and implementing recovery measures."
- s. 38: "In preparing a recovery strategy, action plan or management plan, the competent minister must consider the commitment of the Government of Canada to conserving biological diversity and to the principle that, if there are threats of serious or irreversible damage to the listed wildlife species, cost-effective measures to prevent the reduction or loss of the species should not be postponed for a lack of full scientific certainty."
- s.49(1)(e): "An action plan must include, with respect to the area to which the action plan relates, an evaluation of the socio-economic costs of the action plan and the benefits to be derived from its implementation."
- s.55: "The competent minister must monitor the implementation of an action plan and the progress towards meeting its objectives and assess and report on its implementation and its ecological and socio-economic impacts five years after the plan comes into effect. A copy of the report must be included in the public registry" (personal communication, Sharmane Allen, DFO, July 27 2005).

According to the DFO, SARA "not only prohibits the killing, harming, harassing, capturing or taking of species at risk, but also makes it illegal to destroy their critical habitats." Under SARA, the DFO has to produce recovery strategies and actions plans for aquatic species listed as endangered or threatened. These recovery strategies are supposed to be developed in collaboration with relevant stakeholder groups like the fishing industry and should include a socio-economic impact evaluation component and ongoing monitoring of such impacts. Once a species is listed, fisheries management plans will need to be adapted to comply with SARA, and recovery strategies will have to be developed (http://www.dfo-mpo.gc.ca/species-especes/actMeans/actMeans_commercialFishing_e.asp).

One potential issue is that of who benefits from recovery. Different recovery strategies might impact differently on different fleet sectors or regions. The DFO and the Newfoundland government recently produced a report on cod recovery in Newfoundland and Labrador, but it did not address this issue. A second issue is the data sources for listing. To our knowledge, much of the data used to date in developing listing arguments (for wolffish, for example) have been taken from the RV trawl survey data. In feedback meetings, coastal fish harvesters indicated that at least some species of wolffish are still relatively common in coastal waters. This difference in the basis for observation between scientists listing species, and harvesters, and the very limited knowledge that exists on the ecology of some species, their range, distribution, and so forth, may lead to disagreements between scientists and fishers and in some cases to listings that should apply primarily to offshore areas and to restrictions on trawling being applied across the board.

MARINE PROTECTED AREAS

The Oceans Strategy purports to accept ecosystem-based management, but it represents depleted marine ecosystems on its Integrated Management Board, along with oil and gas, salmon farming, urban waste disposal, and other issues. It does, however, accept conservation measures such as marine protected areas (MPAs). MPAs are regions of the marine environment under various levels of protection from human disturbance. There are quite a few worldwide, ranging in size from hectares to thousands of square kilometres, but most are small and coastal. Protection can vary from minor prohibitions, such as prohibitions on disturbance of habitat, to strict ones, such as prohibitions on entry. Comparative studies have shown that fish populations are often enhanced in the vicinity of fully protected regions, which is valuable for local communities.

In the development of an MPA its objectives should be clearly defined at the outset, and flexibility in the choice of an area should also be highlighted (Day and Roff 2000; NRC 2001). As many species as possible should be protected, and whenever feasible, special attention should be paid to species at risk. The public may need to be educated about the purpose of MPAs to minimize opposition in important ecological areas (Roberts 2000). Usually, a steering committee is established that represents groups with economic, political, or cultural ties to the area and provides a forum through which diverse interests and concerns can be brought together in an integrated management setting to discuss the ongoing operations of the MPA and arrive at consensus-based decisions. Ultimate decision-making authority is usually vested in a government body, but advice from the steering committee is taken into account. Government administrators often also sit on steering committees, playing a mediating and coordinating role.

Of course, the awareness of consequences and the appreciation of the value of ecosystems that exists among First Nations and permanent fishing communities is not reflected to the same extent in central government, nor can it be until knowledge of the dangers of our impact on ecosystems becomes more widely held.

MPAs pose problems for humans in terms both of lost dietary input, and of lost employment, albeit only in the short term while stocks rebuild (see chapters 9 and 10). Fishing quotas and single-species management were and remain the most common methods of managing marine systems. These methods focus on species that are commercially important and tend to ignore other species in the ocean. Therefore, fishes subject to by-catch or ghost-fisheries, but not directed fisheries, rarely have adequate protection. Global by-catch and discards between 1988 and 1990 amounted to approximately one-third of total landed biomass (Murray et al. 1999 and chapter 3). As a part of overall restructuring it has become apparent that other methods of management must be adopted to prevent the marine system from being depleted to the point where recovery is impossible. Although opposition still persists to the "fencing of the sea," MPAs have proven to be a valuable tool to improve fisheries, preserve habitat and biodiversity, and enhance the aesthetic and recreational values of an area (Day and Roff 2000; NRC 2001). MPAs can also be used to protect critical or threatened habitats, in order to foster the restoration of biological communities (NRC 2001).

MPAs at a scale of the continental shelf make a lot of sense in the light of fishery collapses and attempts to restore populations and ecosystems, but the problem is how to decide upon candidate areas for their establishment and how to address the potentially huge social and economic consequences of such MPAs. There is no precedent for this. Just as with protected areas on land, MPAs are often identified on the basis of some clear physical structure like a coral reef or a rocky bank. Offshore, the identification of marks like this becomes difficult, especially over a broad expanse like the Newfoundland continental shelf. In that case, biological characteristics might be used. One logical approach is to seek to protect as many species as possible. We propose using diversity as an index, with areas of consistently high diversity ranked the highest for protection. We can measure species diversity and map "super hotspots" using the ECNASAP database. Super hotspots were considered to be areas where species richness was three standard deviations above the mean. We first substitute the abundance data for each species in the database with presence and absence information, i.e., 1 or 0 respectively. The richness at each station is then calculated by totalling the number of species reported, and the mean richness and standard deviation are determined for each region. The super hotspots were mapped to assess whether distributions are clumped or apparently random. A cluster of hotspots would represent an area with high overall species richness, and thus protection for this area should be a priority.

The areas of species richness were not uniformly scattered throughout the region we studied (figure 14.1) but tended to occur in specific

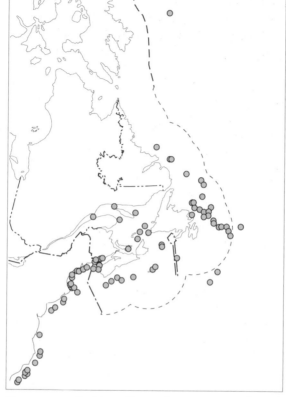

Figure 14.1   Position of super hotspots across the
entire ECNASAP region from Cape Hatteras to
Greenland

regions.[3] The obvious localities of several super hotspots indicate these
are regions of high species richness and not misleading outliers in the
data. Regions, therefore, that should be considered for conservation
include the Bay of Fundy, the Gulf of Maine, the Bonavista Corridor,
Laurentian Channel, and portions of the Scotian Shelf. The super
hotspots were able to represent almost 50 percent of the total species
sampled in a given region.

Although this type of examination identifies regions that are important for conserving a large number of species, there was no attempt to
take the special needs of species at risk into consideration. Species at
risk require recovery and habitat conservation programs to help rebuild
populations. Halpern (2003) found that marine sites protected from
fishing lead to a rapid increase in the biomass, abundance, average
size, and diversity of all functional groups. The density of organisms is

roughly doubled in no-take MPAs, and the biomass of organisms is nearly tripled (Halpern 2003). Thus, protected populations are expected to produce many times more offspring (Roberts et al. 2001) and exhibit increased population trends. If areas important for species at risk are identified and protected, the productivity of the population can be enhanced and the population can begin to rebuild. Although scientists caution that MPAs will not be enough to restore all fish populations (Halpern 2003), they can be viewed as an important tool for species recovery programs (Roberts et al. 2001).

Little is known about critical habitats or potential sites that are important for adult marine fishes. These can be inferred from data gathered by trawl surveys made over extended periods of time to determine significant locations and the nature of these locations. However, areas of particular importance may change over time, and the results were examined to determine if the areas are temporary or relatively permanent.

The ECNASAP database provided information to evaluate regions of high abundance for species at risk. The stations within the top 95 percent of abundance (99 percent for Atlantic cod) were mapped over five-year intervals to determine if the areas of high abundance for individual species were temporary and uniformly distributed. If this was the result, areas of high abundance would not represent potential sites for MPAs. If the alternative was true, the areas should be considered important sites for potential MPAs.

Several high-abundance areas were consistent throughout the period of the sampling, and therefore they are not outliers in the data. These areas were extracted and mapped for all species at risk. There appear to be distinct areas where the species at risk are abundant (figure 14.2), largely corresponding with areas of high diversity. Since they represent regions important for the species, they should be considered important candidate areas for conservation. The northern edge of the Grand Banks of Newfoundland (Bonavista Corridor) appears to be very important for all the species at risk that exhibit northerly ranges. North of Hamilton Bank appears very important for the wolffish species, and the Laurentian channel appears important for Atlantic cod and Atlantic wolffish. A portion of the southwest Scotian Shelf appears very important for cusk. Baker (2003) can be viewed for individual species maps depicting high-abundance areas over five-year intervals.

There are distinct locations where MPAs could be designated to protect diversity and species at risk (figure 14.3). Overlaps in the high-priority areas occur in the Bonavista Corridor, the northern side of Hamilton Bank, the Laurentian Channel, the Bay of Fundy, and a

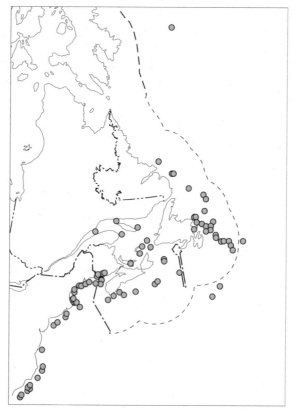

Figure 14.2　　Areas that appear to have static clusters
of highly abundant species at risk

portion of the southwest Scotian Shelf. The Bonavista Corridor should
be considered the highest priority area for an MPA, because it is home
to abundant populations of all the species at risk (except cusk) and
many super hotspots occur in this region.

The objectives of the MPAs being evaluated thus far have been to
conserve diversity and species at risk, and these objectives should
remain the priority of the potential MPAs. Nevertheless, good protection is related to both compliance and enforcement (Roberts 2000). If
at all possible, MPAs should not be suggested in areas where restriction,
termination, or displacement of activities such as fishing and oil development will occur (Murray et al. 1999). To minimize conflicts, MPAs
should be designated only where public acceptance is likely. Roberts et
al. (2001) indicate that while it is important to evaluate the needs of
the stakeholders, the original criteria concerning ecological parame-

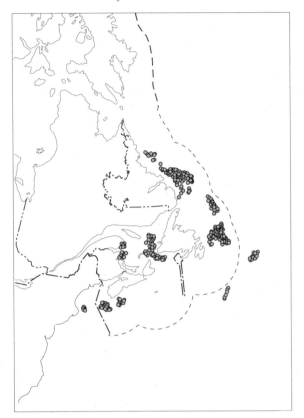

Figure 14.3   High-priority areas for MPAs in the
Northwest Atlantic. Circles represent stations where
species at risk are abundant.

ters should remain the highest prioritized criteria for identifying
potential sites for MPAs. This would ensure that the sites most important for ecological considerations would remain potential sites for
MPAs. The next step to evaluate potential sites for MPAs, however,
should be to view areas important to stakeholders in the region.

Trawl effort data can be used as an index to determine high-quality
fishing grounds on the northwest Atlantic continental shelf and thus to
determine high-cost areas. High-cost areas are regions where there is a
high probability of conflict if an MPA is established. The oil industry is
ever-growing in Nova Scotia and Newfoundland and Labrador and
thus must also be considered a stakeholder in the process of selecting
sites for MPAs.

We obtained data for the top twenty areas trawled in each year
(David Kulka 2003). These data are from the Fisheries Observer Pro-

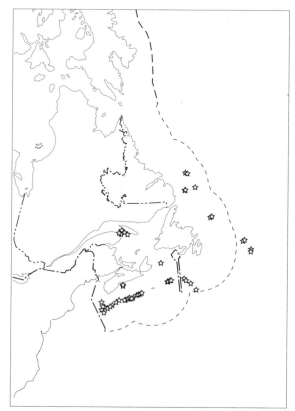

Figure 14.4   Top twenty fishing areas for each year
during the period 1995 to 2000. These areas,
therefore, represent recent high-quality fishing
grounds: from Kulka (2003).

gram and are in the form of geo-referenced fishing set locations
(Kulka and Pitcher 2001). The data illustrating the top twenty areas in
each year from 1995 to 2000 were mapped to represent areas that have
recently been important to trawling (figure 14.4). The results were
overlaid with high-priority areas (figure 14.5). There is an overlap in
areas that are deemed to be of high priority for potential MPAs and
high-quality fishing grounds; however, the overlaps are minimal.
Potential areas of conflict could occur on relatively small portions of
the southwest Scotian Shelf, the Flemish Cap, and Hamilton Bank.

  Sites of potential or current oil and gas exploration (chapter 5) were
also considered high-cost areas. Maps depicting these regions were
found on the Canada-Newfoundland Offshore Petroleum Board

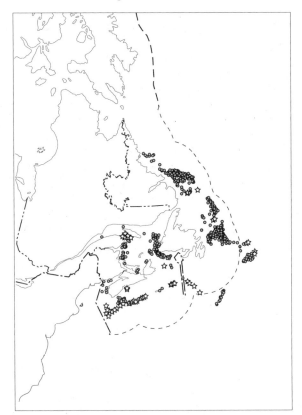

Figure 14.5   High-priority areas (circles) and top
twenty areas of trawl effort for each year during the
period 1995–2000 (diamonds)

(CNOPB) and the Canada–Nova Scotia Offshore Petroleum Board
(CNSOPB) websites (CNOPB 2003; CNSOPB 2003).

The scientific survey data for demersal fishes allow us to develop
maps of biodiversity for areas deeper than one hundred metres. The
importance of areas so identified for commercial species and for other
indicator species in the Newfoundland region, including species at
risk, suggests their potential impact. At the same time, the maps also
identify possible conflicts because of the location of traditional fishing
grounds and oil and gas leases. With this information, we are able to
recommend a few areas of the Newfoundland continental shelf with
the highest promise for designation as MPAs. Because these are based
on the distribution of fish biodiversity, a legislative framework for
implementing them already exists under the Fisheries Act (Canada

Fisheries and Oceans 1985) and the Species at Risk Act (Canada 2002b).

The areas of high cost associated with oil and gas development in Newfoundland and Labrador do not overlap very much with the high-priority MPA areas. There appear to be no plans for the Bonavista Corridor and only slight interest near the Hamilton Bank. Western Newfoundland development may extend into part of the regions deemed high-priority sites for MPAs, but not extensively so. There is extensive oil and gas development on the edge of the Scotian Shelf; however, it is further out on the shelf and slope than the high-priority MPA areas and should not be a concern. There is no oil and gas development in the Bay of Fundy, but there is extensive aquaculture along the coast, which may cause some opposition to an MPA being established there.

This preliminary investigation reveals that there are few overlaps of high-priority MPA areas and areas important for other stakeholders on the shelf. However, this conclusion may not be entirely true. The high-quality fishing grounds were identified by only examining the top areas where trawling occurs. There are other fishing techniques and species, and thus other fishers may depend on the areas that are not trawled to a great extent. Public consultations are required to create more reliable data that would help identify other areas considered important to fishers. Consultation with First Nations has become a legal requirement. Commercial fishers and the oil industry are not the only stakeholders of the marine environment. Other important stakeholders to consider in future research would include the subsistence users, the aquaculture industry, the shipping industry, marine mining interests, tourism and recreation stakeholders, and marine scientists. These groups should also play an important role in establishing MPAs, and they need to be consulted during the process.

Other taxonomic groups should be incorporated into the analysis to determine high-priority areas for, among others, seabirds, marine mammals, and benthic fauna. However, if data are not currently available, this lack should not be used as an excuse for delaying the establishment of MPAs. Once an MPA is established, complementary sites can be identified and examined to further increase the diversity being protected and create a broadly effective MPA network.

Our results indicate that there are minimal overlaps in recent (since 1995) high-cost areas and high-priority areas for potential MPAs on Canada's continental shelf. Using ecological values as the highest priority and recent fishing pressure, the Bonavista Corridor, the northern side of Hamilton Bank, the Laurentian Channel off southwest Newfoundland, the outer Bay of Fundy, and a portion of the southwest Scotian Shelf remain high-priority areas for MPAs.

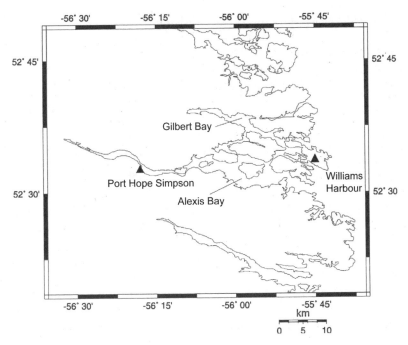

Figure 14.6   Gilbert Bay, Labrador, and surrounding area

Alternative means need to be identified for mapping potential MPA
areas in the bays and coastal/intertidal areas. One way to begin to
develop such maps is through coastal-zone inventories carried out in
recent years by the DFO in collaboration with regional development
organizations around Newfoundland and Labrador and through LEK
research with fish harvesters (Murray, Neis, and Bavington 2005). On
the East Coast, there is a new MPA in Gilbert Bay, which was created as a
result of joint community-CUS work. Gilbert Bay lies between the communities of Port Hope Simpson and Williams Harbour on the southeast coast of Labrador (figure 14.6). Its residents have long utilized the
bay's marine resources for both commerce and sustenance. Local cod
fishers suspected that a stock of golden-brown coloured fish, which was
resident in the bay, was distinct – a belief based on the colour difference from the offshore (grey-countershaded) cod and the timing of
the local fishery, which is a springtime fishery for bay cod, conducted
before the summer fishery for offshore cod migrating to the coast
(Wroblewski 2000).

   This local ecological knowledge prompted us to carry out a study of
the biology and migration behaviour of cod in Gilbert Bay, an analysis

that revealed that these local cod are distinctive in both genetic character and appearance, thus confirming that they are a self-reproducing local population. More specifically, we found that the reddish or golden colouration of the fish is the result of their diet of invertebrates rich in carotenoids (Gosse and Wroblewski 2004). Since, therefore, it would have been possible for offshore cod to migrate into Gilbert Bay and acquire a reddish or golden colour gradually by feeding on invertebrates, we needed the corresponding evidence for distinctness from DNA genotyping, since brown variant colouration alone would not have been a conclusive indicator of inshore stocks. Finally, we confirmed through tracking experiments that Gilbert Bay cod remain in the bay year-round and home to a spawning ground within the bay to breed. This is a good example of fishers' local knowledge (about a bay stock living in Gilbert Bay) being supported by new scientific knowledge. Interestingly, it is only now, when fishers' knowledge is being substantiated by scientific study, that such knowledge is being used by the coastal communities with increased confidence in planning resource development, conservation measures, and co-management.

Concerned about their local resource and about the risk of such a relatively small "bay stock" meeting the same fate as did the Grand Banks cod stocks, the residents of the communities of Port Hope Simpson and Williams Harbour asked the DFO to consider Gilbert Bay for the Marine Protected Areas program under Canada's Oceans Act.[4] Subsequently, Gilbert Bay was identified as an Area of Interest under the MPA program in October 2000. In 2001 the Gilbert Bay Steering Committee (GBSC made up of representatives from local fishing committees, local communities, the Labrador Métis Nation, Memorial University [including a CUS researcher], the provincial Department of Fisheries and Aquaculture, and the federal DFO) was established as part of this marine protected-area initiative. The mandate of the committee makes it responsible for bilateral and public consultations on the MPA initiative and for advising the DFO on establishing an MPA. Finally, in 2003, the GBSC recommended to the federal minister of fisheries that Gilbert Bay be designated an MPA under the Oceans Act. The minister accepted this recommendation and asked the committee to begin management planning. The GBSC continues to function as an advisory body to the DFO in developing regulations and a management plan for Gilbert Bay. Identification, protection, and indeed recovery of local cod stocks such as the one in Gilbert Bay may be essential for the future recolonization of offshore areas (Wroblewski, Neis, and Gosse 2005).

ECONOMIC ISSUES IN THE RESTORATION
OF MARINE ECOSYSTEMS

The Oceans Act does not come to grips with economic issues such as the discount rate, which is often argued to be inexorable. But our new Back to the Future economic analysis can operate in terms of *inter-generational* discounting (Sumaila and Walters 2005; Ainsworth and Sumaila 2005), and it enables us to compare past and present ecosystem states (those of the 1750s, 1900s, 1950s, and the present, which we presented in chapter 2) and to then compute the value of each of these ecosystems, if restored, to future as well as present generations. This approach to restorative marine ecology proposes rebuilding, rather than preserving, the status quo as the proper goal for management (Pitcher and Pauly 1998; Pitcher, Heymans, et al. 2004). It provides a strategic guide for the long-term restoration of depleted marine ecosystems by explicitly quantifying the socio-economic and ecological trade-offs inherent in restoration, using a range of economic, social, and ecological indicators to discriminate and score rebuilding solutions (Pitcher 2005; and see, for example, Ainsworth and Pitcher 2004a; Cheung and Pitcher 2004; Heymans 2003e; Ainsworth and Sumaila 2004a, 2004b; Haggan, Brignall, and Wood 2003). The final decision on what to restore (and how) would, of course, fall to policymakers, but there should be participatory vetting of restoration strategies by local communities (Pitcher, Heymans, and Vasconcellos 2002; Pitcher et al. 2005).

There are two basic components of this work: setting goals for restoration and developing restoration strategies. The first step, basing its restoration goals on historic ecosystems, seeks to restore the optimally productive state directly (Ainsworth and Pitcher 2003; Ainsworth, Heymans, and Pitcher 2004; Heymans 2003e).[5] To assess where that would be, we offer a methodology for calculating the equilibrium biomass configuration that theoretically results from a historic ecosystem after long-term optimal exploitation (Ainsworth, Heymans, and Pitcher 2004). That processed configuration is called the optimal restorable biomass (ORB) (figure 14.7). This should be the actual goal of rebuilding.

The second step is to consider how we may finally achieve this goal. Developing new tools for simulation, we calculate the optimal pattern of fishing mortality that will rebuild the current ecosystem and recapture lost productivity potential (Ainsworth and Pitcher 2004b). Preliminary work suggests that under some circumstances, ecosystem restoration appears to be a win-win situation – the relationship between

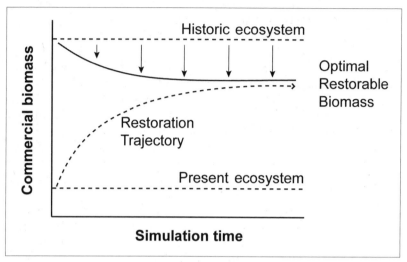

Figure 14.7   Conceptual diagram showing the development of the optimal
restorable biomass (ORB) restoration target and a possible restoration
trajectory. Simultaneity is not implied between ORB determination and
restoration. ORB is the theoretical biomass equilibrium that would result after
long-term optimal harvesting of the historic ecosystem (down arrows).
A restoration plan (dotted line) would see the present ecosystem change to
resemble the ORB state (Ainsworth and Pitcher 2005).

profit and biodiversity is convex (figure 14.8) – implying that there is
some optimal rate of restoration where humans and the environment
can best be served; restoration can be accomplished without forsaking
resource users.

The costs and benefits of this approach can be evaluated for different policy goals, using such criteria as ecological economics, employment opportunities, and impacts on biodiversity. Different policy goals
range from purely economic to purely conservationist through a series
of intermediate weightings, and comparison among them quantifies
the trade-offs that would have to be made to achieve each goal. For
example, one way to evaluate the benefits of alternative restored ecosystems is to compare the economic value of their fisheries, information that is readily estimated from the simulations mentioned above.
But how we determine the economic "success" of restoration will
depend greatly on how we value the expected stream of benefits.

In cost-benefit analysis (CBA), economists use conventional discounting to summarize the expected stream of benefits from an investment
into a single term, the net present value (NPV), which can then be compared among alternative investments. In this calculation, the value that

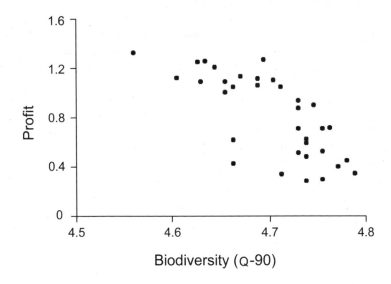

Figure 14.8   Equilibrium trade-offs resulting after twenty-five years of restoration in northern British Columbia. Not shown: baseline (current) system provides profit 0.7 and biodiversity 3.9; goal (1900 ORB) system provides profit 1.2 and biodiversity 8.4

future benefits carry in the present is discounted exponentially through time, reflecting the investor's preference for immediate consumption and delayed payment. Discounting accounts for lost opportunity costs, uncertainty, impatience, and other factors that make a unit of money today worth more than the same unit tomorrow. In this fashion, discounting acts a model of human behaviour; it is an analytical tool to help make our value-based decisions. It is unrelated to the inflation rate.

The discount rate applied in the cost-benefit analysis has a critical influence in determining whether an investment is seen as worthwhile. However, in valuing a long-term conservation plan like an ecosystem restoration project or a sustainable fishery, immediate costs tend to outweigh far-off benefits, so that only "myopic" policies can be viable at any practicable level of discounting (Sumaila and Walters 2005). Sumaila and Waters (2003, 2005) have devised a new form of discounting to evaluate long-term environmental initiatives. It is set apart from other intergenerational discounting techniques (e.g., Fearnside 2002; Weitzman 2001; Chichilnisky 1996) because it allows us to divide explicitly the stream of benefits between ourselves and our children. The method incorporates into each year of the analysis an annual influx of investors who bring a renewed perspective on future earnings; this partially resets the "discounting clock." If this method were to be adopted, it would satisfy national and international requirements to manage for the benefit of future as well as present generations, and it would meet the Aboriginal criterion of planning for the "Seventh Generation" (Clarkson et al. 1992).

The best policy goal is the future scenario that, allowing for the costs and time span of restoration, will produce the highest rewards. The methods are made possible by whole-ecosystem modelling, at present implemented using the Ecopath-with-Ecosim suite of modelling tools (Christensen and Walters 2004). The results may be challenged numerically by the uncertainty of model parameter values and by possible climate fluctuations and changes, in order to evaluate the robustness of each policy goal.[6]

In our work, we have created policy directives that maximized one of five harvest objectives: *ecology* increases the occurrence of long-lived species; *economy* maximizes profit; the concept of the *social* is presently too narrow, being restricted to maximizing the available number of jobs; the *mixed* and *portfolio log-utility objectives* combine these priorities; and the latter considers risk aversion. Using seven valuation indices, we compared the optimal harvest profile and the resulting ecosystem condition, tendered by each period, fleet structure, and harvest objective. Economic valuation considers the net present value of future harvests under conventional and intergenerational modes of discounting; social valuation considers the total number of jobs and employment diversity; and ecological valuation considers the system change in biodiversity evenness, the occurrence of local extinctions, and the change in system resiliency to fishing.

We found that the pre-contact systems (1750 and 1450) have the most to offer as restoration goals, outperforming other time periods in economic, social, and ecological terms. They are the least similar to the present condition, however, and so represent the most ambitious goals for restoration. The five harvest objectives effectively sketch the inherent trade-offs between human consumption and protection of the environment. The social and economic objectives provide wealth and jobs to the detriment of the environment. The ecology and portfolio log-utility objectives preserve the environment but neglect resource users. The mixed objective offers an intermediate solution between these extremes. Having quantified what a restored system may be worth, we provide a benchmark with which to measure and justify the costs of restoration.

This procedure allows us to compare quantitatively possible futures based on restoration with a form of "status quo" represented by fishing the present-day ecosystem in a more sustainable and responsible fashion, which is presumably the policy adopted implicitly or explicitly by most of the world's fisheries management agencies today. Hence it can provide an evaluation of the benefits (or otherwise) of restoration, which may be a painful process, compared to "good management" of today's fisheries and ecosystems. Moreover, we have found that the trade-offs between exploitation and conservation made explicit for the best ORB policy may often be less extreme than the trade-offs faced in the present day: a nice example of a sustainable development policy that addresses a genuine reconciliation of exploitation with conservation.

CONCLUSION

Canada's Oceans Act and associated documents purport to lay out a strategy for the future management of Canada's oceans. It is important that the Canadian state maintain its overall responsibility for management and not effectively hand it over to particular groups of vested interests as some quota management schemes in the past have done. The resulting social-ecological patchwork approach to management and recovery that this can produce is unlikely to ensure sustainable management, let alone recovery. At the same time, monolithic, centralized management based on single-species assessments and policies with inappropriate spatial and temporal management scales also does not work. We could begin to achieve an even-handed and effective management function in the future by laying out strategies that involve multilayered consultation (integrated management planning); attention to habitats, species interactions, and ecology; investing in recovery, rather than sustaining misery; the development of the social and economic structures needed for effective co-management; the strategic identification, development, monitoring, and ongoing adjustment of spatially and temporally appropriate management initiatives; and managing for multiple generations of people and fish, while paying careful attention to the question of recovery for whom and the goal of promoting the health of people, communities, and environments. This will happen if and only if there is genuine representation from all levels of interest and if there are shared goals that go beyond the purely short-term environmental, social, political, and economic. Thus, the goals of conservation, stewardship, ecosystem health, human health, community health, scientific and technological imperatives, environment, and security all need to be interwoven.

We suggest that we can utilize both the knowledge of species-at-risk hotspots and the protective practices of MPAs in the short run, while preparing in the long run to provide a highly flexible method of restoring lost ecosystems, based on the informed choice of stakeholders. In all discussions, issues of access, ownership, and food security for humans will also need to be considered. Managing Canada's oceans means managing Canadian users of the oceans. For this to be effective, it is important that managers understand the goals and interests of various stakeholders and also have a firm grasp of the scientific knowledge of the ecosystems involved, including crucial gaps in that knowledge that can seriously affect stock assessment. Managing all this will involve bringing together the various levels of interest involved, so that they can work together. Only then can we avoid the risk of damaging the integrity of marine ecosystems, and only then can managers ensure that the risks and costs of using Canada's oceans are not paid for by one set of stakeholders while the benefits accrue to another.

Canada's Ocean Act and associated strategies signal growing institutional awareness of the problems of disturbed ecosystems, but much less awareness of the social complexities involved that cannot be considered separately from new forms of ocean and coastal water management. New initiatives will have to consider social-ecological health and will therefore have to use co-management structures of some form. In this chapter we have discussed all these features and identified strengths and weaknesses of the suggested strategies for the future. The approaches we offer here provide ways to evaluate what is "in play" when we recognize that there are rich flows of benefits (spiritual, cultural, existence, social, ecological, and economic) from healthy coastal ecosystems. When we compare these with the cost to future generations of nonsustainable activities, it becomes obvious that we need to ensure that whatever we do does not compromise the productive and regenerative capacity of marine ecosystems in and for the future. Meanwhile, current management practices – often based on the perceived inevitability of the fiscal imperatives justified in conventional discounting – ensure that we continue to deplete marine ecosystems and preclude the level of re-investment in natural and social capital that would be necessary to compensate for the past 150 years of overharvest. In the next chapter, we focus the discussion on the challenges of landward management and of the land/sea interface, using some examples that are being tested on different parts of coastal Canada.

15

New Options for Governance II: The Land and Sea/Land Interface

Finally, we turn to consider some of the resource management options that coastal communities have considered as they seek to come to grips with the changing socio-economic, environmental, and political conditions that have affected their terrestrial and land/water interface ecosystems.[1] Interactive restructuring has occurred at all levels and on multiple dimensions, and now at the national level it is being driven to a significant extent by international global restructuring of various kinds.[2] While international restructuring is beyond the scope of our study, we needed to be constantly aware of it and to recognize it as part of the driving force behind international manifestations of social-ecological distress, as seen in the over-exploitation of the world's oceans or the destruction of the Amazon tropical rainforest, for example, where it has been driven by the industrial and fiduciary agendas of transnational actors, in the form of transnational corporations and banks and their various instruments, such as the International Monetary Fund (IMF) or the World Trade Organization (WTO). There is little in the way of transnational institutional governance capacity or autonomy that can offset the economic agendas of the large banks and corporations, since the United Nations and transnational nongovernmental organizations (NGOs) lack the necessary instruments of adjustment with which to alter any concerted global economic developments, and structures like the WTO are essentially under the influence of First World powers. Particularly since the early years of the 1970s, national governments have become increasingly sensitive to the requirements and needs of the global economy, and much of their politico-economic strategy now falls within the rubric of a global capitalist trajectory (Ommer 2004).

Canadian policy restructuring has affected our study areas in several ways, interacting with industrial, environmental, and social restructuring.

Nationally, in the context of deficit fighting and an ethos of increased reliance on the private sector (both driven by the requirements of globalization), we have seen that policies that traditionally supported rural economies and communities – including, for example, federal-provincial transfer payments, rural development initiatives, and Employment Insurance (EI) – have been significantly eroded. Regional economic development programs have been replaced by community economic development and subprovincial, rather than multiprovince "regional" initiatives, which challenge struggling communities to find their own solutions. Rural communities and industries have been undermined by depleted resources, market shifts and cutbacks in essential transportation services, as we have shown. Political restructuring has reconfigured welfare state entitlements to benefits such as EI, social assistance, and workers' compensation and to services such as health care and education. While health care restructuring has sometimes increased the range of services available in rural areas, centralization and the escalation of user fees for health services have hurt many communities. In practice, most of the restructuring in the governance of health care has been influenced more by the biomedical model than social determinants or eco-health (Tomblin 2004). The consolidation of education has been an ongoing concern in rural communities.

There has also been significant deregulation and re-regulation of resource access. The traditional resource sectors of forestry and fishing have both seen new efforts to change the rules affecting access, often in the name of resource sustainability. In fisheries, licensing costs, the privatization of public infrastructure like wharves and catch monitoring, professionalization programs, and the spread of individual quotas and individual transferable quotas are limiting access, driving up the cost of entry into the industry, and facilitating the concentration of ownership and control over the resource and the industry as a whole. In both fisheries and forestry competing interests between sectors (pulp, sawmills) and between large companies and smaller operators come into play in resource management. Such changes in resource policies both shape, and are influenced by, industrial and environmental restructuring. In turn, the households that are trying to make a living in the resource sectors have had to adjust to a new set of constraints and rigidities, although there are also a few new opportunities. Communities prosper or decline in the process.

Clearly, the top-down model of governance has failed coastal and other rural communities and needs replacing with something that can operate across scales and beyond silos. Piecemeal, highly localized programs are not an appropriate alternative. Some new land-based and coastal governance structures already exist, and their achievements tell us much about what strategies might help to protect coastal social-ecological health in the future. This is where Warren's proposed SES model (2005; chapter 1) is very useful, in that it insists on the need to recognize constraints at all levels, and particularly the institutional constraints that can render local-scale innovations (as spoken of in *panarchy* [Gunderson and Holling 2002]) impotent. With that in mind, we scrutinize some models of collaborative governance that now actually operate at the regional or subregional level, including examples within the realms of watershed management, integrated coastal management, regional economic development, and community-university research and education partnerships. We then turn to the crucial issue of the stewardship of local social-ecological systems by local communities for a further assessment of that concept as a foundation for success in collaborative governance arrangements aimed at the long-term resilience of coastal social-ecological systems. Here too, as always, we are enlightened by our empirical findings, and by questions of social-ecological cross-scale and cross-institution interactions in matters of governance.

COLLABORATIVE GOVERNANCE STRUCTURES AND PROCESSES IN COASTAL ECONOMIC DEVELOPMENT

Policy development and implementation play a significant role in community, social, economic, and environmental change, as we have seen throughout this book, but particularly in chapters 7, 10, and 12. Driven by government, often in partnership with business and/or labour, conventional approaches as identified by us and others have been significant contributors to coastal distress (see, for example, Markey, Pierce, and Vodden 2000). In the course of this book we have identified ways in which such damage could be at least partially redressed, pointing to areas in which bottom-up approaches to governance could be created and connected to and balanced with existing top-down structures. It is important to understand that all parts of the national meta-system must be involved, because the contributions of all of them are necessary (see chapter 3) if our coastal communities and ecosystems are to maintain and improve their often surprising resilience, which they have crafted over time but which is now seriously threatened. The creation of new adaptive strategies and the enhancement of older ones require the support of all levels of the meta-system if Canada is to strengthen coastal social-ecological systems in a globalizing world. In this respect, the creation of marine protected areas (MPAS) is a healthy sign that government is grappling with ways of achieving multiscale governance, and this is very good news indeed, but a few small MPAs cannot substitute for broad-scale, effective strategies for managing the bulk of marine or terrestrial ecosystems.

As we said at the beginning of this book, the view from the coasts is relevant to the rest of the country and, indeed, the world. Attempting to change our approach so that Canada will abjure current (not only coastal) patterns of ecological degradation, social inequity, discord, and community decline will require new ways of governing – new objectives, new institutions, and new relationships. We could become, as a nation, a leader in balancing the well-being of the urban and rural worlds, with their different roles, contributions, and strengths, rather than just one more nation involved in the riddle of worsening ecological degradation and increasing urban poverty. In seeking to find the solutions to coastal community distress, then, we seek also to solve this wider problem. That is why this last part of our work explores what new forms of governance might look like for coastal communities, for the nation, and for the world.

We think of collaborative governance as very similar to the kind of complex adaptive system that we examined in our discussion (in chapter 1) of the work of Gunderson, Holling, Berkes, and others (Gunderson and Holling 2002; Berkes and Folke 1998; Berkes, Colding, and Folke 2003), particularly as it has been extended and (to some degree) recast by Warren (2005) to take cross-scale and multilevel influences into account. In all the recent literature that searches for ways to promote social-ecological health, the necessity of involving multiple and diverse actors within a complex planning and decision-making structure made up of formal and informal relationships is emphasized. Such a system can integrate issues, objectives, and values (seek balance and sustainability) because it is multiscale; and because it is also cross-scale, it recognizes the importance of the "local," and submits itself to an ongoing process of dialogue, struggle, learning, and adaptation over time. That is to say, adaptivity is built into the system (Vodden 2004a).

Throughout this book we have compiled evidence of the inadequate responses over time by higher levels of government to coastal social-ecological problems in which, indeed, they have sometimes been complicit. This, we have shown, is because these levels of government have not been sufficiently in touch with the realities, needs, concerns, and aspirations of local communities, nor have they been sufficiently aware of the realities of local social-ecosystem dynamics. Downloading of responsibilities and costs (such as the responsibility for wharves, which has been handed over to communities) is not multiscale governance. Similarly, offloading the management of resources onto private businesses is also not good governance but dereliction of duty. We need governments to protect and enhance Canadian resources and the health of people and their communities. The resources are the property of the people, and government manages them in an implicit contract with all its citizens, not just its industrial or commercial partners (Ommer 2004).

One of our concerns has been the inadequacy of the linkages between collaborative governance and regional economic development.[3] Post–1960s regional economic development experiments in Canada were often rigid top-down systems that inhibited innovation and adaptation (and thereby often maintained poverty). That said, they did seek to counter the ill effects of mono-staple development (the staple trap) and to buffer the regions against the problems of both the long-term shifts and short-term crisis dynamics of restructuring that have challenged the resilience of individuals and households (see chapter 9). Consequently, decline has continued in many coastal communities.

Efforts at regional economic development evolved differently on both coasts due both to variation in the models applied by the government of Canada in each area and to differences in the regions themselves. By the mid–1980s, federal policy-makers had acknowledged that the solutions to local development problems could often be found at the local level, and they responded by establishing regional development agencies such as the Atlantic Canada Opportunities Agency (ACOA) in the east and western diversification (WD) in the west. Investing $1 billion new money in each over five years they turned to a "localized regional approach" (Savoie 2000). While decisions were still made in regional and provincial offices and federal priorities of entrepreneurship and diversification were emphasized, these agencies turned to locally operated Community Futures Development Corporations (set up in 1986 under the Canadian Jobs Strategy) for assistance.

In western Canada, including British Columbia, the Community Futures model has been maintained. These organizations have a dual mandate of small-business lending and community/regional (multi-community) economic development planning. They are governed by a community-based board of directors and rely primarily on federal funding to deliver their services, thus strengthening this federal funding linkage.[4] In the east, ACOA has taken a more flexible approach and developed different models in different areas. Local and regional development here has also involved greater degrees of federal-provincial cooperation. In Newfoundland and Labrador, the creation in the mid-1990s of twenty Regional Economic Development Boards (community-based institutions with federal and provincial support) provided a new model for the province. The boards were overlaid on a system of more than fifty smaller-scale Rural Development Associations (RDAS), along with other groups (House 1999). Economic Development Boards are responsible for planning and facilitating community and regional economic development, while small-business lending is handled by Community Business Development Centres (also federally funded, but locally governed), among other agencies. Such was the Great Northern Peninsula Development Corporation, which was created by the six RDAS on the Great Northern Peninsula as an experiment in locally controlled development and which to some degree still exists (Felt and Sinclair 1995). Many RDAS still exist and are now one of a number of sets of actors within the new complex Newfoundland and Labrador regional economic development system. Reviews of the effectiveness of the current model in Newfoundland and Labrador have recently been undertaken by federal and provincial governments and by the boards themselves. These evaluations provide an opportunity to learn from experience. Without such adaptation and renewal the system is likely to decline. The boards recognize that monitoring and evaluation is best undertaken early on, on an ongoing, system-wide, and collaborative basis. This kind of monitoring and evaluation, coupled with ongoing and multiscale planning, capacity building, and communication, is a critical governance tool.

Unlike British Columbia and Newfoundland and Labrador, Regional Development Authorities are key players across Nova Scotia and are funded equally by federal, provincial, *and* municipal partners. As elsewhere, actors struggle with their respective roles and relationships, and communication and coordination among them remain challenging.[5] RDA relationships with local development associations, development corporations and the Cape Breton-wide, federally sponsored Enterprise Cape Breton Corporation (ECBC), for example, remain fluid and ill-defined. The Nova Scotia Sustainable Communities Initiative (SCI – formed in 1999) focused initially on the Annapolis River and Bras d'Or Lakes watersheds. In Cape Breton, the SCI has now shifted its focus to the island as a whole and moved from a more holistic approach to one focused on community social and economic development issues. At least in its original form, the SCI was a potential model for holistic planning and development involving municipal, First Nations, provincial, and federal actors, as well as each of the development agencies referred to above. However, there is a distinct, although narrowing, gap between First Nations economic development efforts and other initiatives in their territories, because Aboriginal-controlled economic development agencies are often not well linked into other regional efforts. That said, as First Nations' success in gaining legal recognition of their rights and title in treaty settlements has grown, provincial and federal governments have begun to recognize both First Nations' commitment to land and communities and the lessons to be learned from that commitment (as opposed to market-based short-term focused strategies of mobility and exploitation).

In these matters, it is also important not to underestimate the power and capacity of inherited regime arrangements or their ability to resist or incorporate pressures for transformation, even in periods of crisis, when new popular models appear to have much public support. Nor should we underestimate the role of leadership and other political changes in new experiments in governance (Sinclair and Gray 2005). The Newfoundland model, for example, has significant problems, including the need for renewed federal-provincial cooperation, improved private-sector and community engagement, strengthened reporting and accountability mechanisms, and a revisiting of the relationship between planning and financing for local development. Proponents of the model call for greater government commitment to one regional agency (the zonal boards), but the north island British Columbia case and the Community Futures model both demonstrate that caution is necessary here, particularly in combining the functions of lending and community economic-development planning and facilitation. Measures are needed to ensure regional agencies are supportive of and accountable to local organizations and communities. Bearing in mind that significant community energy has been devoted to building each of these regional institutions, we think that in a time of increasing strain on volunteers, the option of improving existing institutions rather than creating new ones makes sense (Vodden 2004a).

There is unfortunately some evidence to support the suspicion that some forms of regionalization are simply government attempts to appease or (worse still) utilize community energy and commitment to meet central government objectives, and that they are not the decentralization initiative that they are often portrayed as being. There is ongoing tension as local agencies struggle for control over policy-setting agendas and objectives, and senior orders of government in several cases appear to be attempting to move away from a more localized approach by trying to increase the size of subprovincial regions to such a scale that the sense of local identity is lost. Provincially and federally driven coastal land use planning efforts have been expanded, and we suspect that this sense of "boundary creep" can be avoided only by consciously planning a nesting of new layers within, rather than in place of, existing (and often more local) ones.

Of course, some adjustments may be the result of a lack of capacity and resources in rural areas, since small rural regions have fewer resources to draw upon to define problems, produce new mental maps, and then act upon them, and this has worked against these kinds of experiments. On the other hand, whether in the area of health or economic development policy, the creation of bigger regions may end up further weakening local community control, while making it easier to impose universal solutions that do not take into account the needs of small, marginal, coastal communities. The ongoing efforts to strike an appropriate balance between policy unity and community diversity result in constant change. Complicating matters further, attempts to promote common approaches across provinces (for example, through organizations such as the Atlantic Premiers' Conference) have created pressure to reduce diversity, not reinforce it. As a result of these contradictory political policy trends, institutionalizing a new long-term regional vision has encountered a number of political obstacles within provinces.

On the East Coast, renewing governance based on a community-based model has been a constant struggle, in which rhetoric and reality have been very different. In Newfoundland, for example, six Strategic Social Plan (SSP) regions were introduced in 1998 (overlaid on the territories of the twenty zonal boards and formerly more than fifty Rural Development Associations) to address social and economic development. What was described as a crowded institutional landscape when the twenty economic zones were established is now even more congested.[6] Formerly existing institutions have not been well integrated into new structures, nor have they disappeared, as some may have hoped. In the end, the political struggle to effect structural change and renew governance based on a more community-based model has proved incapable of generating major change. Clyde Wells was a reformer who took advantage of the cod crisis to press for new governance and political practices. But with new leaders and changing economic conditions connected with offshore development, bringing about major change has proved difficult.

These factors have had a major impact on the attempt to create new community-based governance approaches and are one reason why old sources of networking and power sharing have survived intact. Premier Tobin dismantled the Economic Recovery Commission, the Enterprise Network, and other institutions (Tomblin 2002). The old community structures and processes had failed to build the kind of political support or the capacity required to effect major political policy change, and neither Tobin nor Danny Williams, the current premier, had political incentives to adopt new measures. Premier Grimes appeared more open to trying new approaches to civic engagement, but his experiment in reinforcing the capacity and autonomy of volunteers at the community level (the SSP) was abandoned when Williams defeated him. Even before this occurred the SSP process, at least in some regions, had come to be dominated by government bureaucrats (another source of power capable of either resisting or facilitating change) and organizational staff rather than volunteers. With better economic prospects and a solid majority, Williams redrew the health regions without public input and in general made it clear that he had a different vision for the province. He also introduced new Rural Secretariat regions, a program to be directed from the Executive Council in St John's. He did so without working with existing community-based structures such as Regional Economic Development Boards, which were still struggling with how to integrate their efforts with those of Tobin's SSP Councils, to bring about this new approach.

In short, as hierarchy theory, complex adaptive systems, and cross-scale innovation thinking all predict, governments can (and do on occasion) act as a brake on the overall state system, being very slow in devolving real power and decision-making authority to regional agencies. It is theoretically and politically significant that, while international and federal forces of change have clearly been influential (through program and policy support enabling initiatives to get off the ground, for example), local actors have often been the primary initiators of change in local development initiatives of many kinds, particularly in the case of watershed management initiatives driven by concern over fisheries and ecosystem decline. Indeed, variations in local context, and the actors operating within these contexts, account to a large extent for differences in how collaborative governance models have developed across the country, although provincial policy variation and a regionalized federal economic development approach have also contributed.

Local, and particularly multi-community regional organizations, are charged with maintaining a very difficult balance of internal and external relations. Multi-community collaboration requires significant effort and resources that are often not available except on a project-by-project basis, but such collaboration is a possible and worthwhile endeavour that could be more effectively supported, benefiting economic and social development, as well as resource management, protection and restoration, and land use planning. With some degree of shared jurisdiction in both fields, provincial-federal collaboration is also a significant challenge. Federal-provincial and federal-provincial-community agreements, memoranda of understanding, and regular contact between the different levels of the system, have been important tools in this regard. The Nova Scotia SCI, which includes local- and senior-level action committees, provides an innovative example of how to link government and community agencies vertically and horizontally and address a range of sustainability and resilience issues. More generally, case studies in collaborative governance demonstrate that the active participation of staff members from senior agencies in initiatives at the subprovincial level allows them to get out of their particular box (or silo) and thus promote cooperation across jurisdictions.

COLLABORATIVE GOVERNANCE IN WATERSHED MANAGEMENT, COASTAL PLANNING, AND THE MARINE/TERRESTRIAL INTERFACE

In Canada, "coastal zone management" (czm) was initiated as early as 1978 and was built on czm efforts ranging from those that were local and community-based to pre-existing provincial, federal, and even international programs.[7] czm aimed to coordinate federal policies and activities and to promote federal and provincial participation in the planning of shore zone areas (Hildebrand 1989; Huggett 2003), but lack of coordination between the multitude of government levels and agencies with jurisdiction in the coastal zone was (and remains) challenging. To be effective, czm must have appropriate legal/legislative authority, be able to affect the activities of all agencies and levels of government that have decision-making authority relative to the coastal zone, be seen as a legitimate and appropriate part of the process, and be capable of making "informed" decisions, which means that it must have access to appropriate scientific and technical expertise and data (Cicin-Sain and Knecht 1998).

Coastal zone management typically involves integrating local research with the zoning and regulation of coastal development in such a way as to reconcile conflict among multiple uses and protect vulnerable areas, using defence mechanisms such as seawalls or safety zones around oilrigs. The czm vision seeks long-term planning for present and future uses of coastal and marine areas, the promotion of economic development through the appropriate use of coastal and marine areas; the creation of stewardship of resources, the protection of the ecological base of coastal and marine areas, and the resolution of conflicts, so as to harmonize and balance existing and potential uses and protect public safety, and thus to generate wise management of government-held areas and resources (Cicin-Sain and Knecht 1998). Missing from this vision is a strong commitment to recovery, wherever possible, of degraded ecosystems and to promoting the health of people and communities.

Governance models that have been created and rendered operational at the level of the subprovincial region represent the crucially important middle ground between local communities and institutions at higher levels in the decision making hierarchy (such as provincial and national governments). It is at this level that cross-scale interactions may most easily and productively be facilitated, as Warren (2005) indicates. What successes and/or problems have these models had for resilient development in coastal communities?[8] What are their key characteristics? Which appear to be most effective, and under what circumstances? What are the appropriate roles to be played by various actors in the decision-making hierarchy? What are the barriers to a shift toward collaborative governance, and how might these barriers be overcome?

We analysed the resource management, stewardship, and economic-development functions of six collaborative governance models in coastal Canada. In northern Vancouver Island (British Columbia), Indian Bay/Kittiwake Coast (Newfoundland and Labrador), and Bras d'Or Lakes (Cape Breton, Nova Scotia), we looked at two governance structures in each area, one focused on community and regional economic development, the other on resource management and land use (specifically watershed management),[9] and we supplemented our work with an extensive review of additional models from the case study literature. For each of the six cases, we considered governing principles, structures (networks, organizations, institutions) and other operational mechanisms (processes, tools), actors and their interests, and values and relationships that shape governance, and we looked at cross-scale linkages, sustainability outcomes, and barriers to change.

All three areas chose community watershed management as their preferred model for community resource management and development, but they exhibited a diversity of approaches, actors, and actor relationships in the details of the application of the model. In two of three cases, First Nations played a leadership role, one that arose naturally from their long culturally embedded experience with ecosystem practices. In the third case, settler communities led the way, with support from a neighbouring Mi'kmaq community.[10] In all three cases, the organizational structures and the levels of community, university, government, and private-sector involvement varied, as did financial and human-resource investment. Importantly, each initiative was motivated by resource decline and associated cultural, economic, and/or human health implications, so the links between ecological and socio-cultural health were well understood. Private sector involvement was driven by legal requirements associated with Aboriginal rights and title and by the desire, particularly among forest companies, to pursue ecological certification in response to environmentally oriented public, market, and policy pressures. All the case studies experience ongoing funding challenges and spend significant amounts of time (human resources) seeking financing for future projects and priorities.

Significantly, in all cases the range of issues addressed by the communities (and therefore the level of integration) has increased over time, while ongoing tensions in community-government relations and in relationships between local institutions and the public they serve or represent demonstrate the importance of acquiring the requisite skill and resources with which to overcome such difficulties. As Warren (2005) indicates, building "social capital" and political awareness within an extensive and complex network of associated actors is a major challenge requiring dedicated strategies and resources. Our social-ecological analysis (Vodden 2004b) tells us that according to sustainability indicators, the overall condition of these watersheds and the human communities within/surrounding them remains similar to, or worse than, it was before the new systems were put in place, with the possible exception of Indian Bay. However, it is also clear that without these organizations in place, the situation would have been significantly worse, since watershed management efforts have helped stem ecosystem decline and restored past damage.

Through experience gained over fifteen years of operation, these organizations have learned the importance of local, traditional, and academic/conventional science linkages of the kind we have demonstrated throughout this book. They understand that different forms of knowledge contribute to one another, help offset each other's weaknesses, and address significant gaps in understanding. Their linkages flourish in collaborative research and extension programs, community dialogue workshops, cross-cultural and scientific training, scholarships, exchanges, community-based inventory and mapping initiatives and new kinds of community-university institutions, especially when these operate both through formal, well-managed organizational structures *and* unstructured informal relationships, fostered over time through communication, dialogue, and trust building. Formal mechanisms are required for accountability, role clarity, and securing financial support, while informal relations provide the "people element" that ultimately makes collaborative models work, although over time programs and organizations become not only more integrated but also increasingly formal and structured.

The scale of the watershed has been shown to be appropriate for community-based ecosystem management, particularly for freshwater and also for terrestrial and land-use planning issues. Watersheds in which people live are the source of their drinking water, their recreation, and often their subsistence and culture, and this is an important consideration in building stewardship and generating and sharing knowledge, since people are more apt to care for something they understand and to which they feel directly connected. Aside from representing a biophysical unit with relatively clear ecological boundaries, then, the scale is one to which communities can relate, especially as environmental awareness increases. By the same token, the watershed scale is of sufficiently significant size to warrant the attention of senior governments and, since watersheds are themselves nested systems, this makes cross-scale linkages in management (Vodden 2004c) relatively simple. It does, however, mean that more than one community is involved, and it requires a broader scale of "community" involvement than single places. That said, issues of tourism and forestry development, along with conflicts between formal and informal economic activities within an area, are now being addressed under a watershed management framework in British Columbia, for example, where the Clayoquot Sound Land Use decision is now being implemented under a series of watershed-level plans.

However, the success of such initiatives depends, among other factors, on an ability to create linkages to broader environmental and political systems (Vodden 2004b). Building awareness of, and the capacity to address and respond to, broader social issues of institutional rigidity, cross-boundary issues (such as air or water pollution), and political constraints are all significant challenges at this scale of governance. They have been met to varying degrees by the close collaboration of watershed-level institutions working with senior levels of government to identify windows of opportunity and potential for flexibility, since cross-scale interactions that facilitate flexibility through feedback are essential to success. Encouragingly, given some of our findings, increasing attention is being paid to the interconnectedness of terrestrial, freshwater, estuarial, and ocean environments, a reality of which coastal communities have long been aware. Initiatives to study, plan, and manage these interactions are both bottom-up (initiated by community-led institutional structures) and top-down (driven by provincial and federal policies and institutions) and frequently involve partnerships among communities, researchers, governments, and others (Vodden 2004c).

Research is an important part of the work in all these watersheds. Federal policy is now influencing the process: the Bras d'Or Lakes watershed has been designated a Coastal Management Area (CMA), and workshops have been held to design an appropriate integrated management (IM) governance structure. A state-of-the-ecosystem report is in preparation as a starting point for future efforts. Recent national policies such as the Oceans Act (and associated integrated management policy directions), along with the National Programme of Action for Protection of the Marine Environment from Land-Based Activities (NPA),[11] recognize the significant impacts of land-based development and activities on coastal and oceans waters and associated implications for human health. The DFO (2002, 15) acknowledges that IM "will require the integration of land use practices within catchment areas of rivers and streams that feed estuaries and coastal waters." Watershed management practices and organizations must therefore become an integral component of federal and provincial programs for effective integrated coastal management in Canada, just as watershed management organizations and programs continue to expand their efforts to incorporate estuarine and marine linkages to freshwater systems. Finally, reviews of CZM programs across the globe suggest that coastal planning can and should occur at various scales, with linkages established between each.

This hierarchical "nested systems" approach, we conclude, is showing positive signs of being able to take management of social-ecological systems forward into a deeper understanding of what wider functions (creative experiment, for example) would be most appropriate for each level. This is a broader geographical scale than many think of community-based management as operating at, but it will empower local communities as they work together and will mean that geography is a constant reminder of the need to work together across scales.

INSTITUTIONAL INNOVATION IN COMMUNITY-UNIVERSITY PARTNERSHIPS

In attempting to balance the three pillars of governance (government, business, and civil society) and thus create a functional (as opposed to dysfunctional) complex adaptive system, the creation of community-based institutions that empower marginalized groups like coastal communities is a vital task. We need new partnerships, new forms of collaboration that will redistribute power between the state and civil society. This includes partnerships between academic researchers and coastal communities that draw upon a combination of "expert" scientific knowledge and local knowledge or observations to address social and ecological problems. "Across the higher echelons of U.S. government," says Sclove (1998, 1283), "the long-standing norm is to invite scientific leaders, but no one else who will be affected or who might have an illuminating alternative perspective." We think that collaborative research is a hugely useful tool for better science and for more effective policy decisions on social and environmental issues, not least because it involves collaborations that will broaden the traditional narrow linkages in academic research that have often been restricted to elite "sets of interests and power relations" (Hall 1999, 35) and place it at the service of society at large. Moreover, if institutional research arrangements reproduce existing power relationships, they ultimately run the risk of eroding the social capital upon which collaborative research depends. How, then, do we facilitate community-university research collaboration at the *institutional* level, rather than merely at the individual project level?

Merely working with local communities to confront social and ecological problems through research is not enough. We think it imperative that we work both with local knowledge (once local people have themselves agreed about knowledge claims) and with formal science, putting that joint knowledge to work in the research and policy realm, particularly since, as we have seen throughout this volume, collaboration with communities leads to more appropriate and effective ways to address the complex, value-laden, contextual problems that have to be faced in local social-ecological systems. This means developing "new skills, including diplomacy and negotiation and a willingness to engage the 'other' in a respectful manner over long periods of time," but "conventional scientific institutions and training have not provided scientists with the training or support necessary for collaboration with communities" (Song and M'Gonigle 2000, 986–7). Building research relationships and mutual understandings with communities are prerequisites to working collaboratively and essential to ethical research practices. However, the need for adequate time and funds to build relationships, develop informed consent, and establish appropriate research goals and processes is often overlooked by universities and funding councils. The consequences, then, are significant tensions for researchers, as a result of entrenched norms in academic reward structures, research timelines, institutional research and ethics policies, and publication practices, as well as, in some cases, entrenched community suspicion of academic researchers, not to mention confusion or distress if the academic funding councils reject research proposals. An appropriate process to facilitate these collaborations is largely lacking at universities and funding councils, making the challenge of moving from the ideals of collaborative research practices into practice itself extremely difficult to overcome.

Institutional innovations to facilitate research collaborations between communities and universities, such as the European "science shops" and the US "community research network" models have existed elsewhere for decades (SCIPAS 2001; Snyder et al. 1997).[12] Within Canada, Dunnett (2004) has identified various contexts and creative arrangements within which community-university collaborations are occurring, but most are time-limited and project-specific, and they rely on individual champions who often operate on the margins of their institution. Processes and institutional structures to facilitate collaborative research, especially over the long term, are largely lacking,[13] although incentives to create these kinds of institutional arrangements in Canada are on the rise, with federal funding initiatives such as the Community-University Research Alliance (CURA) program of SSHRC[14] and the Community Alliance for Health Research (CAHR) program of the Canadian Institutes of Health Research (CIHR).

Working through the Clayoquot Alliance for Research, Education, and Training[15] (a CURA-funded institutional alliance between the University of Victoria and the Clayoquot Biosphere Trust), we examined the underlying philosophies, power dynamics and decision-making structures embedded in community-university research partnerships, and the relevance and usefulness of the science-shops concept in addressing socio-cultural and natural resource issues that may lead to more sustainable, ecosystem-based management systems for coastal communities in the northern Barkley and Clayoquot Sound region of western Vancouver Island. A priority issue, identified by community members and First Nations representatives involved in the Clayoquot Alliance for Research, Education, and Training, was the lack of agreed "protocols" governing equitable and ethical research conduct for projects led by university researchers. The region already has a long history of "being studied" by researchers from various sectors (e.g., government, industry, academia, non-profit organizations), and while there is widespread community interest in promoting research (social, economic, and environmental), there are also concerns about some kinds of research practices related to shortcomings in obtaining appropriate consent or permission, informing and involving local people, sharing the benefits of research by returning results in appropriate forms, and adequately protecting community knowledge or property from misrepresentation or misappropriation.

An iterative process based on a series of open public meetings, workshops, a discussion listserv, and the formation of a working group over a two-year period was used to understand and address these problems and co-develop an agreed set of principles and guidelines for research in the region. Over fifty individuals were involved, and more than twenty local organizations (including First Nations, non-government organizations, government bodies, industry, and universities) were represented in the process of developing the first *Standard of Conduct for Research in Northern Barkley and Clayoquot Sound Communities* (CLARET 2003). Recognizing that projects would encompass a spectrum of community involvement (from low to high), depending on the objectives, subject matter, nature of the study, and numerous logistical factors, the *Standard of Conduct* is meant to encourage adherence to agreed principles, expected processes for permission (particularly for research involving First Nations), and practical ways of involving community members.

LeRoy (1999, 3) points out that "a complex array of new institutions and practices has emerged in the Clayoquot Sound region which are designed to reconcile governmental or corporate management decisions with the needs of First Nations, local communities, and the ecosystem itself" (see also Dobell and Bunton 2001; Dobell 2002). Aside from the obvious practical use of the *Standard of Conduct*, there is the beginning of an institutional innovation – a shop-like science entity – that has contributed to network building and more effective governance through building social capital and participation in local policy formation by those who will be most directly affected by research in the region. Dobell (2002, 8) notes that "the cultural barriers to effective cross-boundary linkage apply not only to the organization and structures for operational management ... but also to a distinct, prior activity of building understanding, shared beliefs about facts and visions or norms, and indeed to an extended investment in building understanding and agreement around process and procedure."

Canada has considerable political experience with building linkages among experts, key interests, and the general public. Royal commissions, for example, have been relied upon as agents of change in Canada in situations where political "parties have not been effective catalysts or carriers of policy innovation" (Bradford 1998). Yet, as with other mechanisms and processes relied upon for knowledge construction and agenda-setting, there are always problems with ensuring that "marginal" interests have an opportunity to first define their problems and then come up with practical solutions based on realistic assessments of community needs and priorities. In the case of the Romanow Commission on the Future of Health, for example, there was little focus on eco-health but much discussion about the problems of rural and remote communities. "People in rural and remote communities have poorer health status than Canadians who live in larger centres. Access to health care also is a problem, not only because of distances, but because these communities struggle to attract and keep nurses, doctors and other health care providers" (Canada 2002a, 159). Unfortunately, despite much discussion about these problems and issues, there has been little action with respect to accomplishing the goals of either improving the health of individuals in coastal communities or dealing with the problems of maintenance and recruitment of health care providers.

Improving the flow of science into policy and knowledge into action will require better understanding of natural systems and better synthesis of local and traditional knowledge with conventional science. It will also depend on a re-allocation of decision-making power to represent different knowledge systems, which then can inform understanding and new directions for collaborative research. The latter has important implications for individual and community interests, for concerns and rights around the well-being of adjacent resources and ecosystems, and for citizen participation in processes of governance.

STEWARDSHIP: A SOCIAL-ECOLOGICAL INNOVATION

If citizens are to participate in governance, then it needs to be remembered that a key characteristic of co-management is a focus on geographic locale. Place matters. Central to place-based policy processes have to be the communities of people who live in, work in, and know intimately the areas that will be affected by management decisions. "Co-management" – a term that has been used for decades to describe management systems that involve sharing decision-making authority among government and non-government interests – really means having a say in what happens to the place in which one is embedded. In the area of central concern to this book, natural resources management, the degree of control held by local communities, levels of formality in working agreements, and mechanisms for funding and knowledge sharing are all part of the governance issues we have been discussing in this chapter. These issues, we suggest, are necessary but not sufficient components of good place-based co-management. For real place-based co-management, the totality of social-ecological well-being must be considered, and that means adding stewardship to the co-management governance mandate.

Stewardship is not a new concept, of course, having entered modern popular consciousness through its application in numerous public policy strategies as a key component in the management, protection, and/or enhancement of the environment and natural resources. But tying it to co-management sharpens its application with the kind of specificity that is lacking in its use in current descriptions of activities as diverse as those initiated through the government of Canada Habitat Stewardship Program for Species at Risk and the Stewardship Action Plan for Canada or those of local activist or "Nimby" groups (Lerner 1993). The key to stewardship, and the way in which it differs fundamentally from legislated measures or land acquisitions, is that it incorporates a notion of *people* caring for or being responsible for a local natural area or resource and, moreover, *local* people, who have the most to gain or lose. We see, that is, concern for the local operating in stewardship and, indeed, we know no better definition than that of

Roach (2000, 71), who describes it as *hands-on daily care*. The implication here is that the stewards' decisions and subsequent actions are undertaken voluntarily (Roach et al. 2006) as a result of a very specific place-based form of enlightened self-interest. Stewardship, in the forms of both the local activist group and of programs initiated from the government level, share the common philosophy that people are motivated to act in the care of their local environment by "ethical knowledge" (Parker 1995) of what aspects of the local environment are of value and why. This is possible only if that knowledge is rooted in place-based local social-ecological knowledge.

We see true stewardship, then, as grounded in people's valuing or prioritizing the preservation or conservation of a natural environment or natural resource (or valuing the outcomes associated with such actions) over exploitative and destructive activities that are not rooted in place but in relatively short-term economic objectives. This kind of stewardship is very different, and much stronger, than top-down legislated measures that rely on frightening people into obeying regulations (which they may not understand or see as wise or equitable) out of fear of punitive sanctions. Considering stewardship as a value-based component of co-management means that shared value judgments (agreement on the *goals* and the vision that underlie good natural environment management) are decisive when formulating any decision, including decisions about the natural environment, whether made by an individual or at a policy-making level. The point here is that nature is valued both in cases where people are motivated by (shorter-term) economic values and in those where they are motivated by (longer-term) environmental preservation values – but the values are very different in the two cases, as is the underlying vision (Hollis 2004). Such value conflicts underlie many environmental battles, and it is values that usually make the practical difference to the outcomes of the ensuing decision or policy (Stenmark 2002). Stewardship in a co-management framework places the formulation of the value-statement that underlies environmental decision-making in the hands of the stewards themselves, working with (rather than under or even against) centralized policy-makers.

To bring such a value-based co-management framework to fruition is not easy. We found intriguing variations in the understanding of environmental health in our work, and these often depended on the level at which interpretation occurred. For example, in our work on wetland conservation in rural communities (which was ranked among the provincial Eastern Habitat Joint Ventures), we found that our task of delivery of a provincial goal had necessarily to be sensitive to variations in local (community) interpretation, which raised questions like the following: Who is involved? How fast will the program be adopted? How will its goal be delivered?

It was important to set this work up so that it was clear that community diversity and the cultural context would be recognized. However, if we set these local interpretations against the same goals in their broader context of, say, North America, we found that local communities' interpretations may be opposed to that wider continental one. A classic example is how the goal of the North American Waterfowl Management Plan (NAWMP) is interpreted. American hunters take it to be the maintenance of waterfowl populations through habitat conservation that is specifically geared to satisfying the demand for hunting, and because the US Fish and Wildlife Service is a major funding source for the NAWMP, it is reasonable to interpret the fundamental goal this way. However, rural communities in Newfoundland delivering the NAWMP goal with NAWMP funds often interpret *local* habitat conservation to mean the prohibition of hunting. At the local scale this interpretation makes sense, and it is, indeed, the best ecological approach, but it results in different understandings of fundamental goals at different scales.

What we need to learn from this example is that it is at the local scale that people and ecosystems interact in ways that bring to light the key properties of a social-ecological system. This is a lesson for the upper levels of partners in a co-management scheme, and appreciation of this will ensure that the kinds of social-ecological damage that was done through, for example, historic logging in British Columbia and Newfoundland and Labrador (chapter 4) will not be repeated. Under stewardship, not only does local-level input into decision making usually consider the needs, aspirations, and values of local people, it is also more likely to be designed to mitigate depleting or damaging consequences, since the decision maker often bears the costs of such consequences. That is, stewardship is by definition and purpose a community-based initiative focused on making local communities not only the stewards but also the beneficiaries of local efforts at environmental protection (Roach et al. 2006). Stewardship utilizes local awareness of the cultural value of a specific environment (ethical knowledge), which is invaluable when assessing the impact *on a community* of the loss of a locally cherished area (a hard thing to state) against the *economic value* of the development of what has been a "natural" area, which can easily be appreciated (Parker 1995). The pursuit of economic gain in the management of the Newfoundland fishery has led to a loss of intangible cultural values such as the preservation and celebration of local knowledge, informal food production strategies, and the quality of life as defined by local people. Such things, although non-quantifiable in relation to economic health, are important to local concepts of community health and are examples of the role of ethical knowledge.

The model of stewardship we used for our wetlands work harnesses local concern for community wetlands and empowers local decision making, but it also takes the step of explicitly seeking to foster and develop values that fulfil its objectives of wetlands' preservation. Education and awareness of the value of waterfowl and critical wetland habitat, most overtly conveyed by trails and interpretive media have, in participating communities and elsewhere, fostered the perception of aesthetic and recreational value, on which locals have been moved from non-environmental to pro-environmental stances. There is evidence, informing the power and potential of local stewardship, that this has made the threat of legislated sanctions or expensive land acquisitions unnecessary (Hollis 2004). Community members speak of the "implicit protection" afforded to local wetlands by the values they now convey to the community and argue that formal protection from development is not necessary, because "there are too many people here that would be against it."

This promotion of wetland values is at the base of the stewardship agreements we examined, but it also privileges preservation values at the expense of use values. Expressions of concern, and even resentment, by hunter groups and business concerns with development agendas in effect articulate the presence of underlying conflicts over use versus preservation at a local scale. The wetlands model is a new form of stewardship in these communities in the way that it introduces new values to many community members – new bases on which to value the natural environment. Trail building is not an explicit objective of the program, but because trails serve to bring people into closer contact with wetlands, they serve to increase awareness of wetland values, or particular wetland values, which in turn motivates protective attitudes and behaviours. Yet one hunter spoke volumes when, referring to these efforts, he talked of how he loved being out on the wetlands, a place where he hunts, traps, cuts wood, and picks berries. But "why," he said, "would I want to walk around one in the evening?" This opinion and the opinions of other hunters we interviewed spoke to a traditional form of stewardship, even though it was never referred to as such: a stewardship developed perhaps over generations of daily, intimate contact with the same local nature environment that has provided key sources of subsistence, if not survival, in harsh economic climates. Why, indeed, would a traditional (as opposed to a recreational) hunter need to walk around a trail in order to *learn* of the importance of wetlands?

Importantly, rather than polarizing use and preservation values, the stories here reflect a form of stewardship that attempts to find a balance, a middle ground that must be found, particularly in communities that continue both to rely heavily on subsistence "hunter gathering" and to experience constrained economic choices. This stewardship has, and probably always has had, a bottom line in terms of preservation, the privileging of the values of which conflict with legitimate concern for community survival: "The grandchildren won't be staying here. Parsons Pond will die; 'The Pond' will die unless the oil wells are started. If everyone has gone to Ontario, what's the point in preserving the marsh?"

It is also a stewardship that relies on local ethics, not merely a far-sighted, informed, anthropocentric view but one that views nature with a moral concern. However informal, it functions in a way comparable to traditional ecological knowledge that incorporates value-based "authority system(s) of rules for resource use" (Berkes 1999, 6), the rationale for the study of which includes the need to find better ethics of resource management, not merely to uncover scientifically compatible local ecological knowledge. It is, after all, impossible to formulate environmental policy unless scientific information is supplemented with certain value judgments (Stenmark 2002).

The net result is that calls for the incorporation of local values into policy formation and decision making are grounded not merely in a social-justice ideology. Part of the process that is required for the local ownership of policies and for them to gain sufficient support among community members is the need for communities to own their values (Parker 1995). The potential power of this process is clear from our work, but – since the values that motivate local community members are not uniform and since the success of the process is constrained by circumstance – its effectiveness may be constrained as well. One of the hurdles that co-management with stewardship faces is that policy or participatory co-management arrangements that retain value judgments also thereby limit choices. There is a risk that realistic potentials for finding the middle ground between use and preservation may be missed. Moreover, values may be in subtle conflict at different levels of the process – the DFO may seek to conserve the ocean ecosystem but not the communities that depend on it in the aftermath of the collapse of the groundfish stocks in the northwest Atlantic, for example. It is important that all levels of a co-management stewardship system understand *and share* the goals of the system, for if local values and the ethics of traditional stewardship are marginalized, there is a real risk that the resilience of the human communities themselves will also be marginalized, and that is not a desirable outcome for social-ecological health. By contrast, sharing social-ecological knowledge can become part of (and is enhanced by) larger processes of sharing power and responsibility in a manner that is more ecologically enlightened and socially just. Most importantly, an apparently democratically idealistic model can be seen to be workable at the community level, because such a model tends to eliminate externalities, unlike what can happen in the decision-making arena of political or corporate interests (Berkes 1999).

CONCLUSION

New approaches evolve from historical ones and take place within a context that is multiscalar in its temporal and spatial dimensions, among others. Acknowledging complexity and focusing on collective ways to understand and better manage complex adaptive systems for social-ecological health (see chapter 1), helps to develop the mindset in which connections between system parts and levels are cultivated. Collaboration is itself an adaptive response for policy-makers who have, over the last several years, had to deal with difficult new circumstances. In this chapter we have examined a range of governance models that hold promise for future ways of managing social-ecological systems, coastal and otherwise: regional economic development structures, coastal-zone management, university institutional change, and local social-ecological stewardship initiatives that can be integrated into co-management structures. Most of the initiatives, things that we think are helpful and productive, are sadly either not linked to, or else not well supported by, existing formal structures. This is deeply troubling, because rural development that is not linked to environmental recovery and stewardship and that does not support a broad co-management model operating across the nested scales from state to community is, as we have shown throughout this section, problematic. What has become clear in Coasts Under Stress is that it is not enough to address integration only in terms of the spatial and ecosystem dimensions of a coastal zone – its marine, terrestrial, freshwater and estuarine, local, regional, and national aspects. It is also vital that intersectoral, intergovernmental, and interdisciplinary aspects are considered – in other words, both the vertical and horizontal, cross-scale and multisectoral dimensions of the coastal social-ecological system (this volume; Sorensen 1997; Meltzer 1998). Considering these dimensions creates the opportunity for leadership at the local level within a national and even an international policy framework that provides coordination, funding, and administrative support.

The link between regional economic development, integrated management, and resource management has yet to be fully developed or established in these models. Greater attention to environment-economy interrelationships and a commitment to the concept of resilience are required. Many local institutions, guided by higher-level policies, have pursued the goal of economic diversification, but "new economy" sectors (such as information technology, tourism, and small business development), while critical to diversification and thus to increased resilience and long-term community survival, must be balanced with traditional sectors that form the foundation of community economies, history, and identity. Development in resource sectors necessarily involves attention to intergenerationally sustainable resource management (see the previous chapter) and land use and an incorporation of the stewardship ethic into regional economic-development policy and practice.

It is clear to us that regional economic-development institutions must fit the needs of local community economic development organizations and activities and that governments need to pay attention to linkages between scales, clarifying relationships and increasing communication and cooperation, while recognizing financial and human-resource realities. We think that CZM should evolve in this direction through a multistaged multiscale process in which time and resources are allocated for building capacity (human, technical, financial, and legal/administrative), cross-scale and cross-sector relationships, and a solid foundation for the process. We should be working towards a more balanced and collaborative approach that creates linkages within the divide of bottom-up endogenous development and top-down government-run economic-development programs for "disadvantaged" regions, while at the same time highlighting still existing "silos" that deny the connectedness of society and ecology in social-ecological systems and the difficulties of local-level change in the midst of higher-level rigidities.

In this respect, "development institutions," both formal and informal, are important vehicles for governance learning that could, with appropriate support and appropriate organizational and institutional frameworks, facilitate the undertaking of increased responsibilities at the local and regional level and lead to resilience in local social-ecological systems within the greater metasystem (panarchy) of the state. There needs to be a cross-scale and cross-sector drive to recovery, which is consciously and structurally combined with the promotion of both human health (employment, income, physical and work environments, gender, social equity – all of the health determinants) and the health of the environment, taken together. Cooperation between provincial and federal actors and other interests such as the private sector and academia is essential here, as are supportive legislative and policy frameworks, strong local institutions, community action, and commitment.[16] Also critical are the informal personal relationships that develop through interactions associated with these institutions. It is these relationships that provide the "glue" that ultimately makes these institutions work, building trust, mutual respect, and understanding over time. At the same time shorter-term concrete achievements, reflection, and learning should be achieved and celebrated along the way.

Significantly, the multiple versions of new organizational structures that exist in Canada speak to the panarchy idea that creative experimentation must take place at the lowest (grass-roots) level of a complex system for it to be at a small enough scale to be testable, and hence adaptive. That creativity needs to then be discussed across scales and adopted at the upper levels when judged appropriate and useful. It is encouraging that this appears to be feasible inside the Canadian federal and provincial systems. With their local focus, regional development institutions exercise significant local influence. They now need to link up (to scales beyond the region), down (to communities within their region), and out (to other regional efforts) in order to be effective and influence broader changes in society that ultimately affect their own efforts. This can be done by encouraging and supporting the continued development of community stewardship and collaborative governance models across the country, utilizing knowledge gained through critical analysis of existing experiences and the sharing of lessons learned. Such an approach must be pursued with awareness that "one size does not fit all," that different regions will require different solutions, and that not all communities and regions may wish or have the capacity to participate. Policies and programs to support local initiatives must be accompanied by realistic evaluations of resource requirements and availability, and conducted with the participation of all actors and the space for creative solutions to emerge.

The twenty-first-century notion of collaborative governance can be conceptualized as a broader form of co-management, extending beyond the management of ecological resources to a broad cross-section of societal affairs ranging from fisheries management and land use planning to education and health care. We therefore suggest, based on local and international experience, that Canada adopt ten key guidelines for integrated coastal management:

1 ensure ecosystem health and integrity;
2 assert community rights and leadership, identifying and working through value differences to reach a mutually satisfactory set of goals based on agreed values and underlying principles;

3 recognize Aboriginal rights and title;

4 create and commit to new ways of working together, including merging stewardship and co-management approaches that enshrine the principles of precautionary management and adjacency;

5 determine the appropriate scales for different parts of the planning and management process;

6 set realistic timelines;

7 practise adaptive management, which includes identifying and learning to work with and around real constraints at any level of the process;

8 make capacity building an integral part of the planning process;

9 integrate local and scientific knowledge, include existing information in the process, and strive to fill critical knowledge gaps; and

10 recognize and incorporate multiple values and uses.[17]

Significant opportunities exist to create and build upon new networks and institutional models, such as we have discussed in this book, in this time of continuing coastal governance reform. For this to happen, however, will require us as a nation to adopt an attitude of open and ongoing examination of innovations and to be free to implement, and if necessary to change to suit place and circumstance, any improvements that may occur. This is an adaptive co-management approach to social-ecological health that reflects a commitment to ongoing social learning. We think that the result will be not only a dramatic increase in social-ecological health and coastal community resilience but also a new leadership role for Canada in the ongoing global striving for coastal community reform.

16

Building a More Resilient Future

An inability to distinguish between the risks to which people expose themselves and the risks to which they expose others appears to be the defining disease of modern capitalism.

George Monbiot (2005)

In the Amazon jungle (what we still have left of it) the canopy is intimately connected to the jungle floor. Big Brazil nuts form in the canopy, birds crack them open, and drop bits that fall to the forest floor, the base of the system. Termites at ground level carry bits of food back up the trunks of the trees, and hence back up the system. Male butterflies flutter from tree to tree at the same level, but go to "salt licks" at river's edge (ground level) to gain the capacity to generate the pheromones that will render them attractive and continue the species. Birds nest at some levels but feed at many. Monkeys swing from branch to branch, feeding themselves, and in the process, the whole system. The system in its undisturbed state can be said to be in balance, at all levels and between them. In the days when humans were in balance with the nonhuman part of the ecosystem, this set of interlocking ecological niches worked both vertically (top-down, bottom-up interdependencies) and horizontally (trophic levels). Today it is endangered because humanity has disturbed that balance in so many different ways that the integrity of the whole system is now imperilled.

The jungle has served as a metaphor for human society before – as the social jungle, the city jungle, the jungle of human emotions, and so on, but it is the imperilled jungle that now serves us best, because it reflects the fundamental truth that we have so disturbed the planetary ecosystem that we are faced, in this twenty-first century, with a major imbalance between people and the global environment that sustains them.[1] Planet Earth is a closed system, and we are wise not to forget that. It is, moreover, a highly complex adaptive system – a panarchy of a kind, but one in which top levels of hierarchical human systems wield power in unpredictable ways and in which uncertainty in its human and natural dimensions leads to both positive and negative "surprises" that mean that we are unlikely ever to be able to predict outcomes or achieve the kind of dynamic equilibria we may seek. That said, just as the "social systems" of butterflies or monkeys or termites need to be in sync with the biological systems on which they all ultimately depend if the metasystem is to function efficiently, so must the planetary system, which includes humans.

We are the top predator in this system and also the first such predator to be capable of destroying it in whole or in part. That is why we carry a huge responsibility to be efficient in the sense of that word that goes well beyond business economics – efficiency is not just about money and profits, it is about the systemic use of resources (including human ones) in such a way as to produce the least waste possible. In combustion engines, this means the maximum use of energy, with the minimum amount of exhaust. In efficient biological systems, the organization of the whole takes into account the interactive requirements of all component parts, as the jungle balances everything from termites to jaguars to the tallest tree. To generate efficiency in a complex system that includes modern humanity in all its technological and social complexity requires extremely good systems of governance that will not favour the productivity of one part over another but will strive to work across scales and parts of the system and move them closer to harmony. The challenge is enormous – meeting it will require much more than liberating market forces on the one side and placating protesting resource extraction industries on the other. We will need more and better science, precautionary approaches, and knowledgeable stewards at every level, along with a much wiser use of communications technology and organizational connectivities and rapprochements than we have yet achieved.

The scale misalignments and asymmetries we have spoken about throughout this book show us how far we are from such an (admittedly utopian) goal. Worse, they are symptoms of serious global systemic inefficiency and dysfunction – the long-term accretion of outcomes that may benefit a part but may also destabilize or destroy other parts of the global social-ecological system. They speak to the use and the misuse of power, whether knowingly or not, that result from inadequate governance structures at many levels: international, national, regional, and local. In Canada, as elsewhere, the response to global restructuring (which is only the most recent set of ongoing global power struggles and associated adjustments) reflects political as well as economic will. In our case, that will is expressed through the overall federal, provincial, municipal, and other governance structures and processes within which the coasts (and the country) have evolved over the last one hundred and fifty years and which still operate.

Our work in Coasts Under Stress, with its focus on the relationship between interactive restructuring and health, has shown that policy-making and industry practices have been silo-based (constructed within the walls of one department or company, without reference to impacts in other parts of the system) and not very skillful at listening to the needs or to the knowledge of the grass-roots levels of the Canadian citizenry. Ecosystems, however, work at multiple levels, and the result of this social-ecological scale misalignment and silo-based approach has been to exacerbate lack of awareness of impending crises such as the destruction of, or serious damage to, marine and terrestrial ecosystems, brought about by the failure to appreciate and remedy the impact of particular industrial practices on other parts of the environment (log booming, for example, on inshore waters, or the loss of groundfish and other marine species). These failures in ecosystem well-being in turn have impacted on resource communities, where cross-sector misalignments have produced remarkably little political awareness of impending multiple-sector failures (fisheries plus forestry plus mining, for example). In other words, cross-scale misalignments have compounded the problem of cross-sector misalignments, with the result that there have been serious failures of policy response to local-level distress in our resource-based coastal communities on both coasts. Such lack of appropriate government response has then exacerbated an already bad situation by channelling the energies of coastal communities into continuing the extraction of natural resources even when those are at risk, rather than channelling energies into alternative uses, such as stewardship and recovery, because the lack of supportive policies for the latter have thrown people back onto the former, of necessity.

The solution, however, is not to hand management of our natural resources over to either local communities or the private sector. Turning public resources into private wealth and/or to power concentrated in the hands of a few will not ensure that the management obligation is met and, indeed, will hamper it. This "solution" is actually a dereliction of duty. We need good governance and, however difficult and challenging it may be, governments have a fundamental role to play in ecosystem management, which cannot be handed over to any other part of the system (the current buzz word would be "stakeholder"), because all other players are neither disinterested nor capable of seeing and managing the whole for the better good of all, nature included. Government, that is, has an ethical as well as a democratic obligation to govern and to protect all the components of the nation state, even the smallest coastal community, the smallest functioning part of the national social-ecological system. That, not merely economic growth, is its primary mandate, and it is an appallingly difficult one. Our work in Coasts Under Stress has forced us to ask how government can create a functional national metasystem that serves the needs of all levels of the state, because coastal communities are one (currently dysfunctional) part of the metasystem. To come to grips with how to bring the state metasystem back into balance is a significant challenge, not least because human-environment interconnections and interdependencies (what we have called social-ecological systems in this book) are poorly understood in our urbanized societies.

This should not surprise us, for much of the world regards the environment ("nature") as something to be exploited – something to be used not just to support people in their habitat but to produce marketable goods for profit. We have seen how in North America at least since colonial times we have inflicted serious damage on the complex checks and balances of our social-ecological systems, without really understanding in most cases what we were doing. Colonial developers and their masters in governments (either in the colonial European metropole or, later, in the colonial capital) operated with quite extraordinary degrees of hubris, assuming that this kind of development improved the well-being of people (including Aboriginal peoples, who were assumed to be in need of salvation, education, and civilization). The continent was replete with resources to be exploited, wilds to be conquered, barriers and challenges to be overcome. We live with the fruits (and, more often, the residue) of such a mindset today, most poignantly in our coastal communities (First Nations and otherwise) whose resource bases, and hence their sustainability, have been consistently undermined and often expropriated over the last century and more.

To try to come to grips with the problem of the impact of restructuring on the social-ecological health of the coastal communities of Canada's East and West Coasts, this volume has covered a great deal of intellectual – as well as geographical and methodological – territory, moving back and forward also across several temporal, geographical, and institutional scales of analysis. We have done this because we have sought to come as close as possible to the lived reality of coastal communities without losing sight of the overall patterns that have shaped and continue to shape, that reality. We have also sought to compare and contrast coastal experiences, to see how much place mattered, when and how it mattered, and if, when, and how social structures and institutional policies overrode geographical differences. Drawing on more traditional disciplinary analyses where possible, but blending them together and adding new research and new interpretations, we have created a social-ecological history of both coasts, their resource bases, their societies, communities, and economies over time, and the way in which the lives of people are, and have always been, embedded in the environments to which they belong and on which they depend. In doing this, we have sought to counteract the effects of the "shifting baseline syndrome" that leads us to wrongly estimate the productive capacity of our natural world, and we have taken note of temporal differences, for the timing of development makes a difference, as does distance, place, culture, technology, and the nature of the resource base that was found on each coast.

What we have been struck by, however, is not the differences, but the similarities between places thousands of miles apart in terms of the difficulties they now face. These similarities are significant and, we think, structural. That means that it is our human systems that created them and that it is, therefore, our human systems that can remedy our own mistakes. We have found that many people on both coasts are suffering poorer individual health as a result of job losses that derive from economic restructuring in natural resource industries (chapters 8 to 10, and most others). We have seen that economic restructuring has been, in turn, driven by environmental restructuring, some of which has been caused over the long run by industrial and postindustrial strategies and related government policies that have been designed to maintain profits in a competitive globalizing world. By this we mean that overfishing (for example) has restructured ecosystems, while "efficient" timber-cutting methods have hewn down the forest community, reducing biodiversity and damaging mountainsides, forests, and coastal waters.

We deliberately use the examples of fishing and forestry here, because these are the so-called renewable resource sectors, which should by definition be the most sustainable. However, as we have seen, fishing and forest industries, faced with declines in the resource and heavy global competition, have changed their business and employment strategies, tightening both their wallets and their wage employment regulations in order to maintain competitiveness, so that the final reverberations of this economic restructuring has then been felt at the local level in social-ecological ill-health of various kinds. This situation exists on both coasts, and closer examination (such as that in the historical and current-day chapters of this book), tells us that it has been brought about not only by market conditions but by poor long-term development and resource management strategies at provincial and federal governmental levels that have left communities and their resource bases highly vulnerable to such conditions, which have then been exacerbated by overharvesting in both sectors and on both coasts.

We have further found not only that governmental environmental concerns have been a poor second to economic ones but that the environmental impact of industrial resource extraction technologies has been deleterious over the long run. Indeed, the most charitable thing that can be concluded is that the environmental impact in renewable resource industries has been poorly understood (and hence not managed with sufficient awareness and concern) for a long time, on both coasts. For example, the wood debris accumulations in British Columbia coastal waters (chapter 6) are damaging a fragile "interface" ecosystem upon which people and biological species depend, while the way in which forests have been harvested has been more mindful of corporate efficiency and competitiveness than community and ecosystem health. By the same token, the way in which the fishery has been prosecuted on the East and West Coasts has been more mindful of corporate efficiency and competitiveness than of community survival and ecosystem health (chapters 2 and 3). The result has been the collapse of the East Coast marine ecosystem, while that on the West Coast is seriously (but not irremediably) damaged at this point. Ecosystem damage, terrestrial or marine, has resulted from inadequate management strategies, which have considered quick profit over long-term and local development (chapters 2 to 5). We have "logged down" the forest ecosystem as we have "fished down" the oceans. We stress the renewable resource sectors in our conclusions because, while environmental concerns have grown in the regulation of nonrenewable resources (such as oil and gas), there seems to be a serious lack of awareness that we are rendering our renewable resource sectors nonrenewable by virtue of our own neglect of the implications of serious damage to living systems.

Added to this neglect and the hardships that are consequent upon it and having a further deleterious effect on human health on both coasts has been an insufficiently sensitive restructuring of social systems such as the EI reviewing process (chapter 9), the restructuring of health care, and the restructuring of education, where funding has been cut back in a cross-scale downloading of policy rigidities (chapters 9 to 11), brought about, once again, by the economic restructuring that is perceived to be required by the exigencies of a nation seeking its competitive place in a globalizing economy without counting the non-urban grass-roots costs of its solutions to what have been seen as fiscal necessities.

For those on both coasts who seek to, or feel forced to, go elsewhere, the adjustment costs of out-migration are potentially horrendous, and the lack of appropriate skills training has made it very difficult for those going elsewhere to get the training they need to compete in an urban job market (chapter 9). For those who wish to stay and develop local resources, we have uncovered many instances where such local development, along with productive efforts at community-based enterprises, has over time been created by local people, only to be ignored or even rejected by provincial governments – the case of British Columbia forest development is particularly striking (chapter 4), as is the case of Newfoundland and Labrador inshore versus offshore fisheries – a federal jurisdiction (chapters 2 and 3). In the neoliberal government and business agendas that have dominated Canada since the 1970s, there has been a great deal of rhetoric about self-support and self-esteem and almost no recognition that coastal communities and coastal ecosystems require a modicum of care if they (and we) are to achieve sustainable recovery. We have found little active institutional support for the start-up requirements for community economic development, very few policies that would operate proactively to assist communities as they try to respond to restructuring, and far too little appreciation of the knowledge that such places embody and could, and would, usefully put at the service of the nation. Indeed, overall, it seems as though the realities of life on both coasts are not penetrating, as they might, into the policy agendas at the provincial or federal levels. There seems to be a sense that small rural coastal communities are, or should be, a thing of the past, communities that at best should be grudgingly supported through their death throes, but without any consideration that remedial care might be of benefit to the regions and the nation as a whole.

What about ecosystems? They are thought about in separate boxes, unrelated to people. The fish are now protected: the Oceans Act and associated legislation protect the fish but not the fishing communities. Governments protect the forest industry but not the logging communities. Both types of communities must be transformed from resource-extraction, single-industry towns to fully fledged, diversified communities tasked with stewardship and the recovery of our natural environments, as well as with production of products and services. It is startling to consider that, taken overall, there has been more social and environmental concern shown in the nonrenewable resource industries of this country (flawed though that concern is) than in its renewable sectors (chapters 5 and 13).

Our findings show that, regardless of which coast, regardless of major environmental differences, coastal communities suffer remarkably similar ills. These, we repeat, cannot be merely the product of geography, location, the particular provincial (or subprovincial) culture, or the particular sector ailments – they are the result of our own systems, institutional, economic, and social. Human and environmental health is poorer on both coasts. Interestingly, just where we have to report very bad news, we can also report very good news – stress is lower than one would be led to expect in these places, because they are populated by people whose social cohesion in the face of significant difficulties is remarkable. We have here people who are adaptive and resilient, even in hard times, and who are capable of ongoing resilience if given half a chance. But in many cases, successful local initiatives remain local and are jeopardized over the longer term because all that is available is pilot-project funding without any support or program for turning successful pilot projects into models for regional or industry-wide change. If Tofino (for example) turns to tourism and has its local health services and other supporting infrastructure removed, how would that affect the tourist dollar? If Labrador develops a series of locally run entrepreneurial experiments whose funding is not sustained long enough for "take-off" and that must replace existing industries rather than diversifying around them, how can we expect these places to become self-supporting again? These are classic examples of misalignments of policy and also of scale asymmetry – where policy is designed at the level of the province as if it were a level playing field, neglecting the way things can play out at smaller scales.

As a nation, we depend on these people as stewards of the coastal environment and its resources and as bearers of history and culture; they in turn depend on this environment and the existence of the places there. Fish does not come from the fish counter in the local store; building timber does not come from the local timber yard; energy comes down a pipe or along electrical wiring – but it comes down and along from somewhere, and often that somewhere is coastal. Preserving the cultures, ways of life, and resilience of Canadian coastal communities preserves our access to, and stewardship of, our natural resources; it protects our shores and provides concrete examples of our sovereignty over those shores, resources, and waters. In an era of Arctic warming, vessels of all nations will be able to sail the Northwest Passage, ply their way through Canadian waters, and challenge Canadian sovereignty over shores and resources that will appear abandoned and inoperative if our coastal communities do not survive. If nothing else, practical requirements for twenty-first-century living, coupled to greed and national pride, should alert us to the need for such communities, even in an urban age, before it is too late. And it is close to becoming too late. The institutional confusion of efficiency with profit and output with the short-term "bang for the buck," has been clogging the arteries of coastal life, impeding natural flows of life-sustaining requirements. The health of our coastal ecosystems and our coastal people – our coastal social-ecological health – is imperilled. And this is not merely a "regional issue," although that should be a good enough reason to take note of what is happening. It is a national, and even an international, issue.

Why is this happening? There is a wealth of evidence that complex adaptation/flexibility used to be the hallmark both of coastal biological ecosystems and the historical and pre-contact coastal communities who are part of them. Charles Wilkinson's book on the American West, for example, spoke of a local who "showed me poor land health, the history that had made it that way, and how we can – we must – both restore the land health and preserve the communities that depend on the land" (1992, 294). As for the American West, so for rural (in this case coastal) Canada. In chapter 1 we spoke of how in "a healthy society, each level is allowed to operate at its own pace, protected from above by slower, larger levels, but invigorated from below by faster, smaller ones" (Gunderson and Holling 2002, 76). When we look at coastal Canada, we see stubborn determination to survive, and survive locally, among a significant number of people in coastal communities, including youth. We see experiments in management and co-ownership, and in economic diversification. We see, that is, a creative base on which government can build.

Sadly, however, that community voice is not listened to often enough. In a globalizing world (see chapter 9), corporations have responded to stressors (global competition) by becoming flexible, at the cost of the environment, community economies, and labour.[2] Government seems all too often to have done the same, privileging one part of the meta-system over another, misguided by the erroneous idea that this will promote national systemic efficiency. The corporate lobby, of course, sees the world this way – and their voices can resound in the corridors of power in ways in which those of coastal community leaders or spokespersons cannot, being a muted small whisper compared to the thunder of corporate demands to be allowed to become efficient and competitive in a global economy. That is a legitimate aim for business, but government needs efficient functioning at all levels of the state metasystem if the state is to be efficient. So governments need to listen or we will lose our rural communities and continue to damage our ecosystems through a continued misunderstanding of goals, misalignment of policies, and lack of comprehension of scale differentials. Moreover, the community adjustment costs involved in government and corporate restructuring are seriously inefficient (consider the cost of a house in Cartwright versus Toronto) and replete with environmental and human distress. Such costs are all too often still seen as externalities for such firms, costs for which corporations are not accountable. Now we need to decide if we value our rural communities in general, and our coastal communities and their ecosystems in particular, enough to redress the balance by restoring to them the flexibility they once had and could have again, albeit under vastly changed circumstances.

This calls, of course, for a change in perspective in the corridors of power. It calls for government to remember what its mandate actually is – to represent all citizens, not just corporate citizens, to be mindful of tomorrow's citizens as well as those of today, and to protect the territory (marine and terrestrial) that is Canada, which it holds on behalf of all Canadians, not just Canadian business. Government has more than an economic role to play in a democracy. Problems of imbalance and increasing poverty allied to resource depletion are now recognized as evils in the underdeveloped world that need to be overcome. However, we forget that such problems also haunt underdeveloped parts of the developed world. In a powerful essay, William Rees speaks of the importance in the developed world of relative income. "Among high income countries it is not the richest societies that have the best individual and population health,[3] but rather those with the smallest income differences between rich and poor" (2002, 12). What is it, he asks, that "compels people so adamantly to defend the goal of unlimited growth when the getting of it apparently sacrifices much they ... value in life, deprives others people of the right to live, and demonstrably threatens the ecological integrity of the planet?"

Staples development has left Canada with a sub- or unconscious export-dependent mentality that, over time, has distracted the nation's policy-makers (key players among whom have been businessmen in sectors based on staples, including their transportation) from reinforcing local diversification and manufacturing around the staple base, and hence research and development (R&D) to move staple growth into domestic development. For R&D since the beginning of the twentieth century, we have increasingly looked to multinational corporations (MNCs), whose branch plants in Canada, however, have failed (not surprisingly) to foster Canadian research, preferring their own in their own home bases. From that long-term Canadian myopia[4] has come much of the weakness Canada faces in today's restructuring for global competitiveness. The core lesson of the staple thesis is as true today as it ever was: the Canadian state remains weak on research and development and strong on export-led growth, trapped in a classic staples mentality (Watkins 1963; Britton and Gilmour 1978; Ommer 1991; Hayter 2000). So long as everything depends on ownership/control and so long as property and ownership regimes support external control, the social-ecological health of a country, region, or community is at the whim of a global market and individual profit-focused investors. We do not control our own destiny.

That said, missed development opportunities may be captured again (Gershenkron 1962) – they come and go, although never in the same combination. It appears as though the neoliberal agenda sees no way to challenge transnational corporations (TNCs) and can only placate them as the drivers of the national and international economy. The flexibility that used to be part of the traditional rural economy has been seriously eroded by wage labour and then by subsequent post-Fordist strategies that have sacrificed worker (and hence community) flexibilities. The question now is how can communities and their environments survive in the face of a staples approach to development that has become destructive?

Businesses have now built flexibility into their operations as a response to globalization. This could be good, but not if they achieve flexibility (as they have done) by passing rigidities down the system to the ground floor. Multi-tasking (which used to be a rural-community strength) is now required of industrial work forces and, combined with piece work, leaves people waiting for a call to work, and unable to become involved in other employment, to take just one example. The firm benefits – it has the numbers of workers it needs when it needs them, but it does not have to pay them a steady wage. The state does not then provide the support they require with flexible EI regulations, so local workers are caught in a double bind.

The challenge to the Canadian state is how to reverse rigidities and restore flexibilities down the system, while protecting flexibility in the business world. It can be done (Cohen and Winn 2007) – and it will be done more easily with restored ecosystems and with healthy, well-educated people living in well-networked, multigenerational, diversified communities spread throughout the country. State institutions also need to do this within themselves. Poverty isn't simple. It involves power structures, fiscal and financial inequities, poor health, access to health care, and so on. The supreme challenge to government today, especially but not solely at the federal level, is to recognize that some activities of the state cannot be handed over to other players in the game. We need good leadership that respects the grass roots and works with all levels of the system to provide functional interconnections that protect the weakest and most vulnerable, who have, nonetheless, a huge amount of social capital and who can teach us all about social cohesion.[5] We need, too, to recognize that those who are most dependent on a particular local healthy ecosystem will do the best job of taking care of it – people whose lives depend on coastal resources are the best bet for protecting them. Footloose corporations, who can (and do) move on when a resource is depleted are not the national best bet for resource protection. The principle of adjacency, as well as that of "precaution," needs to be fundamental to state resource management policy – and coastal communities can be of major assistance to the state, when it comes to protection of Canada's marine and terrestrial inheritance, which belongs to us all.

In Coasts Under Stress, our researchers and coastal communities have come together to collaborate in the generation of new forms of knowledge and strategies that should facilitate the building of the kind of partnerships, vision, and leadership essential for managing complex interdependencies – while discovering new ways for achieving environmental and social benefits in coastal communities. Good governance requires vision. It also, however, requires finding ways to work out differences, share objectives, facilitate community learning and capacity building, and discover new ways for adapting formal institutions and informal networks to the changing needs of coastal communities. Coastal communities have many sensible things to say about how this might happen. They should be consulted, and they cannot be abandoned. Why bother? There is clear empirical evidence that governance matters. In the economic development sphere, for example, there is much evidence to support the claim that good governance is linked to community prosperity and cohesion (Locke and Tomblin 2003; Abdellatif 2003; Higgins and Savoie 1995). Similar findings can be found in the literature on health restructuring and sustainable development.

We are living in an era when there is much uncertainty – political, economic, biogenic, environmental, and climatological. That means that there is less faith in the old way of doing things and thus an opportunity to try other ways. In line with panarchy thinking, we suggest that the best chances lie in reconciling top-down and bottom-up approaches and in doing that in a way that increases local control over Canada's system of coastal governance, improves health outcomes, and facilitates the development of policy recommendations. This would give Canadian decision makers a better understanding of available options, and it would mean that their thinking would be informed by a more critical understanding of the complex interdependencies between coastal communities and their ecosystems. Many of the objectives and critiques associated with new competing models of governance come from a top-down efficiency and business mentality that fails to recognize the complexity of people and their governance requirements. By contrast, we need to build on real efficiencies, adopt the flexible approach that business has itself refined, and provide an opportunity to define changes in the state, civil society, and the international economy in a way that works across scales, enhancing local leadership, capacity, autonomy, social capital, and the knowledge required to participate in governance.

Doing so would remove the conflict zones (and the crossed purposes) between different orders of governance. In this period of rapid political restructuring, it would help governments who are seeking community support to build the kind of partnerships that would help legitimize and strengthen traditional Canadian values, while at the same time pushing the issue of sustainability of Canada's coastal regions onto the public agenda. Now more than ever, experts and coastal communities need to collaborate and work together in constructing shared objectives and a vision for the future. Much of this will depend on the creation of a system of coastal governance that can address complex realities. For it to be successful, responsibilities will need to be sorted out and integration and cooperation among decision makers, local communities, and the public will need to be promoted. How to bring this about is a central question in governance debates.

We believe that other discussions of governance have tended to ignore or undervalue the importance of finding new ways to sustain coastal community and environmental health. We intend to question core ideas often associated with dominant ideas over governance because (as we have demonstrated in many chapters) neoliberalism, globalization, privatization, continentalism, and other redistributions of power and accountability within the Canadian governance system have, particularly since the 1990s, contributed to the problems associated with staple underdevelopment and various kinds of institutional rigidity. This has, we argue, been an organic process in which rigidities have been pushed down the system as responses have been sought to problems in government or industry, problems that have themselves grown out of the changing nature of global capitalism.

Such responses, however unintentionally, have contributed to the decline of coastal communities and the poverty of many coastal peoples. Even more significantly, they have also made it difficult for local people and others (different levels of the national metasystem) to build the networks and the interaction among experts and coastal communities that are required if flexible governance is to become an essential part of the Canadian state and if the various orders of government are to be able to assess the outcomes of strategies that are adopted. We hope that our case studies and subsequent analysis, as contained in this volume, will provide useful examples of what works and what does not work and why, and that they will also produce ideas that will help to increase the opportunities for coastal communities to tell their own stories, build networks, strengthen social capital, and come up with their own strategies for future development based on shared local and expert knowledge. We hope as well that they will do all of that in a manner that can be integrated with the actions of other, broader, levels of government.

There needs to be more discussion of state-society relations, of how power, networking, and so on operate within the state and society, but also of how interactions between the state and society influence future prospects for coping/resisting. We argue in this book that there are many problems with governance, power-sharing, networking, and dominant assumptions (for coastal communities) and that there is still much we need to learn about transformation, knowledge creation, and the policy process to turn this around and give ownership to these various communities. Our interdisciplinary approach is designed to question old ways of thinking, to explore old as well as new forms of resistance, and to understand the intended (and unintended) consequences of previous economic, social, and political systems and choices. It asks how we might broaden actions and strategies based on new assumptions and political actions. We are optimistic in that we believe that there is much to be gained by first better understanding different forms of restructuring (social, political, environmental), contesting old assumptions and paradigms, and doing what we can to create new spaces and opportunities for networking, contesting old ways of thinking, and defining problems.

"History," Anne Michaels once remarked, talking of it as a "gradual instant" (1996, 77), a kind of slow cataclysm, "stalks before it strikes" (159). It is true. Not only are evolutionary and biological processes slow; so are social processes. Society, like the natural world of which it is a part, evolves. Evolution, revolution, and transformation all take time. We (and our governments) need to understand that. Michaels also warns that "Something tolerated soon becomes something good" (159). It is true that we get used to things we don't like, learn to live with them, even find new strengths through them. But it should not be true that we come to see inequity and the disempowerment of people and communities and the destruction of environments as acceptable. So while long-term adaptive processes can render even major errors productive, in the widest possible sense of that word, in that they teach us and we learn, they are painful and they may turn out to be folly over the long run. The collapse of rural society and the global pattern of increasing rural-to-urban migration has been defended today, initially as something that had to be tolerated because of better opportunities in urban areas, and then because it is easier and more efficient to provide people with the benefits of modern living if they are in large "central places." The evidence, however, is mounting globally that many who move to the city end up in slums and an endless cycle of poverty and disease. Even in the West, urban decay has become a problem, and unemployment, bad housing conditions, loss of self-esteem, and other social ills are rising. Along with this, local cultures are being lost, as are ways of life that we can ill afford to lose. This is not wisdom: this is folly.

The loss of Canada's coastal communities is the kind of slow cataclysm to which Michaels refers. The ghost of closure has stalked them for generations. In the twentieth century, many First Nations were relocated, but today, many return and now are claiming the lands that were taken from them by a capitalist settler society. Settler peoples on the coast have also become place-attached over the generations. Many on both coasts love the places they call home. Some do not. Some are content to leave, seeking another lifestyle, and other goals. For those who wish to stay it is, we contend, unwise governance to let their communities die, for these places, these people, encapsulate things that are valuable to, and in, the Canadian state. They are our roots and our history. They are potentially the stewards of our coastal lands – they could play a vital role for a nation with the longest coastline in the world. They are, in commercial terms, a huge "heritage" tourist attraction. Much less trivially, they embody a lifestyle that does not seek as a primary aim the goals of increasing riches but instead seeks and values sufficiency. These are qualities we cannot afford to lose, even if not all (or even many) of us choose to live under such a rubric. They teach resilience, and they teach adaptive innovation. In an age of increasing uncertainty, it is a lesson we need to have to hand.

Canada needs to think hard, to reconsider the political, bureaucratic, and staple-industrial mindset that privileges industrial and bureaucratic flexibilities over community adaptive capacity. It is clear in nature, culture, and society that innovation – true creativity – emerges at the grass-roots level, although it is developed further up the system. The very flexibility that coastal communities once had has been adopted by business management strategies – piece work and shared work and multitasking were all features of traditional community life, furnished out of seasonal marketable resources and the occupational pluralism of settler and First Nation alike. But bureaucratic regulations have frowned upon such pluralism. Censuses do not recognize it; informal economies (the rump of the old system) are misunderstood and seen as problematic. This staple state continues to rely on its resources, semi-processed for export, and has added a service sector, missing out on much of the stability that diversification into manufacturing would have allowed, choosing instead the branch plant economy that has failed to produce the domestic control we need.

And now there is talk of closing down rural communities, or letting them wither away, deprived of services, infrastructure, and youth. At the same time, footloose industrial fleets pillage the seas – and we cannot afford sufficient numbers of coast guard vessels to ensure that the hard-won two-hundred-mile limit is observed. Who will steward the seas if the inshore communities are gone? Where will the experiments in co-management and marine protected areas (MPAs) and cooperative living occur if they are gone? Urban Canada dare not forget that it depends on its hinterland, on the resources that sustain it and that need stewardship. Governments at the federal and provincial level seek revenues for all – but "all" often means urban-industrial cores and industrial businesses. The rural producer communities of this country do not need urban-level wages, but they need the opportunity to provide sustenance for themselves so that they may provide it for the nation. We ignore, reject, or dismiss them at our peril.

This is not, however, a book about blame. It is a book about identifying what has gone wrong, and right, so that we as a nation can make better-informed choices for the future. If we think about the jungle metaphor with which this chapter started, it becomes clear that all parts of the political panarchy must function in an integrated manner. Strategies that privilege one part of the system over another will create stasis down the road for the whole system. Upper levels will starve if the essential intellectual nutrients that come from the grass roots do not reach them, and the whole system will wither if the creative ideas do not reach the top, or if they do but are then not valued and used. Blockages in the pathways that keep the social-ecological state healthy are dangerous. But there are signs that efforts are being made to remove them and to create meaningful communication throughout the political system, not just in its upper layers, and we examined some of them in this book.

The lesson we have learned in Coasts Under Stress is that communication between all levels and all sectors of the state is a necessary prerequisite to identifying and preventing or fixing potential and actual misalignments that affect the overall health of Canada's environments, its people, and its communities. Although the contemporary period is one of managing for scarcity, our capacity to degrade social-ecological health is growing not declining, because current management policy supports short-term technological and organizational efficiency over longer-term social-ecological well-being. Hence, we continue to fish down ecosystems, going from more to less abundant species, from species we know more about to those we know less about – and we still use the technologies whose catching capacity was developed for abundance (Neis and Kean 2003). On the social side we see accelerating change as well, often masked because many people try hard to stay in the places they call home and because there are intergenerational differences in adaptive strategies. This means that what we have witnessed over the past decade in terms of coastal community decline is not the full picture. The decline has been ongoing for a very long time, as we have shown in the earlier chapters of this book, until now there is probably going to be no next generation coming into the fishery – and without a fishery, there will be no, or very few, coastal communities once the elderly are gone.

Many people have talked about the jungle, but looking at coasts has told us about some things that are perhaps less clear in jungles, because the littoral is such a vast and vital interface, as we discussed at the outset of this book. Coasts have relevance globally and for cities; resource-dependent communities are important – the science involved sits at a point of intensive interaction between people and environments. On the coasts, one sees what is hidden in the city, where the impacts of globalization, microlevel and even quite macrolevel shortages, like the collapse of northern cod, are masked by international movements of fish, species replacement, and so on, and they will be masked potentially until there is nothing left. We need our coastal communities for this as well as other reasons of culture, equity, and strategic presence.

We have concluded that with respect to natural resource extraction and the future social-ecological health of our coasts, we must get out of our management and regional silos. Co-management with stewardship is an essential component in finding the best way forward. That in turn will need to be married to other formal and informal structures – and supported by the necessary funding – to support local communities as they seek other forms of regional development, whether as complementary to resource extraction or as replacement for it. Canada's coastal communities have large reserves of social capital and quite remarkable social cohesion. Nature, too, if respected, has significant powers of recovery. Integrated strategies that understand the interdependencies that make up social-ecological health and good governance can turn potential disaster into productive, albeit complex, adaptive strategies for a more resilient future for our coastal communities. The good news, then, is that what we face is challenge, not necessarily defeat.

APPENDICES

Interdisciplinary Team Research: The Coasts Under Stress Experience

CREATING AN INTERDISCIPLINARY RESEARCH TEAM: THE STRUCTURE

Coasts Under Stress is a major experiment in interdisciplinary research, covering many aspects of the natural and social sciences and humanities. The team was organized into five research "arms" – hence our metaphor of the seastar, which we adopted for our logo. These arms each focused on one set of research questions that were related to those in all the other arms. We also had a "centre" that contained both administrative and research areas, the research being that which pertained to the whole. Figure A.1 shows the CUS seastar and its research focus. Each arm had a leader on each coast, a social scientist/humanist with a natural scientist/health scientist, depending on which interdisciplinary groupings were involved. Arm 1 asked, How do different forms of knowledge (scientific, technical, and local-community) contribute to the understanding of ecosystems and the development of local policy with respect to environmental (and hence human) health? Arm 2 asked, How can local ecological and scientific knowledge help us to understand changes in environmental, community, and individual health and to identify strategies for future ecological recovery? Arm 3, which looked at renewable resources, asked: What are the consequences on both coasts of traditional and new strategies in the forestry and fisheries sectors for environmental, community, and human health? Arm 4 looked at nonrenewable resources and asked, What are the risks and benefits to environmental, community, and human health of the development and exploitation of hydrocarbon and mineral resources? Finally, Arm 5 asked, How has social and political change (or lack thereof) affected the health of individuals, families, and their communities? Further, the centre of the seastar incorporates, beyond its administrative functions, three case studies. The task of the first was to synthesize the work of

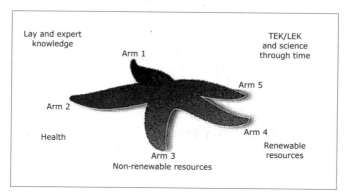

Figure A.1   Seastar structures of Coasts Under Stress

the project. The second examined the ethics of some of our policy recommendations, focusing specifically on the thorny issue of aquaculture. The third assessed the perceptions of the team on the success or failure of its ability to produce interdisciplinary work.

The challenge we faced throughout our work was to integrate all the subcomponents in this complex array of related research, and we evolved various techniques for doing that in the course of the study. The research, like this book, had a common aim: to demonstrate the relationship between different kinds of restructuring and the cascade of effects they each create.

The challenge that faced us was to produce a shared coherent set of analyses that meshed in such a way that it provided an interdisciplinary intellectual synthesis. That is what this book has, we believe, achieved. Our ability to keep ourselves informed about all facets of our work and thus to achieve ongoing integration was facilitated in a number of ways. There was the normal scholarly exchange among people at the same location and telephone calls between those at distance, of course. Beyond those, we met as coastal subteams twice a year, with the project director (PD) and the members of the executive attending each meeting (minuted and seen on both coasts). This aided the flow of ideas between coasts. There were also three joint team meetings, where everyone met on one coast. The first of these was the start-up and the second a midterm, and the third took place one year later to finalize integration and dissemination plans. Between times, there were "arm" teleconferences, and each arm, along with the students (and later our advisory and partners), had a listserv in which there was ongoing discussion. The PD belonged to all listservs and participated in as many of the teleconferences as possible, as well as in all meetings of the team. National-level advisors came to all our meetings and were joined in the coastal meetings by local advisors. We met our partners in the field,

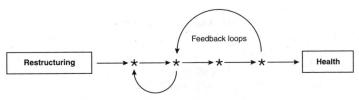

Figure A.2   A simplified portrayal of the CUS concept, where * is the point of entry of a given sub-component into the analysis and → is a pathway identified in that work

in workshops, and before or after team meetings, when they were able to see the coastal context of all our work.

CREATING AN INTERDISCIPLINARY RESEARCH TEAM: THE PEOPLE

One challenge of team research is to get people comfortable with each other, so that a genuine exchange of ideas can occur. When the team is big or when it has to work across large distances, or when (as in this case) both these conditions apply, then team "bonding" requires particular strategies (see appendix 2 for a list of team members). Key to the success of any interdisciplinary research is a shared vision that holds the researchers together. We had that vision in our shared concern for what has been happening to our coastal communities and the environment of which they are a part, but we had many different ways of thinking about the problem. Perspectives varied according to discipline, interest, and even ideology, but in all cases we were aware that we needed to be involved in the examination of interfaces *between* sectors, places, and people, and *between* scales, ranging from the local to the national.

When we were preparing the proposal for funding, the East and West Coast team members met separately to experiment with schematic diagrams that would let each subcomponent identify itself within the whole project (see figure A.2). That exercise in creating diagrams continued throughout the first half of the project, as each arm debated, and then finalized, its own schematic, sharing its development at team meetings with the whole team. This was a key exercise because, in discussing the implications of lines, curves, linkages, and connections, the predilections, assumptions, and perspectives of the various scholars, their disciplines, and even their personal ideologies rapidly became apparent. Natural scientists, on the whole, along with some quantitative social scientists, preferred flow charts; scholars with long experience in community-based research preferred spirals (to get at all those feedback loops); geographers liked things organized spatially, historians by time period; political scien-

tists and those with a policy concern thought in terms of spheres of influence and institutions. Scale was an ongoing issue and the two coasts,
drawing on their particular experiences, were not always in concert.

Initially, we were unable to find a schematic that everyone was comfortable with, but when the funding councils asked all of us from both coasts
to meet together, in April 2000, and organize ourselves as a coherent team
with detailed milestones on which we all agreed, we were faced with the
need to draw our varying perspectives together. In a series of plenary discussions and break-out sessions, the team self-organized into the seastar
that became our trademark.

We agreed to report in verbal presentations every six months to the
whole team, through coastal team meetings that were attended by the PD
and the co-chairs and presentations that followed an agreed format and
were circulated to the whole team. For the team report, members had to
summarize the research they had completed since the previous report,
the milestones that had been met (and if they were not met, what was
being done to remedy that), their expenditures, their results, the anticipated linkages, and (as time went on) the outcomes of the work in terms
of policy guidelines, presentations, publications, community workshops,
videos, and so on. These team reports continued every six months until
the last eighteen months of our work, at which point contributions to this
and other volumes replaced them. They were posted on the "members
only" part of our website, and sent to our advisory board and to SSHRC and
NSERC. They were an invaluable link in such a scattered team, but they
were not enough to generate integration, being primarily reports on progress.

CREATING A TEAM: STRUCTURING INTEGRATED RESEARCH

Integration is a buzz-word in interdisciplinary research that is usually used
without much reflection. It means, at a simplistic level, that the research
should hang together, have a unified goal, and be able to show a coherent
outcome. But in our experience, integration was much more than this. Integration had to reflect the complexity of the coasts and needed different kinds
of ongoing exchanges between team members, often on particular points. By
extension, individual team members would have to have a way of finding out
what was going on among colleagues on both coasts if we were to avoid dupli-
cating findings or constructing contradictory frameworks or failing to capture
the synergies that are the huge reward in research of this kind.

We therefore instigated listservs for each arm, for students, and (later)
for our partners at their request, and we set up arm teleconferences, which
were hugely productive. These mechanisms, along with team meetings,

became the backbone of our integrative communications. The PD participated in all listservs and teleconferences, as well as meetings. The executive held regular (bimonthly) teleconferences, where linkages, budgetary
issues, partnership concerns, and other administrative matters were dealt
with, and all executive members attended all team meetings to identify
potential points of integration as these arose from coastal presentations.
The PD and the chair of the advisory board were also in regular communication. Budgets were issued to each subcomponent yearly and reviewed
every six months by the PD in consultation with the arm leaders: extraordinary budgetary matters were dealt with by arm leaders (if a small amount
was involved) and the executive (if a large amount was involved). The
team met three times as a whole, once at start-up, once midway through,
and once in the middle of year four. Final meetings, for those who contributed to the synthetic volumes we are publishing (of which this is one),
were held in the last year of our work. The result was a coherent organization within which there was relative autonomy inside the agreed research
framework and a capacity for ongoing communication by phone, email,
and text, supplemented by occasional bi-coastal visits by individuals and
regular visits by the PD. Without all this organization, the integration of
research results would have been much weaker, many important cross-
coastal linkages would not have been identified, and several highly productive cross-coastal research partnerships would not have been realized.

With that structure in place, it was not possible for any one part of the
research to operate in a vacuum, but integration still took time.

INTEGRATION — THE RESEARCH

In the early stages of our work, we had a range of ways of indicating ecosystem health (species at risk, abundance, biodiversity, biomarker analysis,
lipids, and so on) social health (community economic health, social well-
being, determinants of health, cultural energy/creativity/liveliness [=
health]), future health (youth, education and infrastructure), and individual health (sickness, risk danger). We also employed various ways of assessing how social-ecological health was protected and promoted (through
academic and government research, stewardship, MPAs, federal/provincial
regulations [their history and application], and TEK/LEK), and we had an
historical component to all this work, since we believe that resilience is a
long-term concept and can be demonstrated only over an extended
period.

It took time for the relationships and the linkages to become apparent
(in some cases) or to begin to appear compatible in terms of joining
together into a seamless whole (figure A.3). At start-up, the link between,
say, the stewardship of eastern wetlands and oil and gas exploration in the

➤ How do we emphasize community context yet make comparisons to other communities in different locales?

➤ How can we use historical data to talk about the present?

➤ How can we extend natural science data on resources to current health, education, and employment issues?

Connections:

• Arm 1/2. Archival information on Aboriginal communities could provide background for Marshall's work in Alert Bay.

• Arm 1/2. Traditional science workshops can be viewed as a career resource for FN youth.

• Arm 3. Aquaculture impacts (Parrish) and employment (Marshall) and health (Taylor).

• Arm 3. Forestry impacts on estuaries and impact on FN health and diet.

• Arm 3. Local economic development and the forestry industry (Sinclair) and community health and consequences for individuals/families (Taylor).

• Arm 4. Annotated healthcare bibliography for northern BC (Hammond) provides comparative data with FN and LA.

Figure A.3   Challenges in research integration

Hecate Strait could only be described as tenuous at best – there is an intellectual linkage, but that is a different matter from a research project linkage. Integration does not happen all at once, nor will it happen with every subcomponent at the same time. After a year, the potential for integration was still nascent rather than real, but links were beginning to emerge, and this was when a second whole-team meeting should have been held, had there been the funds to do that.

By the end of year 2, it was possible to produce a working paper that drew all the research together into a narrative text, in which the linkages and relationships between subcomponents were clearly articulated. At this time also it became clear that arms 1 and 2 were close to integration, and they were combined as arm 1/2.

At year 2.5, building on that narrative, each arm prepared power points that addressed that part of the research as an integrated unit and clearly stated the linkages between arms. The PD wrote, and presented on both coasts, a position paper on "pathways" that explicitly laid out a way to integrate all subcomponents through shared analytical thinking.

As we discussed among ourselves how to identify pathways between environmental and social restructuring and environmental and human

J. Volpe: Effects of Salmon Farms on Inshore Biological
   Communities

- Purpose: to determine the effect of fish farms in Clayoquot
  Sound on intertidal ecology, specifically First Nations food
  gathering grounds.
- Analysis: ANOVA.
- Results: No significant effects.
- Pathways: Possible cumulative impact on environmental and
  community health over long-term, even if there is no detectable
  immediate damage.
- Linkages: None reported to date. But are the following
  possibilities? N. Turner (LEK re sustainable fisheries, e.g., clam
  beds on Haida Gwaii); J. Lutz (welfare/assistance among First
  Nations); C. Parrish (organic waste from fish farms); DFO West
  Vancouver Labs (studying effects of aquaculture on marine
  environment); AquaNet (NSERC Cente of Excellence).
- Dissemination: MSC thesis; refereed papers; reports to
  stakeholders.

Knowledge Pathways

- Fisheries management in relation to environmental, economic,
  and social health (Parzival Copes)
- Ecosystem health and its implications for resource management
  and public policy (R. Haedrich and M. Wernerheim, et al.)
- Facilitating use of local, traditional and Western scientific
  knowledge in policy development (Michael M'Gonigle and Kelly
  Bannister)
- Knowledge transformation and change within communities
  (Wolff-Michael Roth)
- Linking knowledge with management (Dave Schneider and
  Barbara Neis)

Figure A.4   Examples of the integration of CUS research from the September
2003 mid-term review meeting

health, and, even more importantly, as we integrated those pathways into
power point presentations for team meetings (figure A.4), tighter integrative formulations began to appear, and the flexibility with which it was
necessary to treat the concept also emerged. There were important policy implications in that flexibility, as there were in all the surrounding
discussions.

In the process, some parts of our work were adjusted to take the new insights into account. We made a major shift in deciding to study the regionalization of decision making. This involved getting at health more directly by shifting our attention away from regional *economic* policies and structures to regional *health* matters. The regionalization of health administration and service delivery was, we noted, a key change that occurred in almost all Canadian jurisdictions during the 1990s but one that had, to date, not been studied in much detail, either in our two provinces or comparatively. Our work therefore was redesigned to study the regional health authorities in our two study zones and to compare their responses to the challenges of economic, social, and ecological restructuring.

THEORY AND METHODOLOGY

In some respects, the various methodologies adopted by CUS researchers were no different from what would have taken place outside CUS except, perhaps, that the range of methods employed were somewhat wider than might have been possible otherwise. The strategies adopted by the team to ensure coherence across the whole project, however, represent in and of themselves a new interdisciplinary methodology for working with large teams located at a distance and embracing many disciplines. That is why we have discussed them in some detail.

Working in the team involved research at a whole range of scales, but what bound the work together was the conviction – inherent in the problem itself, the methodologies that were employed, and the theories that were used as constructs – that place matters, people matter, and things change over time. Methods ranged across the spectrum of the disciplines involved, of course. We employed surveys (both physical and social), participant observation, snowball interview techniques, systems modelling (both physical and socio-economic), archival research, archaeological "digs," sampling techniques in the marine and geological sciences, and field work in the places where we worked – the schools, the sea, the shoreline, and the islands. We worked at the broad scale of analysis that deals with the impacts of global economies on resource-based economies or with ocean "hotspots" for species at risk. We looked at regional and national health frameworks, at diet and nutrition in the twentieth century, and at plant life and First Nations culture in one small settlement on the northern British Columbia coast. We talked to fishers on and off the water, in kitchens, and in schoolrooms, and we sat in small planes to identify old fishing and First Nation aquaculture sites and in research vessels to scan the ocean floor for such things as fossils that could tell us about oil and gas or the accumulated debris from log booms over the span of a century.

Much of our research used standard methodologies, but in some cases we had to build a new model (chapter 13) or use a new and still-developing marine-life accounting set of methodologies, whose complexities are recorded elsewhere, and referenced where appropriate (chapters 2 and 15). As a general principle we summarize methodologies used in the first footnote for each chapter, where the subcomponent from whom the work is drawn is also recognized.

In effect, we worked with theory and concept at a variety of scales. No one method could apply to our work, and clearly we also needed other specific bits of disciplinary theory in our case studies: they are indicated by footnotes and references in the appropriate parts of the book. More important in this kind of work was the coherence of the framework we employed for thinking as a team about the research problem and for guiding the overall operation of our work. That was the concept of social-ecological health that allowed us to think in terms of interdependence in ecology and society, including the capacity to create and/or withstand stress. And since environment and society are multilayered, we were constantly aware of scale and the problems of scaling up or down, be that from policy at the national level to the ability to obtain good nutrition in remote areas or to changes in ecosystem structure and the capacity of a species (*Homo sapiens* or *Gadhus morhua*, for example) to survive and be healthy.

The result is this volume, which is dedicated to providing an overview of all our work, showing how the components fit together and how in their entirety they cover the crisis in which coastal communities (human and natural) find themselves. Other books "drill down" into details of particular themes and analyses: the movement of knowledge from people to researchers to policymakers to students and back to communities in many interwoven ways; the analysis of power, whether as energy (oil and gas, hydro) or as "power over nature constructs" or as power and agency in nature and human communities; and the history of health, diet, and nutrition – the issue of decreasing food security in places where once-stable food webs have suffered radical shock, as has the kitchen culture of human communities, which has always been interdependent with the food sources that are now endangered.

THE HUMAN FACE OF INTERDISCIPLINARY TEAM RESEARCH

Our administrative and intellectual account of ourselves would not be complete without some reflections on the impact of cus on various team members, including students. Working in a team project like this has enormous creative potential in terms of research attitudes and understanding of the work of other scholars from different disciplines, and CUS has, we think, broken down some significant solitudes. Take this comment from one scientist, for example:

CUS had me recognize, through reading and through the research approach I adopted, that ecological processes themselves can be investigated via a study involving local ecological knowledge (LEK). My prior assumption was that LEK was limited to assisting an ecological study by perhaps initiating it or confirming its results in an anecdotal way. I have newly learned that LEK can be strategically investigated and systematically analyzed to produce its own study of an eco-system.

Another commented that

My research is based on the theory of fisheries ecology. As a scientist, I practise the scientific method in conducting research. Before my involvement in CUS, I used "anecdotal information" from fishers in planning my field studies. Through my collaboration with sociologists and humanists in CUS, I learned that this "anecdotal information" is actually another knowledge base relevant to my research, i.e., fishers' ecological knowledge (FEK). I have been introduced to epistemology. My CUS collaborators and I have constructed a new theoretical framework showing how FEK blends with scientific knowledge, which we have named "closing the loop" [see Gosse, Wroblewski, and Neis 2003]. I have always valued FEK, and used FEK to guide my field research, but since becoming part of CUS, I have learned that LEK is intellectual property and must be respected and protected. Through my collaboration with sociologists and humanists in CUS, I became aware of the ethical issues involved in participatory research with fishers. I found that other researchers are also engaged in participatory research with fishers. This methodology, which is becoming accepted by the fisheries science community, is being formalized and developed in CUS. My research method-ology has changed in other ways also, in that I now obtain ethical clearance from the MUN committee before interviewing fishers. The guidelines from SSHRC and NSERC have made me aware of the risks to fishers for sharing their FEK. Although I was aware of the economic risks, I was oblivious to the sociological risks fishers face in granting interviews with researchers. My research in CUS has involved disseminating new scientific knowledge to coastal communities, empowering them with scientifically validated LEK. With CUS, my interaction with com-munities is now structured, rather than ad hoc. I am more aware of the implications of fisheries research, i.e., the power of new knowledge. I have also become involved with co-management of local marine resources. My research has broadened beyond science into management policy.

CUS also created new academic partnerships, allowing people to work together who might otherwise never have had that opportunity. A social scientist commented as follows.

The sharing of interview guides with other social scientists and scientists, and the input from scientists in terms of the conceptualization of the areas of inquiry, helped to integrate questions/ideas that explored changes in the physical and natural environment and how they impacted youth through various mechanisms. For example, the collapse of the cod fishery has led the community to develop and expand new relationships with the natural environment including an expansion of tourism based on attracting tourists to the physical beauty and historical character of community. The "selling" of the physical or natural setting (i.e., tourism) has interacted with social factors (e.g., resources placed into tourist development), which have also affected youth's relationship to the physical or natural environment and the nature of their leisure activities and lifestyle (e.g., improved outdoor recreational facilities allow for easier access to and use of the natural environment for leisure activities, and alcohol is a key part of the leisure lifestyle).

The sharing of ideas and conceptualizations between scientists and other social scientists has also been key in terms of widening our observational work. The inclusion and integration of an understanding of the role of the physical and natural environment has allowed us, for example, to move our obser-vational work beyond the social context and relationships in the social world, and to be sensitive to and integrate into our research the interactions between youth, and the physical/natural worlds. Further, our work has placed a lens on the changing seasons and the diversity of activities within and between seasons, seeking to understand how such variations influence youth's patterns of behaviour and lifestyles. Participating in CUS has also influenced the methodology in terms of the interactions between East and West Coast research projects, especially a youth-focused project in northern British Columbia. By providing a bi-coastal comparative context we have been able to identify key impacts that are occurring among youth in communities on both coasts, irrespective of differences in historical context and the specific socio-economic location. Commonalities across communities also point to strategies of program and policy development that may be useful in multiple places.

All of us, humanists, social scientists, and natural scientists alike, learned from the work that was done with First Nations in their communities. It informed the integrative thinking of the whole team in terms of the capac-ity for adaptation that human beings possess in the face of significant restructuring, both social and environmental.

Finally, our students were also enormously excited by working in an interdisciplinary project. Indeed, they were a motivating force for interdisciplinarity. Graduate students worked on various aspects of the project from the pre-testing of the interview guides, through natural and social science field and lab work, to the final stages of data analysis and the production of reports/publications. They attended all team meetings. Graduate students have therefore obtained training in all aspects of the research process and have been able to take part in an interdisciplinary program of research. Coasts Under Stress included not only natural and social scientists, but also postdoctoral fellows, research associates, and research assistants, as well as more than fifty graduate students. They asked permission to conduct their own survey of themselves, which we also report on here.

Our students told us that, like Bruhn (2000), they think "interdisciplinarity is a philosophy, an art form – it is a synchronicity of questions, of viewpoints, and of methods focused on a problem of common concern. It is more than sharing views or coordinating methods [among the team members]." While studies have been done of the effectiveness of interdisciplinary teams,[1] the role of graduate students in interdisciplinary research has not been explored. Our graduate students told us that from their perspective, the many facets of interdisciplinary research present both benefits and challenges to graduate studies. On one hand, interdisciplinary research facilitates students by defining the topic and scope of research and by providing funding; on the other hand, it challenges students to define their unique contribution within this interdisciplinary research. For instance, how should individual roles be designated and how can individual contributions to the entire team be recognized? To answer those questions, a student-perspective survey was conducted in December 2003 to allow us to better understand the role of graduate students in the Coasts Under Stress research project for both the natural sciences and the social sciences/humanities. The survey found that for students who are open to participating fully in the project and enthusiastic about the unique opportunity provided for academic growth, the intellectual benefits of team research are considerable.

Students in the natural sciences become accustomed to following the scientific method, in which experiments are designed around a definable problem and hypotheses are tested and proved or disproved – a process that can be somewhat remote from the real world. Involvement with an interdisciplinary program like Coasts Under Stress can be very valuable for a student of the natural sciences, as there are few careers where science is practised in a vacuum. In attending meetings and sharing research ideas with members of different disciplines in Coasts Under Stress, once a slight "language barrier" is overcome, natural-science students gain a broader theoretical perspective through which they will be better able to understand the sociological implications of decisions made in resource-harvesting communities. One of the most important lessons learned from experience with an organization like Coasts Under Stress is that a hypothesis proved true or false according to the scientific method will not necessarily correspond with what resource users will believe to be true or false.

The social sciences and humanities consist of a range of academic disciplines that focus on evaluating the human element of society. Scholars of such fields generally approach their topics of research through intellectual interpretation and critical analysis of human documentation, including statistical data, official documents, oral interviews, and more personal written, visual, and auditory treatments. The results of social sciences and humanities research depend in large part on the initial theoretical and methodological approaches taken in the research. Coasts Under Stress has introduced many students of such fields to the different approaches taken by other disciplines in conducting research and has broadened students' academic growth. The influence of other disciplines (whether in the natural sciences or in other fields of the social sciences and humanities) on the students of the social sciences and humanities is not always evident and at times difficult to evaluate, but it is certainly there. Graduate students are at an important and impressionable growing-stage of their academic careers and Coasts Under Stress has provided financial support and mentorship for students to conduct research in a field that interests them. Furthermore, the unique opportunities provided by the interdisciplinarity of the project cannot be over-emphasized. Whether they realize it now or later in their academic development, the interdisciplinarity of Coasts Under Stress and the intellectual forum it provides with professors is helping to shape the careers of graduate students in a progressive manner.

One final set of reflections. Work of this scope, scale, and distance is new. Its practice is sometimes met with hostility ("this money would have been better spent on individual studies that would be more detailed and reflective"), and sometimes with overly high expectations ("with this money, you will be able to solve the problems of coastal communities and bring scholars of all disciplines together in a seamless research web"). There is a grain of truth in both of these extremes. We think we have brought scholars from many disciplines together in such a way as to make us aware of the strengths and limitations of one another's disciplines, thereby getting rid of inappropriate expectations that can sound the death knell of interdisciplinary scholarship. We have also become clear that interdisciplinary work has many faces – it includes solitary work (in the lab, for example, or at the desk) and also the exhilaration of team problem solving and insights suddenly obtained by thinking outside one's usual disciplinary box (*cultural* refugia, for example). It does show the pathways that come together to create the bigger picture, and it does point both to the intransigence of scale problems and also to some solutions in scaling. On the other hand, there isn't time in five years to absorb all the information from all parts of the research, and so some rough links remain: the web is not seamless. It has to be recognized that some research is not "cutting-edge" – there are contexts that have to be established, baselines that must be filled in, historical backdrops that cannot be continuous, given limits of time and money.

Good research design takes care of these where possible and takes them into account where not. By the time a proposal goes forward for funding, it should be sufficiently integrated that the pathway through the research is clear, necessary gaps are identified and allowed for, and gaps that can realistically be closed have been. A clear unified vision is a sine qua non, and an administrative structure that works towards integration is also essential. Without them, integration is impossible.

This larger model, somewhat like "big science" but much more flexible,[2] is increasingly becoming essential for work in the social sciences and humanities, given the kinds of problems that this complex postindustrial global capitalist world poses for society.[3] In that context, the small, focused, applied study can illustrate, and sometimes forecast, problems in our world, but solutions can be arrived at only when the complexity of the everyday web of economy, ecology, and society, taken together, is considered. Such work costs money, sometimes a great deal of money, although if one were to cost a project like Coasts Under Stress by individual subcomponents, it would be seen that each of these is in fact highly cost-effective, because of the ability to share resources (equipment, administrative costs, ship time, students, accommodations, software, and hardware, for example). The collective "bang for the buck," then, is considerable if the work does actually cohere into one final complex and unified analysis. That is what we have tried to make it do, and the results collected in short, focused form, will permit the reader to judge how far we have managed to achieve our goal.

The Coasts Under Stress Team

Co-investigators	Department	Organization
Chris Barnes	Earth and Ocean Sciences	University of Victoria
Vaughn Barrie	Earth and Ocean Sciences; Pacific Geoscience Centre	University of Victoria; Geological Survey Canada
Brian Bornhold	SEOS	University of Victoria
Stephen Bornstein	Political Science	Memorial University of Newfoundland
Sharon Buehler	Community Health, Medicine	Memorial University of Newfoundland
Sean Cadigan	History	Memorial University of Newfoundland
Patricia Canning	Education	Memorial University of Newfoundland
Ross Chapman	Earth and Ocean Sciences	University of Victoria
Parzival Copes	Institute of Fisheries Analysis, Economics	Simon Fraser University
Harold Coward	Studies in Religion and Society	University of Victoria
Catherine Donovan	Community Medicine	Memorial University of Newfoundland
Barry Glickman	Environmental Health	University of Victoria
Robert Griffin	Human History	Royal British Columbia Museum
Richard Haedrich	Biology	Memorial University of Newfoundland
Nigel Haggan	Fisheries Centre	University of British Columbia
Lorne Hammond	History	University of Victoria; Royal British Columbia Museum
John Harper	Earth and Ocean Sciences	University of Victoria
Carol Harris	Leadership Studies, Education	University of Victoria
Douglas C. Harris	Law	University of British Columbia

Co-investigators	Department	Organization
Craig Hawryshyn	Biology	University of Victoria
Robert Hood	Tourism Management	University College of the Cariboo
Robin J. Hood	Education	University of Victoria
Lois Jackson	Health and Human Performance	Dalhousie University
Ian Joyce	Geography	Douglas College
Greg Kealey	History	University of New Brunswick
Linda Kealey	History	University of New Brunswick
John Kennedy	Anthropology	Memorial University of Newfoundland
John Lutz	History	University of Victoria
Martha MacDonald	Economics	St Mary's University
Quentin Mackie	Anthropology	University of Victoria
Anne Marshall	Educational Psychology and Leadership Studies	University of Victoria
Brian McLaren	Forestry/Biology	Lakehead University
Michael M'Gonigle	Law	University of Victoria
Bill Montevecchi	Psychology, Biology, Ocean Sciences	Memorial University of Newfoundland
Barbara Neis	Sociology	Memorial University of Newfoundland
Rosemary Ommer	Office of the Vice President, Research	University of Victoria
Chris Parrish	Ocean Sciences, Biology, Chemistry	Memorial University of Newfoundland
Tony Pitcher	Fisheries Centre	University of British Columbia
Wolff-Michael Roth	Education	University of Victoria
David Schneider	Ocean Sciences	Memorial University of Newfoundland
John Schofield	Economics	University of Victoria
Peter Sinclair	Sociology	Memorial University of Newfoundland
Shirley Solberg	Nursing	Memorial University of Newfoundland
Martin Taylor	Geography	University of Victoria
Susan Tirone	Health and Human Performance	Dalhousie University
Stephen Tomblin	Political Science	Memorial University of Newfoundland
Peter Trnka	Philosophy	Memorial University of Newfoundland
Nancy Turner	Environmental Studies, Geography	University of Victoria
Michael Wernerheim	Economics	Memorial University of Newfoundland

Co-investigators	Department	Organization
Michael Whiticar	Earth and Ocean Science	University of Victoria
Derek Wilton	Earth Sciences	Memorial University of Newfoundland
Joe Wroblewski	Ocean Sciences, Biology, Physics and Physical Oceanography	Memorial University of Newfoundland

Post Doctoral Fellows, Research Associates, and Collaborators	Department	Organization
Jack Barnes	Earth and Ocean Sciences	University of Victoria
Ingrid Botting	History; Centre for Applied Health Research	Memorial University of Newfoundland
Moyra Brackley	Centre on Aging	University of Victoria
Villy Christensen	Fisheries Centre	University of British Columbia
Gail Davoren	Biopsychology	Memorial University of Newfoundland
Holly Dolan	Geography	University of Victoria
Robyn Forrest	Fisheries Centre	University of British Columbia
John Gibson	Biology	Memorial University of Newfoundland
Theodore Haimberger	Biology	University of Victoria
Sheila Heymans	Fisheries Centre	University of British Columbia
Steven High	History	Memorial University of Newfoundland
Marji Johns	Earth and Ocean Sciences	Pacific Paleoquest
Jahn Petter Johnsen	Centre for Rural Research	University of Tromsø
Russ Jones	Fisheries Centre	University of British Columbia
Pat Kavanagh	Fisheries Centre	University of British Columbia
Susanne Klausen	History	University of Victoria
Lyubov Laroche	Education	University of Victoria
Kaija Metuzals	Biology	Memorial University of Newfoundland
John Munro	Biology	Memorial University of Newfoundland
Grant Murray	Ocean Sciences Centre	Memorial University of Newfoundland
Melanie Power	Centre for Applied Ethics	University of British Columbia
Richard Rajala	History	Royal British Columbia Museum
Rick Rennie	History	Memorial University of Newfoundland
Kristen Rohr	Earth and Ocean Sciences	University of Victoria

Post Doctoral Fellows, Research Associates, and Collaborators	Department	Organization
Gurmit Sandhu	Economics	University of Victoria
Torge Schümann	Earth and Ocean Research	University of Victoria
Marcelo Vasconcellos	Fisheries Centre	University of British Columbia
Carl Walters	Fisheries Centre	University of British Columbia

Graduate Students	Department	Organization
Cameron Ainsworth	Fisheries Centre	University of British Columbia
Erin Alcock	Environmental Science	Memorial University of Newfoundland
Krista Baker	Environmental Science	Memorial University of Newfoundland
Elizabeth Bennett	Environmental Science	Memorial University of Newfoundland
Leanna Boyer	Education and Geography	University of Victoria
Erin Breen	Marine Management	Dalhousie University
Claire Brignall	Fisheries Centre	University of British Columbia
Eny Buchary	Fisheries Centre	University of British Columbia
Diana Cardoso	Environmental Science	Memorial University of Newfoundland
Heather Chaffey	Biopsychology	Memorial University of Newfoundland
Jackie Cheeseman	Environmental Science	Memorial University of Newfoundland
William Cheung	Fisheries Centre	University of British Columbia
Jamie Chippett	Biology	Memorial University of Newfoundland
Patrick Collins	Earth Sciences	Memorial University of Newfoundland
Heidi Coombs	History	University of New Brunswick
Louise Copeman	Biology	Memorial University of Newfoundland
Sulan Dai	Geography	University of Victoria
Reade Davis	Anthropology	Memorial University of Newfoundland
Robert Dennis	History	University of New Brunswick
Jennifer Devine	Biology	Memorial University of Newfoundland
Michael Fleming	Sociology	Memorial University of Newfoundland

Graduate Students	Department	Organization
Robyn Forrest	Fisheries Centre	University of British Columbia
Thomas Good	Earth and Ocean Sciences	University of Victoria
Karen Gosse	Environmental Science	Memorial University of Newfoundland
Chris Grandin	Earth and Ocean Sciences	University of Victoria
Brenda Grzetic	Women's Studies	Memorial University of Newfoundland
Louise Hahn	Biology	University of Victoria
Julie Halliday	Earth and Ocean Sciences	University of Victoria
Tim Hollis	Geography	Memorial University of Newfoundland
Tara Hooper	Chemistry	Memorial University of Newfoundland
Honna Janes-Hodder	Sociology	Memorial University of Newfoundland
Yan Jiao	Biology	Memorial University of Newfoundland
Amanda Karst	Biology and Environmental Studies	University of Victoria
Sheila Keefe	Community Psychology	Wilfred Laurier
Valerie Kendall	Environmental Science	Memorial University of Newfoundland
Lindsay Lawrence	Economics	Memorial University of Newfoundland
Kate Leslie	Environmental Studies	University of Victoria
Hector Lozano	Fisheries Centre	University of British Columbia
Gail Lush	History	Memorial University of Newfoundland
Maria Martinez-Murillo	Biology	Memorial University of Newfoundland
Duncan McLaren	Anthropology	University of Victoria
Mark McLaughlin	History	Memorial University of Newfoundland
Heather Molyneaux	History	University of New Brunswick
Andrew Moss	Environmental Science	Memorial University of Newfoundland
Ann Murrin	Economics	Memorial University of Newfoundland
Roshni Narayan	Earth and Ocean Sciences	University of Victoria
Martin Niemann	Earth and Ocean Sciences	University of Victoria
Chris Picard	Earth and Ocean Sciences	University of Victoria; Department of Fisheries and Oceans Canada
Melanie Power	Fisheries Centre	University of British Columbia

Graduate Students	Department	Organization
Kerri Riggs	Earth Sciences	Memorial University of Newfoundland
Steve Riley	Community School Principal, Prince Rupert	University of Victoria
Leah Robinson	Educational Psychology and Leadership Studies	University of Victoria
Kara Rogers	Biology	Memorial University of Newfoundland
Leah Smith	Political Science	Memorial University of Newfoundland
Nicole Spencer	Marine Studies	Memorial University of Newfoundland
Richard Stanford	Fisheries Centre	University of British Columbia
Iain Stenhouse	Bio-Psychology	Memorial University of Newfoundland
Shelby Temple	Biology	University of Victoria
Judith Thompson	Education and Environmental Studies	University of Victoria
Pablo Trujillo	Fisheries Centre	University of British Columbia
Geert Van Biesen	Environmental Science	Memorial University of Newfoundland
Kelly Vodden	Geography	Simon Fraser University
John Volpe	Biology	University of Victoria
Colleen Walsh	Environmental Science	Memorial University of Newfoundland
Beibei Wang	Economics	University of Victoria
Jennifer Whalen	Geography	Memorial University of Newfoundland
Amy Williams	Environmental Science	Memorial University of Newfoundland
Kevin Wilson	History	University of Victoria
Louisa Wood	Fisheries Centre	University of British Columbia

Research Assistants	Department	Organization
Ellen Adams	Coasts Under Stress	Memorial University of Newfoundland
Gord Adams	Coasts Under Stress	Memorial University of Newfoundland
Erin Alcock	Environmental Science	Memorial University of Newfoundland
Katie Armitage	Law	University of British Columbia
Meghan Atherton	Psychology	University of Victoria
Dwayne Avery	Philosophy	Memorial University of Newfoundland

Research Assistants	Department	Organization
Emerence Baker	Women's Studies	Memorial University of Newfoundland
Jodi Baker	Biopsychology	Memorial University of Newfoundland
Shelley Balshine	Law	University of British Columbia
Jordan Barclay	Earth and Ocean Sciences	University of Victoria
Andrea Barnes	History	Memorial University of Newfoundland
Russell Barnett	Economics	St Mary's University
Katherine Barrett	Law	University of Victoria
Suzanne Batten	Counselling Psychology	University of Victoria
Dean Bavington	Environmental Science	Wilfred Laurier University
Yusuf Baydal	Philosophy	Memorial University of Newfoundland
Erica Beauchamp	Earth and Ocean Sciences	University of Victoria
Robert Blazey	Economics	University of Victoria
Jeff Braun-Jackson	Coasts Under Stress	Memorial University of Newfoundland
Chantelle Burke	Biology	Memorial University of Newfoundland
Turkan Cadgas	Economics	Simon Fraser University
Carol Cantwell	Coasts Under Stress	Memorial University of Newfoundland
Heather Chaffey	Biopsychology	Memorial University of Newfoundland
Keith Chaulk	Biopsychology	Memorial University of Newfoundland
Jackie Cheeseman	Environmental Science	Memorial University of Newfoundland
Milagros Chu	Economics	Simon Fraser University
Dominique Clement	History	Memorial University of Newfoundland
Julianna Coffey	Archaeology/Biology	Memorial University of Newfoundland
William Coffey	Coasts Under Stress	Memorial University of Newfoundland
Megan Coles	Coasts Under Stress	Memorial University of Newfoundland
Christopher Cook	History	University of Victoria
Sarah Corrin	Counselling Psychology	University of Toronto
John Corsiglia	Education	University of Victoria
Lisa Culihall	Coasts Under Stress	Memorial University of Newfoundland
Jill Cumby	Coasts Under Stress	Memorial University of Newfoundland
Tarah Cunningham	Biology	Memorial University of Newfoundland
Lindsay Dalton	Geography	University of Victoria

Research Assistants	Department	Organization
Allan Davidson		Haida Nation
Jennifer Davison	Economics	University of Victoria
Fern Delamere	Coasts Under Stress	Memorial University of Newfoundland
Tanya Demchuk	Earth and Ocean Sciences	University of Victoria
Michel Demyen	Geography	University of Victoria
Mary des Roches	Coasts Under Stress	St Mary's University
Danielle Devereaux	Women's Studies	Memorial University of Newfoundland
Andrea Dunne	Counselling Psychology	Memorial University of Newfoundland
Jodi Durdle	Nursing	Memorial University of Newfoundland
Mark Eddy	Sociology	Memorial University of Newfoundland
Anne English	History	University of Calgary
Aftab Erfan	Fisheries Centre	University of British Columbia
Daniel Ferguson	Philosophy	Memorial University of Newfoundland
John Foster	Earth Sciences	Memorial University of Newfoundland
Steve Gardner	Science Studies and Ecology	Memorial University of Newfoundland
Jennifer Gee	Biology	University of Victoria
Robin Grazely	Coasts Under Stress	University of Calgary
Brenda Grzetic	Women's Studies	Memorial University of Newfoundland
Francis Guenette	Education	University of Victoria
Liam Haggarty	History	University of Victoria
Scott Harris	Earth and Ocean Sciences	University of Victoria
Kathy Harrison	Education	University of Victoria
Dean Hay	Human Kinetics and Recreation	Memorial University of Newfoundland
Courtney Hayden	Ocean Sciences	Memorial University of Newfoundland
Dale Haynes	Coasts Under Stress	Memorial University of Newfoundland
Anapuma Hewa	Coasts Under Stress	Memorial University of Newfoundland
G. Hill	Anthropology	University of Victoria
Alison Hounsell	Coasts Under Stress	Memorial University of Newfoundland
Michelle Howell	Recreation Programme	Dalhousie University
Rebecca Hudson-Breen	Educational Psychology and Leadership Studies	University of Victoria
Sheldon Huelin	Earth Sciences	Memorial University of Newfoundland

Research Assistants	Department	Organization
Danny Ings	Biology	Memorial University of Newfoundland
Jennifer Jackson	Biology	University of Victoria
Michael Januska	Economics	University of Victoria
Stefan Jensen	History	Memorial University of Newfoundland
Heather Johnstone	Anthropology	University of Victoria
Chantelle Jones	Biopsychology	Memorial University of Newfoundland
Jessica Jorna	Earth and Ocean Science	University of Victoria
Robert Kean	Sociology	Memorial University of Newfoundland
Brenda Kitchen	Coasts Under Stress	Memorial University of Newfoundland
Brian Kopach	Biology	University of Victoria
Leanne Kryger	Biology	Memorial University of Newfoundland
Dianne Labonte	History	Memorial University of Newfoundland
Cynthia Lake	Anthropology	University of Victoria
Erin Lane	Biology and Geology	Memorial University of Newfoundland
Yew Jin Lee	Curriculum Studies, Education	University of Victoria
Heather Legge	Psychology, Math, and Statistics	Memorial University of Newfoundland
Wendy Lendrum	Education	University of Victoria
Jason Letto	Earth Sciences	Memorial University of Newfoundland
Tanya Lopez	Sociology	Memorial University of Newfoundland
Ryan MacDonald	Economics	University of Victoria
Debbie Martin	Health Education	Dalhousie University
Emma Mason	Education	University of Victoria
Lani Maxwell	Educational Psychology and Leadership Studies	University of Victoria
Margaret Ann McCarthy	Coasts Under Stress	Memorial University of Newfoundland
Keith Mercer	History	Memorial University of Newfoundland
Peter Milley	Education	University of Victoria
Lori Morgan	Biology	Memorial University of Newfoundland
Kimberley Morrissey	Earth Sciences	Memorial University of Newfoundland
Ian Mosbey	History	University of British Columbia
Heather Nelson	Geography	University of Victoria
Andrew (Jiye) Ni	Economics	University of Victoria

Research Assistants	Department	Organization
Ioan Nichifor	Economics	University of Victoria
Becci Noel	Coasts Under Stress	Memorial University of Newfoundland
Chad O'Brien	Philosophy	Memorial University of Newfoundland
Niall O'Dea	Biology	Memorial University of Newfoundland
Jacqueline O'Driscoll	Earth Sciences	Memorial University of Newfoundland
Chris O'Grady	Fisheries Centre	University of British Columbia
Trevor Orchard	Anthropology	University of Toronto
Peter Panek	Institute of Fisheries Analysis	Simon Fraser University
Grace Parr	Nursing	Memorial University of Newfoundland
Diane Peacock	Geography	University of Victoria
Josh Pennell	Biology	Memorial University of Newfoundland
Brian Penney	Ocean Sciences	Memorial University of Newfoundland
Cindy Phillips	Educational Psychology and Leadership Studies	University of Victoria
Erika Pittman	Coasts Under Stress	Memorial University of Newfoundland
Tim Pittman	Ocean Sciences	Memorial University of Newfoundland
Byron Plant	History	University of Victoria
Elmar Plate	Education	University of Victoria
Victoria Plowman	Coasts Under Stress	Memorial University of Newfoundland
Jason Pollard	Environmental Science	Memorial University of Newfoundland
Andrea Powell	Coasts Under Stress	Memorial University of Newfoundland
Derek Pritchett	Ocean Sciences	Memorial University of Newfoundland
Samuel Ramsden	Biology	University of Victoria
Carolyn Ramsey	Anthropology	University of Sheffield
Mia Reimers	History	University of Victoria
Aaron Retzlaff	Biology	Memorial University of Newfoundland
Claire Rillie	Geography	Memorial University of Newfoundland
Tamara Rozeck-Allen	Counselling Psychology	University of Victoria
Lisa Russell	Political Science	Memorial University of Newfoundland
Sara Russo Garrido	Rhodes Scholar	Oxford/Memorial University of Newfoundland

Research Assistants	Department	Organization
David Sampson	Ocean Sciences	Memorial University of Newfoundland
Gregory Saunders	Coasts Under Stress	Memorial University of Newfoundland
John Scaplen	Earth Sciences	Memorial University of Newfoundland
Torben Schau	History	University of Victoria
Christine Shaw	History	University of Victoria
Colleen Shepherd	Educational Psychology and Leadership Studies	University of Victoria
Trudi Smith	Environmental Studies	University of Victoria
Michael Sooley	Sociology	Memorial University of Newfoundland
Jason Stanley	Rhodes Scholar	Oxford/Memorial University of Newfoundland
Graigory Sutherland	Earth and Ocean Sciences	University of Victoria
Susan Tam	Economics	Simon Fraser University
Susannah Taylor	Health Education	Dalhousie University
Alkanani Thamir	Ocean Sciences	Memorial University of Newfoundland
Frances Thorsen	Education	University of Victoria
Jann Ticknor	Health and Human Performance	Dalhousie University
Sandra Umpleby	Educational Psychology and Leadership Studies	University of Victoria
Kim Valenta	Anthropology	University of Calgary
Gunita Wadhwa	Education	Memorial University of Newfoundland
Katherine Waller	Education	University of Victoria
Mary Walter	Economics	University of Victoria
David Warner	Education, Psychology, and Leadership Studies	University of Victoria
Theresa Watson	Ocean Sciences	Memorial University of Newfoundland
Jeff Webb	Sociology	Memorial University of Newfoundland
Jeanette Wells	Ocean Sciences	Memorial University of Newfoundland
Laura Wheeler	Ocean Sciences	Memorial University of Newfoundland
Veryan Wolsak	Education	University of Victoria
Louisa Wood	Fisheries Centre	University of British Columbia
Zhiliang Xing	Ocean and Naval Architectural Engineering	Memorial University of Newfoundland
Sheila Yeomans	History	University of Victoria
Sylvania Yuen	Economics	University of Victoria
Danielle Zietsma	Economics	University of Victoria

Other Staff	Department	Organization
Catherine Baur	Senior Clerk	St Mary's University
Rhonda Bowring	Transcriber	Memorial University of Newfoundland
Paula Cowx	Interviewer (from research community)	Memorial University of Newfoundland
Patsy Cull	Interviewer (from research community)	Memorial University of Newfoundland
Angela Drake	Project Coordinator (East Coast)	Memorial University of Newfoundland
Sandra Estabrooks	Interviewer (from research community)	Memorial University of Newfoundland
Vivian Gaulton	Transcriber (from research community)	Memorial University of Newfoundland
Pamela Harris	Transcriber	Memorial University of Newfoundland
Carrie Holcapek	Project Coordinator (West Coast)	University of Victoria
Scott Hutchings	Interviewer (from research community)	Memorial University of Newfoundland
Coleen Hynes	Interviewer (from research community)	Memorial University of Newfoundland
Cathy King	Secretary (East Coast)	Memorial University of Newfoundland
Marjorie Lavers	Interviewer (from research community)	Memorial University of Newfoundland
Vanessa Lavers	Interviewer (from research community)	Memorial University of Newfoundland
Kari Marks	Accounts Administrator (West Coast)	University of Victoria
Cathy Matthews	Transcriber	Memorial University of Newfoundland
Joan Muir	Transcriber	Memorial University of Newfoundland
Nicole Offrey	Interviewer (from research community)	Memorial University of Newfoundland
Janet Oliver	Administrator (East Coast)	Memorial University of Newfoundland
Patsy Pilgrim	Interviewer (from research community)	Memorial University of Newfoundland
Lillian Russell	Interviewer (from research community)	Memorial University of Newfoundland
Bonnie Ryan	Interviewer (from research community)	Memorial University of Newfoundland
Joan Sheppard-Wells	Transcriber	Memorial University of Newfoundland
Kristopher Smith	Interviewer (from research community)	Memorial University of Newfoundland
Maureen Smith	Interviewer (from research community)	Memorial University of Newfoundland

Other Staff	Department	Organization
Moira Wainwright	Secretary (West Coast)	University of Victoria
Kathy Watson	Secretary to West Coast Accounts Administrator	University of Victoria
Susan Williams	Transcriber	Memorial University of Newfoundland

Glossary of Technical Terms

By-catch   All species captured other than target species. The by-catch is
often discarded because it is the illegal size of a regulated species or a
species of low market value, or because the quota has been reached or
the fish are damaged or of poor quality (Crean and Symes 1994). The
by-catch encompasses discards along with non-targeted catch or incidental catch or capture (Alverson 1977; Alverson et al. 1994; Hall,
Alverson, and Metuzals 2000; chapter 3).

Data fouling   Fishers deliberately under-reporting catches in order to
avoid detection.

Discarding   The selective removal of harvested fish (by size), which usually
occurs for economic reasons such as marketability, although ecological
value and the sensitivity of public opinion have also been cited
(Alverson et al. 1994, Hall 1996). Discard rates in most fisheries are
high, variable, and very difficult to estimate (Alverson et al. 1994;
Borges et al. 2001). In the recent past, discards were not regulated and
often went unreported (chapter 3).

Discards   Fish harvested but returned to the sea. Discards might include
both targeted and non-targeted species and by-catches.

Gap disturbance   Any event or force – e.g., windthrow, root-rot, fire – that
removes individual trees or groups of trees within a forest, leaving gaps
in the canopy and creating opportunities for understory species or tree
seedlings requiring higher light intensity to grow. Spatial and temporal
variation in the gap disturbance regime can result in differing degrees
of biodiversity.

High-grading   Discarding of fish of lesser quality (size or condition).

Illegal landings   Catches of fish that are landed or sold in contravention of
fisheries regulations.

Misreporting Deliberate and willful distortion of true catch statistics: this practice includes illegal activities and in this sense, both under- and overreporting of catches.

Population health studies The study of the complex interactions (biological, social, cultural, and environmental) that determine the health of individuals, communities, and global populations; and the application of that knowledge to improve the health of both populations and individuals (adapted from the mandate of the CIHR Institute for Population and Public Health).

Price dumping Discarding fish to obtain a better price: nonselective process of returning whole or partial (unculled) catches to the sea. This dumping occurs when the quantity of fish exceeds quotas or processing capacity.

Quota busting Catching a larger quantity of fish than the individual quota allows.

Resilience The capacity to survive under stressful conditions by employing strategies for adaptation or change.

Unreported catches This category includes unreported discards, which may or may not be legal; unmandated catches, which are catches that a given agency is not mandated to record; and illegal catches.

Unreported fishing Fishing which is unreported, misreported, or underreported.

Glossary of Species Mentioned in the Text, by Scientific Name

Scientific Name	Organism
Acipenser spp.	sturgeons
Alces alces	moose
Alosa pseudoharengus	alewife
Anarhichas denticulatus	northern wolffish
Anarhichas spp.	wolffishes
Anarhichas lupus	Atlantic wolffish
Anarhichas minor	spotted wolffish
Anoplopoma fimbria	sablefish, Alaska cod, black cod
Antimora rostrata	blue antimora, blue hake
Arctogadus glacialis	Arctic cod
Argentina egedii	Pacific silverweed
Artediellus uncinatus	snowflake hookear sculpin
Balaena mysticetus	bowhead whale
Bankia setacea	teredo worm, shipworm
Branta bernicla	brant goose
Brosme brosme	cusk
Caligus clemensi	sea lice
Callorhinus ursinus	northern fur seal
Cancer magister	Dungeness crab
Castanea dentata	American chestnut
Chionoecetes opilio	snow crab
Clupea harengus	Atlantic herring
Clupea pallasii	herring
Cornus canadensis	bunchberry
Cottidae (family)	sculpins
Crassostrea gigas	Pacific oyster
Enhydra lutris	sea otter
Eubalaena glacialis	right whale
Eumetopias jubatus	Steller's sea lion

Scientific Name	Organism
Fucus spp.	marine brown algae
Gadus morhua	Atlantic cod, northern cod
Gadus ogac	Greenland cod
Gaultheria shallon	salal
Halichoerus grypus	Atlantic, gray seal
Haliotis spp.	abalones
Hemitripterus americanus	sea raven
Hippoglossoides platessoides	American plaice
Hippoglossus stenolepsis	Pacific halibut
Homo sapiens	human
Hydrodamalis gigas	Steller's sea cow
Lebbeus groenlandicus	lobster
Lepeophtheirus spp.	sea lice
Lepus spp.	rabbits
Lophius americanus	goosefish
Lottia alveus	eelgrass limpet
Macrocystis integrifolia	kelp
Mallotus villosus	capelin
Melanogrammus aeglefinus	haddock
Metridium spp.	anemones
Mirounga angustirostris	northern elephant seal
Mitchella repens	partridgeberry
Morone saxatilis	striped bass
Munida quadrispina	squat lobster
Myoxocephalus octodecemspinosus	longhorn sculpin
Mytilus edulis	blue mussel
Nereocystis luetkeana	bull kelp
Odobenus rosmarus	walrus
Oncorhynchus spp.	Pacific salmon
Oncorhynchus gorbuscha	pink salmon
Oncorhynchus keta	chum salmon
Oncorhynchus kisutch	coho or silver salmon
Oncorhynchus nerka	sockeye salmon
Oncorhynchus tshawytscha	chinook salmon
Ophiodon elongates	ling cod
Orcinus orca	orca, killer whale
Oreamnos americanus	mountain goat
Pandalus borealis	northern prawn
Panopea abrupta	geoduck clam
Parastichopus spp.	holothurians
Peprilus triacanthus	butterfish
Phoca groenlandica	harp seal

Scientific Name	Organism
Phoca vitulina	harbour seal
Phoebastria sp.	albatrosses
Picea sitchensis	sitka spruce
Pinus monticola	western white pine
Pinus strobus	eastern white pine
Pollicipes polymerus	gooseneck barnacle
Porphyra abbottiae	red laver seaweed
Pseudopleuronetes americanus	winter flounder
Pycnopodia helianthoides	sunflower sea star
Rangifer tarandus	caribou
Reinhardtius hippoglossoides	turbot, Greenland halibut
Rubus chamaemorus	bakeapple, cloudberry
Salmo salar	Atlantic salmon
Saxidomus gigantea	butter clam
Scomber scombrus	Atlantic mackerel
Sebastes spp.	rockfishes
Sebastes faciatus	redfish, ocean perch
Sebastes marinus	redfish, ocean perch
Sebastes mentella	redfish, ocean perch
Shepherdia canadensis	soapberry
Somateria mollissima	eider duck
Spartina spp.	cord grasses
Squalus acanthias	dogfish
Strongylocentrotus spp.	sea urchins
Strongylocentrotus droebachiensis	green sea urchin
Strongylocentrotus franciscanus	red sea urchin
Symplocarpus foetidus	skunk cabbage
Thaleichthty pacificus	eulachon, ooligan, oulachon
Thuja plicata	western red cedar
Tricholoma magnivelare	pine mushroom
Triglops murrayi	moustache sculpin
Tsuga spp.	hemlocks
Tsuga heterophylla	western hemlock
Ulmus americana	American elm
Urophycis tenuis	white hake
Ursus spp.	bears
Ursus americanus	black bear
Ursus americanus kermodei	kermode, or spirit bear
Ursus arctos	brown bear
Venerupis philippinarum	Manila clam, Japanese littleneck
Viburnum edule	highbush cranberry
Zalophus californianus	California sea lion
Zostera marina	eelgrass

Usage note. Capital letters have been used only for names associated with places, for example, Pacific silverweed. Plurals have been used when more than one species was indicated.

Notes

1 See also Harold Lasswell's notion that "values" such as wealth and power tend to stick together (Lasswell 1977).

2 But we refer readers to a thorough and excellent review of the literature to 1999 by I. Scoones (1999), from which we draw, while extending the discussion to deal with more recent work of significance to CUS.

3 Other resilience thinking – actor network theory (Law and Hassard 1999; Latour 1987) and political ecology (M'Gonigle 1999; Dalby and Keil 2003) – avoids the pitfall that is present in so much systems theory, which assumes that systems will tend to maintain or restore equilibrium without essentially changing the relationships among the parts. Actor network theory in particular recognizes that systemic change will alter such relationships, but it fails to deal with issues of choice and responsibility.

4 Preamble to the Constitution of the World Health Organization as adopted by the International Health Conference, New York, 19–22 June, 1946; signed on 22 July 1946 by the representatives of sixty-one states (WHO 1948, 100) and entered into force on 7 April 1948. The definition has not been amended since 1948.

5 See Sinclair and Ommer (2006): differences of power, which underlie the processes we discuss in this book, are so important that we have devoted this separate volume to a close examination of power.

6 Borrowing from Wackernagel and Rees (1996.)

7 The recent *Proceedings of the World Summit on Salmon* (Gallaugher and Wood 2004) is a case in point. The authors' goal is laudable, their concern demonstrable, and the science important. Their logo reads "Speaking for the Salmon," which is a fine motivation ... but the findings are sadly constrained by very restricted social science input. The economist contributor there (William Rees), however, presented a world history perspective that (while it

lumps all humans together as "inherently biased against sustainability" [150])
– has very important things to say about the negative effects of globalization.

8 Appendix 2 names the team. This book is team written: we use the collective
"we," "us" and "our(s)" throughout, because the work of individual researchers and research groups is reported in this text, which has been brought
together through team contribution and discussion.

9 For a more concentrated discussion of the implication of our work for the
creation and flow of knowledge see Lutz and Neis (forthcoming), particularly
chapter 2: Ommer, Coward, and Parrish. We have also considered the different ways in which knowledge is transmitted and have ourselves disseminated
our work using a range of transmission vehicles, some traditional or formal
(for example, academic journal articles and books) and some less so (for
example, brochures, posters, a website, and film).

10 Warren's model, like ours, draws on complex systems theory, especially
panarchy, and adds Sewell's (1992) adaption of Giddens' structuration theory in order to bring out the mutual interdependence of micro and macro
action, to use a scale metaphor, or agency and structure, as emphasized by
Giddens. The emerging and controversial character of this approach (and of
our own too) is reflected in the critical responses to Warren, especially by Bell
(2005), who challenges the system metaphor and the peripheral position of
power in the analysis as he sees it.

CHAPTER TWO

1 This chapter draws on the work of Alcock (2%), D. Harris (10%), Ings (5%),
Kennedy (2%), Murray (10%), Neis (30%), Vodden (9%), Rogers (2%),
Ryan (2%), and Whelan (2%). It also draws from the work of Pitcher,
Ainsworth, Sumaila, and Haggan, and other members of the Back to the
Future team at the Fisheries Centre at UBC (24%), and the jointly undertaken
work of Mackie and Fedje, the latter of Parks Canada (2%).

2 The trophic level is the relative position occupied by an organism in a food
chain or web, with primary producers (e.g., phytoplankton) constituting the
first trophic level.

3 Back to the Future (BTF) uses collaborative reconstruction of past ecosystems
to set benchmarks of trophic structure, abundance, and biodiversity. Knowledge of the past is then used to inform policy goals for the restoration and
sustainable management of fisheries: "to determine where we were, how that
relates to where we are, and where we could be" (Pitcher 2005). Full details
of data sources, methodology, collaborative workshops, and construction of
the eight models for this project are published in a series of six reports
(Pitcher, Vasconcellos, Heymans, Brignall, and Haggan 2002; Ainsworth,
Heymans, Pitcher, and Vasconcellos 2002; Pitcher, Heymans, and

Vasconcellos 2002; Pitcher, Power, and Wood 2002; Heymans 2003d; Pitcher, 2004d).

4 This is, of course, our central research question.

5 A fine-detailed history, focused on power relations, can be found in D.C. Harris (2006). While we were unable to carry out a detailed study of traditional ecological knowledge (TEK) on that coast, some TEK work was done along with the mass balance modelling, in order to ground-truth the models against First Nations experience and memory.

6 We have drawn here from the papers published in the six reports mentioned in note 3 above, notably Orchard and Mackie (2004), Pitcher (2004a, 2004b, 2004c), and Heymans and Pitcher (2004).

7 The interview database is searchable at www.fisheries.ubc.ca/projects/btf.

8 See the analysis of Lekwiltok expansion in Galois (1994).

9 The report of the First Nations Panel on the Fisheries recommended that, as a starting point, First Nations should hold 50 percent of the fisheries (First Nations Panel on Fisheries 2004).

10 There is some discrepancy between this information, commonly believed to be true by Alert Bay residents, and Cicely Lyons' (1969) history of BC Packers, which states that the saltery was established by two settlers named Mack and Neill in the early 1860s (rather than in 1870), who later sold the operation to Spencer and Earle of Victoria.

11 We have drawn here on the Fisheries Centre Reports once again: Bundy (2002); Burke et al. (2002); Lilly (2002); Vasconcellos and Heymans (2002); Vasconcellos, Heymans, and Pitcher (2002); Vasconcellos et al. (2002); Heymans and Pitcher (2002a, 2002b). Note also the local extinction of the walrus, grey seal, and Atlantic grey whale, and the global extinction of the great auk (the original penguin), which had considerable effects on the ecosystem structure (Pitcher 2004c).

12 Total system throughput expresses flow across the whole food web in mass-balanced Ecopath models.

13 The literature is extensive. See references in Ommer (1994, 2002).

14 We use the problematic term "settler" here to refer to Labrador's permanent population, most of whom were of mixed European-Inuit ancestry. Although the term "settler" was most commonly used by Moravian missionaries in northern Labrador, it was also used in central, and to a lesser extent, southern Labrador, where the term "liveyere" (a corruption of "to live here") was more common. In central Labrador, the Hudson's Bay Company invariably used the word "planter" to refer to Labrador's first and succeeding generations of permanent residents. All these terms are problematic because they often referred to European founders and to their descendants of mixed ancestry. Today, many of the mixed-ancestry descendants of early "settlers," "liveyeres," or "planters" call themselves Métis.

15 Ecopath modelling approaches suggest that critical decreases in biomass actually occurred throughout the 1970s, after the period when landings peaked. Reduced maturity in the system at the time of the killer spike in the late 1960s suggests the onset of the ecological shift.

16 For a detailed ecological and demographic history of the Strait of Belle Isle (both shores) see Thornton (1979).

CHAPTER THREE

1 This chapter is drawn from the work of Bavington (5%), Copes (6%), Davis (7%), Haedrich (10%), Lutz (5%), Metuzals (12%), Murray (11%), Murrin (8%), Neis (10%), Vodden (16%), and Wernerheim (10%).

2 Precision is a measure of the variability of data around its mean value, whereas accuracy is a measure of the closeness of a measured or computed value to its true value (NRC 2000).

3 It fell in 1987, however, due in part to the use of larger meshes and of square webbing and in part to the observer program.

4 Discarding of major species such as redfish and flatfish doubled in the early 1980s. Atkinson (1984) described discarding of young redfish in the shrimp fishery.

5 A few brief definitions are in order: *discarding* is the selective removal of fish (by size) and usually occurs for economic reasons such as marketability, although ecological value and a sensitivity of public opinion have also been cited (Alverson et al. 1994; Hall 1998). Discard rates in most fisheries are high, variable, and very difficult to estimate (Alverson, Sweeting, and McKeown 1994; Borges et al. 2001). In the recent past, discards were not regulated, and these discards often went unreported. *By-catch* is the non-targeted catch, often discarded because it is the illegal size of a regulated species or a species of low market value or because the quota has been reached or the fish are damaged or of poor quality (Crean and Symes 1994). *By-catch* encompasses discards plus non-targeted catch or incidental catch or capture (Alverson 1977; Alverson, Sweeting, and McKeown 1994; Hall, Alverson, and Metuzals 2000).

6 It is not known exactly how much illegal fishing is taking place, but we know that for some important fisheries, fishing accounts for a large percentage of total catches, and the amount of it world-wide appears to be increasing (FAO 2002). Evans (2000) estimates that fish stocks are underreported by 75 percent and high-sea stocks by 100 percent.

7 For more detailed discussion of the results of this study see Metuzals (forthcoming). We thank the many fishers for their participation and time that they offered in this study. Further thanks to Michael Ryan and Danny Ings who also helped out. Many thanks to Martin Hall who provided many helpful comments on the questionnaire.

8 Illegal landings and discards can be described as being almost like a moral hazard that arises from individual catches that are unobservable to society and hence are private information (Jensen and Vestergaard 2002).

9 When fishing for crab, most fishers will use a combination of mesh sizes in their pots (e.g., 50 percent of the pots with 5¾ inch mesh and 50 percent with 5½ inch mesh). The pots are designed with the view that the larger animals get trapped and that the females or smaller ones get out. The mesh size regulation is 5¼ inches, which ensures that the minimum legal size of 95 millimetre carapace width (cw) is respected. This regulation excludes females from the fishery while the adult males in the population remain available for reproduction.

10 Fisheries Resource Conservation Council, Ottawa. http://www.frcc-ccrh.ca.

11 See http://eng.msc.org.

CHAPTER FOUR

1 This chapter draws on the work of Cadigan (37%), MacDonald (8%), McLaren and Pollard (9%), and Rajala (46%).

2 One of British Columbia's early Japanese union movements began with a strike at Swanson Bay in 1919 (Campbell 1993, 71).

3 Throughout the summer of 1900, for example, people who lived near Mount Moriah (in what is now a suburb of Corner Brook) turned out periodically to fight fires caused by cinders from Reid railway locomotives (Cadigan 1999).

4 The Newfoundland government initially negotiated with the International Paper Company of New York for the development of the Corner Brook mill. The American parent company operated locally as the International Power and Paper Company of Newfoundland, Limited. The company acted in partnership with Quebec mills owned by the Canadian International Paper Company, and a mill at Dalhousie, New Brunswick, operated by the New Brunswick International Paper Company (Great Britain 1933, 4).

5 Student Mark McLaughlin, with Sean Cadigan, interviewed eleven loggers from White Bay South in June 2004. These interviews are important evidence for McLaughlin 2004.

6 Hereafter, we refer to the company as Kruger. In 1994, when the mill was in financial difficulty, the provincial government repurchased some company lands at an inflated price ($17 million, compared with an estimated market value of $3 million), including areas on the Northern Peninsula. Kruger retained logging rights, including first right to purchase wood cut by independent contractors.

7 In July 2005, Abitibi Consolidated announced the closure of the Stephenville mill and their plans to shut down the Number 7 paper machine in Grand Falls-Windsor, with a loss of four hundred jobs province-wide (Jackson 2005).

8 See Rajala (2006), the major study conducted in partnership with the Royal British Columbia Museum (and published jointly with them), which spells out this history in political, legal, industrial, human, and environmental terms. This whole section draws from that manuscript in its draft form.

9 For details of the operation, government activities, and working and living conditions, see Rajala, chap. 3.

10 Sloan had been chair of the 1845 commission that implemented sustained yield and the Forest Management Licences system.

11 The results are based on a 1977 survey of residents of Ocean Falls, Gold River, and Tasu by Robert F. Wilson and John Campbell, cited in Ocean Falls Corporation (1980, 78).

CHAPTER FIVE

1 This chapter draws on the work of Dennis and G. Kealey (17%), Rennie (25%), Griffin (31%), Schofield and Davison (3%), and Wilton and Riggs (24%).

2 For some of the formative statements on this debate in the Canadian literature, see Innis (1936) and Careless (1954). Clement (1980, 15–26) interprets industrial relations at INCO as shaped largely by a combination of Innis's staples theory and Marxist concepts of capital accumulation and worker control. On underdevelopment and dependence in the Atlantic region, see Brym and Sacouman (1979); Acheson, Frank, and Frost (1985); Samson (1994); and Burill and McKay (1987).

3 For most historical information we had to rely upon two often contradictory written records from Dr Moses Harvey, a physician and prominent natural scientist (and promoter) who was a widely read and respected commentator in Newfoundland during the latter half of the nineteenth century, and from James Howley, the well-known chief geologist with the Newfoundland Geological Survey, protégé of Alexander Murray (founder of the survey) and brother of the Catholic archbishop of St John's. For information on the methodologies and long-term environmental effects of the operations, we had to derive our own geochemical data.

4 In this chapter "ton" is the imperial ton, also called the UK ton or long ton, and is equal to 1.016 tonnes, or 2,240 pounds (not to be confused with the US, or short, ton at 2,000 pounds; the tonne used elsewhere in this volume, is 1,000 kilograms, or 2,205 pounds).

5 The value depended, of course, on the degree of refinement. Figures quoted in the customs returns are usually average prices.

6 Some coal was quarried by the Reid Company, from deposits near the railway line, and for a very brief period coal from Reid properties in South Branch and Grand Lake helped supply the railway. See Martin (1983, 38–9) and Snelgrove (1953, 5).

7 Most of the supervisory staff, as well as other personnel, were drawn from Phoenix and Grand Forks.

8 The electric railway in the mine used 3-foot gauge, 56-pound steel rails, and haulage was performed by 6-ton electric locomotives. The cars were of 3.7-ton capacity, self-dumping steel construction.

9 A very large proportion of the 1869 population was between twenty and forty years old or under ten years old, and in all categories except the one-to nine-year-olds, there were nearly twice as many males as females (Census of Newfoundland 1869).

10 Of the 768 people in Tilt Cove in 1869, 667 had been born in Newfoundland and 101, or 13 percent, in England or elsewhere in the British Isles. The proportion for all of Newfoundland in the same census year was just 0.05 percent (Report on the Census of Newfoundland 1921).

11 What data are available suggest that Copper Cliff in Ontario had a similarly ethnically diverse community. Very clearly, however, Copper Cliff was a much more regimented society than the one at Anyox. For example, the company-sponsored recreation facilities at Copper Cliff were restricted to Anglo-Saxons (Goltz 1989, 158, 162) but not at Anyox.

12 Both 1869 figures are from the Census of Newfoundland for that year. The Bett's Cove figure for 1878 is based on Harvey's estimate.

13 Unfortunately no plans exist for the buildings, but clearly the cottages were built to a set architectural design for this type of accommodation. Certainly this was the approach adopted by the company four years later at Cassidy, Vancouver Island, when a planned community was created to support the company's coal-mining operation.

14 Anyox was clearly a company town, despite being referred to as a model town in the *Mining and Scientific Press* (1913, 163): "A model town is being laid out with all the conveniences."

15 Margaret Crawford (1991, 53) notes: "Not all company towns employed such dramatic forms of social control. Industries that needed to attract skilled workers found it more to their advantage to create towns where their power was less evident."

16 Topography has been used in other locations, but the seemingly clear distinctions that appear in Anyox are not so readily apparent. In fact, most company towns seemed to tend toward homogenizing the landscape rather than intentionally using it to create distinct enclaves of social groups. Goddard found that it was Steppoe, the ancillary open community, that tended to allow topography to dictate community dispersion, rather than McGill, the associated company town (Goddard 1999, 72, 77).

17 Loudon (1973, 57) notes that young men who made the trip across the slag pile to visit the small community were struck off the list of those permitted to call on their daughters.

18 Goltz has also pointed this out for Copper Cliff (1989, 21).

19 Death statistics compiled by Robert Griffin and Susan Johnson from data in Vital Statistic Records at the BC Archives and from Coroner Reports at the BC Archives.

20 A siemen (s) is the International System of Units (si) unit of conductance. One siemen is the conductance at which a potential of one volt causes a current of one ampere to flow; conductivity is measured in siemens per centimetre.

21 As a result of this work, Huelin (2004) developed a new analytical technique (Laser Ablation – Inductively Coupled Plasma – Mass Spectrometry (LAM-ICP-MS) for the analysis of base-metal distributions and environmental monitoring of iron-manganese (Fe-Mn) oxide coatings on stream pebbles, which should be useful in future analyses of this kind.

22 Goltz (1989, 21) has also pointed this out for Copper Cliff.

23 In December 1870, in fact, the same Elias Marett who had predicted great things for Tilt Cove only six months earlier noted that the slump in copper demand had caused Tilt Cove operations to be "in a great measure suspended, and a large number of the operatives dismissed" (Marett 1870).

24 The prohibition against non-French settlement and resource exploitation on a portion of the Island's west coast had long been a source of controversy in the mining and exploration sectors.

25 This coincides with a study of post–World War II company towns in British Columbia. Individual hardship is less when the company maintains property ownership (Western Economic Consulting Ltd and Clayton Resources Ltd n.d.).

26 For a more comprehensive account of the events and negotiations leading up to signing the Atlantic Accord, see Nemeth (1997), Crosbie (2003), and Dennis (2005).

27 This appears surprising considering that the Accord was signed just three years after the sinking of the oil rig Ocean Ranger and the loss of eighty-four lives on the Hibernia field. It should be noted, however, that a royal commission inquiring into the Ocean Ranger tragedy had yet to conclude its work.

28 Note 2 of schedule 2 of the federal Petroleum and Natural Gas Act stated that "no drilling licences will be issued within the Provincial Inland Marine Zone defined in BC regulation 237/81" (British Columbia 1989, schedule 2, footnote 2).

29 Two important conclusions regarding the moratorium on oil and gas activities include (1) providing an adequate regulatory regime is put in place, there are no science gaps that need to be filled before lifting the moratoria on oil and gas development; and (2) the present restriction on tanker traffic in transit along the West Coast of North America from entering the coastal zone should be maintained for the time being. See Royal Society of Canada (2004, xix).

30 The Coasts Under Stress website has a guide containing step-by-step descriptions of all the regulatory approval processes for the Newfoundland offshore industry.

31 The EIS for Hibernia was completed in 1985, based on field data gathered in 1980–81. It became the basis for an EPP through a process that evolved over time.

32 For example, the effects of normal operation at Hibernia and Terra Nova on seafloor life are predicted to attenuate with distance out to two kilometres. At Whiffen Head, the prediction is that there will be no effect of normal operations beyond the boat basin. The same design is used for all East Coast oil and gas projects that have point-source effects.

CHAPTER SIX

1 This chapter draws on the work of B. McLaren (10%), Turner (20%), Bornhold and Harper (40%), Picard (8%), Wroblewski (15%), and Neis and Murray (7%).

2 We also conducted a thorough review of siting, monitoring, and abandonment of log-handling leases policies in Canada and other jurisdictions, including guidelines from the Department of Fisheries and Oceans (G3 Consulting Ltd 2003). We hope our work will result in policy improvements that will lead to better siting of log booms such that environmental impacts will be fewer and more comprehensive approaches to monitoring their effects through the life of the lease will be established.

3 Bornhold (1978), for example, describes carbon:nitrogen (C:N) ratios from a wide range of British Columbia marine environments – in general, fjord sediments ranged from about 6.85, for Saanich Inlet and Quatsino Sound, to 14.3 near Kitimat (Douglas Channel). The latter reflects a contribution of about 18 percent terrestrial plant material.

4 The reduction of infaunal prey species (those living *within* the sediments on the seafloor) in wood-dominated habitats (previously documented by several authors such as Pease [1974], Conlan and Ellis [1979], and Jackson [1986]) probably accounts for low predator abundance in these habitats.

5 This section draws substantially on the PhD dissertation of Reade Davis, Memorial University of Newfoundland (Davis forthcoming).

6 During our taxonomic interviews with retired fish harvesters, we asked them to provide us with local names for eighty-three different species of fish, marine invertebrates, birds, and seals. From thirty interviews we compiled a list of twenty-one names for cod. A maximum of seven other names was given for the other eighty-two species. Generally, names for cod referred to size and appearance, but activities such as feeding (e.g., herring cod, capelin scull) were common as well.

CHAPTER SEVEN

1 This chapter is based on the work of Bornstein (21%), Dalton (21%), Dolan (17%), Kean (20%), and Tomblin (21%).

2 The history of Newfoundland and Labrador's health care system has been studied by CUS team member Linda Kealey in work to be published elsewhere.

3 As the Newfoundland Hospital and Nursing Home Association (NHNHA) put it, "Each new advance that gives hope to another category of sufferers converts a latent need into an immediate and continuing demand" (NHNHA 1993, 3).

4 It should be noted that only $12 billion of this money will be transferred directly to the provinces; most of the rest of the $34.8 billion has been allocated to a reform fund for primary health care reform, home care, and catastrophic drug coverage.

5 "People no longer want the health care system to deliver basic services but now expect to have all of the options, all of the time, wherever they live" (Provincial and Territorial Health Ministers 2000, 47).

6 The numbers for the western region are, again, quite similar: 23 percent of people in the western region smoked daily, 52 percent were overweight or obese, and 6 percent reported having diabetes.

7 The figures cited here do not include the three territories.

8 House noted that "it is becoming increasingly difficult to finance the needs of the system as presently structured."

9 Unable to find additional savings, CEO George Tilley called upon the department to bring in "a fresh set of eyes."

10 Rurality is assessed using six weighted variables: remoteness from a basic referral centre, remoteness from the closest advanced referral centre, the drawing population, the number of general practitioners, the number of specialists, and the presence of an acute hospital. Rural areas are defined as communities of up to ten thousand people where most medical care is provided by a small number of general practitioners with limited or distant access to specialist services and high technology health care facilities. Rural remote areas are rural communities about eighty to four hundred kilometres, or about one to four hours transport in good weather, from a major regional hospital (Provincial Co-ordinating Committee for Remote and Rural Health Services 1999, 25–6).

11 Based on the World Health Organization definition of health care, the committee stated that primary care is "essential health care based on practical, scientifically sound, and socially acceptable methods and technology made universally accessible to individuals and families in the community through their full participation and at a cost that the community and the country can afford to maintain at every stage of self-reliance and self-determination. It is the first level of contact of the individual, the family, and the community with

the national health system, bringing health care as close as possible to where people live and work, and constitutes the first element of a continuing care process" (Provincial Co-ordinating Committee for Remote and Rural Health Services 1999, 25).

12 The Northwest Health Authority represents Bulkley Valley, Kitimat, Snow Country, Upper Skeena, North Coast, Stikine, Terrace, Queen Charlotte Islands, and the Nisga'a Health Authority. The steering committee consisted of people from the nursing and medical professions, each of the Health Councils, the North West Community Services, the North West Regional Hospital District, the Aboriginal population, and a senior member of the MOH (North West Health Authority 2000).

13 The NLHNA placed this figure closer to $90 million.

14 These facilities include Curtis Memorial Hospital in St Anthony, the Mary's Harbour coastal nursing station, and the Community Health Centres in Forteau, Flower's Cove, and Roddickton. There are also four other nursing stations in this region.

15 We have arrived at these numbers by comparing the figures on page 43 of the 1984 *Report of the Royal Commission on Hospital and Nursing Home Costs* with figures from William Nycum and Associates (1994, D5–D16).

16 The four facilities in the western health region that fall within our study area – Western Memorial Hospital, Sir Thomas Roddick Hospital, the Rufus Guinchard Health Centre in Port Saunders, and the Bonne Bay Health Centre in Norris Point – also lost a substantial number of beds between 1983 and 2001, though it is impossible to say exactly when these reductions occurred. Sometime over this period, the combined number of beds in these four facilities declined from 411 to 247 (Royal Commission on Hospital and Nursing Home Costs 1984, 43; Newfoundland and Labrador Department of Health and Community Services 2001b, 5).

17 The patterns for licensed practical nurses (LPNs) are virtually the same – the number of LPNs per 100,000 population in Newfoundland and Labrador increased by 14 percent between 1993 and 2001, at which time there were about 541.7 LPNs for every 100,000 citizens, and 2,889 in total. In the country as whole the corresponding number was 234.9, representing an 18 percent decline since 1993.

18 The Bay St George Status of Women's Council has argued that de-listing discourages some people from seeking medical advice or from staying home and taking care of themselves when they are sick.

19 In 1993 there were 8,142 physicians and 27,879 practising RNs. By 1997, the number of physicians had grown by 8 percent to 8,636. The number of RNs increased by 4 percent to 29,116. In 1993 there were 8,522 practising LPNs and RPNs; by 1997 there were only 7,462, a decrease of 10 percent (Penning et al. 2002).

20 Similar doubts were also raised about the advisability of separating cancer care from all other types of treatment.

21 The projected NITA/POP expenditures for 1998/99 were $800,000.

22 The number of communities eligible for a Northern Isolation Allowance has increased from 69 in 1989–90 to 103 in 1999–2000; the total number of doctors living and practising in these communities and eligible for NIA was 386 for 1999–2000, up from 212 in 1989–90.

23 Doctors may access up to twenty-eight days annually in five-day minimum blocks. Long-term locum coverage is provided to communities that were temporarily without resident physicians. In 1998–99, 1,670 days of locum coverage was provided in 165 assignments. The 1998–99 projected expenditures were $700,000. The external stakeholders include representatives from health authorities, the College of Physicians and Surgeons, the College of Family Physicians, the Union of BC Municipalities, the BC Medical Association, the BC Chamber of Commerce, the Registered Nurses Association (RNABC), the Central Interior Native Health Society, and the Professional Association of Residents of BC.

CHAPTER EIGHT

1 This chapter draws on the work of Taylor (40%), Solberg, Bornstein, Canning, and Sinclair (30%), and Dai, Nelson, and Dolan (30%).

2 In our review of demographic characteristics, we find that no published compilations of census data fit our study areas exactly. We have therefore chosen to report data for the 2003 health regions in Newfoundland and Labrador, while using the census regional districts as the unit of analysis for British Columbia, because these are the best fits we can achieve. Because the British Columbia data cover more than the communities we studied, we also report information for individual centres as appropriate.

3 The Grenfell region runs from New Ferrole on the Great Northern Peninsula to St Anthony to Harbour Deep and includes coastal Labrador south of Black Tickle.

4 Earlier research for the Great Northern Peninsula documents how out-migration of youth has been a constant part of life at least since the 1950s (Felt and Sinclair 1995; Palmer and Sinclair 2000). However, the decade 1991–2001 was the worst since the 1950s in this respect. This conclusion is based on cohort analysis of the relevant census data for 1991, 1996, and 2000, from http://www.statscan.ca/.

5 In the western region the population declined by 14 percent between 1991 and 2001. Between 1991 and 1996 net out-migration accounted for the loss of 6.4 percent of the population, the vast majority of whom were under thirty-five. The percentage of the population over forty increased from 35 percent to 42 percent over this period. Regional data were taken from the Community Accounts database at http://www.communityaccounts.ca/communityaccounts/onlinedata/default.htm.

6 We are unable to provide evidence of migration patterns for small places (those under about 3,000) because the Statistics Canada privacy requirement leads to potentially serious errors when small absolute numbers are rounded up or down, as is the case for many five-year age categories.

7 See also the discussion of health in chapter 1, and in the health literature more generally.

8 http://www.communityaccounts.ca.

CHAPTER NINE

1 This chapter draws on the work of Dai (15%), Dolan (7%), Grzetic (10%), Lutz (14%), MacDonald (22%), Neis (10%), Nelson (7%), Sinclair (10%), and Taylor (5%). It summarizes results presented elsewhere, including Bavington, Grzetic, and Neis (2004); Dolan et al. (2005); MacDonald, Grzetic, and Neis (2004); MacDonald, Neis, and Grzetic (2005); MacDonald, Neis, and Murray (2005); Power et al. (2005); and Sinclair et al. (forthcoming).

2 Specific East and West Coast examples throughout this chapter are drawn from case studies' fieldwork. On the East Coast we worked in the Labrador Straits, Port au Choix/Hawke's Bay and White Bay South. On the West Coast, we worked in Hartley Bay, Alert Bay, Port Hardy, Prince Rupert, Tofino, and Ucuelet.

CHAPTER TEN

1 This chapter is based on the work of C. Harris (15%), L. Kealey (35%), Shepard (5%), Shepherd (5%), Solberg and Parr (25%), and Turner (15%).

2 We adapted our interview and spatial-mapping research tools for gathering fish harvesters' ecological knowledge, and ethnobotanical tools developed in British Columbia and elsewhere, to develop this work.

3 See Ommer et al. (forthcoming) for details of food security and the informal economy in Newfoundland. See also Ommer and Turner (2004).

4 On the modernization of the fishery, see Wright (2001).

5 Swanson was hired after Anderson retired and had saved some of the latter's files. We are grateful to Eleanor Swanson for access to these files at the Confederation Building, St John's, and for her sharing of information with us.

6 Over the period from 1974 to 1975, 248 nutritional interviews were undertaken to examine changing food patterns.

CHAPTER ELEVEN

1 This chapter is based on the work of Canning (5%), Donovan (1%), C. Harris (20%), Harrison (2%), Robert Hood (1%), Jackson (25%), Marshall (25%), Maxwell (1%), Shepard (5%), Solberg (5%), and Tirone (10%).

2 It should be noted that the three districts mentioned here were part of the old system of ten boards, which the Williams government has targeted for further consolidation. The department has yet to post the redrawn district map on its website.

3 No school-level data were available by sex.

4 Data on the East Coast were obtained from four main sources: eleven focus groups with youth, observational work in the community during four different time periods and over various seasons, nineteen one-on-one semi-structured key informant interviews, and many informal conversations with community members. In British Columbia, data were obtained in five communities from six main sources: individual semi-structured interviews with youth, group interviews with youth, individual interviews with adult key informants, community observations, and informal conversations during multiple site visits over three years (see also Jackson et al. forthcoming; Marshall 2002; and Marshall et al. forthcoming). Our investigation follows the Participatory Action Research paradigm of significant and on-going participant and community engagement. An ethnographic narrative approach was utilized to explore young people's meaning-making regarding their experiences.

5 Youth and adults in the East Coast community did not report concerns about youth drinking and driving cars, but drinking and snowmobiling was raised as a health issue.

CHAPTER TWELVE

1 This chapter draws on the work of Dai (5%), Hahn (5%), Hawryshyn, S. Lee, and Temple (10%), Kennedy (15%), Neis (10%), Parrish (15%), Russo Garrido and Stanley (15%), Turner (5%), Volpe (5%), and Wroblewski (15%).

2 Details about the funding arrangement for and purpose of the Trans Labrador Highway (or Labrador Road) are also available online, under www.gov.nf.ca/releases.

3 One resident said, for example, that "the first vehicle to come through when the road finally opened was the beer truck ... but the road meant jobs."

4 During the 2000 construction season, eight Lodge Bay men worked on the road, most in menial labour positions, although one former fisher operated a piece of heavy equipment; in Mary's Harbour local employment on the road was provided for between twenty and thirty people, and at Port Hope Simpson, forty worked on the road. The workers, mostly men, generally worked twelve hours each day, seven days a week, transforming hinterland barrens and forest into road at the rate of around one kilometre per week. Local development was not much helped: nearly all construction companies were located on the Island of Newfoundland, and little consideration appears to have been given to encouraging the ability of companies located in southeastern Labrador to build or to maintain the road.

5 For a careful and detailed report on our work, see the occasional paper by S. Russo Garrido and J. Stanley (2002), especially 8–25.

6 Their previous access, by airplane, was more secure.

7 The fish species of major interest were brook trout (*Salvelinus fontinalis*), Atlantic salmon (*Salmo salar*), and arctic char (*Salvelinus alpinus*).

8 On 4 May 2004 Michael Wernerheim interviewed Mr Terry McCarthy, senior coordinator for Trans Labrador Highway, Department of Works, Services, and Transportation, Government of Newfoundland and Labrador, concerning economic details of the new road.

9 http://www.pointamourlighthouse.ca/heritage.php.

CHAPTER THIRTEEN

1 This chapter is drawn from the work of Barnes, Johns, and Narayan (25%), Chapman, Rohr, and Halliday (20%), Schofield, Sandhu, and Davison (25%), and Whiticar, Schümann, and Niemann (30%). Further details on the economic content of the chapter can be found in Davison and Schofield (2003) and Sandhu and Schofield (2001).

2 The concepts include backstripping, the evolution of thermal parameters, temperature distribution within the sedimentary column, temperature distribution within crust and mantle, and models for the kinetics of petroleum generation.

3 Vitrinite reflectance is an indicator of the maturity of the organic material, which acts as a control parameter for the burial/thermal history. Vitrinite reflectance is defined as the percentage of light reflected by a given vitrinite particle. On the one hand it can be measured, using the organic material collected in wells, and on the other hand vitrinite reflectance is calculated during basin modelling, according to the widely used Easy%R_o algorithm of vitrinite reflectance.

4 To create petroleum systems models in this study, IES Petromod version 8.0 software was used.

5 See note 3.

6 The acoustic classifications are highly correlated with independent information about sediment type from sediment core samples.

7 For a more detailed methodology please see Sandhu and Schofield (2001).

8 For example, in Newfoundland and Labrador: see Community Resource Services Ltd (2003).

CHAPTER FOURTEEN

1 The work in this chapter is drawn from Baker (25%), Davis (24%), Haedrich (11%), Hamilton (8%), Pitcher, Heymans, Sumaila, and Ainsworth (17%), Vodden (10%), and Wroblewski (5%).

2 Much of this section is drawn from Davis (2004).

3 Baker (2003) can be viewed for more detailed maps of each region.

4 A Marine Protected Area has its own management plan, developed jointly by the resource users and the DFO.

5 Fisheries science used to inform us that the most productive state of a stock is not when it is in its pristine condition but when older animals have been removed. We, however, would not choose to restore the past system only to fish it down to a more productive state, although this is increasingly debated as the role of older fish in the survival of a species (through learned behaviour and superior breeding) becomes better understood.

6 The most recent summary of these procedures may be found in Pitcher (2005).

CHAPTER FIFTEEN

1 This chapter is based on the work of MacLaren and Roach (10%), Vodden (60%), Hollis (10%), Bannister (12%), and Tomblin (8%), with the section on collaborative governance structures and processes (below) drawn 100% from the PHD work of Kelly Vodden.

2 See the interesting volume *Communities of Work. Rural Restructuring in Local and Global Contexts*, which deals with similar problems as experienced throughout the rural United States (Falk, Schulman, and Tickamyer 2003).

3 Regional Economic Development Boards (NL), Rural Development Authorities (NS) and Community Futures Development Corporations (BC).

4 This is due to limited and/or inconsistent provincial involvement in, and commitment to, local and regional economic development in British Columbia.

5 For further details on the challenges of Cape Breton Governance see Locke and Tomblin (2003).

6 To further complicate matters, nine rural secretariat regions were created in 2004–5.

7 Indeed, this is true of several of the "new" governance models we look at here: their current form is often new, but they actually have evolved out of earlier approaches or earlier governance experiments.

8 Paquet (2000) and others use the term "collaborative governance." Other terms used with similar meaning include multilevel governance (Canada 1999), shared governance (Brown 1999), distributed (Paquet 1999), and multicentric (Wolfish and Smith 2000).

9 This work is still under way and will be published in Vodden's (forthcoming) PHD thesis. Here, we provide only a brief survey of the work.

10 The original Beothuk peoples of Indian Bay were killed and/or displaced (Ommer 2002, 22–5) in the early twentieth century.

11 See chapter 14. The NPA was established to meet Canada's commitments under the 1995 Global Programme of Action for the Protection of the

Marine Environment from Land-based Activities (GPA). The Global
Programme of Action calls on countries to develop regional and national programs of action to prevent, reduce, and control land-based activities that contribute to the degradation of the marine environment.

12 Science shops are a model of community-based research that originated in
the Netherlands in the 1970s. They are defined as facilities that provide independent, participatory research support in response to concerns experienced
by civil society (SCIPAS 2001).

13 Canadian examples are discussed further in Vodden and Bannister (forthcoming).

14 http://www.sshrc.ca/web/apply/program_descriptions/cura_e.asp.

15 http://www.clayoquotalliance.uvic.ca/.

16 We need to link local and regional institutions into larger efforts, creating
cross-scale interaction, through networks of local and regional efforts such as
Coastal Community Networks in both Nova Scotia and British Columbia, provincial Regional Economic Development Boards, and other organizations
that create communication and cooperation across scales. Coalitions like these
can be expanded to national and even international levels, and initiatives such
as Coasts Under Stress and Canada's Ocean Management Research Network
can help to support and encourage networking at these broader levels.

17 See detailed discussion in Vodden, Penikett, and Berry (2003).

CHAPTER SIXTEEN

1 The World Millennium Project: draft preliminary report. Sadly, the authors
of this report, almost all natural scientists, offered solutions naively based on
market principles that, as we have shown throughout this book, fail to consider the complexities involved.

2 This has been ongoing for some time: see Cohen and Winn (2007).

3 Or that provide the greatest humanitarian aid, which speaks to a concern for
people over wealth.

4 But see Britton and Gilmour (1978), who warned of this myopia a long time
ago.

5 See, for some insightful comments on this, the "Working Group on Seasonal
Employment and Seasonal Economies in the Twenty-first Century," discussion paper for the national roundtable (2005, 8).

APPENDIX ONE

1 Schofield and Amodeo (1999) drawing on more than 2,200 abstracts and
224 articles, aimed to answer questions about the benefits and drawbacks of
interdisciplinary teams, as well as about what the relationship is among interdisciplinary teams.

2 Fundamental, basic, curiosity-driven, discipline-based research is not being impugned here. Quite the contrary: without it, there would be no applied research, because there would be no disciplines on which to draw, and no wisdom on which to build.

3 There also appears to be a preference for this kind of "useful" research, given the patterns of monies flowing to the academic funding councils from the Department of Industry, Ottawa.

Bibliography

Abbott, J.G. 2002. *Labrador Coastal Marine Service Changes: Impacts Assessment Report.* Report commissioned by government and conducted by the Institute for the Advancement of Public Policy, Inc., St John's, Newfoundland, 29 October.

Abdellatif, A. 2003. "Good Governance and Its Relationship to Democracy and Economic Development." Paper presented at the Global Forum III on Fighting Corruption and Safeguarding Integrity, Seoul, Korea, May.

Acheson, T.W., David Frank, and James Frost, eds. 1985. *Industrialization and Underdevelopment in the Maritimes, 1880–1930.* Toronto, ON: Garamond Press.

Adamson et al. 1945. "Medical Survey of Nutrition in Newfoundland." *Canadian Medical Association Journal* 52:227–50.

Adger, W. 2000. "Social and Ecological Resilience: Are They Related?" *Progress in Human Geography* 24(3): 347–64.

Adler, N.E., and K. Newman. 2002. "Socioeconomic Disparities in Health: Pathways and Policies." *Health Affairs* 21(2): 60–76.

AGBC (Auditor General of British Columbia). 2002–3. "Information Use by the Ministry of Health in Resource Allocation Decisions for the Regional Health Care System." Report 6. http://www.bcauditor.com/.

– 1997/98. "A Review of Governance and Accountability in the Regionalization of Health Services." http://bcauditor.com.

AGC (Auditor General of Canada). 1999. "Fisheries and Oceans: Pacific Salmon: Sustainability of the Fisheries." Report of the Auditor General of Canada. http://www.oag-bvg.gc.ca/domino/reports.nsf/html/9920ce.html.

– 1997. "Fisheries and Oceans Canada: Sustainable Fisheries Framework: Atlantic Groundfish." Chap. 14. http//:www.oag-bvg.gc.ca.

Ainsworth, C. 2004a. "How We Carried Out 'Back-to-the-Future' Community Interviews." In *Back to the Future: Advances in Methodology for Modelling and*

Evaluating Past Ecosystems as Future Policy Goals, ed. T.J. Pitcher, 116–24. Fisheries Centre Research Reports 12(1).

– 2004b. "Estimating the Effects of Prey-Predator Vulnerability Settings on Ecosim's Dynamic Function." In *Back to the Future: Advances in Methodology for Modelling and Evaluating Past Ecosystems as Future Policy Goals*, ed. T.J. Pitcher, 45–7. Fisheries Centre Research Reports 12(1).

Ainsworth, C., and T.J. Pitcher. Forthcoming. "Back-to-the-Future in Northern British Columbia: Evaluating Historic Marine Ecosystems and Optimal Restorable Biomass as Restoration Goals for the Future." In *Reconciling Fisheries with Conservation: Proceedings of the 4th World Fisheries Congress*, ed. J. Nielson. Bethesda, MD: American Fisheries Society.

– 2004a. "Modifying Kempton's Biodiversity Index for Use with Dynamic Ecosystem Models." In *Back to the Future: Advances in Methodology for Modelling and Evaluating Past Ecosystems as Future Policy Goals*, ed. T.J. Pitcher. Fisheries Centre Research Reports 12(1).

– 2004b. "Back to the Future in Northern BC: Evaluating Historic Goals for Restoration." Oral presentation, 4th World Fisheries Congress, American Fisheries Society. 2–6 May.

– 2003. "Evaluating Goals for Restoration in the Marine Ecosystem of British Columbia." Oral presentation. 21st Lowell Wakefield Fisheries Symposium. Alaska Sea Grant. 22–5 October.

Ainsworth, C., J.J. Heymans, and T.J. Pitcher. 2004. "Policy Search Methods for Back to the Future." In *Back to the Future: Advances in Methodology for Modelling and Evaluating Past Ecosystems as Future Policy Goals*, ed. T.J. Pitcher. Fisheries Centre Research Reports 12(1).

Ainsworth, C., J.J. Heymans, T.J. Pitcher, and M. Vasconcellos. 2002. *Ecosystem Models of Northern British Columbia for the Time Periods 2000, 1950, 1900 and 1750*. Fisheries Centre Research Reports 10(4).

Ainsworth, C., and U.R. Sumaila. 2005. "Intergenerational Valuation of Fisheries Resources Can Justify Long-Term Conservation: A Case Study in Atlantic Cod (*Gadus morhua*)." *Can J Fish Aquat Sci* 62:1104–10.

– 2004a. "Economic Valuation Techniques for Back to the Future Optimal Policy Searches." In *Back to the Future: Advances in Methodology for Modeling and Evaluating Past Ecosystems as Future Policy Goals*, ed. T.J. Pitcher. Fisheries Centre Research Reports 12(1).

– 2004b. "An Employment Diversity Index Used to Evaluate Ecosystem Restoration Strategies." In *Back to the Future: Advances in Methodology for Modelling and Evaluating Past Ecosystems as Future Policy Goals*, ed. T.J. Pitcher. Fisheries Centre Research Reports 12(1).

Albo, G., and C. Roberts. 1998. "The MAI and the World Economy." In *Dismantling Democracy: The Multilateral Agreement on Investment (MAI) and Its Impact*, ed. Andrew Jackson and Matthew Sanger, 283–318. Ottawa: Canadian Centre for Policy Alternatives.

Alcock, A., D. Ings, and D.C. Schneider. Forthcoming. "From Local Knowledge to Science and Back: Evolving Use of Local Ecological Knowledge in Fisheries Science." In *Making and Moving Knowledge*, eds. J. Lutz and B. Neis. Montreal: McGill-Queen's University Press.

– 2003. "From Local Knowledge to Science and Back: Evolving Use of Local Ecological Knowledge in Fisheries Science." In *Ecosystem Models of Newfoundland and Southeastern Labrador: Additional Information and Analyses for Back to the Future*, ed. J.J. Heymans, 20–39. Fisheries Centre Research Reports 11(5).

Alexander, G., R. Sweeting, and B. McKeown. 1994. "The Shift in Visual Pigment Dominance in the Retinae of Juvenile Coho Salmon (*Oncorhynchus kisutch*): An Indicator of Smolt Status." *Journal of Experimental Biology* 195:185–97.

Alverson, D.L. 1998. *Discarding Practices and Unobserved Fishing Mortalities in Marine Fisheries: An Update*. Washington Sea Grant Program, 98–06. Seattle, Washington.

– 1977. *Report of the Advisory Committee and Marine Resources Research Working Party on Marine Mammals*. FAO, Rome. FAO Fisheries Report No. 194.

Alverson, D.L., M.H. Freeberg, S.A. Murawski, and J.G. Pope. 1994. *A Global Assessment of Fisheries Bycatch and Discards*. FAO Technical paper no. 339. Rome: FAO.

Ambers, V. 2003. Personal communication, Alert Bay, British Columbia.

Amilien, V. 2001. "What Do We Mean by Traditional Food? A Concept Approach." Working paper no. 4. Norway: National Institute for Consumer Research.

Amorosi, T., T. McGovern, and S. Perdikaris. 1994. "Bioarchaeology and the Cod Fisheries: A New Source of Evidence." *ICES, Marine Science Symposium* 198:31–48.

Amott, T. 1993. *Caught in the Crisis: Women and the U.S. Economy Today*. New York, NY: Monthly Review Press.

Anderson, O. 1967. "Three Day Food Habits Study of 135 Expectant Mothers." *Canadian Nutrition Notes* 23 (7) (August-September): 73–5.

Anderson, R. 1972. *Hunt and Deceive: Information Management in Newfoundland Deep-Sea Trawler Fishing, North Atlantic Fishermen. Anthropological Essays on Modern Fishing*, ed. R. Anderson and C. Wadel. Newfoundland Social and Economic Papers, no. 5, Memorial University, NL.

Andrews, E., P. Gill, O. Janzen, L.L. Hollett, and E. Hollett. 2000. *Putting the Hum on the Humber: The First 75 Years*. St John's, NL: Robinson-Blackmore.

Angel, J.R., D.L. Burke, R. O'Boyle, F.G. Peacock, M. Sinclair, and K. Zwanenburg, eds. 1994. *Report of the Workshop on Scotia-Fundy Groundfish Management from 1977 to 1993*. Can. Tech. Report Fish and Aquatic Sciences 1979.

Antle, Rob. 2005. "Strategic Investments." *Evening Telegram*, 31 January.

Arlien-Søberg, P. 1992. "Styrene." In *Solvent Neurotoxicity*, ed. P. Arlien-Søberg, 129–53. Boca Raton: CRC Press.

Armstrong, P. 1995. "The Feminization of the Labour Force: Harmonizing Down in a Global Economy." In *Invisible: Issues in Women's Occupational Health*, ed. K. Messing, B. Neis, and L. Dumais, 368–92. Charlottetown: Gynergy.

Atkinson, B. 1984. "Discarding of Small Redfish in the Shrimp Fishery off Port au Choix, Newfoundland, 1976–80. *J Northwest Atlantic Fishery Science* 5(1): 99–102.

Aykroyd, W.R. 1930. "Beriberi and Other Food Deficiency Diseases in Newfoundland and Labrador." *Journal of Hygiene* 30:357–86.

Bailey, I. 2001a. "BC Nurses Ready to Quit, Union Says: Province's Health Ministry in a Standoff after Offering $300 Million Less than Demanded." *National Post*, 27 July, A4.

– 2001b. "BC to Pass Cooling-off Law in Health Strike. *National Post*, 19 June, A4.

Baird, Edna. 1945. *Food and Your Health*. Newfoundland Government Bulletin 19.

Baker, K. 2003. "Potential Sites for Marine Protected Areas in Canada's North Atlantic." Master of Environmental Science project report, Memorial University.

Baker, N. 2002. "Health Care Restructuring in Acute Care Settings: Implications for Registered Nurses' Attitudes." Unpublished master's thesis, Memorial University, St John's, Newfoundland.

Balmford, A., R.E. Green, and M. Jenkins. 2003. "Measuring the Changing State of Nature." *Trends in Ecology and Evolution* 18:326–30.

Barrett, B., D. Gregory, C. Way, G. Kent, J. McDonald, A. Batstone, M. Doyle, B. Curtis, L. Twells, C. Negrijn, S. Jelinski, S. Kraft, D. O'Reilly, S. Smith, and P.S. Parfrey. 2003. *The Impact of Restructuring on Acute Care Hospitals in Newfoundland*. Ottawa, ON: Canadian Health Services Research Foundation.

Barrie, J.V., and B.D. Bornhold. 1989. "Surficial Geology of Hecate Strait, British Columbia Contintental Shelf." *Can J Earth Sci* 26:1241–54.

Barrie, J.V., J.L. Luternauer, and K.F. Conway. 1991. "Surficial Geology and Geohazards of the Queen Charlotte Basin, Northwestern Canadian Continental Shelf, British Columbia." In *Evolution and Hydrocarbon Potential of the Queen Charlotte Basin, British Columbia*, ed. G.L. Woodsworth, Geological Survey of Canada, Paper 90–10, 507–12.

Batch, R.A. 2000. "Finding Stability in a Company Town: A Community Study of Slickville Pennsylvania, 1916–1943." PhD thesis, University of Pennsylvania, AAT 9965442.

Bavington, D., B. Grzetic, and B. Neis. 2004. "The Feminist Political Ecology of Fishing Down." *Studies in Political Economy* 73 (spring/summer): 159–82.

BC Federationist. 1916. "A Typical 'Company' Town." 8 September, 3. Quoted in Rajala 2006, ch. 3:74.

BCMA (British Columbia Medical Association). 2004. "BCMA Issue Backgrounder: Federal Transfer Payments for Health Care," May 2004. http://www.bcma.org/public/news_publications/publications/ policy_backgrounders/pdfs/federal%20transfer%20payments.pdf.

BCNU (British Columbia Nurses Union). 2004. "Provincial Cuts." http://bcnu.org/provincial_cuts.html.

Beaudry, M., A-M. Hamelin, and H. Delisle. 2004. "Public Nutrition: An Emerging Paradigm." *Canadian Journal of Public Health* 95:375–7.

Belec, Bonnie. 1993. "Some Fear Cutting Hospital Boards Could Adversely Affect Patient Care." *Evening Telegram,* 4 March, 3.

Bell, M. 2005. "The Vitality of Difference: Systems Theory, the Environment, and the Ghost of Parsons." *Society and Natural Resources* 18 (5): 471–8.

Beneria, L., and A. Lind. 1995. *Engendering International Trade: Concepts, Policy, and Action.* GSD Working Paper Series No. 5, July. Gender, Science and Development Programme and United Nations Development Fund for Women.

Bergeron, Y., Alain Leduc, Brian Harvey, and Sylvie Gauthier. 2002. "Natural Fire Regime: A Guide for Sustainable Management of the Canadian Boreal Forest." *Silva Fennica* 36:81–95.

Berkes, F. 1999. *Sacred Ecology: Traditional Ecological Knowledge and Resource Management.* Philadelphia, PA: Taylor and Francis.

Berkes, F., J. Colding, and C. Folke, eds. 2003. *Navigating Social-Ecological Systems: Building Resilience for Complexity and Change.* New York, NY: Cambridge University Press.

Berkes, F., and C. Folke, eds. 1998. *Linking Social and Ecological Systems: Management Practices and Social Mechanisms for Building Resilience.* New York, NY: Cambridge University Press.

Birch, L.L. 1999. "Development of Food Preferences." *Annual Review of Nutrition* 19:41–62.

Bird, A.L. 1997. "Earthquakes in the Queen Charlotte Region: 1982–1996." MSc thesis, University of Victoria.

Black, W.A. 1965. *Forest Utilization, Great Northern Peninsula, Newfoundland.* Ottawa: Queen's Printer.

Bloxam, C.L. 1882. "Extraction of Copper from Its Ores: The Welsh Copper Process." In *Metals.* London: King's College.

Bobbitt, J., and S. Akenhead. 1982. "Influence of Controlled Discharge from the Churchill River on the Oceanography of Groswater Bay, Labrador." *Can Tech Rept Fish Aquat Sci* 1097.

Bonnano, A., L. Busch, W. Friedland, L. Gouveis, and E. Mingioni, eds. 1994. *From Columbus to ConAgra: The Globalization of Agriculture and Food.* Lawrence, KS: University of Kansas Press.

Borges, T.C., K. Erzini, L. Bentes, M. Costa, J. Goncalves, P. Lino, C. Pais, and J. Ribiero. 2001. "By-catch and Discarding Practices in Five Algarve (Southern Portugal) Metiers." *J Appl Ichthyol* 17:104–14.

Bornhold, B.D. 1978. *Carbon/Nitrogen (C/N) Ratios in Surficial Marine Sediments of British Columbia.* Geological Survey of Canada, Paper 78-c, 108–12.

Botting, Ingrid. 2000. *Health Care Restructuring and Privatization from Women's Perspective in Newfoundland and Labrador.* St John's: the author.

Bowering, W.R., and G.R. Lilly. 1992. "Greenland Halibut (*Reinhardtius hippoglossoides*) off Southern Labrador and Northeastern Newfoundland (Northwest Atlantic) Feed Primarily on Capelin (*Mallotus villosus*)." *Neth J Sea Res* 29:211–22.

Boyd, Robert. 1999. *The Coming of the Spirit of Pestilence: Introduced Infectious Diseases and Population Decline among Northwest Coast Indians, 1774–84.* Vancouver and Washington: UBC Press and University of Washington Press.

Bradford, Neil. 1998. *Commissioning Ideas.* Toronto, ON: Oxford University Press.

Bradstreet, M.S.W., and W.E. Cross 1982. "Trophic Relationships at High Arctic Ice Edges." *Arctic* 35:1–12.

Bradstreet, M.S.W., K.J. Finley, A.D. Serak, W.B. Griffiths, C.R. Evans, M.F. Fabijan, and H.E. Stallard 1986. "Aspects of the Biology of Arctic Cod (*Boreogadus saida*) and Its Importance in Marine Food Chain." *Can Tech Rep Fish Aquat Sci* 1491.

Brander-Smith, D., D. Therrien, and S. Tobin. 1990. *Public Review Panel on Tanker Safety and Marine Spills Response Capability.* Final Report.

Brett, Ella. 1950. "Good Food for Good Health." *Atlantic Guardian* 5, 7 (May): 54–5 and 7, 8 (August): 61.

British Columbia. 2000. *The Barometer, Economic and Labour Market Review,* May.

– 1989. *Petroleum and Natural Gas Act.* Victoria, BC: Government of British Columbia. http://www.qp.gov.bc.ca/statreg/stat/P/96361_01.htm.

– 1875. *Papers Connected with the Indian Land Question, 1850–1875.* Victoria: Richard Wolfenden.

British Columbia. Department of Mines. 1919. *Annual Report.* Victoria, BC: King's Printer.

– 1918. *Annual Report.* Victoria, BC: King's Printer.

– 1917. *Annual Report.* Victoria, BC: King's Printer.

– 1915. *Annual Report.* Victoria, BC: King's Printer.

– 1914. *Annual Report.* Victoria, BC: King's Printer.

British Columbia. Ministry of Energy and Mines, Scientific Review Panel. 2002. *British Columbia Offshore Hydrocarbon Development: Report of the Scientific Review Panel.* Victoria, BC: Ministry of Energy and Mines.

– Ministry of Health and Ministry Responsible for Seniors. 1997. "Health Goals for British Columbia." Victoria, BC, December. http://www.hlth.gov.bc.ca (accessed May 15, 2004).

- Ministry of Health. 1993. *A Guide for Developing Community Health Councils and Regional Health Boards: Meeting the Challenge – Action for a Healthy Society.* Victoria, BC.
- Ministry of Health Planning. 2002. "The Picture of Health: How We Are Modernizing British Columbia's Health Care System." http://www.healthservices.gov.bc.ca/bchealthcare/publications.
- Ministry of Health Planning. 2001. "A New Era for Patient-Centred Health Care: Building a Sustainable, Accountable Structure for Delivery of High-Quality Patient Services." http://www.healthservices.gov.bc.ca/socsec/pdf/new_era_sustain.pdf.
British Columbia. Ministry of Health Services. 2004. 2004/5–2006/7 Service Plan. http://www.bcbudget.gov.bc.ca/sp2004/crownagency/fph/cc.pdf(accessed May 30th, 2004).
- Ministry of Labour and Citizens' Services. 2001. BC Stats. http://www.bcstats.gov.bc.ca.
- Royal Commission on Health Care and Costs (Seaton Commission). 1991. "Closer to Home" [Report]. 3 vol. Victoria, BC: The Province of British Columbia.
Britton, J.N.H., and J.M. Gilmour. 1978. *The Weakest Link: A Technological Perspective on Canadian Industrial Underdevelopment.* Background study 43. Ottawa, ON: Science Council of Canada.
Brody, B.L., H.J. Simon, and K. Stadler. 1997. "Closer to Home (or Home Alone?): The British Columbia Long-Term Care System in Transition." *Western Journal of Medicine* 167:336–42.
Bronfenbrenner, U., and G.W. Evans. 2000. "Developmental Science in the 21st Century: Emerging Questions, Theoretical Models, Research Findings, and Empirical Findings." *Social Development* 9:115–25.
Bronfenbrenner, U., and P.A. Morris. 1998. "The Ecology of Developmental Processes." In *Handbook of Child Psychology,* ed. R.M. Lerner, 993–1028. 5th ed., vol. 1, series editor, W. Damon. New York, NY: Wiley.
Brown, D. 1999. *The Impact of Global and Regional Integration on Federal Political Systems: A Baseline Study.* Institute of Intergovernmental Relations.
Brown, S.K., R. Mahon, K.C.T. Zwanenburg, K.R. Buja, L.W. Claflin, R.N. O'Boyle, B. Atkinson, M. Sinclair, G. Howell, and M.E. Monaco. 1996. *East Coast of North America Groundfish: Initial Explorations of Biogeography and Species Assemblages.* Silver Springs, MD: NOAA and Dartmouth, NS: DFO.
Bruhn, J.G. 2000. "Interdisciplinary Research: A Philosophy, Art Form, Artifact or Antidote?" *Integrative Physiological and Behavioral Science* 35 (1): 58–66.
Bruntland, G., ed. 1987. *Our Common Future: The World Commission on Environment and Development.* Oxford: Oxford University Press.
Brym, Robert J., and R. James Sacouman, eds. 1979. *Underdevelopment and Social Movements in Atlantic Canada.* Toronto, ON: New Hogtown Press.

Bundy, A. 2002. "Adaptation of a Newfoundland-Labrador Ecopath Model for 1985–1987 in Statistical Areas 2J3KLNO to the Area 2J3KL." In *Information Supporting Past and Present Ecosystem Models of Northern British Columbia and the Newfoundland Shelf*, eds. T.J. Pitcher, M. Vasconcellos, J.J. Heymans, C. Brignall, and N. Haggan. Fisheries Centre Research Reports 10(1): 13–21.

Burill, Gary, and Ian McKay, eds. 1987. *People, Resources, and Region*. Fredericton, NB: Acadiensis Press.

Burke, C., G.K. Davoren, W.A. Montevecchi, and I.J. Stenhouse. 2002. "Winging Back to the Future: An Historic Reconstruction of Seabird Diversity, Distribution and Abundance in the Northwest Atlantic, 1500–2000." In *Information Supporting Past and Present Ecosystem Models of Northern British Columbia and the Newfoundland Shelf*, eds. T.J. Pitcher, M. Vasconcellos, J.J. Heymans, C. Brignall, and N. Haggan. Fisheries Centre Research Reports 10(1): 27–37.

Burleson, Ron (Senior Project Manager, BC Offshore Oil and Gas Team). 2003. Interview by R. Dennis. Thursday, 11 December. Cited on page 56.

Bustin, R.M. 1997. "Petroleum Source Rocks, Organic Maturation and Thermal History of the Queen Charlotte Basin, British Columbia." *Bulletin of Canadian Petroleum Geology* 45 (3): 255–78.

– 1995. "Organic Maturation and Petroleum Source Rock Potential of Tofino Basin, Southwestern British Columbia." *Bulletin of Canadian Petroleum Geology* 43 (2): 177–86.

Bustin, R.M., and M. Mastalerz. 1995. "Organic Petrology and Geochemistry of Organic-Rich Rocks of the Late Triassic and Early Jurassic Sandilands and Ghost Creek Formations, Queen Charlotte Islands, British Columbia." *Marine and Petroleum Geology* 12:70–81.

Cadigan, Sean. 2006. "Restructuring the Woods: Governance and Management Issues in a Newfoundland Case Study." In *Power and Restructuring: Shaping Coastal Society and Environment*, ed. Peter Sinclair and Rosemary Ommer. St John's, NL: ISER Books.

– 2003. "The Moral Economy of Retrenchment and Regeneration in the History of Rural Newfoundland." In *Retrenchment and Regeneration in Rural Newfoundland*, ed. Reginald Byron, 14–42. Toronto, ON: University of Toronto Press.

– 2000. "Community Regulation of Coastal Forests in Newfoundland, 1875–1945." Paper presented to the Atlantic Canada Studies Conference XIII, Halifax, May 2000.

– 1999. "A Smoke in the Woods: The Moral Economy of Forest Fires in Outport Newfoundland, 1875–1946." Paper presented to the annual meeting of the Canadian Historical Association, Sherbrooke, 1999; and to the 15th International Seminar on Marginal Regions, Looking to the Sea, Looking to the Land: Survival Strategies in the 21st Century. St John's and Terra Nova, NF, July 1999.

Cail, Robert E. 1974. *Land, Man and the Law: The Disposal of Crown Lands in British Columbia, 1871–1913*. Vancouver, BC: UBC Press.

Caledon Institute of Social Policy. 2002. "A New Era in British Columbia: A Profile of Budget Cuts across Social Programs." http://www.caledoninst.org.

Cameron, B.E.B. 1980. *Biostratigraphy and Depositional Environment of the Escalante and Hesquiat Formations (Early Tertiary) of the Nootka Sound Area, Vancouver Island, British Columbia.* Geological Survey of Canada Paper 78–9.

Campagna D., D. Mergler, G. Huel, S. Belanger, G. Truchon, C. Ostiguy, and D. Drolet. 1995. "Visual dysfunction among styrene-exposed workers." *Scandinavian Journal of Work and Environmental Health* 21:382–90.

Campbell, C.M., and Community Resource Servives Ltd. 2002. *Analysis of Potential Services to the BC Offshore Oil and Gas Industry.* Report prepared for Pacific Offshore Energy Group, Nanaimo, BC. 6 August.

Campbell, K. 1993. *North Coast Odyssey: The Inside Passage from Port Hardy to Prince Rupert.* Victoria, BC: Sono Nis Press.

Canada. 2004. "Developing an Oceans Action Plan." Presentation for Strategic Planning Working Group, 8 June 2004. http://www.omrn.ca/documents/Ocean%20Action%20Plan.ppt.

– 2002a. Commission on the Future Health of Health Care in Canada. *Final Report.* November.

– 2002b. "Species at Risk Act." (2002, c. 29) http://www.sararegistry.gc.ca/the_act/default_e.cfm.

– 1987. *Canada-Newfoundland Atlantic Accord Implementation Act.* Ottawa: Government of Canada.

– 1985. *The Atlantic Accord: Memorandum of Agreement between the Government of Canada and the Government of Newfoundland and Labrador on Offshore Oil and Gas Resource Management and Revenue Sharing.* Ottawa: Government of Canada.

– 1903. *Annual Report of the Department of Indian Affairs for the Year Ended June 30th 1902.* Ottawa: Department of Indian Affairs, 257, http://www.collectionscanada.ca/indianaffairs/020010–119.01-e.php?uidc = ID&uid = 15444&queryString. Quoted in Rajala 2006.

– Committee on the Status of Endangered Wildlife in Canada. 2003. *Assessment and Update Status Report of the Atlantic Cod.* Ottawa, ON: Department of Fisheries and Oceans.

– Environment. 1999. "The New World of Environment Policy Making: Theory and Practice." ESAC Annual Meeting, Sherbrooke, Quebec.

– Fisheries and Oceans. 2004. *A Policy Framework for the Management of Fisheries on Canada's Atlantic Coast.* Ottawa, ON: Fisheries and Oceans Canada.

– Fisheries and Oceans. 2002a. *Canada's Oceans Strategy.* Ottawa: Oceans Directorate. http://www.cos-soc.gc.ca/doc/publications_e.asp.

– Fisheries and Oceans. 2002b. *Policy and Operational Framework for Integrated Management of Estuarine, Coastal and Marine Environments in Canada.* Ottawa: Oceans Directorate. http://www.cos-soc.gc.ca/doc/im-gi/index_e.asp.
– Fisheries and Oceans. 1998. *Canadian Code of Conduct for Responsible Fishing Operations.* Fisheries and Aquaculture Management, Reports and Publications. http://www.dfo-mpo.gc.ca/communic/fish_man/code/cccrfo-cccppr_e.htm.
– Fisheries and Oceans. 1996. *Oceans Act.* S.C. 1996, c. 31. http://www.pac.dfo-mpo.gc.ca/oceans/OceansAct/oceansact_e.htm.
– Fisheries and Oceans. 1985. *Fisheries Act* (R.S. 1985, c. F–14) http://laws.justice.gc.ca/en/F-14/p.22.
– Health Canada. 2002. *Canada's Food Guides from 1942 to 1992.* Catalogue H39–651/2002E-IN. Ottawa: Health Canada.
– Health Canada. 1999. *Diabetes in Canada.* Ottawa: Health Canada. http://www.hc-sc.gc.ca/pphb-dgspsp/publicat/dic-dac99/.
– House of Commons Standing Committee on the Environment. 1990. Transcript, 12 September.
– National Health and Welfare, Nutrition Canada. 1975. *The Newfoundland Survey Report.* Ottawa: National Health and Welfare.
– Natural Resources. 2004. *Report of the Public Review Panel on the Government of Canada Moratorium on Offshore Oil and Gas Activities in the Queen Charlotte Region, British Columbia.* Ottawa: Natural Resources.
– Statistics Canada. 2004. "Self-rated Health by Sex, Canada, Provinces, Territories, Health Regions and Peer Groups." Health Indicators 2004 (1). http://www.statcan.ca/english/freepub/82-221-XIE/free.htm.
– Statistics Canada. 2001. "Community Profiles." http://www.statcan.ca/english/profil101/.
– Statistics Canada. n.d. http:// www.statscan.ca.
Canning, P., and C. Strong. 2002. "Children and Families Adjusting to the Cod Moratorium." In *The Resilient Outport,* ed. R.E. Ommer, 342–8. St John's, NL: ISER Books.
Cannon, Aubrey. 1991. *The Economy Prehistory of Namu: Patterns in Vertebrate Fauna.* Burnaby, BC: Archaeology Press, Simon Fraser University.
Cardoso, D. 2004. "Modelling Variations of the Seasonal Cycle of Plankton Production: The Labrador Sea, Labrador Shelf, and Hamilton Inlet." MSc thesis. Environmental Science Graduate Program, Memorial University.
Careless, J.M.S. 1954. "Frontierism, Metropolitanism, and Canadian History." *Canadian Historical Review* 35 (March).
Carlton D., J. Carlton, E. Dudley, D. Lindberg, and G. Vermeij. 1991. "The First Historical Extinction of a Marine Invertebrate in an Ocean Basin: The Demise of the Eelgrass Limpet, *Lottia alveus.*" *Biol Bull* 180:72–80.

Carscadden, J.E., K.T. Frank, and W.C. Leggett. 2001. "Ecosystem Changes and the Effects on Capelin (*Mallotus villosus*), a Major Forage Species." *Can J Fish Aquat Sci* 58:73–85.

CBC News. 2004. "Air Labrador Pulls Service from Isolated Towns." 15 December. http://stjohns.cbc.ca/regional/servlet/View?filename=nf-air-labrador-20041215.

Cellabos, G., and P.R. Ehrlich. 2002. "Mammal Population Losses and the Extinction Crisis." *Science* 296:904–7.

Census of Canada. 1991, 1996, 2001.

Census of Newfoundland. 1845, 1857, 1869, 1884, 1891, 1911.

Chamberlayne, R., B. Green, M.L. Barer, C. Hertzman, W.J. Lawrence, S.B. Sheps. 1998. "Creating a Population-Based Linked Health Database: A New Resource for Health Services Research." *Canadian Journal of Public Health* 89 (4): 270–3.

Cherry, N., H.A. Waldron, G.G. Wells, R.T. Wilkinson, H.K. Wilson, and S. Jones. 1980. "An Investigation of the Acute Behavioral Effects of Styrene on Factory Workers." *British Journal of Industrial Medicine* 37:234–49.

Cheung, W.L., and T.J. Pitcher. 2004. "An Index Expressing Risk Of Local Extinction for Use with Dynamic Ecosystem Simulation Models." In *Back to the Future: Advances in Methodology for Modelling and Evaluating Past Ecosystems*, ed. T.J. Pitcher. Fisheries Centre Research Reports 12(1).

Chichilnisky, G. 1996. "An Axiomatic Approach to Sustainable Development." *Social Choice and Welfare* 13:231–57.

Christensen, V., and C.J. Walters. 2004. "Ecopath with Ecosim: Methods, Capabilities and Limitations." *Ecological Modelling* 172:109–39.

Cicin-Sain, B., and R. Knecht. 1998. *Integrated Coastal and Ocean Management – Concepts and Practice*. Washington, DC and Covelo, California: Island Press.

CIHI (Canadian Institute for Health Information). "Health Spending – Macro Spending." http://secure.cihi.ca/cihiweb/dispPage.jsp?cw_page=statistics_results_topic_macrospend_e&cw_topic = Health %20Spending&cw_subtopic=Macro%20Spending.

– 2004a. "Health Personnel Trends in Canada, 1993–2002." Ottawa, ON: CIHI.

– 2004b. "Improving the Health of Canadians." Ottawa, ON: CIHI.

– 2002. "Supply and Distribution of Registered Nurses in Rural and Small Town Canada, 2000." Ottawa, ON: CIHI.

Clague, M. 1997. "Thirty Turbulent Years: Community Development and the Organization of Health and Social Services in British Columbia." In *Community Organizing: Community Experiences*, ed. B. Wharf and M. Clague, 91–112. Toronto, ON: Oxford University Press.

Clapp, C.H., and H.C. Cooke. 1917. *Sooke and Duncan Map-Areas, Vancouver Island*. Canada Department of Mines, Geological Survey, Memoir 96, Geological Series no. 80.

Clapp, L.R. 1923. "Notes on Anyox Mining and Smelting Practice." *Monthly Bulletin of the Canadian Institute of Mining and Metallurgy* 140 (December): 682–700.

Clapp, R.A. 1998. "The Resource Cycle in Forestry and Fishing." *Canadian Geographer* 42 (2): 129–44.

CLARET (Clayoquot Alliance for Research, Education and Training). 2003. *Standard of Conduct for Research in Northern Barkley and Clayoquot Sound Communities.* Version 1.0. Developed through the Protocols Project. Victoria and Tofino: CLARET. Available at http://www.clayoquotalliance.uvic.ca/.

Clarke, B.L., and R. Arnold. 1923. "Fauna of the Sooke Formation, Vancouver Island." *Bulletin of the Department of Geological Sciences,* University of California Publications, 14 (5): 123–234, pls. 15–42.

Clarkson, L., V. Morrissette, and G. Régallet. 1992. *Our Responsibility to the Seventh Generation: Indigenous Peoples and Sustainable Development.* Winnipeg, MB: International Institute for Sustainable Development.

Clayoquot Sound Scientific Panel. 1995. "First Nations' Perspectives Relating to Forest Practices Standards in Clayoquot Sound."

Clayton, L., and D. Smith. 2003. "West Coast Vancouver Island Cultured Shellfish Initiative – Moving Forward." Prepared for Alberni-Clayoquot Economic Development Commission.

Cleary, Ryan. 2001. "Little Time to Balance Health Care Corp. Books." *Evening Telegram,* 5 December, 1–2.

Clement, Wallace. 1980. "The Subordination of Labour in Canadian Mining." *Labour/Le Travailleur* 5 (spring): 133–48.

– 1981. *Hard Rock Mining: Industrial Relations and Technological Changes at INCO.* Toronto, ON: McClelland and Stewart.

Clowes, R.M., D.J. Baird, and S.A. Dehler. 1997. "Crustal Structure of the Cascadia Subduction Zone, Southwestern British Columbia, from Potential Field and Seismic Studies." *Canadian Journal of Earth Sciences* 34:317–35.

CNOPB (Canada-Newfoundland Offshore Petroleum Board). 2003. "Canada-Newfoundland Offshore Petroleum Board." http://www.cnopb.nfnet.com.

CNSOPB (Canada-Nova Scotia Offshore Petroleum Board). 2003. Canada-Nova Scotia Offshore Petroleum Board. http://www.cnsopb.ns.ca.

Coakes, Edward. 1973. "Sawmilling in Newfoundland." MA thesis, Department of Geography, Memorial University.

Cohen, B., and M.I. Winn. 2007. "Market Imperfections, Opportunity and Sustainable Entrepreneurship." *Journal of Business Venturing* 22 (1): 22–49.

Coleman, F.C., W.F. Figueira, J.S. Ueland, and L.B. Crowder. 2004. "The Impact of United States Recreational Fisheries on Marine Fish Populations." *Science* 305 (5692): 1958–1960.

Colinvaux, P. 1986. *Ecology.* New York, NY: John Wiley and Sons.

Community Resource Services Ltd. 2003. *Socio-economic Benefits from Petroleum Industry Activity in Newfoundland and Labrador.* Report prepared for Petroleum Research Atlantic Canada, November 2003.

Conlan, K.E. 1977. "The Effects of Wood Deposition from a Coastal Log Handling Operation on the Benthos of a Shallow Sand Bed in Saanich Inlet, British Columbia." MSc thesis, University of Victoria.

Conlan, K.E., and D.V. Ellis. 1979. "Effects of Wood Waste on a Sand-Bed Benthos." *Marine Pollution Bulletin,* 10:262–7.

Connors, M., C.A. Bisogni, J. Sobal, and C.M. Devine. 2001. "Managing Values in Personal Food Systems." *Appetite* 36:189–200.

Copeman, L.A., and C.C. Parrish. 2004. "Lipid Classes, Fatty Acids, and Sterols in Seafood from Gilbert Bay, Southern Labrador." *J Agricultural Food Chem* 52:4872–81.

Copes, P. 2003. Personal communications.

– 2000. "Adverse Impacts of Individual Quota Systems on Conservation and Fish Harvest Productivity." Simon Fraser University Institute of Fisheries Analysis Discussion Paper 00–2.

– 1986. "A Critical Review of the Individual Transferable Quota as a Device in Fisheries Management." *Land Economics* 62:278–92.

Corbett, M. 2007. *Learning to Leave: The Irony of Schooling in a Coastal Community.* Halifax: Fernwood Press.

COSEWIC 2003. (Committee on the Status of Endangered Wildlife in Canada, Environment Canada). "Canadian Species-at-Risk, May 2003." http://www.cosewic.gc.ca.

Costanza, R., R. d'Arge, R. de Groot, S. Farber, M. Grasso, B. Hannon, K. Limburg, S. Naeem, R.V. O'Neill, J. Paruelo, R.G. Raskin, P. Sutton, and M. vandenBelt. 1997. "The Value of the World's Ecosystem Services and Natural Capital." *Nature* 387:253–60.

Coward, H., R.E. Ommer, and T.J. Pitcher, eds. 2000. *Just Fish: Ethics and Canadian Marine Fisheries.* St John's, NL: ISER Books.

Cox, S.P., T.E. Essington, J.F. Kitchell, S.J.D. Martell, C.J. Walters, C. Boggs, and I. Kaplan. 2002. "Reconstructing Ecosystem Dynamics in the Central Pacific Ocean, 1952–1998. II. A Preliminary Assessment of the Trophic Impacts of Fishing and Effects on Tuna Dynamics." *Can J Fish Aquat Sci* 59:1736–47.

Cranmer, Roy 2003. Personal communication, Alert Bay, British Columbia.

Crawford, Margaret Lee. 1991. "Designing the Company Town, 1910–1930." PHD dissertation, University of California, Los Angeles.

Crean, K., and D. Symes. 1994. "The Discards Problem: Towards a European Solution." *Marine Policy* 18 (5): 422–34.

Crosbie, John C. 2003. "Overview Paper on the 1985 Canada-Newfoundland Atlantic Accord." In *Royal Commission on Renewing and Strengthening Our*

Place in Canada, Newfoundland and Labrador. St John's: Royal
Commission.

Cruikshank, D. 1991. *A Commission of Inquiry into Licensing and Related Policies
of the Department of Fisheries and Oceans: The Fisherman's Report.* Vancouver:
United Fishermen and Allied Worker's Union.

Curtis, Kent Alexander. 2001. "An Ecology of Industry: Mining and Nature in
Western Montana, 1860–1907." PHD thesis, University of Kansas.

Cuthbertson, D.P. 1947. *Report of Nutrition in Newfoundland.* London:
Dominion's Office.

Dai, Sulan. 2006. "Socio-economic Restructuring and Health: A Multi-
Method Study of Coastal Communities in British Columbia." PHD thesis,
Department of Geography, University of Victoria.

Dalby, S., and R. Keil, eds. 2003. "Political Ecology." Special issue, *Studies in
Political Economy* 70.

Davidson, A.R. 1999. "British Columbia's Health Reform: 'New Directions'
and Accountability." *Canadian Journal of Public Health* 90 (Suppl 1): s35–8.

Davis, A., B. Wilson, and B. Compton. 1995. *Salmonberry Blossoms in the New
Year: Some Culturally Significant Plants of the Haisla Known to Occur within the
Greater Kitlope Ecosystem.* The Kitamaat Village Council. Kitamaat, BC:
Nanakila Press and Ecotrust Canada.

Davis, Reade. Forthcoming. Doctoral Thesis, Department of Anthropology,
Memorial University of Newfoundland.

– 2004. "Catching the Wave: Citizens, Stakeholders and Sustainable
Development in Canadian Oceans Policy." Paper presented to the Annual
Meetings of the Canadian Anthropology Society (CASCA), London, ON, 5–9
May.

Davis, Reade, J. Whalen, and B. Neis. 2007. "From Orders to Borders: Efforts
toward a Sustainable Co-managed Lobster Fishery in Bonavista Bay,
Newfoundland." *Human Ecology* 34(6): 851–67.

Davis, W. 2001. *Light at the Edge of the World: A Journey through the Realm of
Vanishing Cultures.* Vancouver, BC: Douglas & McIntyre, and Washington,
DC: National Geographic Society.

Davison, J., and J.A. Schofield. 2003. "Offshore Energy Development in BC:
Lessons from Elsewhere and Implications for Research." CUS working
paper, University of Victoria.

Day, A. 2004. "West Coast Vancouver Island Aquatic Management Board."
Paper presented at The Future of Coastal Communities in Canada: some
case studies workshop; All Within One Ocean: Cooperation in Sustainable
Coastal and Ocean Management Coastal Zone Canada Conference, St
John's, Newfoundland and Labrador.

Day, J.C., and J.C. Roff. 2000. *Planning for Representative Marine Protected Areas:
A Framework for Canada's Oceans.* World Wildlife Fund, Toronto.

Dayle, J.B., and L. McIntyre. 2003. "Children's Feeding Programmes in Atlantic Canada: Some Foucauldian Theoretical Concepts in Action." *Social Science and Medicine* 57:313–25.

Deimling, E.A., and W.J. Liss 1994. "Fishery Development in the Eastern North Pacific: A Natural-Cultural System Perspective." *Fisheries Oceanography* 3:60–77.

DeLong, S.C., and D. Tanner. 1996. "Managing the Pattern of Forest Harvest: Lessons from Wildfire." *Biodiversity Conservation* 5:1191–1205.

Dennis, Robert H. 2005. "Understanding Accord and Moratoria: Offshore Oil and Gas Development on the East and West Coasts of Canada, 1949–1989." Master's thesis, University of New Brunswick.

Deur, Douglas, and N.J. Turner, eds. 2005. *"Keeping it Living": Indigenous Plant Management on the Northwest Coast.* Seattle, WA: University of Washington Press, and Vancouver: UBC Press.

Diamond, Jared. 2005. *Collapse: How Societies Choose to Fail or Succeed.* New York, NY: Viking Press.

Dibsdall, L.A., N. Lambert, and L.J. Frewer. 2002. "Using Interpretative Phenomenology to Understand the Food-Related Experiences and Beliefs of a Select Group of Low-Income UK Women." *Journal of Nutrition and Education Behaviour* 34:298–309.

Dietrich, J.R. 1995. "Petroleum Resource Potential of the Queen Charlotte Basin and Environs, West Coast Canada." *Bulletin of Canadian Petroleum Geology* 43 (1): 20–34.

Dietrich, J.R., R. Higgs, K.M. Rohr, and J.M. White. 1993. "The Tertiary Queen Charlotte Basin: A Strike-Slip Basin on the Western Canadian Continental Margin." In *Tectonic Controls and Signatures in Sedimentary Successions,* ed. L. Frostick and R. Steel. Special publication number 20, International Association of Sedimentologists, 161–9.

Dobbin, Lucy. 1993. *Report on the Reduction of Hospital Boards.* March.

Dobell, Rod. 2002. "Devolution and Discretion: Building Community-Based Resource Management into Contemporary Governance." In *New Players, Partners and Processes: A Public Sector without Boundaries?* ed. Meredith Edwards and John Langford, 108–25. Victoria, BC: Centre for Public Sector Studies.

Dobell, Rod, and Martin Bunton. 2001. "Sound Governance: The Recent Emergence of Collaborative Networks and Institutions in the Clayoquot Sound Region." Background paper for the Clayoquot Sound Regional Workshop on Adaptive Management and Sustainable Communities, 8–11 May.

Dolan, A.H., S.M. Taylor, B. Neis, J. Eyles, R.E. Ommer, D.C. Schneider, and W.A. Montevecchi. 2005. "Restructuring and Health in Canadian Coastal Communities: A Social-Ecological Framework of Restructuring and

Health." *EcoHealth* 2 (3): 195–208. http://dx.doi.org/ 10.1007/
   s10393–005–6333–7.
Drewnowski, A. 1997. "Taste Preferences and Food Intake." *Annual Review of
   Nutrition* 17:237–53.
Drushka, Ken. 1995. *HR: A Biography of H.R. MacMillan.* Madeira Park:
   Harbour Publishing, 191–203. Quoted in Rajala, 2006, ch. 4.
Dunnett, Janet. 2004. "University and Community Linkages at the University
   Of Victoria: Towards a New Agenda for Community Based Research."
   Unpublished MPA thesis. School of Public Administration, University of
   Victoria, Victoria.
Dunning, Paula. 1995. *Who's Running Our Schools: Education Governance in the
   90s: Provincial/Territorial Summaries.* Ottawa, ON: Canadian School Boards
   Association.
Duval, W.S., and F.F. Slaney, et al. 1980. *A Review of the Impacts of Log Handling
   on Coastal Marine Environments and Resources.* Report prepared for the
   Review Panel of COFI/Government, Estuary, Foreshore and Water
   Handling, and Transportation Study.
Dyer, A.K., and A.W. Robertson. 1984. "Forestry." In *Encyclopedia of
   Newfoundland and Labrador,* vol. 2, 297–312. St John's, NL: Joseph R.
   Smallwood Heritage Foundation, Inc.
Dyke, A.P. 1969. *Community Inventory of Coastal Labrador.* St John's, NL:
   Department of Labrador Affairs.
Ecotrust Canada. 1997. *Seeing the Ocean through the Trees: A Conservation-Based
   Development Strategy for Clayoquot Sound.* Vancouver, BC: Ecotrust Canada.
Ecotrust Canada, and Ecotrust. 2004. *Catch–22: Conservation, Communities and
   the Privatization of BC Fisheries.* Vancouver, BC: Ecotrust Canada.
Elder, G.H., Jr. 1974. *Children of the Great Depression.* Chicago, IL: University of
   Chicago Press.
Engineering and Mining News, 13, 17, and 20 October 1917. Cited in BC
   Department of Mines, 1917, F46.
Ennis, Gerry, and J. Whalen. 2005. "Presentation to St John Bay Fish
   Harvesters." Port au Choix, spring.
Epstein, A.G., J.B. Epstein, and L.D. Harris. 1977. "Conodont Color
   Alteration – An Index to Organic Metamorphism." United States
   Geological Survey, Professional Paper 995.
Errington, John Charles. 1975. "Natural Revegetation of Disturbed Sites in
   British Columbia." PHD, Forestry, University of British Columbia.
Estes, J.A., M.T. Tinker, T.M. Williams, and D.F. Doak. 1998. "Killer Whale
   Predation on Sea Otters Linking Oceanic and Nearshore Ecosystems."
   Science 282:473–5.
Evans, D. 2000. *The Consequences of Illegal, Unreported and Unregulated Fishing for
   Fishery Data and Management: Expert Consultation on Illegal, Unreported and
   Unregulated Fishing.* FAO; IUU/2000/12.

Evans, R.G., M.L. Barer, and T.R. Marmor, eds. 1994. *Why Are Some People Healthy and Others Not? The Determinants of Health of Populations.* New York, NY: Walter de Gruyter.

Evening Telegram. 1993. "Trade Board Says Reducing Debt Must Be Government's Priority." 2 March, 9.

– 1983. "Hospital, Nursing Home Costs to Be Studied." 16 April, 1.

– 1984. "New Health Act Isn't Realistic." 18 February, 3.

Falk, William W., Michael D. Schulman, and Ann R. Tickamyer, eds. 2003. *Communities of Work: Rural Restructuring in Local and Global Contexts.* Athens, OH: Ohio University Press.

FAO (Food and Agriculture Organization of the United Nations). 2005. "Forest Area Statistics, Canada." http://www.fao.org/forestry/site/22030/en/can.

– 2002. *Stopping Illegal, Unreported and Unregulated Fishing.* Rome: FAO.

– 1995. *The Code of Conduct for Responsible Fisheries.* Rome: FAO. http://www.fao.org/fi/agreem/codecond/ficonde.asp.

Fearnside, P.M. 2002. "Time Preference in Global Warming Calculations: A Proposal for a Unified Index." *Ecological Economics* 41:21–31.

Felt, L.F., and P.R. Sinclair, eds. 1995. *Living on the Edge: The Great Northern Peninsula of Newfoundland.* St John's, NL: ISER Books.

Field, Albert E. 1993. *Fields of Endeavour.* Kelowna, BC: Ehmann Printing.

Finley, K.J., and C.R. Evans. 1983. "Summer Diet of the Bearded Seal (*Erignathus barbatus*) in the Canadian High Arctic." *Arctic* 36: 82–9.

Firestone, Melvin M. 1967. *Brothers and Rivals: Patrilocality in Savage Cove.* St John's, NL: ISER Books.

Fisher, M., D. Pastore, and M. Schneider. 1994. "Eating Attitudes in Urban and Suburban Adolescents." *International Journal of Eating Disorders* 16:67–74.

Fodor, J.G. 1980. "Epidemiology of Hypertension." *Medicine North America* 1 (3): 1–4.

Folke, C., F. Berkes, and J. Colding. 1998. "Ecological Practices and Social Mechanisms for Building Resilience and Sustainability." In *Linking Social and Ecological Systems: Management Practices and Social Mechanisms for Building Resilience,* eds. F. Berkes and C. Folke, 414–36. Cambridge, England: Cambridge University Press.

Foster, D.R. 1983. "The History and Pattern of Fire in the Boreal Forest of Southeastern Labrador." *Canadian Journal of Botany* 61: 2459–71.

Foulkes, R.G. 1974. *Health Security for British Columbia.* Vol. 2, *Management of the British Columbia Health Programme.* Victoria: Health Security Programme Project.

– 1973. *Health Security for British Columbians,* vol. 1. Victoria: Health Security Programme Project.

Fowler, M.G., L.R. Snowdon, P.W. Brooks, and T.S. Hamilton. 1988. "Biomarker Characterization and Hydrous Pyrolysis of Bitumen from

Tertiary Volcanics, Queen Charlotte Islands, British Columbia." *Organic Geochemistry* 13:715.

Frankish, C.J., B. Kwan, P.A. Ratner, J.W. Higgins, and C. Larsen. 2002. "Social and Political Factors Influencing the Functioning of Regional Health Boards in British Columbia (Canada)." *Health Policy* 61 (2): 125–51.

Frechet, A., J. Gauthier, P. Schwabb, H. Bourdages, D. Chabot, F. Collier, F. Gregoire, Y. Lambert, G. Moreault, L. Pageau, and J. Spingle. 2003. *The Status of Cod in the Northern Gulf of St Lawrence.* Canadian Science Advisory Secretariat: Research Document 2003/065.

Friedland, W.H. 1991. "Introduction: Shaping the New Political Economy of Advanced Capitalist Agriculture." In *Towards a New Political Economy of Agriculture*, eds. W.H. Friedland, L. Busch, F.H. Buttel, and A.P. Rudy, 1–34. Colorado: Westview.

Fuller, S., C. Fuller, and M. Cohen. 2003. *Health Care Restructuring in BC.* Vancouver, BC: Canadian Centre for Policy Alternatives.

G3 Consulting Ltd. 2003. *Guidebook: Environmentally Sustainable Log Handling Facilities in British Columbia.* Report Prepared for Fisheries and Oceans Canada, Pacific and Yukon Region, Habitat and Enhancement Branch.

Gallaugher, P., and L. Wood, eds. 2004. *Proceedings of the World Summit on Salmon, June 10–13, 2003.* Burnaby, BC: Simon Fraser University. http://www.sfu.ca/cstudies/science/summit.htm (accessed June 16, 2005).

Galois, Robert. 1996. "Measles, 1847–1850: The First Modern Epidemic in British Columbia." *BC Studies* 108: 31.

– 1994. *Kwakwaka'wakw Settlements, 1777–1920: A Geographical Analysis and Gazetteer*, 233–5. Vancouver, BC: UBC Press.

Garibaldi, A., and N. Turner. 2004. "Cultural Keystone Species: Implications for Ecological Conservation and Restoration." *Ecology and Society.* http://www.ecologyandsociety.org/volxx/issYY/artzz.

George, Earl Maquinna. 2003. *Living on the Edge: Nuu-Chah-Nulth History from an Ahousaht Chief's Perspective.* Winlaw, BC: Sono Nis Press.

Gershenkron, Alexander. 1962. *Economic Backwardness in Historical Perspective.* Harvard, MA: Belknap Press.

Gezelius, S. 2003. "The Morality of Compliance in Coastal Fisheries: Cases from Norway and Newfoundland." Paper presented at the IASCP Northern Polar Regional Meeting, Anchorage, Alaska. 17–21 August.

Gibson, R.J., R.J. Luther, and R.L. Haedrich. 2002. "Labrador Road Study: A Survey of Water Crossings on a Section of the Trans Labrador Highway, 26 June–10 July." http://www.coastsunderstress.ca/publications.php.

Gislason, G. 1998. *Fishing for Money: Challenges and Opportunities in the BC Salmon Industry.* Victoria, BC: The ARA Consulting Group Inc., for the BC Job Protection Commission.

Gislason, G., E. Lam and M. Mohan. 1996. *Fishing for Answers: Coastal Communities and the BC Salmon Fishery*. Victoria, BC: ARA Consulting Group Inc. for the BC Job Protection Commission.

Goddard, Richard. 1999. "On the Edge: The Historical Ethnography of a Satellite Settlement." PHD diss., University of Nevada.

Goltz, Eileen Alice. 1989. "The Exercise of Power in a Company Town: Copper Cliff, 1886–1960." PHD diss., University of Guelph.

Gosling, W.G. 1910. *Labrador – Its Discovery, Exploration and Development*. London, England: Alston Rivers.

Gosse, K., and J.S. Wroblewski. 2004. "Variant Colourations of Atlantic Cod (*Gadus morhua*) in Newfoundland and Labrador Nearshore Waters." *ICES Journal of Marine Science* 61:752–9.

Gosse, K., J.S. Wroblewski, and B. Neis. 2003. "Closing the Loop: Commercial Fish Harvesters' Local Ecological Knowledge and Science in a Study of Coastal Cod in Newfoundland and Labrador." Paper presented at the Putting Fishers' Knowledge to Work Conference, Fisheries Centre, The University of British Columbia, Vancouver, BC, 27–30 August 2001. Fisheries Centre Research Reports 11 (1): 25–35.

Goudzwaard, B., and H. DeLange. 1995. *Beyond Poverty and Affluence: Towards a Canadian Economy of Care*. Toronto, ON: University of Toronto Press.

Gowgaia Institute. 2003. "Haida Gwaii: What's Been Logged?" *SpruceRoots Magazine*. http://www.spruceroots.org/Gallery/Logging.html.

Great Britain. Economic Advisory Council. Committee on Nutrition in the Colonial Empire. 1939. *First Report – Part 1, Nutrition in the Colonial Empire*. CMD 6050 London.

Great Britain. 1933. *Newfoundland Royal Commission 1933*. London: King's Printer. (http://www.heritage.nf.ca/law/amulree/chap7_4.html).

Greenfield, T.B. 1978. Reflections on Organization Theory and the Truths of Irreconcilable Realities. *Educational Administration Quarterly* 14 (2): 1–23.

Groot, A., S. Gauthier, and Y. Bergeron. 2004. "Stand Dynamics Modelling Approaches for Multicohort Management of Eastern Canadian Boreal Forests." *Silva Fennica* 38(4): 437–48.

Grzetic, Brenda. 2004. *Women Fishes These Days*. Halifax, NS: Fernwood Publishing.

Guardian Weekly, 8–14 April 2005, 12.

Gunderson, L., and C.S. Holling. 2002. *Panarchy: Understanding Transformations in Human and Natural Systems*. Washington, DC: Island Press.

Hache, J.E. 1989. *Report of the Scotia-Fundy Groundfish Task Force*. DFO, Halifax, NS.

Haedrich, R.L., N.R. Merrett, and N.R. O'Dea. 1998. "Can Ecological Knowledge Catch Up with Deep-Water Fishing?" CM *ICES* 98/O:37, Deepwater Fish and Fisheries.

Haggan, N., C. Brignall, and L. Wood, eds. 2003. *Putting Fisher's Knowledge to Work*. Fisheries Centre Research Reports 11 (1).

Hall, Budd. 1999. "Looking Back, Looking Forward – Reflections on the International Participatory Research Network." *Forests, Trees and People Newsletter* 39:33–6.

Hall, M. 1998. "An Ecological View of the Tuna-Dolphin Problem: Impacts and Trade-Offs." *Reviews in Fish Biology and Fisheries* 8:1–34.

Hall, M., D. Alverson, and K.I. Metuzals. 2000. "Bycatch: Problems and Solutions." In *Seas at the Millennium*, ed. C. Sheppard, chap. 116. Amsterdam: Elsevier.

Hall, W. 1937. "Moresby Forest: Survey and Preliminary Management Plan." British Columbia Forest Branch (unpublished report), 19, 33. Quoted in Rajala, 2006, chap. 4.

Halpern, B.S. 2003. "The Science of Marine Reserves – The Impact of Marine Reserves: Do Reserves Work and Does Reserve Size Matter?" *Ecological Applications* 13 (1): S117–39.

Hamilton, L.C., and R.L. Haedrich. 1999. "Ecological and Population Changes in Fishing Communities of the North Atlantic Arc." *Polar Research* 18 (2): 383–8.

Hamilton, T.S., and B.E.B. Cameron. 1989. "Hydrocarbon Occurrences on the Western Margin of the Queen Charlotte Basin." *Bulletin of Canadian Petroleum Geology* 37 (4): 443–66.

Hammond, Lorne, and Robert Griffin. 2002. "Resource Communities of BC's Northern Coast Living Landscapes." CUS Project Mid-term Report, September 2002.

Hancock T. 2002. "Indicators of Environmental Health in the Urban Setting." *Canadian Journal of Public Health* 93(s1):45–56.

– 2000. "Healthy Communities Must Also Be Sustainable Communities." *Public Health Reports* 115 (March/April and May/June): 151–6.

– 1993. "Health, Human Development, and the Community Ecosystem: Three Ecological Models." *Health Promotion International* 8:41–7.

Hannigan, P.K., J.R. Dietrich, P.J. Lee, and K.G. Osadetz. 2001. "Petroleum Resource Potential of Sedimentary Basins on the Pacific Margin of Canada." *Geological Survey of Canada*, Bulletin 564.

– 1998. Petroleum Resource Potential of Sedimentary Basins on the Pacific Margin of Canada. *Geological Survey of Canada*, Open File 3629.

Harper, John. 2004. "The Clam Gardens of the Broughton Archipelago: A Case for Pre-contact, Large-Scale Mariculture in Queen Charlotte Strait." Paper presented at Malaspina College, Science and Technology Lecture Series, January 28.

Harris, C.E. 1998. "Time, Place, and the Rationality of Rural School Restructuring: Visiting the Politics of Site-Based Management?" In *Centering the Margins: The Evaded Curriculum*, ed. J. Epp, 113–20. Ottawa: International Institute Proceedings of the Canadian Association for the Study of Women and Education, University of Ottawa, 31 May–1 June.

Harris, C.E., and C. Shepherd. 2007. "Food Security, Learning, and Culture: Bridging the Responsibility Gap in Public Schools." In *Resetting the Kitchen Table: Food Security, Culture, Health and Resilience in Coastal Communities*, eds. C. Parrish, N.J. Turner, and S.M. Solberg. Hauppauge, NY: Nova Science Publishing.

Harris, C.E., and S. Umpleby. Forthcoming. "Students as Community Participants: Knowledge through Engagement in the Coastal Context." In *Making and Moving Knowledge,* eds. J. Lutz and B. Neis, Montreal: McGill-Queen's University Press.

Harris, Cole. 2002. *Making Native Space: Colonialism, Resistance, and Reserves in British Columbia.* Vancouver, BC: UBC Press.

– 1997/98. "Social Power and Cultural Change in Pre-Colonial British Columbia." *BC Studies* 115, 116: 45.

– 1973. *The Resettlement of British Columbia: Essays on Colonialism and Geographical Change.* Vancouver, BC: UBC Press.

Harris, D.C. 2006. "Colonial Territoriality: The Spatial Restructuring of Native Land and Fisheries on the Pacific Coast." In *Power, Agency and Nature*, eds. P. Sinclair and R.E. Ommer, 135–53. St John's: ISER Books.

– 2004. "Land, Fish, and Law: Indian Reserves and Native Fisheries in British Columbia, 1850–1927." PHD thesis, York University.

– 2001. *Fish, Law, and Colonialism: The Legal Capture of Salmon in British Columbia.* Toronto, ON: University of Toronto Press.

Harris, L. 1990. *Independent Review of the State of the Northern Cod Stock.* Ottawa: Communication Directorate, DFO.

Harris, M. 1998. *Lament for an Ocean: the Collapse of the Atlantic Cod Fishery.* Toronto, ON: McClelland and Stewart.

Harrison, K. 2005. "Changing Communities, Changing Goals and Changing Dreams: Youth Perceptions of Present and Future Possibilities in Coastal British Columbia." Unpublished MA thesis, University of Victoria.

Harrison, P. 2005. *Budget 2005: The Oceans Action Plan.* http://www.omrn.ca/documents/FINAL_OCEANS_BUDGET_2005.pdf.

– 2004. "Canada, UNCLOS, and the Future of the Hydrographic Sciences." Presentation to the Canadian Hydrographic Conference, Ottawa, May.

Hart, Michael. 1996. *A Multilateral Agreement on Foreign Direct Investment – Why Now?* Occasional Paper in International Trade Law and Policy, no. 37. Centre for Trade Policy and Law, Norman Paterson School of International Affairs, Carleton University.

Harter, S. 1999. *The Construction of the Self: The Developmental Perspective.* New York, NY: The Guilford Press.

Harvey, Reverend M. 1879. *Across Newfoundland with the Governor: A Visit to Our Mining Region.* St John's, NL.

– 1878. "Our Newfoundland Letter." Correspondent, unknown newspaper, 12 September 1878.

Hawryshyn, C.W., T.J. Haimberger, et al. 2001. "Microspectrophotometric Measurements of Vertebrate Photoreceptors Using CCD-Based Detection Technology." *Journal of Experimental Biology* 204:2431–8.

Hay, D. 2002. "School-Based Feeding Programmes: A Good Choice For Children?" Policy paper commissioned by Childhood and Youth Division, Health Canada, Ottawa.

Haynes, Dale, and Stephen White. 2002. *Dynamics of Rural Areas: A Comparative Study of Economic Performance in Two Rural Newfoundland Areas.* St John's: Memorial University of Newfoundland, Public Policy Research Centre. http://www.mun.ca/policycentre/pdf/DORA.pdf.

Hayter, Roger. 2000. *Flexible Crossroads: The Restructuring of British Columbia's Forest Economy.* Vancouver, BC: University of British Columbia Press.

Heart and Stroke Foundation of Canada. 2003. *The Growing Burden of Heart Disease and Stroke in Canada.* Ottawa, ON: Heart and Stroke Foundation of Canada.

Hertzman, C., and A. Siddiqi. 2000. "Health and Rapid Economic Change in the Late Twentieth Century." *Social Science and Medicine* 51:809–19.

Hewes, G.W. 1973. "Indian Fisheries Productivity in Pre-contact Times in the Pacific Salmon Area." *Northwest Anthropological Research Notes* 7 (2): 133–55.

Heymans, J.J. 2003a. "Revised Models for Newfoundland and Southern Labrador (2J3KLNO) for the Time Periods 1985–87 and 1995–97." In *Ecosystem Models of Newfoundland and Southeastern Labrador: Additional Information and Analyses for "Back to the Future,"* ed. J.J. Heymans. Fisheries Centre Research Reports 11(5): 40–62. Vancouver, BC.

– 2003b. "Fitting the Newfoundland Model to Time Series Data." In *Ecosystem Models of Newfoundland and Southeastern Labrador: Additional Information and Analyses for "Back to the Future,"* ed. J.J. Heymans. Fisheries Centre Research Reports 11 (5): 72–8.

– 2003c. "First Nations Impact on the Newfoundland Ecosystem during Pre-contact Times." In *Ecosystem Models of Newfoundland and Southeastern Labrador: Additional Information and Analyses for "Back to the Future,"* ed. J.J. Heymans. Fisheries Centre Research Reports 11 (5): 4–11.

– ed. 2003d. *Ecosystem Models of Newfoundland and Southeastern Labrador: Additional Information and Analyses for "Back to the Future."* Fisheries Centre Research Reports 11(5).

– 2003e. "Comparing the Newfoundland-Southern Labrador Marine Ecosystem Models Using Information Theory." In *Ecosystem Models of Newfoundland and Southeastern Labrador (2J3KLNO): Additional Information and Analyses for "Back to the Future,"* ed. J.J. Heymans, 62–71. Fisheries Centre Research Reports 11(5).

Heymans, J.J., and T.J. Pitcher. 2004. "Synoptic Methods for Constructing Models of the Past." In *Back to the Future: Advances in Methodology for*

Modelling and Evaluating Past Ecosystems *as Future Policy Goals*, ed. T.J. Pitcher, 11–17. Fisheries Centre Research Reports 12(1).

– 2002a. "A Picasso-esque View of the Marine Ecosystem of Newfoundland and Southern Labrador: Models for the Time Periods 1450 and 1900." In *Ecosystem Models of Newfoundland for the Time Periods 1995, 1985, 1900 and 1450*, eds. T.J. Pitcher, J.J. Heymans, and M. Vasconcellos. Fisheries Centre Research Reports 10 (5): 44–71.

– 2002b. "A Model of the Marine Ecosystem of Newfoundland and Southern Labrador (2J3KLNO) in the Time Periods 1985–1987 and 1995–1997." In *Ecosystem Models of Newfoundland for the Time Periods 1995, 1985, 1900 and 1450*, eds. T.J. Pitcher, J.J. Heymans, and M. Vasconcellos. Fisheries Centre Research Reports 10 (5): 5–43.

Higgins, B., and D.J. Savoie. 1995. *Regional Development Theories and Their Application*. New Brunswick, NJ, and London: Transaction Publishers.

Higgs, R. 1991. "Sedimentology, Basin-Fill Architecture, and Petroleum Geology of the Tertiary Queen Charlotte Basin, British Columbia." In *Evolution and Hydrocarbon Potential of the Queen Charlotte Basin, British Columbia*, ed. G.J. Woodsworth, 337–71 Geological Survey of Canada, Paper 90–10.

Hilborn, R., T.A. Branch, B. Ernst, A. Magnusson, C.V. Minte-Vera, M. Scheuerell, and J. Valero. 2003. "State of the World's Fisheries." *Annual Review of Environment and Resources* 28:359–99.

Hildebrand L.P. 1989. *Canada's Experience with Coastal Zone Management*. Halifax, NS: Oceans Institute of Canada.

Hiller, J.K. 1967. "The Establishment and Early Years of the Moravian Missions in Labrador, 1752–1805." Unpublished MA thesis, Department of History, Memorial University of Newfoundland.

Hilton-Taylor, C., compiler. 2000. *2000 IUCN Red List of Threatened Species*. Gland, Switzerland, and Cambridge, England: IUCN.

Holling, C.S. 1986. "The Resilience of Terrestrial Ecosystems, Local Surprise and Global Change." In *Sustainable Development of the Biosphere*, ed. W.C. Clark and R.E. Munn, 292–317. Cambridge: Cambridge University Press.

– 1973. "Resilience and Stability of Ecological Systems." *Annual Review of Ecology and Systematics* 4:1–23.

Hollis, T.I. 2004. "Stewardship of Local Wetlands: Environmental Ethics and Traditional Ecological Knowledge in Four Rural Newfoundland Communities." MA dissertation, Memorial University of Newfoundland.

Hood, R.J., and Ben Fox, producers and directors. 2003. *Gitga'ata Spring Harvest – Traditional Knowledge of Kiel*. A film co-produced by the Gitga'at First Nation and the Coasts Under Stress research project.

Horne, G. 2004. *British Columbia's Heartland at the Dawn of the 21st Century: 2001 Economic Dependencies and Impact Ratios for 63 Local Areas*. BC Ministry of Management Services, Victoria, BC.

House, J.D. 2000. "Myths and Realities about Petroleum-Related Development: Lessons from Atlantic Canada and the North Sea." Paper presented at Exploring the Future of Offshore Oil and Gas Development in BC: Lessons from Atlantic, Simon Fraser University, Burnaby, BC, 17–19 May.

– 1999. "Does Community Really Matter in Newfoundland and Labrador? The Need for Supportive Capacity in the New Regional Economic Development."

House, J.D., Sheila M. White, and Paul Ripley. 1989. *Going Away and Coming Back. Economic Life and Migration in Small Canadian Communities.* ISER Report no. 2.

Howse, D., D. Gautrin, B. Neis, A. Cartier, L. Horth-Susin, M. Jong, and M. Swanson. 2006. "Gender and Snow Crab Occupational Asthma in Newfoundland and Labrador, Canada." *Environmental Research* 101:163–274.

Hudec, A. 2003. *Developing BC's Offshore Oil and Gas Resources.* Vancouver, BC: Davis and Company.

Huelin, S. 2004. "Determination of Trace Elements in Iron-Manganese Oxide Coatings by Laser Ablation ICP-MS for Environmental Monitoring/Mineral Exploration." MSc thesis, Memorial University of Newfoundland, St John's, NL.

Huggett, D. 2003. *The Role of Federal Government Intervention in Coastal Zone Planning and Management.* Bedfordshire: Coastal Policy Officer, Royal Society for the Protection of Birds.

Hughes, J.B., G.C. Daily, and P.R. Ehrlich. 1997. "Population Diversity: Its Extent and Extinction." *Science* 278: 689–91.

Hume, Mark. 2004. "You Don't Know What You've Got Till It's Gone." *Globe and Mail,* 24 February, A15.

Hunter, M.L. 1993. "Natural Disturbance Regimes as Spatial Models for Managing Boreal Forests." *Biological Conservation* 65: 115–20.

Husky Energy. 2003. "Whiterose Project Diversity Plan. St John's, NL." http://www.huskyenergy.ca/operations/canadaseastcoast/projects/ whiterose.asp.

Hussey, G. 1981. *Our Life on Lear's Room.* St John's: Robinson-Blackmore Printing.

Hutchings, J.A. 2000. "Collapse and Recovery of Marine Fishes." *Nature* 406 (6798): 882–5.

Hutchings, J. 1996. "Spatial and Temporal Variation in the Density of Northern Cod and a Review of Hypotheses for the Stock's Collapse." *Canadian Journal of Fisheries and Aquatic Sciences* 53:943–62.

Hutchings, J., and M. Ferguson. 2000. "Temporal Changes in Harvesting Dynamics of Canadian Inshore Fisheries for Northern Atlantic Cod, *Gadus morhua.*" *Canadian Journal of Fisheries and Aquatic Sciences* 57: 805–14.

Hutchings, Jeffrey A., and R. Myers. 1995. "The Biological Collapse of Atlantic Cod off Newfoundland and Labrador: An Exploration of

Historical Changes in Exploitation, Harvesting Technology, and Management." In *The North Atlantic Fisheries: Successes, Failures, and Challenges,* ed. Ragnar Arnason and Lawrence Felt, 37–93. Charlottetown, PEI: The Institute of Island Studies.

Hutchings, Jeffrey A., C. Walters, and R. Haedrich 1997. "Is Scientific Inquiry Incompatible with Government Information Control?" *Can J Fish Aquat Sci* 54(5): 1198–1210.

Hutchings, Oswald J. n.d. Anyox, Old Mss. Series G AN9 H97, BC Archives.

Hyndman, R.D. 1995. "The Lithoprobe Corridor across the Vancouver Island Continental Margin: The Structural and Tectonic Consequences of Subduction." *Canadian Journal of Earth Sciences* 32:1777–1802.

Hyndman, R.D., C.J. Yorath, R.M. Clowes, and E.E. Davis. 1990. "The Northern Cascadia Subduction Zone at Vancouver Island: Seismic Structure and Tectonic History." *Canadian Journal of Earth Sciences* 27 (3): 313–29.

Ings, D.W., D.C. Schneider, R.S. Gregory, and V. Gotceitas. Forthcoming. "Density of Small Fish Depends on Coastline Complexity and Local Habitat Structure." *Mar Ecol Prog Ser.*

Innis, Harold A. 1956. *Essays in Canadian Economic History,* ed. Mary Q. Innis. Toronto, ON: University of Toronto Press.

– 1954. *The Cod Fisheries: The History of an International Economy.* Revised ed. Toronto, ON: University of Toronto Press.

International Classification of Disease. 9th version. http:// icd9cm.chrisendres.com/.

Ireland, D. 2002. "Facing Budget Crunch, BC Restructures Health System." *Canadian Medical Association Journal* 166:1582.

Jackson, Craig. 2005. "'Devastation' for the Region." *The Telegram,* Thursday, 28 July.

Jackson, J.B.C., M.X. Kirby, W.H. Berger, K.A. Bjorndal, L.W. Botsford, B.J. Bourque, R.H. Bradbury, R. Cooke, J. Erlandson, J.A. Estes, T.P. Hughes, S. Kidwell, C.B. Lange, H.S. Lenihan, J.M. Pandolfi, C.H. Peterson, R.S. Steneck, M.J. Tegner, and R.R. Warner. 2001. "Historical Overfishing and the Recent Collapse of Coastal Ecosystems." *Science* 293 (5530): 629–38.

Jackson, L., S. Tirone, C. Donovan, and R. Hood. Forthcoming. "Community Restructuring and the Health of Youth: A Qualitative Study of a Small Coastal Community in Newfoundland, Canada." *Canadian Journal of Community Mental Health.*

Jackson, R.G. 1986. "Effects of Bark Accumulation on Benthic Infauna at a Log Transfer Facility in S.E. Alaska." *Marine Pollution Bulletin* 17:258–62.

Jangaard, P. 1974. "The Capelin *(Mallotus villosus)*: Biology, Distribution, Exploitation, Utilization and Composition." *Bull Fish Res Board Can,* no. 186.

Jeletzky, J.A. 1954. *Tertiary Rocks of the Hesquiat-Nootka Area, West Coast of Vancouver Island, British Columbia.* Canada Department of Mines and Technical Surveys, *Geological Survey of Canada,* paper 53–17.

Jensen, F., and N. Vestergaard. 2002. "Moral Hazard Problems in Fisheries Regulation: The Case of Illegal Landings and Discard." *Resources and Energy Economics* 24:281–99.

Jenson, J. 1998. *Mapping Social Cohesion: The State of Canadian Research.* Study no. F–03. Ottawa: Canadian Policy Research Network.

JHA (Journal of the Newfoundland House of Assembly). 1880.

– 1879. *Customs Returns: Exports from the Colony of Newfoundland for the Year 1879.*

Johns, M.J., C.R. Barnes, and Y.R. Narayan. Forthcoming. "Biostratigraphic Interpretation of Mesozoic and Cenozoic Ichthyolith Assemblages from the Tofino Basin, British Columbia." *Canadian Journal of Earth Sciences.*

– 2005. "Catalogue of Cenozoic ichthyoliths from the Tofino Basin and Western Vancouver Island, British Columbia, Canada." *Palaeontologia Electronica* 8(2). http://palaeo-electronica.org/2005_2/icth/issue2_05.htm.

Johns, M.J., C.R. Barnes, Y.R. Narayan, and J.A. Trotter. Forthcoming. "New Interpretation of the Tofino Basin Crescent Terrane and Stratigraphy: Information for Hydrocarbon Exploration." *Bulletin of Canadian Petroleum Geology.*

Johnson, S.Y., M.E. Tennyson, W.S. Lingley, Jr., and B.E. Law. 1997. *Petroleum Geology of the State of Washington.* U.S. Geological Survey Professional Paper 1582.

Johnston, D.M., and E.N. Hildebrand. 2001. *BC Offshore Hydrocarbon Development: Issues and Prospects.* Victoria, BC: Maritime Awards Society.

Karst, Amanda. 2005. "The Ethnoecology and Reproductive Ecology of Bakeapple (*Rubus chamaemorus* L.) in Southern Labrador." MSc thesis, Department of Biology, University of Victoria.

Kealey, L. 2003. "Nutrition and Dietary Deficiency Diseases: Discourses of Empire, Region and Gender in 20th Century Newfoundland." Paper presented to the Canadian Society for the History of Medicine, Halifax, NS, 1 June.

Keating, D.P., and C. Hertzman, eds. 1999. *Developmental Health and the Wealth of Nations: Social, Biological and Educational Dynamics.* New York, NY: Guilford Press.

Keats, D., D. Steele, and J. Green. 1986. *A Review of the Recent Status of the Northern Cod Stock (NAFO Div. 2J, 3K and 3L) and the Declining Inshore Fishery.* Report of Memorial University of Newfoundland, St John's.

Keller, B.C., and R.M. Leslie. 1996. *Sea-Silver: Inside British Columbia's Salmon Farming Industry.* Victoria, BC: Horsdal and Shubart Publishers.

Kennedy, J.C. 2001. "Environmental Change, Fisheries Restructuring, Transportation Policies, and Differentiation: Coastal Labrador 2000." CUS Occasional Paper.

– 1997. "Labrador Métis Ethnogenesis." *Ethnos* 62 (3–4): 5–23.

– 1996. *Labrador Village, Prospect Heights.* Illinois, MI: Waveland Press.

– 1995. *People of the Bays and Headlands: Anthropological History and the Fate of Communities in the Unknown Labrador.* Toronto, ON: University of Toronto Press.

Ker, J.W. 1967. *Newfoundland Forestry Report.* Revised. Fredericton.

Kerstetter, Steve. 2001. *BC Home to Greatest Wealth Gap in Canada.* Vancouver, BC: Canadian Centre for Policy Alternatives.

Kispiox Valley Community Association. 1975. "Submission to the Royal Commission on Forest Resources." BC Ministry of Forests Library. Quoted in Rajala 2006, chap. 8.

Kitchell, J.F., T.E. Essington, C.H. Boggs, D.E. Schindler, and C.J. Walters. 2002. "The Role of Sharks and Longline Fisheries in a Pelagic Ecosystem of the Central Pacific." *Ecosystems* 5:202–16.

Kleivan, H. 1966. *The Eskimos of Northeast Labrador.* Oslo: Norsk Polarinstitutt.

Knight, Rolf. 1996. *Indians at Work: An Informal History of Native Indian Labour in British Columbia, 1858–1930.* Vancouver, BC: New Star Books.

Krebs-Smith, S.M., J. Heimendinger, B.H. Patterson, A.F. Subar, R. Kessler, and E. Pivonka. 1995. "Psychosocial Factors Associated with Fruit and Vegetable Consumption." *American Journal of Health Promotion* 10:98–104.

Kuhnlein, H.V., 1984. "Traditional and Contemporary Nuxalk Foods." *Nutrition Research* 4:789–809.

Kuhnlein, H.V., and O. Receveur. 1996. "Dietary Change and Traditional Food Systems of Indigenous Peoples." *Annual Review of Nutrition* 16:417–42.

Kuhnlein, H.V. and N.J. Turner. 1991. *Traditional Plant Foods of Canadian Indigenous Peoples. Nutrition, Botany and Use.* Vol. 8, *Food and Nutrition in History and Anthropology*, ed. Solomon Katz. Philadelphia, PA: Gordon and Breach Science Publishers.

Kulka, D. 1986. "Estimates of Discarding by the Newfoundland Offshore Fleet in 1984 with Reference to Trends over the Past Four Years." NAFO SCR Doc. 86/12. Serial no. N1120.

Kulka, D.W., and D.A. Pitcher. 2001. "Spatial and Temporal Patterns in Trawl Activity in the Canadian Atlantic and Pacific." *ICES CM.*

Kulka, David. 2003. Personal communication.

Kvitek, R.G., J.S. Oliver, A.R. DeGange, and B.S. Anderson. 1992. "Changes in Alaskan Soft-Bottom Prey Communities along a Gradient in Sea Otter Predation." *Ecology* 73:413–28.

Lade, Fred. n.d. "The Miner's Boy in the Lardeau Gold Rush." Reminiscences of life in the Lardeau, and in Stewart and Anyox, told by Fred Lade and written by Ina B. Rawson. Photocopy, MS–0457, BC Archives.

Larson, P. 1997. "Navigating the Implementation of 'New Directions' in a
    Region of British Columbia: Who's at the Helm? A Study in the Social
    Organization of Knowledge." Unpublished Master of Nursing Thesis,
    University of Victoria, Victoria, BC.

Lasswell, H.D. 1977. *Harold D. Lasswell on Political Sociology*. Chicago, IL:
    University of Chicago Press.

Latour, B. 1987. *Science in Action*. Milton Keynes: Open University Press.

Laurel, B.J., R.S. Gregory, and J.A. Brown 2003a. "Predator Distribution and
    Habitat Patch Area Determine Predation Rates on Age–0 Juvenile Cod
    Gadus spp." *Mar Ecol Prog Ser* 251:245–54.

– 2003b. "Settlement and Distribution of Age–0 Juvenile Cod, *Gadus morhua*
    and *G. ogac*, Following a Large Scale Habitat Manipulation." *Mar Ecol Prog
    Ser* 262:241–52.

Law, J., and J. Hassard. 1999. *Actor Network Theory and After*. Oxford: Blackwell.

Lawson, J.W., J.T. Anderson, E.L. Dalley, and G.B. Stenson 1998. "Selective
    Foraging by Harp Seals *Phoc groenlandica* in Nearshore and Offshore Waters
    on Newfoundland, 1993 and 1994." *Mar Ecol Prog Ser* 163:1–10.

Laxer, Gordon. 1995. "Introduction." *Canadian Review of Sociology and
    Anthropology* 32 (3): 247–51.

Leach, B., and A. Winson. 1995. "Bringing 'Globalization' Down to Earth:
    Restructuring and Labour in Rural Communities." *Canadian Review of
    Sociology and Anthropology* 32 (3): 341–64.

Lee, Y.J. 2005. "Working Out Work: Learning, Identity and History." PHD
    diss., University of Victoria.

Legall, F.D., C.R. Barnes, and R.W. MacQueen. 1981. "Thermal Maturation,
    Burial History and Hotspot Development, Paleozoic Strata of Southern
    Ontario-Quebec, from Conodont and Acritarch Colour Alteration
    Studies." *Bulletin of Canadian Petroleum Geology* 29:492–539.

Lerner, S.C. 1993. "The Importance of Active Earthkeeping." In
    Environmental Stewardship: Studies in Active Earthkeeping, ed. S.C. Lerner, 3–8.
    Waterloo, ON: University of Waterloo, Department of Geography
    Publication.

LeRoy, S. 1999. "Aboriginal Land Claims, Co-operative Management and the
    Creation of Parks in Canada." Undergraduate directed study, Department
    of Geography. University of Victoria, Victoria, BC.

Levin, S.A. 1999. *Fragile Dominion*. Perseus Books.

Levy, D.A., T.G. Northcote, and R.M. Barr. 1982. *Effects of Estuarine Log Storage
    on Juvenile Salmon*. Technical Report no. 28. Westwater Research Centre,
    University of British Columbia.

Levy, D.A., T.G. Northcote, K.J. Hall, and I. Yesaki. 1989. "Juvenile Salmonid
    Responses to Log Storage in Littoral Habitats of the Fraser River Estuary
    and Babine Lake." *Canadian Special Publication of Fisheries and Aquatic Science*
    105.

Levy, Ryna. 1993. "Perspective 2000: Eating Healthier in Newfoundland and Labrador." 19 July.

Lilly, G.R. 2002. "Ecopath Modelling of the Newfoundland Shelf: Observations on Data Availability within the Canadian Department of Fisheries and Oceans." In *Information Supporting Past and Present Ecosystem Models of Northern British Columbia and the Newfoundland Shelf*, eds. T.J. Pitcher, M. Vasconcellos, J.J. Heymans, C. Brignall, and N. Haggan. Fisheries Centre Research Reports 10(1): 22–6.

Lindsay, J. 2003. "'Partying Hard', 'Partying Sometimes' or 'Shopping': Young Workers' Socializing Patterns and Sexual, Alcohol and Illicit Drug Risk Taking." *Critical Public Health* 13(1): 1–14.

Lindsay, W.R., and R.L. Healy. 1929. "Mining Methods at Hidden Creek Mine." *Canadian Mining Journal* (26 April): 383–5.

Linehan, J.E., R.S. Gregory, and D.C. Schneider 2001. "Predation Risk of 0-group cod (*Gadus*) Relative to Depth and Substrate in Coastal Waters." *J Exp Mar Biol Ecol* 263:25–44.

Living Oceans Society. n.d. "*The Offshore Oil and Gas Industry.*" http://www.livingoceans.org.

Locke, Wade, and Stephen Tomblin. 2003. "Good Governance, a Necessary but Not Sufficient Condition for Facilitating Economic Viability in a Peripheral Region: Cape Breton as a Case Study." Commissioned by Cape Breton Regional Municipality and presented 21 October.

Loney, Bretton. 1993. "Debt Must Be Tackled." *Evening Telegram*, 4 March, 2.

Longerich, H.P. 1995. "Analysis of Pressed Pellets of Geological Samples Using Wavelength-Dispersive X-Ray Fluorescence Spectrometry." *X-Ray Spectrometry* 24:123–36.

Lorimer, C.G. 1977. "The Presettlement Forest and Natural Disturbance Cycle of Northeastern Maine." *Ecology* 58:139–48.

Loudon, Pete. 1973. *Anyox: The Town That Got Lost*. Sidney, BC: Gray's Publishing.

Lucas, Chief S. 2004. "Aboriginal Values." In *Back to the Future: Advances in Methodology for Modelling and Evaluating Past Ecosystems as Future Policy Goals*, ed. T.J. Pitcher. Fisheries Centre Research Reports 12 (1): 114–16.

Ludwig, D., R. Hilborn, and C. Walters. 1993. "Uncertainty, Resource Exploitation and Conservation: Lessons from History." *Science* 260:17–26.

Lush, Gail. 2007. "Nutrition, Health Education, and Dietary Reform: Gendering the 'New Science' at the Grenfell Mission, 1893–1928." Master's thesis, Department of History, Memorial University.

Lutz, J. 2007. *Makuk: Work and Welfare in Aboriginal non-Aboriginal Relations*. Vancouver, BC: UBC Press.

Lutz, J., and B. Neis, eds. Forthcoming. *Making and Moving Knowledge*. Montreal, QC: McGill-Queen's University Press.

Lyman, R.L., and K.P. Cannon, eds. 2004. *Adding Prehistory to Conservation Biology: Zooarchaeological Studies from North America.* Salt Lake City: University of Utah Press.

Lyons, C. 1969. *Salmon, Our Heritage.* Vancouver, BC: BC Packers.

MacDonald, M. 1999. "Restructuring, Gender and Social Security Reform in Canada." *Journal of Canadian Studies* 34(2): 57–88.

MacDonald, M., B. Grzetic, and B. Neis 2004. "Getting By in a Very Fragile World: Household Livelihood 'Strategies' and Gender." Paper presented at the XI World Congress of Rural Sociology, Trondheim, Norway, 25–29 July and at the International Association for Feminist Economics Conference, Oxford, 5–7 August.

MacDonald, M., B. Neis, and B. Grzetic. 2005. "Making a Living: The Struggle to Stay." People and the Sea Conference III: New Directions in Coastal and Maritime Studies, Amsterdam, 7–9 July.

MacKenzie, B.R., J. Alheit, D.J. Conley, P. Holm, and C.C. Kinze. 2002. "Ecological Hypotheses for a Historical Reconstruction of Upper Trophic Level Biomass in the Baltic Sea and Skagerrak." *Canadian Journal of Fisheries and Aquatic Sciences* 59:173–90.

Mackenzie, S., and G. Norcliffe. 1997. "Restructuring in the Canadian Newsprint Industry." *The Canadian Geographer* 41 (1): 2–6.

Mackie, Richard. 1997. *Trading beyond the Mountains: The British Fur Trade on the Pacific, 1793–1843.* Vancouver, BC: UBC Press.

Madill, M. 2004. Personal communications. 'Namgis Aquatic Resources Dept. July.

Magnússon, J.V. 2001. "Distribution and Some Other Biological Parameters of Two Morid Species *Lepidon eques* and *Antimora rostrata* in Icelandic Waters." *Fisheries Research* 51:267–281.

Mandel, Charles. 2004. "The Fight for Fair Share." *Atlantic Business* 15 (1)(February/March): 40.

Mannion, J.J. 1977. *The Peopling of Newfoundland: Essays in Historical Geography.* St John's, NL: Memorial University of Newfoundland.

Marchak, Patricia, Scott Aycock, and Deborah Herbert. 1999. *Falldown: Forest Policy in British Columbia.* Vancouver, BC: Ecotrust Canada and the David Suzuki Foundation.

Marett, Elias. 1870. "On the Copper and Nickel Mines at Tilt Cove, Newfoundland." St John's, Newfoundland. Files of the Newfoundland and Labrador Department of Mines and Energy, file 002E–13/0107.

Markey, S., J. Pierce, and K. Vodden. 2000. "Resources, People and the Environment: A Regional Analysis of the Evolution of Resource Policy in Canada." *Journal of Regional Science* 23 (3): 427–54.

Marshall, A. 2002. "Life-Career Counselling Issues for Youth in Coastal and Rural Communities: The Impact of Economic, Social and Environmental

Restructuring." *International Journal for the Advancement of Counselling*
24 (1): 69–87.

Marshall, A., L. Jackson, B. Shepard, S. Tirone, and C. Donovan.
Forthcoming. "Knowledge Flows around Youth: What Do They 'Know'
about Human and Community Health?" In *Making and Moving Knowledge*,
ed. J. Lutz and B. Neis. Montreal, QC: McGill-Queen's University Press.

Marshall, J.M. 1990. "Report of the Working Group on the 4R, 3PN Cod
Fishery." DFO, unpublished report. Moncton, NB.

Marshall, W.H., ed. 1975. "The West Coast Health Survey: Progress,
Prospects, and Plans: A Workbook for a Multidisciplinary Investigation in
the Province of Newfoundland." Meeting Report. St John's (SN).

Martin, Hans-Peter, and H. Schumann. 1998. *The Global Trap: Globalization
and the Assault on Democracy and Prosperity*. New York, NY: Zed Books.

Martin, W. 1983. "Once upon a Mine: Story of pre-Confederation Mines on
the Island of Newfoundland." CIM Special Volume 26.

Matthew, W. 2001. "Regarding Native Title Agreements Pursuant to Application
for Petroleum Exploration Licences in the Cooper Basin, South Australia."
Ministry of Minerals and Energy. http://www.beachpetroleum.com.au.

Maxwell, L. 2003. "Teachers Speak: Impact of Restructuring on the Daily
Lives of Elementary Educators." MA thesis, University of Victoria.

Mayhew, C., and M. Quinlan. 1999. "The Effects of Outsourcing on
Occupational Health and Safety: A Comparative Study of Factory-Based
Workers and Outworkers in the Australian Clothing Industry." *International
Journal of Health Services* 29 (1): 83–107.

McCarthy, J. 2001. "Gap Dynamics of Forest Trees: A Review with Particular
Attention to Boreal Forests." *Environmental Reviews* 9:1–59.

McDaniel, N.G. 1973. "A Survey of the Benthic Macroinvertebrate Fauna and
Solid Pollutants in Howe Sound." Fisheries Research Board of Canada
Technical Report No. 385.

McDevitt, E., M. Dove, R. Dove, and I.S. Wright. 1944. "Vitamin Status of the
Population of the West Coast of Newfoundland with Emphasis on Vitamin
C." *Annals of Internal Medicine* 20:1–11.

McDonnell, M.J., et al. 1997. "Ecosystem Processes along an Urban-to-Rural
Gradient." *Urban Ecosystems* 1:21–36.

McGrail, K.M., R.G. Evans, M.L. Barer, S.B. Sheps, C. Hertzman, and A.
Kazanjian. 2001. "The Quick and the Dead: 'Managing' Inpatient Care in
British Columbia." *Health Services Research* 35:1319–38.

McGreer, E.R., D.M. Moore, and J.R. Sibert. 1984. *Study of the Recovery of
Intertidal Benthos after Removal of Log Booms, Nanaimo Estuary, British
Columbia*. Canadian Technical Report of Fisheries and Aquatic Sciences,
1246.

McIlwraith, T.F. 1948. *The Bella Coola Indians*. 2 vols. Toronto, ON: University
of Toronto Press.

McIntyre, L., K. Raine, and J.B. Dayle. 2001. "The Institutionalization of Children's Feeding Programmes in Atlantic Canada." *Canadian Journal of Dieticians Practical Research* 62(2): 53–7.

McIntyre, L., K. Travers, and J.B. Dayle. 1999. "Children's Feeding Programmes in Atlantic Canada: Reducing or Reproducing Inequities?" *Canadian Journal of Public Health* 90(3): 196–200.

McLaughlin, Mark. 2004. "Power Tools as Tools of Power: Mechanization in the Tree Harvest of the Newfoundland Pulp and Paper Industry." Unpublished MA major paper (History), Memorial University of Newfoundland.

McRae, Donald M., and Peter H. Pearse. 2004. *Treaties and Transition: Towards a Sustainable Fishery on Canada's Pacific Coast.* Ottawa: Department of Fisheries and Oceans.

Mellin, A., D. Neumark-Sztainer, M. Story, M. Ireland, and M. Resnick. 2002. "Unhealthy Behaviors and Psychosocial Difficulties among Overweight Adolescents: The Potential Impact of Familial Factors." *Journal of Adolescent Health* 31:145–53.

Meltzer, E. 1998. *International Review of Integrated Coastal Zone Management: Potential Application to the East and West Coasts of Canada.* Department of Fisheries and Oceans.

Metcoff, J., et al. 1945. "Nutritional Survey in Norris Point, NF." *Journal of Laboratory and Clinical Medicine* 30:475–87.

Metuzals, K.I., C.M. Wernerheim, R.L. Haedrich, P. Copes, and A. Murrin. Forthcoming. "Data Fouling in Marine Fisheries: Findings and a Model for Newfoundland." In *Making and Moving Knowledge*, eds. J. Lutz and B. Neis. Montreal, QC: McGill-Queen's University Press.

M'Gonigle, R.M. 1999. "Ecological Economics and Political Ecology: Towards a Necessary Synthesis." *Ecological Economics* 28:11–26.

Michaels, Anne. 1996. *Fugitive Pieces.* Toronto: McClelland & Stewart.

Millennium Ecosystem Assessment. 2005. *Ecosystems and Human Well-being: Synthesis.* Washington, DC: Island Press. http://www.millenniumassessment.org/en/products.synthesis.aspx#.

Milne, L.J., and M.J. Milne. 1951. "The Eelgrass Catastrophe." *Scientific American* 184:52–5.

Mining and Scientific Press. 26 July. 1913, 107.

Monbiot, George. 2005. *Guardian Weekly* 22–8 April, 6.

Morland, K., S. Wing, and A. Diez Roux. 2002. The Contextual Effect of the Local Food Environment on Residents' Diets: The Atherosclerosis Risk in Communities Study. *American Journal of Public Health* 92:1761–7.

Moulton, D. 2002. "BC Takes Axe to Budget of a Health System 'in Danger.'" *British Columbia Medical Association Journal* 19:166.

Mulcahy, Dennis. 1999. *Critical Perspectives on Rural Education Reform.* St John's: Dennis Mulcahy. http://www.mun.ca/educ/faculty/mwatch/win99/mulcahy.htm.

Munn, W.A. 1922. "Annual Migration of Codfish in Newfoundland Waters." *Newfoundland Trade Review*, 23 December, 21–4.

Murata K., S. Arski, and K. Yokoyama. 1991. "Assessment of the Peripheral, Central, and Autonomic Nervous System Function in Styrene Workers." *American Journal of Industrial Medicine* 20:775–84.

Murray, G.D., and B. Neis. 2004. "Lessons Learned from Reconstructing Interactions between Local Ecological Knowledge, Fisheries Science and Fisheries Management in the Commercial Fisheries of Newfoundland and Labrador, Canada." Paper presented at the Tenth Biennial Conference of the International Association for the Study of Common Property (IASCP): The Commons in an Age of Global Transition: Challenges, Risks and Opportunities. Oaxaca, Mexico, 9–13 August.

Murray, G.D., B. Neis, and D. Bavington. 2005. "Local Ecological Knowledge, Science, and Fisheries Management in Newfoundland and Labrador: A Complex, Contested and Changing Relationship." In *Participation in Fisheries Governance*, ed. T.S. Gray. Reviews, Methods and Technologies in Fish Biology and Fisheries 4. New York, NY: Springer.

Murray, G.D., B. Neis, and J.P. Johnsen. 2006. "Lessons Learned from Reconstructing Interactions between Local Ecological Knowledge, Fisheries Science and Fisheries Management in the Commercial Fisheries of Newfoundland and Labrador, Canada." *Human Ecology* 34:549–71. http://springerlink.metapress.com/link.asp?id=t41736u272981224.

Murray, S.N., R.F. Ambrose, J.A. Bohnsack, L.W. Botsford, M.H. Carr, G.E. Davis, P.K. Dayton, D. Gotshall, D.R. Gunderson, M.A. Hixon, J. Lubchenco, M. Mangel, A. MacCall, D.A. McArdle, J.C. Ogden, J. Roughgarden, R.M. Starr, M.J. Tegner, and M.M. Yoklavich. 1999. "Opinion, No-Take Reserve Networks: Sustaining Fishery Populations and Marine Ecosystems – A Scientific Panel Convened by the Pacific Ocean Conservation of Fisheries." *American Fisheries Society* 24 (11): 11–25.

Murray, T.P., J. Kay, et al. 2002. "Linking Human and Ecosystem Health on the Amazon Frontier: An Adaptive Ecosystem Approach." In *Conservation Medicine: Ecological Health in Practice*, A. Aguirre, R. Ostfeld, C. House, G. Tabor, and M. Pearl, eds. Oxford, England: Oxford University Press.

Myers, R., J. Hutchings, and N. Barrowman. 1997. "Why Do Fish Stocks Collapse?" *Ecol Appli* 7(1): 91–106.

Nabhan, Gary Paul. 2002. *Coming Home to Eat. The Pleasures and Politics of Local Foods*. New York: W.W. Norton.

Narayan, Y.R. 2003. "Taxonomy, Biostratigraphy and Paleoecology of Cenozoic Foraminifers from Shell Canada Exploration Wells, Tofino Basin, Offshore Vancouver Island, British Columbia." MSc thesis, School of Earth and Ocean Sciences, University of Victoria.

Narayan, Y.R., C.R. Barnes, and M.J. Johns. 2005. "Taxonomy and Biostratigraphy of Cenozoic Foraminifers from Shell Canada Wells, Tofino

Basin, Offshore Vancouver Island, BC, Canada." *Micropaleontology* 51 (2):
101–67.

Nass Valley Communities Association. 1975. *Logging and Small Communities.*
Report to the Royal Commission on Forest Resources. BC Ministry of
Forests Library. Quoted in Rajala 2006, chap. 8.

Neim, A.R., and P.D. Snavely Jr. 1991. "Geology and Preliminary
Hydrocarbon Evaluation of the Tertiary Juan de Fuca Basin, Olympic
Peninsula, Northwest Washington. *Washington Geology* 19(4): 27–34.

Neis, B., and B. Grzetic. 2000. *From Fishplant to Nickel Smelter: Policy
Implications, Policy Development and the Determinants of Women's Health in an
Environment of Restructuring.* Report on the workshop Industrial
Restructuring and Rural Health: Gender-Based Analysis, Issues, and Policy
Lessons and Alternatives, St John's, 2–5 February.

Neis, B., and R. Kean. 2003. "Why Fish Stocks Collapse: An Interdisciplinary
Approach to Understanding the Dynamics of 'Fishing Up.'" In *Retrenchment
and Regeneration in Rural Newfoundland,* ed. R. Byron. Toronto, ON:
University of Toronto Press.

Neis, B., and M. Morris. 2002. "Fishers' Ecological Knowledge and Stock
Assessment: Understanding the Capelin (*Maillotus villosus*) and Capelin
Fisheries in the Bonavista Region of Newfoundland." In *The Resilient
Outport: Ecology, Economy and Society in Rural Newfoundland,* ed. R.E. Ommer,
205–40. St John's, NL: ISER Books.

Nemeth, Tammy Lynn. 1997. "Pat Carney and the Dismantling of the
National Energy Program." Master's thesis, University of Alberta.

Newell, Dianne. 1993. *Tangled Webs of History: Indians and the Law in Canada's
Pacific Coast Fishery.* Toronto, ON: University of Toronto Press.

Newfoundland and Labrador. 1986a. *Background Report to Building on
Our Strengths.* St John's, NL: Government of Newfoundland and
Labrador.

– 1986b. *Building on Our Strengths: Report of the Royal Commission on Employment
and Unemployment.* St John's, NL: Government of Newfoundland and
Labrador.

– Health and Community Services. 2002. *Healthier Together: A Strategic Health
Plan for Newfoundland and Labrador.* St John's, NL: The Department.

– Health and Community Services. 2001a. *Reaching Consensus and Planning
Ahead: Health Forums 2001.* Discussion Document. St John's, NL: The
Department.

– Health and Community Services. 2001b. *Health Forums 2001 – Regional
Profile Health and Community Services: Health and Community Services – Western
Region.* St John's, NL: The Department.

– Health and Community Services. 1999. "Health Ministers Announce
Primary Health Care Project for Three Rural Communities." 14
September. http://www.gov.nf.ca/releases/1999/health/0914n02.htm.

– Education. 2004a. "Chapter s–12.2 – *An Act to Revise the Law Respecting the Operation of Schools in the Province.*" http://www.gov.nl.ca/hoa/statutes/s12–2.htm.

– Education. 2004b. *K–12 School Profile System Website.* http://www.education.gov.nf.ca/sch_rep/pro_year.htm

– Fisheries and Aquaculture. 2002. *Seafood Industry Year in Review.* St John's, NL: Fisheries and Aquaculture.

– Health Resource Centre. 1910, 1912. "Some Facts on Public Health Nutrition in Newfoundland." *Newfoundland Gazette*, 31 March, 28 June, 16 August, 23 July 1912.

– PANL. 1945. GN 38 s 6–1–8, file 3, J. McGrath to F. Stare, M.D., 20 April.

– 1943. GN 38 s 6–5–2, file 1, Council on Nutrition of the Newfoundland Medical Association to Commissioner for Public Health and Welfare, 23 March.

NHNHA (Newfoundland Hospital and Nursing Home Association). 1993. *Presentation to the Minister of Health and Government of Newfoundland re: 1993–94 Budgets of Health Organizations.* St John's: the Association.

Nicholson, J.R. 1975 *Shetland and Oil.* London, England: William Luscombe Publishers.

Niemelä, J. 1999. "Management in Relation to Disturbance in the Boreal Forest." *Forest Ecology and Management* 115:127–34.

Norcliffe, G. 1999. "John Cabot's Legacy in Newfoundland: Resource Depletion and the Resource Cycle." *Geography* 84:97–109.

North West Health Authority. 2000. Health Service Plan Executive Summary, *Bridge the Gap, Close the Gap.* July.

Northern Health Authority. 2004. "About Northern Health." http://northernhealth.ca/nha/about.

– 2002. "Redesign Update." http://northernhealth.ca/context/articlefiles/86-redesign%20Update.pdf.

NRC (National Research Council). 2001. *Marine Protected Areas: Tools for Sustaining Ocean Ecosystems.* National Research Council Committee on the Evaluation, Design, and Monitoring of Marine Reserves and Protected Areas in the United States. Washington, DC: National Academy Press.

– 2000. *Improving the Collection, Management and Use of Marine Fisheries Data.* Washington, DC: National Academy Press.

Nuxalk Food and Nutrition Program. 1984. *Nuxalk Food and Nutrition Handbook.* Bella Coola, BC.

Ocean Falls Corporation. 1980. *Ocean Falls Corporation: Review of Operations 1973–1980.* Ocean Falls Corporation.

O'Clair, C.E., and J.L. Freese. 1988. "Reproductive condition of Dungeness Crabs, *Cancer magister,* at or near Log Transfer Sites in Southeastern Alaska." *Marine Environmental Research* 26:57–91.

O'Connor, J.S. 1998. "Social Justice, Social Citizenship, and the Welfare State: Canada in Comparative Context." In *The Vertical Mosaic Revisited,* ed. R. Helmes-Hayes and J. Curtis, 180–231. Toronto, ON: University of Toronto Press.

O'Dea, N., and R. Haedrich. 2002. "A Review of the Status of the Atlantic Wolffish, *Anarhichas lupus,* in Canada." *Canadian Field-Naturalist* 116(3): 423–32.

Odhiambo, Ben Kisila. 1995. "Sediment Accumulation, Transport and the Fate of Mine Tailings Disposed of in a Coastal Fjord, Alice Arm, BC." MSc thesis, Earth and Ocean Sciences, University of Victoria.

Ommer, R.E. 2006. *Coasts Under Stress: Restructuring and Social-Ecological Health, Policy Reflections.* St John's, NL: ISER Books.

– 2004. "Nature and Community in the Global Marketplace." In *The Twenty-first Century Confronts Its Gods,* ed. David J. Hawkin. New York, NY: SUNY Press.

– 2000. "The Ethical Implications of Property Concepts in a Fishery." In *Just Fish: Ethics and Canadian Marine Fisheries,* ed. H. Coward, R.E. Ommer, and T.J. Pitcher, 117–39. St John's, NL: ISER Books.

– 1994. "One Hundred Years of Fishery Crises." *Acadiensis* 2 (spring 1994): 5–20.

– 1991. *From Outpost to Outport: The Jersey-Gaspé Cod Fishery, 1778–1886.* Montreal, QC: McGill-Queen's University Press.

– 1985. "What's Wrong with Canadian Fish?" *Journal of Canadian Studies* 20(3): 130.

Ommer, R.E., Harold Coward, and C.C. Parrish. Forthcoming. "Knowledge, Uncertainty and Wisdom." In *Making and Moving Knowledge,* ed. J. Lutz and B. Neis, chap. 2. Montreal, QC: McGill-Queen's University Press.

Ommer, R.E., C.L. Holcapek, R.J. Hood, and the Coasts Under Stress team. 2006. *Voices on the Edge.* Victoria, BC: Coasts Under Stress research project.

Ommer, R.E., ed. 2002. *The Resilient Outport.* St John's: ISER Books.

Ommer, R.E., and N.J. Turner. 2004. "The Informal Economy in History." *Labour/Le Travail* 53 (spring 2004): 125–55. http://www.historycooperative.org/journals/llt/53/ommer.html.

Ommer, R.E., N.J. Turner, M. MacDonald, and P.R. Sinclair. 2007. "Food Security and the Informal Economy." In *Resetting the Kitchen Table: Food Security in Canadian Coastal Communities,* eds. C.C. Parrish, N.J. Turner, and S.M. Solberg, chap. 8. Hauppauge, NY: Nova Science Publishing.

Omohundro, John T. 1994. *Rough Food: The Seasons of Subsistence in Northern Newfoundland.* St John's, NL: ISER Books.

One Ocean. 2003. *The Newfoundland and Labrador Fishing and Petroleum Industries: Operating in One Ocean.* Report submitted to the 3rd Annual Atlantic Oil and Gas Summit, St John's, Newfoundland, 15–16 September.

O'Neill, Robert V., and James R. Kahn. 2000. "*Homo economus* as a Keystone Species." *Bioscience* 50: 333–7.

Orchard, T.J., and Q. Mackie. 2004. "Environmental Archaeology: Principles and Case Studies." In *Back to the Future: Advances in Methodology for Modelling and Evaluating Past Ecosystems as Future Policy Goals,* ed. T.J. Pitcher. Fisheries Centre Research Reports 12(1): 64–73.

Osborne, D., and T. Gaebler. 1992. *Reinventing Government: How the Entrepreneurial Spirit Is Transforming the Public Sector.* New York: Plume.

Ostry, A., S. Marion, L. Green, K. Teshke, R. Hershler, S. Kelly, and C. Hertzman. 2002. "The Relationship Between Unemployment, Technological Change and Psychosocial Work Conditions in British Columbia Sawmills." *Critical Public Health* 10(2): 179–92.

Ostry, A.S., R. Hershler, S. Kelly, P. Demers, K. Teschke, and C. Hertzman. 2001. "Effects of De-industrialization on Unemployment, Re-employment, and Work Conditions in a Manufacturing Workforce." *BMC Public Health* 1(15): 1–11.

Pacific Accord. 1987. Draft agreement, 23 June.

Palmer, C.T., and P.R. Sinclair. 2000. "Expecting to Leave: Attitudes to Migration among High School Students on the Great Northern Peninsula of Newfoundland." *Newfoundland Studies* 16(1): 55–78.

– 1997. *When the Fish are Gone: Ecological Disaster and Fishers in Northwest Newfoundland.* Halifax, NS: Fernwood Publishing.

Paquet, G. 2000. "The New Governance, Subsidiarity and the Strategic State." Paper presented at 21st Century Governance conference, Hanover, Germany.

– 1999. Tectonic Changes in Canadian Governance. In *How Ottawa Spends,* ed. Leslie Pal. Oxford, England: Oxford University Press.

– 1968. "Some Views on the Pattern of Economic Development." In *Growth and the Canadian Economy,* ed. T.N. Brewis, 34–64. Carleton Library no. 39. Toronto, ON: McClelland and Stewart.

Parker, J. 1995. "Enabling Morally Reflective Communities: Towards a Resolution of the Democratic Dilemma of Environmental Values in Policy." In *Values and the Environment: A Social Science Perspective,* ed. Y. Guerrier, N. Alexander, J. Chase, and M. O'Brien, 33–50. Chichester, England: Wiley.

Parkes, K.R. 2001. *Psychological Aspects of Work and Health in the North Sea Oil and Gas Industry.* Oxford Contract Research Department, England: Oxford University. http://www.HSE.gov.uk.

Parrish, C.C., N.J. Turner, and S.M. Solberg, eds. 2007. *Resetting the Kitchen Table: Food Security, Culture, Health and Resilience in Coastal Communities.* Hauppague, NY: Nova Science Publishing.

Pauly, D., V. Christensen, J. Dalsgaard, R. Froese, and F. Torres, Jr. 1998. "Fishing Down Marine Food Webs." *Science* 279:860–3.

Pauly, D., and J. Maclean. 2002. *In a Perfect Ocean.* Washington, DC: Island Press.

Pauly, D., Ma. Lourdes Palomares, Rainer Froese, Pascualita Sa-a, Michael Vakily, David Preikshot, and Scott Wallace. 2001. "Fishing Down Canadian Aquatic Food Webs." *Canadian Journal of Fisheries and Aquatic Sciences* 58(1): 51–62.

Payne, L. 2000. In Neis and Grzetic 2000. Report of Workshop on Industrial Restructuring and Rural Health: Gender-Based Analysis, Issues, and Policy Lessons and Alternatives. St John's, 2–5 February, 34–5.

Pease, B.C. 1974. *Effects of Log Dumping and Rafting on the Marine Environment of Southeast Alaska.* USDA Forest Service, General Technical Report PNW–22.

Penning, M.J., D.E. Allan, L.L. Roos, N.L. Chappell, N.P. Roos, and L. Ge. 2002. *Health Care Restructuring and Community-Based Care: Three Regions in British Columbia.* Victoria, BC: Centre On Aging, University of Victoria.

Perry, I., and R. Ommer. 2003. "Scale Issues in Marine Ecosystem and Human Interactions." *Fisheries Oceanography* 12(4/5): 513–22.

Peterson, C.J.G. 1918. "The Sea Bottom and Its Production of Fish. A Survey of the Work Done in Connection with Valuation of the Danish Waters from 1883–1917." *Rep Danish Biol Sta* 25:1–82.

Picot, G., and T. Wannell. 1993. "Permanent Layoffs and Displaced Workers: Cyclical Sensitivity, Concentration, and Experience Following the Layoff." *Journal of Income Distribution* 3(2): 181–230.

Pimm, S.A. 2000. *The World According to Pimm.* New York, NY: McGraw Hill.

Pinkerton, E. 2000. "Directions, Principles, and Practice in the Shared Governance of Canadian Marine Fisheries." In *Fishing Places, Fishing People. Traditions and Issues in Canadian Small-Scale Fisheries*, eds. D. Newell and R.E. Ommer. Toronto, ON: University of Toronto Press

– 1983. "Taking the Minister to Court: Changes in Public Opinion About Forest Management and Their Expression in Haida Land Claims." *BC Studies* 57 (spring 1983).

Piore, M., and C. Sabel. 1984. *The Second Industrial Divide: Possibilities for Prosperity.* New York, NY: Basic Books.

Pitcher, T.J. 2005a. Personal communication.

– 2005b. "'Back To The Future': A Fresh Policy Initiative for Fisheries and a Restoration Ecology for Ocean Ecosystems." *Philosophical Transactions of the Royal Society. B* 360:107–121.

– 2004a. "Introduction to the Methodological Challenges in 'Back-to-the-Future' Research." In *Back to the Future: Advances in Methodology for Modelling and Evaluating Past Ecosystems as Future Policy Goals*, ed. T.J. Pitcher. Fisheries Centre Research Reports 12(1): 4–10.

– 2004b. "What Was the Structure of Past Ecosystems That Had Many Top Predators?" In *Back to the Future: Advances in Methodology for Modelling and Evaluating Past Ecosystems as Future Policy Goals*, ed. T.J. Pitcher. Fisheries Centre Research Reports 12(1): 18–20.

– 2004c. "The Problem of Extinctions." In *Back to the Future: Advances in Methodology for Modelling and Evaluating Past Ecosystems as Future Policy Goals*, ed. T.J. Pitcher. Fisheries Centre Research Reports 12(1): 21–8.

– ed. 2004d. *Back to the Future: Advances in Methodology for Modelling and Evaluating Past Ecosystems as Future Policy Goals*. Fisheries Centre Research Reports 12(1).

– 2001. "Fisheries Managed to Rebuild Ecosystems: Reconstructing the Past to Salvage the Future." *Ecological Applications* 11(2): 601–17.

Pitcher, T.J., C.H. Ainsworth, E.A. Buchary, W.L. Cheung, R. Forrest, N. Haggan, H. Lozano, T. Morato, and L. Morissette. 2005. "Strategic Management of Marine Ecosystems Using Whole-Ecosystem Simulation Modelling: The Back-to-the-Future Policy Approach." In *The Strategic Management of Marine Ecosystems*, eds. E. Levner, I. Linkov, and J-M. Proth, 199–258. NATO Science Series: 4, Earth and Environmental Sciences. New York, NY: Springer.

Pitcher, T.J., and Forrest, R. 2004. "Challenging Ecosystem Simulation Models with Climate Change: The 'Perfect Storm.'" In *Back to the Future: Advances in Methodology for Modelling and Evaluating Past Ecosystems as Future Policy Goals*, ed. T.J. Pitcher, 29–38. Fisheries Centre Research Reports 12(1).

Pitcher, T.J., J.J. Heymans, C.H. Ainsworth, E.A. Buchary, U.R. Sumaila, and V. Christensen. 2004. "Opening the Lost Valley: Implementing a 'Back to the Future' Restoration Policy for Marine Ecosystems and Their Fisheries." In *Sustainable Management of North American Fisheries*, ed. E.E. Knudsen, D.D. MacDonald, and J.K. Muirhead. American Fisheries Society 43:165–93.

Pitcher, T.J., J.J. Heymans, and M. Vasconcellos, eds. 2002. *Ecosystem Models of Newfoundland for the Time Periods 1995, 1985, 1900, and 1450*. Fisheries Centre Research Reports 10(5).

Pitcher, T.J., and D. Pauly. 1998. "Rebuilding Ecosystems, Not Sustainability, as the Proper Goal of Fishery Management." In *Reinventing Fisheries*, eds. T. Pitcher, P. Hart, and D. Pauly, 311–29. London: Chapman and Hall.

Pitcher, T.J., M. Power, and L. Wood, eds. 2002. *Restoring the Past to Salvage the Future: Report on a Community Participation Workshop in Prince Rupert, BC*. Fisheries Centre Research Reports 10(7).

Pitcher, T.J., M. Vasconcellos, J.J. Heymans, C. Brignall, and N. Haggan, eds. 2002. *Information Supporting Past and Present Ecosystem Models of Northern British Columbia and the Newfoundland Shelf*. Fisheries Centre Research Reports 10(1).

Plaice, E. 1990. *The Native Game: Settler Perceptions of Indian/Settler Relations in Central Labrador*. Social and Economic Studies 40. St John's, NL: ISER Books.

Pollard, Jason. 2004. "The Influence of Logging Technology on Employment and on the Boreal Forest Landscape of Newfoundland, Canada." MES thesis, Environmental Science Programme, Memorial University of Newfoundland.

Post, E. 2003. "Climate-Vegetation Dynamics in the Fast Lane." *Trends in Ecology and Evolution* 18:551–3.

Powell, B.W., Sr. 1987. *The Letter That Was Never Read (A History of the Labrador Fishery)*. St John's, NL: Good Tidings Press.

Power, C., and C. Hertzman. 1999. "Health, Well-Being and Coping Skills." In *Developmental Health and the Wealth of Nations*, ed. D.P. Keating and C. Hertzman, 41–54. New York, NY: The Guilford Press.

Power, M.D., N. Haggan, and T.J. Pitcher. 2004. "The Community Workshop: How We Did It and What We Learned from the Results." In *Back to the Future: Advances in Methodology for Modelling and Evaluating Past Ecosystems as Future Policy Goals*, ed. T.J. Pitcher, 125–8. Fisheries Centre Research Reports 12(1).

Power, N., M. Binkley, B. Neis, and S. Brennan. 2005. "Newfoundland Fishers' Perceptions of Risk." Canadian Association for Research on Work and Health Conference, 15–17 May, Vancouver.

Praxis Research and Consulting. 2005. *Setting a New Course – Phase II Human Resources Sector Study for the Fish Harvesting Industry in Canada*. Ottawa, ON: Canadian Council of Professional Fish Harvesters.

PRFD (Prince Rupert Forest District). 1949. *Annual Management Report*. GR1441, reel B9899, file 027637, BC Archives. Quoted in Rajala 2006, chap. 6.

– 1940. Annual Management Report, GR1441, reel B9899, file 027637, BC Archives. Quoted in Rajala 2006, chap. 4.

Prince Rupert Economic Development Commission. 2002. *A Socio-economic Impact Analysis of Skeena Cellulose Inc*. Prince Rupert, BC: Prince Rupert Economic Development Commission.

Provincial and Territorial Ministers of Health. 2000. *Understanding Canada's Health Care Costs*. www.scics.gc.ca/pdf/850080012e.pdf

Provincial Co-ordinating Committee for Remote and Rural Health Services. 1999. *Enhancing Health Services in Remote and Rural Communities of British Columbia*. Victoria: Ministry of Health.

Rajala, Richard. 2006. *Up-Coast: Forests and Industry on British Columbia's North Coast, 1870–2005*. Victoria, BC: Royal British Columbia Museum and Coasts Under Stress research project.

Raphael, Dennis. 1999. "The Full Monty: From Increasing Poverty to Societal Disintegration: Economic Inequality and the Future Health of Canada." Lecture given as part of the series Philosophy and Contemporary Thought, University of Toronto, School of Continuing Studies. 29 January.

Rapport, D.J., R. Costanza, and A.J. McMichael. 1998. "Assessing Ecosystem Health." *Trends in Ecology and Evolution* 13:397–402.

Recalma-Clutesi, Kim. 1993. Personal communication with N.J. Turner.

Rees, William. 2004. "Net-Pen Salmon Farming: Failing on Two Fronts (and Why This Is Just the Latest Stage in Humanity's Terminal Ravaging of the Seas)." In *Proceedings of the World Summit on Salmon*, ed. P. Gallaugher and L. Wood, 139–52 (Simon Fraser University, Burnaby, BC, 10–13 June).

– 2002. "Globalization and Sustainability: Conflict or Convergence?" *Bulletin of Science, Technology and Society* 22(4): 249–68.

Reeves, William G. 1987. "'Our Yankee Cousins': Modernization and the Newfoundland-American Relationship, 1898–1910." PHD thesis, History, University of Maine (Univ. 1987.R257).

Rejebian, V.A., A.G. Harris, and J.S. Huebner. 1987. "Conodont Colour and Textural Alteration: An Index to Regional Metamorphism, Contact Metamorphism, and Hydrothermal Alteration." *Geological Society of America Bulletin* 99:471–9.

Report on the Census of Newfoundland. 1921.

Report re Strike of Employees of the Granby Consolidated Mining, Smelting and Power Company Limited at Anyox BC, 1933. Submitted to W.A. Mackenzie, Minister of Labour, Victoria, BC. Pattullo Papers 53/20 Add Mss 3, BC Archives.

Richer, Shawna. 2004. "Williams Vows to Take Dispute with Martin across Country." *Globe and Mail*, 4 November, A7.

Riedman, M.L., and J.A. Estes. 1990. "The Sea Otter (*Enhydra lutris*): Behaviour, Ecology and Natural History." US *Fish and Wildl Serv Biol Rep* 90(14).

Riggs, K. 2002. "Heavy Metal Particulates in Soil in the Vicinity of the Tilt Cove Smelter, Newfoundland." Unpub. BSC (Hons.) thesis, Memorial University of Newfoundland, St John's, NL.

Right, Ronald. 2004. *A Short History of Progress.* Toronto, ON: Anansi.

Roach, Catherine. 2000. "Stewards of the Sea: A Model for Justice?" In *Just Fish: Ethics and Canadian Marine Fisheries*, ed. H. Coward, R.E. Ommer, and T.J. Pitcher. St John's, NL: ISER Books.

Roach, Catherine M., Tim I. Hollis, Brian E. McLaren, and Dean L.Y. Bavington. 2006. "Ducks, Bogs and Guns: Stewardship Ethics in Newfoundland." *Ethics and the Environment* 11 (1): 43–70.

Roberts, C.M. 2000. "Selecting Marine Reserve Locations: Optimality versus Opportunism." *Bulletin of Marine Science* 66 (3): 581–92.

Roberts, C.M., J.A. Bohnsack, F. Gell, J.P. Hawkins, and R. Goodridge. 2001. "Effects of Marine Reserves on Adjacent Fisheries." *Science* 294:1920–23.

Roebothan, B.V. 2004. *Nutrition Newfoundland and Labrador: The Report of a Survey of Residents of Newfoundland and Labrador, 1996.* St John's, NL: Department of Health and Community Services, Province of Newfoundland and Labrador.

Rogers, Kara. Forthcoming. "Harvester-Resource Interactions: Historical Reconstruction and Biological Assessment of the Bonne Bay, Newfoundland, Lobster Population." MSc thesis, Biology, Memorial University of Newfoundland.

– 2003. "Micro-level Historical Reconstruction of the Newfoundland Fisheries between 1891–2000: Findings and Issues." In *Ecosystem Models of Newfoundland and Southeastern Labrador: Additional Information and Analyses for "Back to the Future,"* ed. J.J. Heymans. Fisheries Centre Research Reports 11(5): 12–19.

Rohr, K.M.M. 2002. *Seismic Reflection Data in Seismically Active Areas of Hecate Strait.* Centre for Earth and Ocean Research, CEOR Report, 2002–3.

Rohr, K.M.M., and G. Sutherland. 2002. *A Reconnaissance AVO Study of Queen Charlotte Basin.* Centre for Earth and Ocean Research, CEOR Report, 2002–2.

Rose, G.A. 2004. "Reconciling Overfishing and Climate Change with Stock Dynamics of Atlantic Cod (*Gadus morhua*) over 500 Years." *Canadian Journal of Fisheries and Aquatic Sciences* 61:1553–7.

Rose, G.A., and R.L. O'Driscoll. 2002. "Capelin Are Good for Cod: Can the Northern Stock Rebuild without Them?" *ICES J Mar Sci* 59:1018–26.

Rosenberg, Andrew A., W. Jeffrey Bolster, Karen E. Alexander, William B. Leavenworth, Andrew B. Cooper, and Matthew G. McKenzie. 2005. "The History of Ocean Resources: Modeling Cod Biomass Using Historical Records." *Frontiers in Ecology and the Environment* 3 (2): 84–90.

Royal Commission on Hospital and Nursing Home Costs. 1984. *Report of the Royal Commission on Hospital and Nursing Home Costs – Executive Summary.* St John's: s.n.

Royal Society of Canada. 2004. Report of the Expert Panel on Science Issues Related to Oil and Gas Activities, Offshore British Columbia. Ottawa.

Russo Garrido, S., and J. Stanley 2002. "Labrador Road Study: Local Knowledge on the Social and Environmental Impacts of the Newly Constructed Trans-Labrador Highway in South-Eastern Labrador." Coasts Under Stress Occasional Paper. http://www.coastsunderstress.ca/publications.php.

Ruzzante, D.E., J.S. Wroblewski, C.T. Taggart, R.K. Smedbol, D. Cook, and S.V. Goddard. 2000. "Bay-Scale Population Structure in Coastal Atlantic Cod in Labrador and Newfoundland, Canada." *Journal of Fish Biology* 56:431–47.

Saila, S. 1983. *Importance and Assessment of Discards in Commercial Fisheries.* FAO Fisheries Circular No. 765.

Samson, Daniel, ed. 1994. *Contested Countryside: Rural Workers and Modern Society in Atlantic Canada.* Fredericton, NB: Acadiensis Press.

Sandhu, G., and J. Schofield 2001. "Impacts of Restructuring in a Coastal Sub-Region of British Columbia: A Computable General Equilibrium

Analysis." http://www.coastsunderstress.ca/pubs/
   arm4_sandhu_schofield_april2001.pdf.

Savoie, D. 2000. "Community Economic Development in Atlantic Canada:
   False Hope or Panacea?" Moncton: The Canadian Institute for Research
   on Regional Development.

Scheffer, M., and S.R. Carpenter. 2003. "Catastrophic Regime Shifts in
   Ecosystems: Linking Theory to Observation." *Trends in Ecology and Evolution*
   18:648–56.

Schneider, R.H. 1984. "The Formation of Attitudes toward Development in
   Southern Labrador." PHD thesis, McGill University.

Schofield, R.F., and M. Amodeo. 1999. "Interdisciplinary Teams in Health
   Care and Human Services Settings: Are They Effective?" *Health and Social
   Work* 24(3): 210–19.

Schrank, W., R. Arnason, and R. Hanneson. 2003. *The Cost of Fisheries
   Management*. Hants, England: Ashgate.

Scientific Panel for Sustainable Forest Practices in Clayoquot Sound. 1995a. *A
   Vision and Its Context: Global Context for Forest Practices in Clayoquot Sound.*
   Report 4. Victoria, BC: Cortex Consulting.

– 1995b. *First Nations' Perspectives on Forest Practices in Clayoquot Sound.* Report
   3. Victoria, BC.

SCIPAS (Study and Conference on Improving Public Access to Science
   through Science Shops.) 2001. Reports nos. 1–7. In *SCIPAS Project Technical
   Reports Compendium, The Netherlands: Living Knowledge – The International
   Science Shops Network.*

Sclove, Richard E. 1998. "Better Approaches to Science Policy." *Science* 279
   (5355): 1283.

Scoones, I. 1999. "New Ecology and the Social Sciences: What Prospects for a
   Fruitful Engagement?" *Annual Review of Anthropology* 28:479–507.

Seaton, Peter D. 1989. *Toward a Better Age*. Report of the Seaton Royal
   Commission on Health Care. Ottawa: The Queen's Printer.

Sewell, William H., Jr. 1992. "A Theory of Structure: Duality, Agency and
   Transformation." *American Journal of Sociology* 98 (1): 1–29.

Sforza-Roderick, Michelle, Scott Nova, and Mark Weisbrot. 1998. "Writing
   the Constitution of a Single Global Economy: A Concise Guide to the
   Multilateral Agreement on Investment-Supporters' and Opponents'
   Views." Preamble, Center for Public Policy. http://www.rtk.net/preamble/
   mai/maioverv.html (accessed 16 June 2005).

Shank, C.C. 1999. "The Committee on the Status of Endangered Wildlife in
   Canada (COSEWIC): A Twenty-One-Year Retrospective." *Canadian
   Field-Naturalist* 113:318–41.

Shell Canada Ltd. 1968a. *Well History Report, Shell Anglo Apollo J-14.*

– 1968b. *Well History Report, Shell Anglo Cygnet J-100.*

– 1968c. *Well History Report, Shell Anglo Sockeye B-10.*

- 1967a. *Well History Report, Shell Anglo Pluto 1–87.*
- 1967b. *Well History Report, Shell Anglo Prometheus H–68.*
- 1967c. *Well History Report, Shell Anglo Zeus D–14.*
- 1967d. *Well History Report, Shell Anglo Zeus 1–65.*

Sheps, S.B., R.J. Reid, M.L. Barer, H. Krueger, K.M. McGrail, B. Green, R.G. Evans, and C. Hertzman. 2000. "Hospital Downsizing and Trends in Health Care Use among Elderly People in British Columbia." *Canadian Medical Association Journal* 163:397–401.

Shouldice, D.H. 1971. "Geology of the Western Canadian Continental Margin." *Bulletin of Canadian Petroleum Geology* 19:405–36.

Shrimpton, M., and K. Storey. 2001. *The Effects of Offshore Employment in the Petroleum Industry: A Cross-National Perspective.* Herndon, VA: US Department of the Interior, Minerals Management Services, Environmental Studies Program.

Sibert, J.R. 1979. "Detritus and Juvenile Salmon Production in the Nanaimo Estuary: II, Meiofauna Available as Food to Juvenile Chum Salmon (*Oncorhynchus keta*)". *Journal of the Fisheries Research Board of Canada* 36:497–503.

Sibert, J.R., and V.J. Harpham. 1979. "Effects of Intertidal Log Storage on the Meiofauna and Interstitial Environment of the Nanaimo River Delta." *Fisheries and Marine Science Technical Report* 883.

Simenstad C.A., J.A. Estes, and K.W. Kenyon. 1978. "Aleuts, Sea Otters, and Alternate Stable-State Communities." *Science* 200:403–11.

Sinclair, A.F., and S.A. Murawski. 1997. "Why Have Groundfish Stocks Declined?" In *Northwest Atlantic Groundfish: Perspectives on a Fishery Collapse,* ed. J. Boreman, B.S. Nakashima, J.A. Wilson, and R.L. Kendall, 71–93. Bethesda, MD: American Fisheries Society.

Sinclair, P.R. 2003. "Becoming Global: Capitalist Development and Restructuring in the World's Forest Products Industry." http://www.coastsunderstress.ca/publications.
- 1985. *From Traps to Draggers: Domestic Commodity Production in Northwest Newfoundland, 1850–1982.* St John's, NL: ISER Books.

Sinclair, P.R., and I. Gray. 2005. "Local Leaders in a Global Setting: Dependency and Resistance in Regional New South Wales and Newfoundland." *Sociologia Ruralis* 45 (1/2): 37–52.

Sinclair, P.R., M. MacDonald, and B. Neis. 2006. "The Changing World of Andy Gibson: Restructuring Forestry on Newfoundland's Great Northern Peninsula." *Studies in Political Economy* 78:177–99.

Sinclair, P.R., and R.E. Ommer, eds. 2006. *Power and Restructuring: Shaping Canada's Coastal Society and Environment.* St John's, NL: ISER Books.

Sloan, Gordon. 1956. *Report of the Commissioner Relating to the Forest Resources of British Columbia.* Vol. 1. Victoria, BC: Queen's Printer. Quoted in Rajala, 2006, chap. 6: 143–4.

Smallwood, J.R. 1931. *The New Newfoundland.* New York, NY: Macmillan.

Snavely, P.D. Jr, and K.A. Kvenvolden. 1989. "Preliminary Evaluation of the Petroleum Potential of the Tertiary Accretionary Terrane, West Side of the Olympic Peninsula, Washington." *U.S. Geological Survey Bulletin* 1892:1–17.

Snelgrove, A.K. 1953. *Mines and Mineral Resources of Newfoundland.* St John's, NL: Department of Mines and Resources.

Snowdon, L.R., M.G. Fowler, and T.S. Hamilton. 1988. "Sources and Seeps: Organic Geochemical Results from the Queen Charlotte Islands." In *Some Aspects of the Petroleum Geology of the Queen Charlotte Islands,* 37–43. Canadian Society of Petroleum Geologists, Field Trip Guide.

Snyder, Will, Madeleine Scammell, and Phil Shepard. 1997. *Building a Community Research Network: Report of the 1996 Community Research Network Conference, Amherst, Massachusetts.* Amherst, MA: The Loka Institute and University of Massachusetts Extension.

Song, Samantha, and R. Michael M'Gonigle. 2000. "Science, Power, and System Dynamics: The Political Economy of Conservation Biology." *Conservation Biology* 15:980–9.

Sorensen, J. 1997. "National and International Efforts at Integrated Coastal Management: Definitions, Achievements, Lessons." *Coastal Management* 25 (1): 3–41.

Southard, F.E. 1982. "Salt Cod and God: An Ethnography of Socio-economic Conditions Affecting Status in a Southern Labrador Community." MA thesis, Memorial University of Newfoundland.

Speck, D. 1987. *An Error in Judgement.* Vancouver, BC: Talon Books.

Spence, G.D., R.D. Hyndman, S. Langton, C.J. Yorath, and E.E. Davis. 1991. "Multichannel Seismic Reflection Profiles across the Vancouver Island Continental Shelf and Slope." Geological Survey of Canada, Open File 2391.

Stanhope, M.J., and C.D. Levings. 1985. "Growth and Production of *Eogammarus confervicolus* (Amphipoda: Anisogammaridae) at a Log Storage Site and in Areas of Undisturbed Habitat within the Squamish Estuary, British Columbia." *Canadian Journal of Fisheries and Aquatic Sciences* 42:1733–40.

Stanhope, M.J., D.W. Powell, and E.B. Hartwick. 1987. "Population Characteristics of the Estuarine Isopod, *Gnorimophaeroma insulare*, in Three Contrasting Habitats, Sedge Marsh, Algal Bed and Wood Debris." *Canadian Journal of Zoology* 65:2097–2104.

Stanley, Hal (Chairman and Chief Executive Officer of the Canada-Newfoundland Offshore Petroleum Board). 2004. Personal communication, Tuesday, 23 March.

Steele, John. 1998. "Regime Shifts in Marine Ecosystems." *Ecological Applications* 8(1): s33–6.

Stenmark, M. 2002. *Environmental Ethics and Policy Making.* Aldershot, England: Ashgate.

Stenson, G.B., M.O. Hammill, and J.W. Lawson. 1997. "Predation by Harp Seals in Atlantic Canada: Preliminary Consumption Estimates for Arctic Cod, Capelin and Atlantic Cod." *J Northwest Atl Fish Sci* 22:137–54.

Stephenson, Peter, Jennifer Hopkinson, and Nancy J. Turner. 1995. "Changing Traditional Diet and Nutrition of Aboriginal Peoples of Coastal British Columbia." In *A Persistent Spirit: Towards Understanding Aboriginal Health in British Columbia*, eds. Peter H. Stephenson, Susan J. Elliott, Leslie T. Foster, and Jill Harris, 129–65. University of Victoria, Western Geographic Series, No. 31.

Stewart, Hilary. 1977. *Indian Fishing: Early Methods on the Northwest Coast.* Vancouver: Douglas & McIntyre.

Story, G.M., W.J. Kirwin, and J.D.A. Widdowson. 1990. *Dictionary of Newfoundland English.* 2d ed., with supplement. Toronto: University of Toronto Press and St John's: Breakwater.

Sumaila, U.R., and C.J. Walters. 2005. "Intergenerational Discounting: A New Intuitive Approach." *Ecol Econ* 52:135–42.

– 2003. "Intergenerational Discounting." In *Three Essays on the Economics of Fishing*, ed. U.R. Sumaila, 19–25. Fisheries Centre Research Reports 11(3).

Swanson, Eleanor. n.d. "Study of Adult Eating Habits in Four Newfoundland Communities." (Survey October 1967 – September 1968), Eleanor Swanson files, "Nutritional Studies."

Symons, Brenton. 1910. "The Mineral Resources of Newfoundland." *Engineering and Mining Journal* (August): 360–2.

Tengo, M., and M. Hammer. 2003. "Management Practices for Building Adaptive Capacity: A Case of Northern Tanzania." In *Navigating Social-Ecological Systems: Building Resilience for Complexity and Change*, ed. F. Berkes, J. Colding, and C. Folke, 132–62. Cambridge, England: Cambridge University Press.

Tennant, Paul. 1990. *Aboriginal People and Politics: The Indian Land Question in British Columbia, 1849–1989.* Vancouver: UBC Press.

The Courier, 24 November 1866.

The Newfoundlander, 4 January 1876.

The Royal Society of Canada. 2004. *Report of the Expert Panel on Science Issues Related to Oil and Gas Activities, Offshore British Columbia.* Ottawa: Royal Society of Canada.

Thompson, F.F. 1961. *The French Shore Problem in Newfoundland: An Imperial Study.* Toronto, ON: University of Toronto Press.

Thompson, J.C. 2004a. "Gitga'at Plant Project: The Intergenerational Transmission of Traditional Ecological Knowledge Using School Science Curricula." MSc thesis, School of Environmental Studies, University of Victoria, Victoria, BC.

– 2004b. "Traditional Plant Knowledge of the Tsimshian Curriculum: Keeping Knowledge in the Community." *Canadian Journal of Native Education* 28 (1/2): 61–5.

Thornton, Patricia A. 1979. "Dynamic Equilibrium: Population, Ecology and Economy in the Strait of Belle Isle, Newfoundland, 1840–1940." PHD diss., Aberdeen University, Scotland.

Tomblin, Stephen. 2004. "Creating a More Democratic Health System: A Critical Review of Constraints and a New Approach to Health Restructuring." In *The Governance of Health Care in Canada,* ed. Tom McIntosh et al., 280–311. Toronto: University of Toronto Press.

– 2002. "Newfoundland and Labrador at the Cross-roads: Reform or Lack of Reform in a New Era." *Journal of Canadian Studies* 37 (1): 104.

Townson, M. 1999. *Health and Wealth: How Social and Economic Factors Affect Our Well Being.* Toronto, ON: James Lorimer.

Tuck, J.A., and R. Grenier. 1989. *Red Bay, Labrador: World Whaling Capital AD 1550–1600.* St John's: Atlantic Archaeology.

Tunstell, G. 1956. *The Forests of the Island of Newfoundand.* St John's: Government of Newfoundland, Department of Mines and Resources.

Tupper, M., and R.G. Boutilier 1995. "Effects of Habitat on Settlement, Growth and Post-Settlement Survival of Atlantic Cod (*Gadus morhua*)." *Canadian Journal of Fisheries and Aquatic Sciences* 52:1834–41.

Turner, N.J. 2005. *The Earth's Blanket. Traditional Teachings for Sustainable Living.* Vancouver: Douglas & McIntyre and Seattle: University of Washington Press.

– 2004. *Plants of Haida Gwaii. Xaadaa Gwaay guud gina k'aws (Skidegate), Xaadaa Gwaayee guu giin k'aws (Massett).* Winlaw, BC: Sono Nis Press.

– 2003. "Passing on the News: Women's Work, Traditional Knowledge, and Plant Resource Management in Indigenous Societies of NW North America." In *Women and Plants: Case Studies on Gender Relations in Local Plant Genetic Resource Management,* ed. P. Howard. UK: Zed Books.

– 1995. *Food Plants of Coastal First Peoples.* Royal British Columbia Museum Handbook. Victoria, BC: Royal British Columbia Museum (revised from 1975 edition, *Food Plants of British Columbia Indians,* part 1, "Coastal Peoples." Vancouver: University of British Columbia Press).

Turner, Nancy J., and J.T. Jones. 2005. "'A Fine Line between Two Nations': Ownership Patterns for Plant Resources among Northeast Coast Indigenous Peoples – Implications for Plant Conservation and Management." In *"Keeping and Living": Traditions in Plant Use and Cultivation on the Northwest Coast of North America,* ed. D. Deur and N.J. Turner. Seattle: University of Washington Press and Vancouver: UBC Press.

Turner, N.J., I.J. Davidson-Hunt, and M. O'Flaherty. 2003. "Living on the Edge: Ecological and Cultural Edges as Sources of Diversity for Social-Ecological Resilience." *Human Ecology* 31 (3): 439–63.

Turner, N.J., Marianne B. Ignace, and Ronald Ignace. 2000. "Traditional Ecological Knowledge and Wisdom of Aboriginal Peoples in British Columbia." *Ecological Applications* 10 (5): 1275–87.

Turner, Robert D. 1990. *Logging by Rail: The British Columbia Story*. Victoria, BC: Sono Nis Press.

Umpleby, S. 2003. Personal communication with N.J. Turner, 16 October.

Van Biesen, G. 2003. "Lipid Class and Fatty Acid Composition of Plankton and Settling Particles at a Fish Enclosure, and of Cod (*Gadus morhua*) in Gilbert Bay, Labrador." MSc thesis, Environmental Science Programme, Memorial University.

Vancouver Island Health Authority. 2003. Health Services Redesign Plan 2003/04–2005/06. http://www.viha.ca/pdf/mandate_page/health_services_redesign_plan_03_06.pdf.

Vancouver Sun. 2001. "No More Money, Liberals Tell Nurses: Government Is Willing to Force an End to Limits on Overtime; Union Says Nurses Won't Be Bullied." A1.

Vardy, M.C., and W.A. Dickson 1967. "Bowaters Newfoundland Timber Resources and Future Wood Supply." Corner Brook: Memorandum, Woodlands Department, Bowaters Newfoundland Limited.

Vasconcellos, M., and J.J. Heymans. 2002. "System Boundaries for East Coast Ecosystem Models." In *Information Supporting Past and Present Ecosystem Models of Northern British Columbia and the Newfoundland Shelf*, eds. T.J. Pitcher, M. Vasconcellos, J.J. Heymans, C. Brignall, and N. Haggan. Fisheries Centre Research Reports 10(1).

Vasconcellos, M., J.J. Heymans, and T.J. Pitcher. 2002. "Historical Reference Points for Models of Past Ecosystems in Newfoundland." In *Information Supporting Past and Present Ecosystem Models of Northern British Columbia and the Newfoundland Shelf*, eds. T.J. Pitcher, M. Vasconcellos, J.J. Heymans, C. Brignall, and N. Haggan. Fisheries Centre Research Reports 10(1):7–12.

Vasconcellos, M., and T.J. Pitcher. 2002. "Historical Reference Points For Models of Past Ecosystems in Northern BC." In *Information Supporting Past and Present Ecosystem Models of Northern British Columbia and the Newfoundland Shelf*, eds. T.J. Pitcher, M. Vasconcellos, J.J. Heymans, C. Brignall, and N. Haggan. Fisheries Centre Research Reports 10(1):60–7.

Vasconcellos, M., M. Power, T.J. Pitcher, G. Stenson, B. Sjare, M. Hammill, G.K. Davoren, W.A. Montevecchi, I.J. Stenhouse, J.J. Heymans, E. Dawe, D. Orr, D.C. Schneider, and E. Dalley. 2002. "Workshop Notes and Sources on the Ecosystem Groups." In *Information Supporting Past and Present Ecosystem Models of Northern British Columbia and the Newfoundland Shelf*, eds. T.J. Pitcher, M. Vasconcellos, J.J. Heymans, C. Brignall, and N. Haggan. Fisheries Centre Research Reports 10(1):38–59.

Visser, Margaret. 1992. *The Rituals of Dinner: The Origins, Evolution, Eccentricities, and Meaning of Table Manners*. Toronto, ON: Harper Collins.

- 1988. *Much Depends upon Dinner: The Extraordinary History and Mythology, Allure and Obsessions, Perils and Taboos, of an Ordinary Meal.* Toronto, ON: Harper Collins.

Vodden, K. Forthcoming. "Collaborative Governance and the Search for Sustainability in Canadas Coastal Regions." PHD thesis. Simon Fraser University.

- 2004a. "Collaborative Governance and Sustainable Development in Canada's Coastal Regions." Presented to the Canadian Rural Revitalization Foundation (CRRF) Think Tank, Local Governance of Rural-Urban Interactions: New Directions. April, College of New Caledonia, Prince George, BC.

- 2004b. "Watershed Management in Canada: Lessons and Implications for Institutional Design." Presented to the United Nations Environment Programme Hilltops to Oceans Partnership Conference, Cairns, Australia, May.

- 2004c. "Knowledge Sharing and Collaborative Watershed Governance – Coast to Coast." Canadian Water Resources Association Conference. Regina, SK, November.

- 2004d. "Opportunity Knocks: The Shellfish Industry in Alberni-Clayoquot." Alberni-Clayoquot Shellfish Industry Report/Processing Opportunity Package. Prepared for Ecotrust Canada and the Alberni-Clayoquot Shellfish Task Force.

- 2002. "Sustainable Community Development in a Coastal Context: The Case of Alert Bay, British Columbia." *Canadian Journal of Aboriginal Economic Development* 3 (1).

- 1999a. "Cormorant Island Community Profile." Inner Coast Natural Resource Centre, Alert Bay, BC.

- 1999b. "Nanwakola: Co-management and Sustainable Community Economic Development in a BC Fishing Village." MA thesis. Simon Fraser University, Department of Geography. http://www.sfu.ca/coastalstudies.

Vodden, K., and K. Bannister. Forthcoming. "Circularising Knowledge Flows: Institutional Structures, Policies and Practices for Community-University Partnerships." In *Making and Moving Knowledge,* eds. J. Lutz and B. Neis. Montreal, QC: McGill-Queen's University Press.

Vodden, K., and J. Kennedy. 2006. "From Resignation to Renewal: First Nations' Strategies for Resilience." In *Power, Agency and Nature: Shaping Coastal Society and Environment,* eds. P. Sinclair and R.E. Ommer. St John's: ISER Books.

Vodden K., and B. Kuecks. 2003. "The Clayoquot Green Economic Opportunities Project: Taking Steps towards a Conservation Economy. Part Two: Sectoral Analysis." Prepared for Friends of Clayoquot Sound, Ahousaht First Nation and Clayoquot Biosphere Trust. http://www.focs.ca/.

Vodden, K., J. Penikett, and M. Berry. 2003. Examining Best Practices in Coastal Zone Planning: Lessons and Applications for BC's Central Coast. Burnaby, BC: Simon Fraser University Centre for Coastal Studies. http://www.sfu.ca/coastalstudies.

Vodden K., J. Pierce, and D. House. 2004. "Offshore Oil and Gas and the Quest for Sustainable Development: A Rural Development Perspective." In *The Structure and Dynamics of Rural Territories: Geographical Perspectives*, ed. D. Ramsay and C. Bryant. Brandon, MB: University of Brandon.

Wackernagel, M., and W. Rees. 1996. *Our Ecological Footprint: Reducing Human Impact on Earth.* Gabriola Island, BC: New Society Publishers.

Wald, G. 1960. "The Distribution and Evolution of Visual Systems." In *Comparative Biochemistry*, vol. 1, eds. M. Florkin and H. Mason, 311–45. New York: Academic Press.

Walters, C.J. 1995. *Fish on the Line: The Future of Pacific Fisheries.* Suzuki Foundation Report. Vancouver, Suzuki Foundation.

Walters, C.J., and J.J. Maguire. 1996. "Lessons for Stock Assessment from the Northern Cod Collapse." *Rev Fish Biol* 6:125–37.

Warren, William A. 2005. "Hierarchy Theory in Sociology, Ecology, and Resource Management: A Conceptual Model for Natural Resource or Environmental Sociology and Socioecological Systems." *Society and Natural Resources* 18 (5): 447–66.

Wathne J.A., T. Haug, and C. Lydersen 2000. "Prey Preference and Niche Overlap of Ringed Seals *Phoca hispida* and Harp Seals *P. groenlandica* in the Barents Sea." *Mar Ecol Prog Ser* 194:233–9.

Watkins, M.H. 1963. "A Staple Theory of Economic Growth." *Canadian Journal of Economics and Political Science* 29:141–58.

Watkins, M.H., and H.M. Grant. 1993. *Canadian Economic History: Classic and Contemporary Approaches.* Ottawa, ON: Carleton University Press.

Watson, J.C. 2000. "The Effects of Sea Otters (*Enhydra lutris*) on Abalone (*Haliotis* spp.) Populations." In *Workshop on Rebuilding Abalone Stocks in British Columbia*, ed. A. Campbell. *Can Spec Publ Fish Aquat Sci* 130:123–32.

WCVI Aquatic Management Society. 2000. *Current Status and Potential of West Coast Vancouver Island Fisheries.*

Weissbourd, B. 2000. "Supportive Communities for Children and Families." *Public Health Reports* 115 (3): 167–73.

Weitzman, M.L. 2001. "Gamma Discounting." *The American Economic Review* 91 (1): 260–72.

Western Economic Consulting Ltd and Clayton Resources Ltd. n.d. *The Impact of Mine Closure in British Columbia.* Victoria, BC: Province of British Columbia.

Whalen, Jennifer. 2005. "Using Harvesters Knowledge to Develop a Computer Simulation Model of the St John Bay, Newfoundland, Lobster Fishery." MSc thesis, Geography, Memorial University of Newfoundland.

White, Louise, and Frank Johns. 1997. *Marine Environmental Assessment of the Estuary and Gulf of St Lawrence, 1997*. Ottawa, ON: Fisheries and Oceans Canada.

White, Richard. 1991. *"It's Your Misfortune and None of My Own": A New History of the American West*. Norman, OK: University of Oklahoma Press. Quoted in Curtis 2001.

Whitehorse Daily Star. 2001. "BC Nurses Walk Out As Government Prepares to Impose Contract." August 8, 11.

Whiticar, M.J., T. Schuemann, M. Niemann, K. Rohr, and M. Johns. 2003. *Analysis of Petroleum Potential in Queen Charlotte Basin: Phase I Report Broad-Scale Basin Characterization*. Victoria, BC: Biogeochemistry Facility, University of Victoria. http://www.energybc.ca/basin_rep.html.

Whiticar, M.J., T. Schuemann, and K.M.M. Rohr. 2004. *Analysis of Petroleum Potential in Queen Charlotte Basin: Phase II Report 2D Petroleum System Modeling*. Victoria, BC: Biogeochemistry Facility, University of Victoria. http://www.energybc.ca/basin_rep.html.

Whitson, D. 2001. "Nature as Playground: Recreation and Gentrification in the Mountain West." In *Writing off the Rural West: Globalization, Governments and the Transformation of Rural Communities*, ed. R. Epp and D. Whitson, 141–64. Edmonton, AB: University of Alberta Press and Parkland Institute.

WHO (World Health Organization). 1948. *Proceedings and Final Acts of the International Health Conference Held in New York from 19 June to 22 July 1946*. World Health Organization Official Records, No. 2.

Wilkinson, Charles F. 1992. *Crossing the Next Meridian: Land, Water, and the Future of the West*. Washington, DC: Island Press.

Wilkinson, Richard, and Michael Marmot, eds. 2003. *Social Determinants of Health: The Solid Facts*. 2d ed. October 26. Copenhagen, Denmark: WHO Regional Publications.

William Nycum and Associates. 1994. *Discussion Paper for Board Restructuring – Northern Newfoundland and Labrador*. St John's: Newfoundland and Labrador, Department of Health.

Williams, P.L., L. McIntyre, J.B. Dayle, and K. Raine. 2003. "The 'Wonderfulness' of Children's Feeding Programmes." *Health Promotion International* 18 (2): 163–70.

Williams, W.A. 1919. "The By-Product Coke Ovens of the Granby Consolidated Mining, Smelting and Power Company, Limited." *Transactions of the Canadian Institute of Mining and Metallurgy* 22:179–89.

Williamson, C.J., C.D. Levings, J.S. MacDonald, E. White, K. Kopeck, and T. Pendray. 2000. *A Preliminary Assessment of Wood Debris at Four Log Dumps on Douglas Channel, British Columbia: Comparison of Techniques*. Canadian Manuscript Report of *Fisheries and Aquatic Sciences*, no. 2539.

Wilson, Barbara (Ḵii7lljuus) and Nancy J. Turner (Ḵ'ii7lljuus NaanGa). 2004. "*K'aaw k'iihl*: A Time-Honoured Tradition for Today's World." In *Cycle of*

Life/Recycle Handbook for Educators, National Edition, ed. Holly Arntzen, Chris Fisher, Stephen Foster, and Bruce Whittington, 257–9. Victoria, BC: Artists' Response Team (ART).

Wingfield-Bonnyn, William. 1890. *The Colonisation Handbook, Containing a General Account of the Agricultural and Mineral Lands*. St John's: Bowden and Sons.

Winson, Anthony. 1992. *The Intimate Commodity: Food and the Development of the Agro-Industrial Complex in Canada*. Toronto, ON: Garamond.

Witherill, J.W., and J. Kolak. 1996. "Is Corporate Re-Engineering Hurting Your Employees?" *Professional Safety* (May): 28–32.

Wolfish, D., and G. Smith. 2000. "Governance and Policy in a Multi-centric World." *Canadian Public Policy* 26: Supplement.

Woodsworth, G.J., ed. 1991. *Evolution and Hydrocarbon Potential of the Queen Charlotte Basin, British Columbia*. Geological Survey of Canada, Paper 90–10.

Working Group on Seasonal Employment and Seasonal Economies in the Twenty-first Century. Discussion paper for the national roundtable, Crossing Boundaries National Council, Ottawa, 2005.

Wright, Miriam. 2001. *A Fishery for Modern Times: The State and the Industrialization of the Newfoundland Fishery, 1934–68*. Toronto, ON: Oxford.

Wroblewski, J. 2000. "The Color of Cod: Fishers and Scientists Identify a Local Cod Stock in Gilbert Bay, Southern Labrador." In *Finding Our Sea Legs*, ed. B. Neis and L. Felt, 72–81. St John's, NL: ISER Books.

Wroblewski, J.S., W.L. Bailey, and J. Russell. 1998. "Grow-out Cod Farming in Southern Labrador." *Bull Aquacul Assoc Canada* 98–2: 47–9.

Wroblewski, J.S., B. Neis, and K. Gosse. 2005. "Inshore Stocks of Atlantic Cod are Important for Rebuilding the East Coast fishery." *Coastal Management* 33:411–32.

Wyllie-Echeverria, Sandy, Chief Adam Dick, Kim Recalma-Clutesi, and Nancy J. Turner. 2003. Case Study 19.2. "The Link between the Seagrass *Zostera marina* [ts'áts'ayem] and the Kwakwaka'wakw Nation, Vancouver Island, Canada." In *The Seagrasses of the Pacific Coast of North America*, ed. S. Wyllie-Echeverria and J. Ackerman, 217–24. World Atlas. UNEP/WCMC.

Yalnizyan, Armine. 1998. *The Growing Gap Report*. Toronto, ON: Centre for Social Justice. Summary: http://www.socialjustric.org.

Yorath, C.J., R.M. Clowes, R.D. MacDonald, C. Spencer, E.E. Davis, R.D. Hyndman, K. Rohr, J.F. Sweeney, R.G. Currie, J.F. Halpenny, and D.A. Seemann. 1987. "Marine Multichannel Seismic Reflection, Gravity and Magnetic Profiles – Vancouver Island Continental Margin and Juan de Fuca Ridge." Geological Survey of Canada, Open File 1661.

Zimmerly, D.W. 1975. *Cain's Land Revisited: Culture Change in Central Labrador, 1771–1972*. St John's, NL: ISER Books.

Index